International Handbook of Workplace Trauma Support

International Handbook of Workplace Trauma Support

Edited by Rick Hughes, Andrew Kinder, and Cary L. Cooper

A John Wiley & Sons, Ltd., Publication

This edition first published 2012

Wiley-Blackwell is an imprint of John Wiley & Sons, formed by the merger of Wiley's global Scientific, Technical and Medical business with Blackwell Publishing.

Registered Office
John Wiley & Sons Ltd, The Atrium, Southern Gate, Chichester, West Sussex, PO19 8SQ, UK

Editorial Offices
350 Main Street, Malden, MA 02148-5020, USA
9600 Garsington Road, Oxford, OX4 2DQ, UK
The Atrium, Southern Gate, Chichester, West Sussex, PO19 8SQ, UK

For details of our global editorial offices, for customer services, and for information about how to apply for permission to reuse the copyright material in this book please see our website at www.wiley.com/wiley-blackwell.

Library of Congress Cataloging-in-Publication Data applied for

A catalogue record for this book is available from the British Library.

Wiley also publishes its books in a variety of electronic formats. Some content that appears in print may not be available in electronic books.

Set in 10/12.5pt Galliard by Thomson Digital, Noida, India

1 2012

Contents

About the Editors ix
List of Contributors xi

A The Evolution and Development of Workplace Trauma Support 1

1 The Evolution of Models of Early Intervention for Adults: From Inspired Help Giving toward Evidence-based Pragmatism 3
Roderick J. Ørner

2 Evidence-based Trauma Management for Organizations: Developments and Prospects 17
Jo Rick and Rob Briner

3 Large-scale Trauma: Institutionalizing Pre- and Post-trauma Prevention, Intervention, and Treatment 30
Joyce A. Adkins and Bryan M. Davidson

4 Commonalities and New Directions in Post-trauma Support Interventions: From Pathology to the Promotion of Post-traumatic Growth 48
Stephen Regel and Atle Dyregrov

B The Legal and Business Imperatives to Manage Trauma Effectively 69

5 The Trauma Impact on Organizations: Causes, Consequences, and Remedies 71
Ronald J. Burke

6 ASSIST: A Model for Supporting Staff in Secure Healthcare Settings after Traumatic Events That Is Expanding into Other European Territories 87
Annette Greenwood, Carol Rooney, and Vittoria Ardino

7 SAV-T First: Managing Workplace Violence 105
Kate Calnan, E. K. Kelloway, and Kathryne E. Dupré

8 The Occupational Implication of the Prolonged Effects of Repeated Exposure to Traumatic Stress 121
Alexander C. McFarlane

9 The Challenge for Effective Interventions in a Violent Society: Boundaries and Crossovers between Workplace and Community 139
Merle Friedman and Gerrit van Wyk

10 Adversity: Reconceptualizing the Post-trauma Response 154
Kevin Friery

C New Understandings on Models of Trauma Support 165

11 The Role and Nature of Early Intervention: The Edinburgh Psychological First Aid and Early Intervention Programs 167
Chris Freeman and Patricia Graham

12 An Organizational Approach to the Management of Potential Traumatic Events: Trauma Risk Management (TRiM) – the Development of a Peer Support Process from the Royal Navy to the Police and Emergency Services 181
Neil Greenberg and Marilyn Wignall

13 Evidence-Based Support for Work-related Trauma: The Royal Mail Group Experience 199
Jo Rick, Andrew Kinder, and Steven Boorman

14 The Development of a Practice Research Network and Its Use in the Evaluation of the "Rewind" Treatment of Psychological Trauma in Different Settings 213
William Andrews and Scott Miller

15 The Emergency Behaviour Officer (EBO): The Use of Accurate Behavioral Information in Emergency Preparedness and Response in Public and Private Sector Settings 227
Mooli Lahad, Ruvie Rogel, and Steven Crimando

16 Trauma-related Dissociation in the Workplace 240
Onno van der Hart, Xiao Lu Wang, and Roger M. Solomon

D The Theory and Practice of Post-trauma Support 257

17 Utilization of EMDR in the Treatment of Workplace Trauma 259
Roger Solomon and Isabel Fernandez

18 Trauma Inoculation: Mindful Preparation for the Unexpected 274
Gordon Turnbull, Rebekah Lwin, and Stuart McNab

19 How Employee Assistance Programs (EAPs) Respond to Trauma Support and Critical Incident Management: An International Focus 295
Mandy Rutter

20 Training Resilience for High-risk Environments: Towards a Strength-based Approach within the Military 313
Sylvie Boermans, Roos Delahaij, Hans Korteling, and Martin Euwema

E The Organizational Response to Trauma Support 331

21 Preparing for and Managing Trauma within Organizations: How to Rehabilitate Employees Back to Work 333
Andrew Kinder and Jo Rick

22 Healing the Traumatized Organization: An Exploration of Post-trauma Recovery and Growth in the Workplace Setting Using the Metaphor of the Nervous System as a Template to Highlight Collective Learning 350
Tony Buckley and Alison Dunn

23 The Management of Emotionally Disturbing Interventions in Fire and Rescue Services: Psychological Triage as a Framework for Acute Support 368
Erik L. J. L. De Soir

24 Working with Tsunami Survivors in South India: The Problem Lies in a Four-letter Word 384
Sue Santi Ireson and Hash Patel

25 Turning Training into Reality: Considerations When Training Teams for Deployment to Disasters 401
Gail Rowntree and Mark Akerlund

26 Combating the Effects of Post-traumatic Stress and Other Trauma Associated with the Theatre of War 416
Walter Busuttil

27 Trauma Counseling and Psychological Support in the People's Republic of China (PRC) 436
Xiaoping Zhu, Zhen Wang, and Tony Buon

28 How Professionals can Help the Traumatized Organization 447
Pauline Rennie Peyton

29 Military Veterans' Mental Health: Long-term Post-trauma Support Needs 458
Walter Busuttil

30 Post-trauma Support: Learning from the Past to Help Shape a Better Future 474
Rick Hughes, Andrew Kinder, and Cary Cooper

Index 487

About the Editors

Rick Hughes is Lead Advisor, Workplace for the British Association for Counselling and Psychotherapy, the lead body for the therapeutic community in the United Kingdom with over 35 000 members. Rick campaigns for, supports, and champions best practice and effective employee support provision including trauma support. He has worked with most of the major UK employee assistance program (EAP) providers. Rick managed the trauma support for a group of UK train operators whilst employed as a specialist at a major EAP provider. He has supported individuals, teams, and organizations in a range of circumstances including after 9/11, the London bombings, and the Asian tsunami. Rick is a former Deputy Chair of the Association for Counselling at Work (ACW), now BACP Workplace, and has been their journal editor since 2003. His MPhil looked at the impact of emotions in the workplace and this led to his receiving an Honorary Research Fellowship from the University of Strathclyde, Glasgow.

Andrew Kinder is a Chartered Counselling and Chartered Occupational Psychologist, the Past Chair of the Workplace Division of the British Association for Counselling and Psychotherapy (BACP), and an Associate Fellow of the British Psychological Society and was recently made Fellow of the BACP for his contributions to counseling in the workplace. He has published widely and is particularly interested in the management of stress and trauma within an occupational health context. Andrew is currently Chief Psychologist for a large employee assistance and occupational health organization (Atos Healthcare, http://www.atoshealthcare.com) and specializes in delivering training, counseling, and coaching services to increase the psychological health of individuals and organizations. He has been instrumental in the introduction of early intervention programs in a number of large organizations relating to employee engagement and employee wellbeing. He has been active as a researcher and has collaborated with other leading organizations, including the British Occupational Health Research Foundation, which was gathering evidence for organizational interventions used following a work-related trauma. He co-edited *Employee Wellbeing Support: A Workplace Resource*, which was published in March 2008, with Cary Cooper and Rick Hughes. He has also co-written with Rick Hughes *Guidelines for Counselling in the Workplace*, which was published by the BACP. He has also carried out numerous assessments for reality TV and provided advice on many high-profile shows. More information on him is available at http://www.andrewkinder.co.uk.

Cary L. Cooper, CBE, is the author of more than 100 books and is one of Britain's most quoted business gurus. He is Distinguished Professor of Organizational Psychology and Health at Lancaster University Management School. He is a founding President of the British Academy of Management, a Companion of the Chartered Management Institute, and one of only five UK Fellows of the (American) Academy of Management. He was the Founding Editor of the *Journal of Organizational Behavior*, and is the Editor (with Professor Chris Argyris of Harvard Business School and Professor Bill Starbuck of New York University as Associate Editors) of the *Blackwell Encyclopedia of Management*. He has been an advisor to the World Health Organization, International Labour Organization, and European Union in the field of occupational health research and wellbeing, was Chair of the Global Agenda Council on Chronic Disease of the World Economic Forum, and is Chair of the Academy of Social Sciences (comprising 43 learned societies in the social sciences and over 87 000 members). He was awarded the CBE by the Queen in 2001 for his contributions to organizational health and safety.

List of Contributors

Joyce A. Adkins, PhD, MPH, is an occupational health psychologist. Colonel Adkins has served in the US Air Force for more than 28 years in clinical and health psychology, organizational health and occupational stress, human factors, policy, and program development capacities. She received her PhD from Peabody College of Vanderbilt University and her MPH from the Harvard School of Public Health. She has served on the editorial review board of three journals and served as primary investigator for multisite, multi-agency research protocols. She deployed to Iraq and Afghanistan and served as Program Director for combat and operational stress programs for the US Department of Defense.

Mark Akerlund, LCSW, MSW, works as a psychotherapist and consultant in Houston, Texas, and also deploys with Kenyon International Emergency Services (KIES) as needed. Mark received his Master's in Social Work (MSW) degree from the University of Houston in 1998. Mark also works as a Staffing Coordinator for Social Work prn. Mark has attended domestic and international disasters including 9/11, the Asian tsunami, and the Haiti earthquake, and has provided both onsite and call center mental health support for numerous aviation incidents.

William Andrews has a private therapy practice in Sheffield, United Kingdom. Following a 22-year career as a dentist, Bill resigned from active practice following a prolonged period of mental distress. He was diagnosed as bipolar in 1994, and this led him into an acute interest in the field of mental health service delivery. He trained as a human givens therapist, graduating with distinction in 2004, and since then has dedicated his time to the active encouragement of the use of service provider feedback in the delivery of psychological treatment. He is a senior associate with the International Centre for Clinical Excellence, a new worldwide initiative designed to explore empirical findings around excellence in the delivery of behavioral health (http://www.centerforclinicalexcellence.com). As well as providing independent consultancy and supervision in outcome-informed practice, he is an accredited supervisor with the Human Givens Institute and lectures internationally on the subject of feedback-informed treatment. He passionately believes in a movement toward a more pragmatic approach to treatment that de-emphasizes reliance on specific models of therapeutic orientation. He has a special interest in treating psychological trauma.

Vittoria Ardino is the Director of the Trauma Research and Treatment Center at the Italian Red Cross. She is President of the Italian Society for Traumatic Stress Studies and a board member of ESTSS. She has extensive academic experience in the United Kingdom and Italy, and her research interests are in clinical and forensic psychology with a focus on cognitive aspects of post-traumatic stress disorder (PTSD). She edited two books on PTSD in children and adolescents and published several articles in international journals.

Sylvie M. Boermans is working on her PhD in organizational psychology at the University of Leuven, Belgium. Her PhD seeks to understand what enables soldiers to respond with resilience. She specifically focuses on the role of morale and leadership. In 2008 she received her MSc in social psychology (with honors) and was rewarded for her master's thesis at the Vrij Universiteit in the Netherlands. She has recently presented a systematic literature review on "military resilience" at the European Work and Organizational Psychology Congress. Sylvie is currently working with TNO Defence, Security and Safety in the Netherlands on the development of a resilience model aimed at enhancing resilience in military organizations for the Netherlands Defense Force.

Steven Boorman, MBBS, MRCGP, FFOM, FRCP, FRCN, is an experienced specialist in occupational medicine, now leading Abermed's UK Occupational Health Services. Prior to this, he had over 20 years of experience in Royal Mail, becoming Chief Medical Adviser and Director of Health and Safety. He is an honorary senior clinical lecturer to the University of Birmingham and an ex-President of the Royal Society of Medicine's Section of Occupational Medicine. In 2009 he led the review of NHS Workforce Health and Wellbeing, demonstrating the linkage between good staff health and improved organizational and patient outcomes. His work in Royal Mail included a particular interest in developing improved post-trauma support.

Rob B. Briner is Professor of Organizational Psychology in the School of Management, University of Bath. He previously worked at Birkbeck College, University of London for 19 years after completing his PhD at the Social and Applied Psychology Unit (now the Institute of Work Psychology) at the University of Sheffield. His research interests including well-being, emotions, stress, ethnicity, the psychological contract, absence from work, motivation, and everyday behaviour at work. One of his current main interests is in evidence-based practice in organizational psychology, HRM and management more generally. He also has a strong interest in writing for practitioner and more popular publications and has published pieces in many HR magazines and newspapers and was a regular columnist for HR magazine *People Management*.

Tony Buckley is Manager of the Counselling and Trauma Service within the Occupational Health Department at Transport for London. In his role, Tony manages a team of Counselling and Trauma Practitioners in delivering therapeutic support services for company employees. This team also provides psycho-education, stress-reduction group work, and response support following critical incidents. His previous, 20-year therapy career experience included supervision, private practice, and counseling management in both a university setting and an adolescent counseling service within the voluntary sector. Originally Gestalt trained, Tony is also a qualified sensorimotor psychotherapist and is on the teaching faculty of the sensorimotor psychotherapy Institute. He is chair of the UK Association of Sensorimotor Psychotherapists. He holds a BA Hons degree in Counselling and a Diploma in Supervision. Tony has a particular developing interest in somatic psychology and the application of knowledge from the fields of neurobiology and psychobiology to trauma theory and treatment interventions.

Tony Buon is a psychologist and an associate lecturer with the Aberdeen Business School at the Robert Gordon University in Scotland. Tony holds graduate and postgraduate degrees in Psychology, Behavioural Science, and Workplace Education. He is also a partner with the Buon Consultancy in the United Kingdom. Tony works extensively throughout Europe, the Middle East and Africa. He is also a qualified Mediator and runs accredited training for workplace mediators in Europe. Along with Xiaoping Zhu, he established one of the first EAPs in China in 1997. In 2008 he ran training programmes in the PRC for Psychologists providing trauma counselling to the survivors of the Wenchuan Earthquake.

Ronald J. Burke's work has focused on the relationship between the work environment and individual and organizational health, and over the past 40 years he has written articles for numerous academic and professional journals. In addition to his research and teaching activities, Professor Burke was the Founding Editor of the *Canadian Journal of Administrative Sciences.* Burke has served on the editorial board of two dozen journals and has reviewed manuscripts for a dozen more journals. He has participated in research conferences in North and South America, the United Kingdom, Europe, Asia, and Australia. He is a Fellow of the Canadian Psychological Association.

Burke has published several journal articles and book chapters and presented numerous papers at academic conferences around the world. He has also edited or co-edited 39 books to date with McGraw-Hill, Kluwer, Blackwell, Routledge, the American Psychological Association, Sage, Edward Elgar, Cambridge University Press, Emerald, Gower, and Elsevier. In addition, he serves as co-editor of the Gower Publishers series on the psychological and behavioral aspects of risk in organizations.

Walter Busuttil is a consultant psychiatrist who was appointed Medical Director to the national charity Combat Stress in 1997. During his time at Combat Stress, he has worked to upgrade all clinical services for veterans throughout the United Kingdom. In 2011 his clinical services were awarded national specialized commissioning from the Department of Health for the delivery of intensive rehabilitation programs for sufferers of chronic PTSD presenting with co-morbid depression and alcohol problems. He served for 16 years in the Royal Air Force where he was instrumental in setting up mental health rehabilitation services for service members returning from the first Gulf War. He was also part of the clinical team that rehabilitated the released British Beirut hostages. After retiring from the RAF in 1997, for 10 years he worked setting up tertiary services for sufferers of complex PTSD in a general adult setting and within a medium-secure forensic women's service. He has published and lectured internationally about the treatment and rehabilitation of chronic and complex presentations of post-traumatic stress disorder. He is the current Chair of the UK Trauma Group and is a board member of the UK Psychological Trauma Society.

Kate Calnan is a PhD student in industrial/organizational psychology and part-time professor of occupational health psychology at Saint Mary's University, Halifax, NS, Canada. Kate has been awarded many research grants including a doctoral scholarship from the Social Sciences and Humanities Research Council. Kate is also a member of the CN Centre for Occupational Health and Safety where she maintains an active role in research and consulting projects. Her current research interests focus on organizational functioning and employee well-being, with a specialized focus on workplace violence, conflict, and positive occupational health psychology. Kate has presented her research at several conferences as well as contributed to several edited books.

Steven M. Crimando, MA, BCETS, is an internationally known consultant and educator specialized in the application of the behavioral sciences in homeland security, violence prevention, and crisis management. He is the Managing Director of Extreme Behavioral Risk Management (XBRM), a division of ALLSector Technology Group, Inc., a New York City–based consulting firm. Mr. Crimando is a Board-Certified Expert in Traumatic Stress (BCETS), and holds Diplomate status with the National Center for Crisis Management, the American Academy of Experts in Traumatic Stress, and the International College of the Behavioral Sciences, where he serves on the Board of Directors. He is a Certified Trauma Specialist (CTS) and holds Level-5 Certification in Homeland Security (CHS-V). Mr. Crimando served as a disaster field operations supervisor for mental health response to the September 11, 2001, World Trade Center attacks and coordinated onsite psychological operations at New Jersey's Anthrax Screening Center, as well as at many instances of both interpersonal and mass violence in corporate, community, and campus settings.

Bryan Davidson, PhD, ABPP, is the Director of Psychological Health and Chief of the Traumatic Stress Response Team for Langley Air Force Base, VA, US. He received his PhD from Fuller Theological Seminary Graduate School of Psychology and is a Diplomate of the American Board of Professional Psychology in Clinical Health Psychology (CHP). His work as a psychologist in the USAF included full-time work in integrated primary care, directing a CHP service, and deployments to Iraq and Afghanistan. Presently, he is an officer in the US Public Health Service providing leadership in expanding initiatives that target organizational health.

Roos Delahaij, born in 1978, obtained her Master's Degree in social psychology (with honors) at the University of Amsterdam in the Netherlands in 2004. From 2005 to 2009 she was employed by the Faculty of Social Sciences of Tilburg University in the Netherlands. She studied the relationship between individual characteristics and performance under acute stress in a military population. In addition, she studied the role of organizational culture in resilience development. This work resulted in a PhD degree in 2010. She has published in international scientific journals such as *Personality and Individual Differences* and the *International Journal of Stress Management*. Since 2009 she has been working as a researcher at TNO Defence, Security and Safety. She has been involved in studies investigating resilience in military organizations and in the development of resilience-enhancing training programs for the Netherlands Defense Force. Roos is currently a member of the NATO Task Group "Mental Health Training."

Erik De Soir graduated from the Royal Military Academy in 1988 with a Master's Degree in Social and Military Sciences. After Infantry School, he was assigned to the Liberation Battalion of the 1st Mechanized Brigade as a platoon commander. In 1991, he was called back at the department of Behavioral Sciences of the Royal Military Academy, Chair of Psychology, to teach courses in general and social psychology. Major De Soir is currently the Commander of the Psychosocial Support Section at the Well-being Department of the Belgian Defence.

De Soir has a Special Master's degree in disaster medicine & disaster management (Catholic University of Leuven [KU Leuven], 1991), a Master's degree in clinical psychology (KU Leuven, 1995), a psychotherapy training in systemic marital, family and sex therapy (KU Leuven, 1998), a psychotherapy training in hypnotherapy (Scientific Flemish Hypnotherapy Association) and extensive training in psychotrauma therapy and counseling (creative arts therapy, EMDR, sensorimotor trauma therapy, cognitive therapy, etc.). He is studying for a PhD on the theme of peri-traumatic reactions and post-trauma memories under the guidance of Prof. Dr. Rolf Kleber (Utrecht University, the Netherlands), Prof. Dr. Onno Van der Hart (Utrecht University, the Netherlands) and Prof. Dr. Jacques Mylle (Royal Military Academy).

De Soir created the Belgian Model for Psychosocial Support for Peacekeeping Operations (for soldiers and their significant others) and elaborated basic principles for traumatic stress management in the military and rescue services. As a crisis psychologist, he regularly participated in peace support operations in Somalia, Rwanda, Croatia, and Bosnia to study the different problems of deployed soldiers and their significant others.

As a leading expert in fire and rescue psychology, De Soir is also a certified fire fighter and a paramedic and serves as a volunteer fire psychologist in the Regional Fire and Ambulance Brigade of Leopoldsburg. He created a European-wide counseling and support network for the management of traumatic stress in fire brigades, emergence medical services, and emergency departments of hospitals, currently known as Fire Stress Teams (FiST). In 2003, he created the European Association of Fire and Rescue Psychologists – Association Européenne des Psychologues Sapeurs-Pompiers (AEPSP).

De Soir is one of the founding members of the Belgian Society for Psychotraumatology (Société Belge de Psychotraumatologie) and the Revue Francophone du Stress et du Trauma. Between 2001 and 2010, he has been the Vice President of the Association de Langue Française pour l'Etude du Stress et du Traumatisme, Board Member of ESTSS and NtVP and he co-chaired the International Structure and Affiliations Committee within the International Society of Traumatic Stress Studies (ISTSS).

Alison Dunn is Head of Treatment Services in the Occupational Health Department at Transport for London, which provides a range of services for London Underground and the Transport for London group. In this role she oversees the management of the counselling and trauma service, the physiotherapy service, and the drug and alcohol assessment and treatment service. Alison's background is in social work and then counselling – she was awarded a Master's degree in psychological counselling and psychotherapy in 2000.

Alison has thorough experience of providing counseling in an organizational setting, managing a proactive workplace counseling service, and responding to critical incidents. She has coordinated the counseling team's response to several incidents that have taken place during past years as well as the response to the events on July 7, 2005. Alison has previously written about a four-stage model for trauma aftercare in *Trauma: A Practitioner's Guide to Counselling*, edited by Thom Spiers. She has also written about critical incident planning in *Employee Well-Being Support: A Workplace Resource*, edited by Andrew Kinder *et al.*

Kathryne Dupré is an associate professor of organizational behavior and human resource management in the Faculty of Business Administration at Memorial University of Newfoundland. She received her PhD in management from Queen's University, her MSc in Industrial/Organizational Psychology from Saint Mary's University, and her honors BA in psychology from Queen's University. Kathryne's research interests focus on employee well-being, with particular emphasis on workplace aggression, harassment and safety, leadership, occupational stress, and young employees' experiences in the workplace. Kathryne has published her work in a number of journals including the *Journal of Applied Psychology*, *Human Resource Management*, and the *Journal of Occupational Health Psychology*. She has presented her research at numerous national and international conferences, and contributed chapters to several edited books.

Atle Dyregrov, PhD, Director of the Centre for Crisis Psychology, Bergen, Norway, is a clinical and research psychologist and the author of numerous publications and journal articles as well as more than 15 books. He has been closely involved in crisis support following many national and international events, including the 2011 tragedy in Norway. In addition, he has conducted research on various subjects relating to bereavement, trauma, and crisis situations. His clinical work has covered diverse areas such as grief reactions in parents, grief and trauma in children, and organizing psychosocial disaster assistance to victims, families, and first responders. He is one of the founding members of the European Society for Traumatic Stress Studies and the Children and War Foundation. He has worked extensively as a consultant to various UN organizations, especially UNICEF and UNHCR.

Martin C. Euwema is Full Professor in Organizational Psychology at the University of Leuven, Belgium, and chair of the research group Work, Organization and Personnel Psychology. Martin has been working for many years at Utrecht University, the Netherlands, teaching and investigating conflict management, peacekeeping, and human resources management. He has been involved in research and development and conducting training for deployment in operations other than war (OOTW) of military officers in the Netherlands, Finland, and Germany, and for nongovernmental organizations (NGOs) in crisis areas in Denmark, Germany, and the Netherlands. He has published in international scientific journals, such as the *Journal of Personality and Social Psychology*, the *Journal of Applied Psychology*, *Work and Stress*, the *Journal of Organizational Behavior*, *Group and Organization Management*, the *International Journal for Conflict Management*, and others, and has published several books (in Dutch) for conflict professionals and textbooks. Martin is President of the International

Association for Conflict Management and co-director of the Leuven Center for Cooperative Management.

Isabel Fernandez is a clinical psychologist, working in Milan. She has been trained in cognitive-behavioral therapy and has been in the faculty of the Italian School of Cognitive Behavior Therapy for 12 years, providing specialization training in psychotherapy. She has worked as a consultant psychologist at the psychiatric ward of Niguarda Hospital, conducting research projects in the clinical field. She is a lecturer at the Catholic University of Milan and Rome. At the present time, she is the Director of the Psychotraumatology Research Center of Milan and has published many papers, articles, and books on trauma, on EMDR, and on research projects in this field. She is an EMDR Europe-Approved trainer and Chair of the Italian Association of EMDR, and belongs to the Board of Directors of the Italian Federation of Scientific Psychological Societies. She is also a Board Member of the European Society for Traumatic Stress Studies (ESTSS) and of the Standing Committee of Disasters, Trauma and Crisis of the European Federation of Psychologists' Association. She has directed and organized interventions of disaster psychology in natural and incidental disasters (e.g., the air crash on Milan's Pirelli building and Molise's earthquake) and has worked in cooperation with civil defense workers and fire fighters for debriefing and psychological support on stress in emergency workers. She has been training psychology graduates, post-graduates, and clinicians on trauma and crisis interventions.

Chris Freeman is a consultant psychiatrist and psychotherapist based in Edinburgh. He was the Director of the Rivers Centre for Traumatic Stress in Edinburgh until 2007. He is currently President of the UKPTS (United Kingdom Psychological Trauma Society), is on the ESTSS Board, and is leading an ESTSS task force on Trauma Informed Services. His current post is as Regional Consultant for Eating Disorders for the southeast of Scotland. He is Vice Chair for Ethics for the Royal College of Psychiatrists.

Merle Friedman, PhD, is a registered clinical psychologist. She has served on the board of the International Society for Traumatic Stress Studies for eight years and is currently on the board of Psychology Beyond Borders and Foundation for a Safe South Africa. Merle is a co-founder and director of the South African Institute for Traumatic Stress, and a founder of the Wits Trauma Clinic in South Africa. She was a member of the US National Institute of Mental Health Violence and Traumatic Stress Working Group, and of the ISTSS International Working Group on Traumatic Stress. She has been a member of TRT (To Reflect and Trust) since 1997, engaging in issues of conflict and distress between ethnic and racial groups.

Merle is also a corporate psychologist, and runs a consultancy, Psych-Action, working in the areas of leadership, stress, traumatic stress, resilience, team development, and strategic risk management. She is a contributing author of the *KING III Report on Corporate Governance* in South Africa (2009). The focus of her current interest and research is behavioral finance, the psychology of decision making. She is an executive coach, working at senior levels in organizations. She travels extensively nationally and internationally for speaking engagements, conference presentations, and lecture tours to Australia, Europe, Israel, and the United States, has published a number of articles and book chapters focused mainly on aspects of stress and resilience, and appears on national TV and radio in South Africa.

Kevin Friery is Clinical Director at Right Management. A graduate in psychology from Exeter University he has worked in the NHS, local government, the voluntary sector, and private industry. As a counsellor and psychotherapist he has worked for a number of years with people seeking answers to some of the questions life has thrown at them, and is a great believer in the concept that it is not events that change people so much as the sense they make of them.

As Clinical Director, he has worked with a large number of organizations helping to understand the factors that influence stress and well-being and thereby to develop a more resilient workforce. He has authored a number of papers on workplace wellness and for three years was Chair of BACP Workplace, the specialist workplace division of the British Association for Counselling and Psychotherapy. Kevin is frequently asked to speak at conferences and has chaired a number of major national events; he is also a regular contributor to professional workplace media.

Patricia Graham is a consultant clinical psychologist based in East Lothian, Scotland. She worked both clinically and in research at the Rivers Centre for Traumatic Stress in Edinburgh until 1999 before specializing in the treatment of severe eating disorders. Her current post is Head of Adult Mental Health Psychology in East Lothian. She is a member of the IPT Scotland Training Committee, and is a BABCP-accredited practitioner in both CBT and CBASP. She is also a member of the Joint Mental Health Planning Group for East Lothian CHP (Community Health Partnership).

Neil Greenberg is the Defence Professor of Mental Health at King's College London and is a consultant psychiatrist. He has carried out research in a number of hostile environments including Afghanistan and Iraq. Greenberg studied medicine at Southampton University, graduating in 1993. He served as a doctor in a variety of warships and submarines, and with two Royal Marines Commando units. During his time with the Royal Marines, he achieved his arctic warfare qualification and completed the all-arms commando course, earning the coveted Green Beret.

Greenberg has specialized in psychiatry, completing a Master's degree in clinical psychiatry and a doctorate in mental health, and is a Fellow of the Royal College of Psychiatrists. He is a specialist in general adult, forensic and liaison psychiatry. Since 1997 he has been part of the team at the forefront of developing Trauma Risk Management (TRiM). Greenberg has provided psychological input for Foreign Office personnel after the events of September 11, 2001, and in Bali after 12 October 2002, and a number of other significant incidents including assisting the London Ambulance Service in the wake of the London Bombings in 2005.

In 2008 he was awarded the Gilbert Blane Medal by the Royal Navy for his work in supporting the health of naval personnel through his research work. He has published more than 100 scientific papers and book chapters, and has presented to national and international audiences on matters concerning the psychological health of the UK Armed Forces, the organizational management of traumatic stress, and occupational mental health.

Annette Greenwood is an HPC-registered counselling psychologist who specializes in trauma and staff well-being. She has worked at the consultant level for the last 13 years in the NHS and now as the trauma advisor for St Andrews Healthcare. She has led psychological incident response at both international and national levels. At St Andrews she has set up and leads the trauma response service to support staff working in secure mental hospitals. Trauma, stress, and the impact of mindfulness on the well-being of healthcare professionals are her areas of clinical and research practice. More recently she was commissioned by the European Union (EU) to co-write the new psychosocial guidelines for psychological support following a CRBN major incident (Berlin, 2011). The ASSIST model described in Chapter 6 will form the training for volunteers at the new Trauma Centre in Milan, Italy.

E. Kevin Kelloway is the Canada Research Chair in Occupational Health Psychology and the Director of the CN Centre for Occupational Health and Safety at Saint Mary's University, Halifax, NS, Canada. He also holds appointments as Professor of Psychology and Professor of Management at Saint Mary's. A prolific researcher, Dr. Kelloway is a Fellow of the Association for Psychological Science, the Canadian Psychological Association, and the Society for Industrial/Organizational Psychology. He currently serves as Associate Editor of *Work and Stress* and as Section Editor for *Stress and Health* as well as serving on several editorial boards. He is past Chair of the Canadian Society for Industrial and Organizational Psychology. He has authored or edited 10 books, and authored over 100 research articles and chapters, on a wide range of topics in organizational psychology. His current research interests focus on the effects of organizational leadership on individual well-being, positive occupational health psychology, and the prediction and consequences of workplace violence. He also maintains an active consulting practice, working with a variety of public and private sector clients in these areas.

J.E. (Hans) Korteling is a research scientist at TNO Defence, Security and Safety. He received his MSc degrees in psychology from the University of Amsterdam, where his main interests included visual perception and neuropsychology. Until 1998, he worked as a research psychologist in the Traffic Behavior Group and as research group leader in the Vehicle Control Group of TNO Human Factors, the Netherlands. Hans received his PhD in the Psychological, Pedagogical, and Sociological Sciences in 1994 (with honors) at the University of Groningen. Hans has been in charge of the Simulation Research Group of TNO Human Factors which focused on the specification and validation of research- and training simulators. Since 2002 he has been program manager of training and instruction research for the Dutch Defense. From 2005 he has also been senior research scientist of the Training and Instruction Department of TNO Human Factors.

Mooli Lahad, PhD, is a senior medical and education psychologist, Professor of Psychology at Tel-Hai College, Tel Hai, Israel, and the Founder and President of the Community Stress Prevention Center (CSPC), established some 30 years ago. He is one of the world's leading experts on the integration of the arts form therapies for psychotrauma and coping with disasters. Following the terrorist attacks on the United States in September 2001, he trained, in both New York and New Jersey, mental health staff and medical care teams. He was the director of an international initiative to help Sri Lanka following the tsunami of 2004. He is also a former advisor to NATO and UNICEF.

Professor Lahad is the author and co-author of 30 books and many articles on the topics of the use of an integrative approach to treat PTSD and grief, communities under stress, and coping with life-threatening situations. He is the recipient of three professional prizes: the Israeli Psychology Association – Bonner Prize for outstanding contributions to education in Israel, the Adler Institute for the Welfare of the Child Prize, Tel Aviv University, and the Israeli Lottery Prize for innovations in medicine for developing a telepsychology services. For more information, please see http://www.icspc.org.

Rebekah Lwin, PhD, is a senior lecturer at the University of Chester, UK, where she teaches a master's-level program on psychological trauma. She is also a consultant clinical psychologist at Alder Hey Children's Hospital in Liverpool, UK, and previously at Great Ormond Street Hospital for Children and the Institute of Child Health, London, UK. She trained in Edinburgh, Scotland, and has worked in child and adolescent mental health throughout her career. Her clinical and research interests have explored the experiences of children and families within a context of chronic illness and acute physical trauma, and latterly she has developed a particular interest in the application of compassion-focused therapeutic approaches to building psychological resilience in young people. She joined the University of Chester in 2008 where she has helped, with colleagues Dr. Stuart McNab and Professor Gordon Turnbull, to establish the Centre for Research and Education in Psychological Trauma.

Alexander McFarlane, AO, MBBS (Hons), MD, FRANZCP, Dip Psychother, is the Head of the University of Adelaide Centre for Traumatic Stress Studies. He is an international expert in the field of psychological trauma and disasters and an active clinician. He is a past president of both the International Society for Traumatic Stress Studies and the Australasian Society for Traumatic Stress Studies. He has been the Senior Adviser in Psychiatry to the Australian Defence Force, and the Department of Veterans Affairs. He holds the rank of Group Captain in the RAAF specialist reserve.

His research has focused on the epidemiology and longitudinal course of posttraumatic stress disorder, as well as the neuroimaging of the cognitive deficits in this disorder. He has a particular interest in the impact of childhood trauma on adult adjustment. He has directed several major studies on the health of members of the entire Australian Defence Force that are currently being completed. He is the recipient of a number of honors and awards for his research and contributions to the field. He has had extensive experience as an expert witness in a range of jurisdictions in Australia and internationally. He has published over 300 articles and chapters in various refereed journals and has co-edited three books.

Stuart McNab is the Director of the Centre for Research and Education at the University of Chester. He is a chartered counselling psychologist and registered analytic psychotherapist who has worked clinically with trauma for over 20 years. Following his original training in the psychoanalytic approach, he trained in EMDR and trauma-focused cognitive-behavioral therapy (CBT). He has a particular interest in working with trauma through mindfulness and compassionate mind training, and has recently brought his love of horses into his therapeutic work with riders who have experienced traumatic accidents. He designed and launched the first master's program in psychological trauma in 1996. The program now recruits from a wide variety of occupations including psychologists, psychotherapists, social workers, general practitioners, police officers, and military personnel and is attracting students both nationally and internationally.

Scott D. Miller, PhD, is the founder of the International Center for Clinical Excellence, an international consortium of clinicians, researchers, and educators dedicated to promoting excellence in behavioral health services. Dr. Miller conducts workshops and training in the United States and abroad, helping hundreds of agencies and organizations, both public and private, to achieve superior results. He is one of a handful of "invited faculty" whose work, thinking, and research are featured at the prestigious Evolution of Psychotherapy Conference. His humorous and engaging presentation style and command of the research literature consistently inspire practitioners, administrators, and policy makers to make effective changes in service delivery. Scott is the author of numerous articles and books, including *Escape from Babel: Toward a Unifying Language for Psychotherapy Practice* (with Barry Duncan and Mark Hubble [1997]), *The Heart and Soul of Change* (with Mark Hubble and Barry Duncan [1999, 2010]), *The Heroic Client: A Revolutionary Way to Improve Effectiveness through Client-Directed, Outcome-Informed Therapy* (with Barry Duncan [2000] and Jacqueline Sparks [rev., 2004]), *Staying on Top and Keeping the Sand Out of Your Pants: The Surfer's Guide to the Good Life* (with Mark Hubble and Seth Houdeshell [2003]), and the forthcoming *Achieving Clinical Excellence in Behavioral Health: Empirical Lessons from the Field's Most Effective Practitioners* (with Mark Hubble and William Andrews).

Roderick Ørner is consultant clinical psychologist in private practice and Visiting Professor of Clinical Psychology at the University of Lincoln which awarded his PhD in 2005. His clinical experience comprises specialist assessment and therapy clinics for adult patients especially following exposure to traumatic events in the maritime industries (see http://www.ForceMajeureMaritime.com). He is a Fellow of the British Psychological Society, and his trauma-related research interests range from British Falklands War veterans, European war veterans, and the provision of psychological support services for emergency responder groups. Alone, he hosted the First European Conference on Traumatic Stress in Lincoln, UK, in 1988 and for a number of years maintained a leading role within the European Society for Traumatic Stress Studies of which he was President between 1997 and 1999. He is senior editor of *Reconstructing Early Intervention after Trauma* published in 2003 by Oxford University Press.

Hash Patel, BSc(Hons), PhD, was born in India and came to England when he was 7 years old. He started his career in the field of counseling in 1981. As Head of Teacher Education and Counselling Studies at Sandwell College in the West Midlands, he was responsible for developing and delivering an extensive range of training programs for counseling at various levels for different professional groups as well as developing a successful Student Counselling Service. Alongside his academic career, Hash has dedicated much time to working actively in the voluntary sector. From his innovative and visionary ideas, he secured private and public finding for the creation of a large, multipurpose community center for the Gujarati community. His involvement in charitable work stems from the belief in helping others to fulfill their potential both individually and as a community group.

Hash is an accredited humanistic counselor with an extensive experience of the unique fusion of Eastern and Western counseling philosophies. His counseling interests and specialisms include PTSD, child abuse, family dynamics, and ethnicity. Hash is a supervisor and trainer as well as maintaining a private practice. He has been involved with a range of committees and work groups within BACP and was Chair of BACP's Division for Independent Practitioners (then PRG) from 2000 to 2002. Hash is currently a lecturer at Birmingham Metropolitan College.

Pauline Rennie Peyton, D. Psych, is a chartered psychologist who runs a clinical practice in central London and an organizational consultancy country-wide. She is the author of *Dignity at Work – Eliminate Bullying and Create a Positive Working Environment* (2003), a book that was her response to the growing problem, and consequences, of workplace bullying. Pauline specializes in working with interpersonal and occupational relationships in which individuals need help in dealing with difficult relationships. Her focus with couples is about enabling them to bring out the best in each other and to abandon the behaviors that caused them to seek therapy in the first place. But the word "couples" can have a wider context: Pauline also works with mothers and daughters, fathers and sons, or various combinations of people in families or other structures, to encourage them to think in terms of role reversal and reach an understanding of the other's point of view.

As a trained mediator with many years of experience, she works in organizations as a trainer and facilitator, giving people the skills to work together with dignity and respect. Another area of specialism for Pauline is the field of trauma, in which she counsels people who have been involved in traumatic incidents – whether as victim, witness, or even perpetrator.

Stephen Regel is Principal Psychotherapist/Co-director of the Centre for Trauma, Resilience and Growth, Nottinghamshire Healthcare NHS Trust, Special Associate Professor in the School of Sociology and Social Policy, Nottingham University, and a Senior Fellow of the Institute of Mental Health, Nottingham. Since 2002, he has been visiting therapist/consultant at the Family Trauma Centre in Belfast, Northern Ireland. His time is divided between clinical, teaching, and research activities. He consults and trains extensively with UK police forces on the provision of post-trauma peer support. He is also consultant/trainer to the International Committee of the Red Cross (ICRC) peer support initiative. He is a consultant to the International Federation of Red Cross and Red Crescent Societies (IFRC) Reference Centre for Psychosocial Support. Since 2005, he has been part of the British Red Cross Psychosocial Support Team, assisting UK nationals affected by incidents abroad.

Jo Rick, PhD, is a work and organisational psychologist with over 20 years experience of applied research and consultancy in work related mental health and well-being. Jo's research interests include well-being, stress, absence, rehabilitation and organizational culture. Jo has led or been involved in a number of large scale evaluations of workplace and community mental health initiatives. She has authored many research reports for a variety of audiences including organisations, Government departments, practitioners and academics. Jo is particularly interested in evidence based approaches and using evidence, she had undertaken evidence reviews to underpin national policy in the UK on areas such as rehabilitation/return to work and management standards for stress as well as to shape specific workplace practices.

Ruvie Rogel is Deputy CEO in charge of professional and international development at the Community Stress Prevention Center (CSPC), Tel Hai College, Israel. He is an international expert on emergency management and community resiliency. Dr. Rogel holds a PhD in Educational Leadership from Leicester University, UK, where he had also obtained his MSc in Management and Training. His first degree is in psychology (Tel Aviv University). He has a faculty appointment at Wright State University in Dayton, Ohio, and teaches in universities and colleges in Israel, as well as in workshops on resilience and psychosocial coping in emergency events.

His professional experience was gained in community emergency and preparedness work and interventions in Israel through government, public, business, and private sectors. During the 2005 evacuation of Israeli citizens from the Gaza strip, Dr. Rogel served as a special consultant to the Israeli Prime Minister's Office, on the psychosocial aspects of the "Disengagement." Dr. Rogel participated in international aid delegations in Sri Lanka, Mississippi, Haiti, Ethiopia, Uganda, and India, and is involved with international organizations' projects. Dr. Rogel attends international conferences on emergency management as a speaker and as a participant and is a member of the International Association of Emergency Management (IAEM).

Carol Rooney is a registered mental nurse specializing in clinical risk and aggression management. She has worked at the senior management level in the NHS and in the independent sector, and is currently Head of Clinical Risk Management for St Andrews Healthcare, and is the professional lead for prevention and management of violence aggression strategy and training. Her area of research interest is in staff responses to disturbed behavior, and she has presented internationally and published on prosecution issues in mental health settings, staff experience of observations, and use and retention of breakaway skills.

Gail Rowntree studied at Kings College, London (Institute of Psychiatry), and currently splits her time between being a senior lecturer for Buckinghamshire New University, UK, lecturing in organizational psychology and disaster management and resilience to post-graduate students as well as being part of the first responders team for Kenyon International Emergency Services (KIES). Her research interests are around understanding the narrative of mass disasters for deploying teams; Gail is currently undertaking extensive research in this area, and she has published several articles on this subject. Gail has deployed to a variety of disasters from the Thailand Tsunami, airline crashes, and civil unrest to other natural disasters, all in the role of family assistance and team welfare.

Mandy Rutter is a qualified psychologist, counsellor, mediator, and trainer. She began her career working with teenagers and their families whose lives had been disrupted by trauma. In 1994 she joined ICAS, an independent, international EAP provider and experienced a wide range of roles working in both the clinical and commercial departments. In 1999, Mandy joined the ICAS Crisiscall team where the main focus was assisting and guiding organizations through accidents, injuries, death, suicide, and distressing unexpected traumas. Mandy led the Crisiscall team in supporting customers after three major train crashes, the Bali bombings, and 7 July 2005 bombings in London. More recently, Mandy has broadened her role and now consults on the design and delivery of strategic employee-focused crisis management issues. She has been part of a national working group who published British Standard Guidelines on human aspects of business continuity, and she regularly contributes articles for professional journals. Mandy currently works alongside financial institutions, oil and gas companies, retail organizations, and government departments assisting in the development and facilitation of "psychological first aid" training programs. Mandy is also developing services to encourage personal and managerial resilience before and after trauma.

Sue Santi Ireson, BEd(Hons), MACouns, has been working in the counseling field since 1976 when she trained as a volunteer youth counselor. She was then appointed as Director of a Youth Counselling Service in Maidenhead, where she developed an extensive training program of ongoing counseling training and focused workshops. A major focus of her work is in the field of child sexual abuse and its long-term effects, and from this she developed an interest in PTSD and the long-term effects of unacknowledged trauma in the individual, the family, and the community. She has published various articles on these issues and continues to offer training in these fields. Her passionate commitment to enabling people to move on from traumatic past experiences has always informed her work.

Sue is deeply committed to developing and maintaining cross-cultural links. She made contact with Montfort College in South India in 2002 and subsequently became a consultant and trainer, developing a successful program for UK counseling trainers to offer short training programs to the college. In 2002 she was made a Fellow of BACP in recognition of this work. She has been involved with a range of committees and work groups within BACP and was Chair of BACP's Division for Independent Practitioners (then PRG) from 1996 to 2000. As well as maintaining a small private practice, Sue currently works part-time as a counselor and trainer in two GP practices, and continues to offer supervision, counseling, and training both nationally and internationally.

Roger M. Solomon, PhD, is a clinical psychologist who specializes in treatment of trauma and grief. He is co-director of the Buffalo Center for Trauma and Loss in Buffalo, NY. He is on the Senior Faculty of the EMDR Institute and provides EMDR training internationally. Formally a police psychologist with the Colorado Springs Police Department and Washington State Patrol, he has consulted with the FBI, Secret Service, Bureau of Alcohol, Safety, and Firearms, US State Department, CIA, and other government agencies. Currently Dr. Solomon consults with NASA, the US Senate, the South Carolina Department of Public Safety, and Polizia di Stato (Italy).

Gordon Turnbull, a graduate of Edinburgh University in 1973, entered psychiatry at the Neuropsychiatric Centre, Royal Air Force Hospital Wroughton, in Wiltshire, UK, in 1980. Previous post-graduate experience had been in general medicine, expedition medicine, and neurology. Appointed Consultant in 1986, his focus turned to psychological trauma after the Lockerbie Air Disaster in 1988 and active service in the Gulf War of 1991 as RAF psychiatric advisor in the field, and first-ever debriefings of British prisoners of war and released British hostages from Lebanon. He developed new treatment strategies for trauma in the RAF, and post-RAF has concentrated on trauma services for police officers, emergency service personnel, and military veterans. Currently, he is Consultant Psychiatrist in Trauma at Capio Nightingale Hospital, London, and the Ridgeway Hospital in Wiltshire, Adviser in Psychiatry to the Civil Aviation Authority (CAA), and Visiting Professor to the University of Chester, UK.

Onno van der Hart, PhD, Prof. Dr. Jacques Mylle is Honorary Professor of Psychopathology of Chronic Traumatization at the Department of Clinical and Health Psychology, Utrecht University, and a psychologist/psychotherapist at the Sinai Center for Mental Health, Amstelveen, the Netherlands. He specialized in the diagnostics and treatment of clients with complex trauma-related disorders, including the dissociative disorders. Both nationally and internationally, he is a clinical consultant and presenter on the diagnosis and treatment of complex trauma-related disorders. He is a past president of the International Society for Traumatic Stress Studies. Currently Vice President of the Institut Pierre Janet, Paris, he is a scholar in Pierre Janet studies. With colleagues Ellert Nijenhuis and Kathy Steele, he wrote *The Haunted Self: Structural Dissociation and the Treatment of Chronic Traumatization* (2006). A new book, *Coping with Trauma-Related Dissociation: Skills Training for Patients and Therapists* (co-authored with Suzette Boon and Kathy Steele), appeared in March 2011. His website, http://www.onnovdhart.nl, contains a number of articles that may be of further interest.

Gerrit C.B. van Wyk, MA (Clinical Psychology), is a clinical psychologist who has practiced in South Africa since 1975. His professional focus has been on the traumatic effects of high levels of violent crime in South African society. He is the founder and director of Traumaclinic Emergency Counselling Network, a national network of trauma practitioners offering early trauma intervention and support to occupational and civilian trauma victims in South Africa. The network has delivered consultative, intervention, and training services to a large number of companies and corporations in South Africa as well as medical aid schemes, NGOs, and government departments.

Gerrit is the recipient of the Merit Award of the SA Society for Clinical Psychology for Outstanding Service to Clinical Psychology. He has served as Affiliate Member for Africa on the Board of the International Society for Traumatic Stress Studies and of the European Society for Traumatic Stress Studies, and he is member of the Diversity Committee of ISTSS. He is also a founder member of the Continuous Trauma Interest Group in South Africa which focuses mainly on traumatic stress in marginalized, under-resourced communities.

He is the author and co-author of a number of book chapters and journal articles, and has presented a substantial number of papers and workshops at national and international conferences, mainly on the problems of early intervention and the Traumaclinic Trauma Support model developed in response to the high levels of violence in South Africa. With the objective of making information and training in the field of trauma more easily and generally available, Gerrit has developed a website, http://www.traumatrainingonline.com, where course material can be freely downloaded.

Xiao Lu Wang, PhD (Industrial and Organizational Psychology), BSc (Psychology), is a Post-doctoral Research Fellow, Department of Social Work and Social Administration Honorary Research Associate, Centre on Behavioral Health of the Faculty of Social Sciences, University of Hong Kong. Dr. Wang's expertise is in industrial and organizational psychology and organization behavior. Her main research interests are workplace wellness and third-sector management, focusing on leadership and management, work motivation and engagement, and staff well-being and development. Her work has been published in the fields of management, psychology, psychiatry, medicine, and social work. She has authored and co-authored 33 peer-reviewed academic publications including international journal articles, book chapters, and international conference papers and presentations, as well as another four under review or forthcoming. Dr. Wang actively participates in trainings and intervention research targeting workplace wellness and NGO management. She has also led and has been involved in various management and research consulting programs for local NGOs (e.g., Caritas, Hong Kong Christian Services, Sheng Kung Hui, Li Ka Shing Foundation, Harmony House, and Hong Kong Society of Rehabilitation) and government departments (e.g., Hong Kong's Hospital Authority).

Zhen Wang is a psychiatrist at the Shanghai Mental Health Center and an Associate Professor of Psychiatry at the Shanghai Jiao Tong University School of Medicine. He established and is managing the Stress and Trauma Research Program in Shanghai.

Marilyn Wignall (aka MacQueen), M Phil., BASOS (Police), Cert Prac NLP, LNCP, Cert.Ed., MIfL., works with the Devon and Cornwall police service as the Force TRiM Co-ordinator and with March on Stress as a TRiM instructor delivering TRiM training to other emergency services. A retired police officer, Wignall enjoyed a diverse police career before completing her service in 2006. During her service she completed two degrees, the latter in the area of stress management in the police. In 1990 she embarked upon a period of research, being awarded a Bramshill Fellowship which resulted in the force appointing Post Incident Colleague Supporters, in-house training, and force policy. In 2003 she trained in the Royal Marines model of Trauma Risk Management (TRiM) and persuaded Devon and Cornwall Constabulary to adopt it as their model of stress management. Since her retirement Wignall has helped to pioneer the use of TRiM in several UK police forces and other emergency services. She gives presentations to a variety of national audiences on TRiM-related matters and in April 2011 received a national award for the Force TRiM program, which was recognized by the Emergency Services Awards as outstanding in policing excellence and innovation.

Xiaoping Zhu is a Principle Consultant and President of China EAP Service Center. He also holds several positions as a Visiting Professor for universities in China, including Shanghai Institute of Foreign Trade, Shanghai Teacher's University, and Graduate University of Chinese Academy of Science.

Part A

The Evolution and Development of Workplace Trauma Support

1

The Evolution of Models of Early Intervention for Adults: From Inspired Help Giving toward Evidence-based Pragmatism

Roderick J. Ørner

First of All: Do No Harm

Edward Munch's painting *The Scream* enjoys iconic status throughout the Western world. It is not difficult to explain why. With consummate skill this Norwegian expressionist artist created a visual image that literally screams silently at us. The face depicted exerts a pull on those who stand before it. We feel engaged by an image of suffering and at the same time repelled by a depiction of danger and threat. Something awful has happened, human suffering is on display, and a powerful impulse to help dominates while something else warns us of inherent dangers. Through such dynamic and primitive processes, this image becomes a vehicle for evoking an unfolding of profound and yet simple interactions between viewers and images. Notably, all happens without a need for words.

All this powerfully demonstrates our capacity for engagement and attachment to suffering in the early aftermath of trauma. Such considerations alerts us to a possibility that for all their positive aspects, impulses to act may in fact be driven more by helpers' self-interests or organizational imperatives within caring professions. Of the many lessons learned during the last 30 years none, is more important than the recognition that urgency of action carries a risk that outcomes may not necessarily be conducive to favorable outcomes for survivors. The legacy of early intervention would have conferred enduring credit on modern psychotraumatology had proponents and advocates of various acute care models reminded themselves not only of the need to act but also to take account of the traps and pitfalls of engaging such primitive processes. As it stands, the legacy is somewhat equivocal and aspects of current practice remain controversial.

Cautious, reflective, and evidence-based approaches may be less thrilling for helpers in the short-term aftermath of trauma but may facilitate improved survivor coping and

International Handbook of Workplace Trauma Support, First Edition. Edited by Rick Hughes, Andrew Kinder, and Cary L. Cooper.

adjustment over the intermediate and longer term. Surely, the defining criterion for instigating early intervention should be a professional and humanitarian aspiration to act in accordance with evidence that is conducive to best possible survivor outcomes. Far too often the impulse to act has been a pretext for questionable initiatives, ill-considered practices, unsubstantiated claims, and short-lasting whims.

The Primitive Dynamics of Early Intervention

Exposure to expressions of distress in others by crying, audible calls for help, photos, television footage, and so on can evoke our keen attention and interest. An aspect of the dynamic interactive processes engaged by suffering is that continued or repetitive exposures become intolerable. Signals of suffering therefore typically and compellingly prompt action to help. Such impulses are immediate and can be difficult to resist, especially if sufferers are relatives, friends, or those with whom we identify.

While modern history teaches us that such altruism and compulsion to care are shaped by circumstance and are by no means universal, it is reasonable to argue that humans are "hard-wired" to attune and respond to others who are in distress, all with the purpose of ending that which is intolerable. By the same token we hope others will be sensitive to our own suffering.

In the case of caring professionals, a core standard of good practice is to step forward into the fray of distress and needs at times of crises. To do otherwise would generally amount to a failure of duty-of-care obligations. Therefore, when statutory or voluntary services are confronted with challenges of delivering early intervention for trauma survivors, the pattern of response has often been powerfully driven by impulses to act immediately. Thus it came to be that actions taken were often urgent and impulsive in the presumed service of survival with dignity.

Standards of practice have far too often been a function of the primitive and reflexive reactions in which they are rooted. Cognitive elaborations, meticulous planning, and careful consideration of survivors' needs have been lower order priorities.

Impulsive imperatives have tended to rule over more cerebral pursuits such as acting after careful reflection about actual needs, formulating testable theories about evoked responses to trauma, being explicit about the etiology of evoked responses and the consequent models that could inform early intervention practices, and so on. Had this occurred, a gain might have been that actions taken would have been reasoned and rationalized so their impact might have been monitored through rigorous investigations of critical components of effective help as well as their relationship to outcomes.

Models, Methods, Techniques, and Theories

This chapter reviews models of early intervention for adult trauma survivors. Examples chosen trace changes in practice that have occurred over time and place models in their contemporaneous historical setting. No attempt is made to debate, argue, or win the case for use of particular techniques or how to render advice about overcoming practical challenges of delivering acute post-trauma services. Giving detailed consideration to merits and limitations of each model is an enormous undertaking also beyond this chapter's scope, and anyway is covered elsewhere in this handbook. Examples chosen for review highlight key themes or constructs that underpin ways of thinking about early interventions, their objectives as well as their aims.

The importance of models for the broad perspectives offered by this book is that they occupy positions between extremes of pure theory on the one hand and action-oriented pragmatism on the other. In their own particular way, models are a potential force for reconciliation and integration of opposites. They offer a prospect of engendering greater tolerance of pluralism in practice while also staking out a claim for which objectives and aims of early intervention are realistic and realizable.

The method of the chapter is to present an historical overview to illustrate the diverse bases from which various perspectives on early intervention have evolved. It will stake out a position that clarifies the foundations of many middle positions between opposites. These models are not necessarily reconcilable in all or even any of their constituent elements and presumptions. The review describes a diverse range of models of early intervention but will not endeavor to extricate putative elements that comprise a "pure" approach. Instead, the emphasis is to illustrate how different models are largely a product of concurrent historical developments and prevailing notions of professional or political correctness.

None of the models considered can claim a monopoly on veracity or effectiveness. Such an aspiration is but a search for a holy grail of omnipotence. Instead, what emerges from the review is that each model defines an idiosyncratic, time-specific perspective on the predicament of trauma survivors. From each are derived different formulations of the purposes of early interventions, outcomes to be achieved, and criteria for monitoring change attributable to help, advice, and support given in the early aftermath of major events.

Remembering, Forgetting, and Reminders: Emergent Models and Historical Context

To understand the trajectory of evolving models of early intervention through history, it is essential to consider the broader socio-political and military contexts in which they arose. Very clear illustrations of how models and historical context mirror one another emerge from a review of early intervention practices adopted by military services at times of war and peace (Weisaeth, 2003). Shepard (2000) has also pointed out that the process of mapping historical trends, and linking these to models adopted for the types and extent of psychological care, reveals recurrent cycles of acting, accumulating experience, refining practices, learning from survivors, and then dismissing it all by forgetting.

Repetitive, oscillating phases of bringing back past practices and then forgetting accumulated experience occur in conjunction with shifts from denial of psychosocial care needs to transient phases of exaggerated emphasis on troops requiring urgent help. In the former stages, presumptions underpinning models of early intervention are of soldiers being passively dependent on access to treatments delivered by professional experts in whose power is uniquely invested a capacity to deliver resolution of evoked reactions. In subsequent transformative phases, a sense starts to prevail that a more moderate perspective fits observations and accumulated experience. A dispassionate weighing up of experience and practice-based evidence invariably recognizes the healing powers of survivors' own strengths and resourcefulness as well as help delivered through psychosocial support networks.

Whenever evoked reactions are appropriately addressed through early interventions complemented by longer term follow-up care, it is as if assumptions take hold to the effect that a problem has been permanently resolved or eliminated. At their evolutionary zenith, models of acute trauma care fall victim to their own success. Early intervention is duly dismissed as redundant and forgotten.

The consequent amnesia and denial persist until circumstances change. A new war may kick-start the cycle again. This engenders the extraordinary spectacle of product champions vying for the "most privileged position" as if their proposals are new, refreshing, and original. Truth told, they are none of what is claimed. Advocates are but actors who replay scenes with many historical precedents. Even at this juncture in time, and probably in years to come, we do well to remind ourselves that models of early intervention have their roots in precedents and traditions that extend into the distant past. Little is new, and claims to be genuinely innovative are illusory.

So it comes to be that models of early intervention transform and transmute in response to changing political, medical, psychological, social, and military circumstances. Other pragmatic considerations, especially those that derive from fluctuating moral and ethical priorities at times of war and peace, exert additional influence. In this sense emergent models and practices to which they give rise are products of the times and circumstances in which they are formulated.

Also crucial for the process of reinventing models of early intervention are prevailing etiological considerations that account for clinical conditions putatively evoked by trauma. Survivor responses are rarely simple, resulting in a wide variety of postulated etiologies of questionable veracity (McFarlane, 2003). In consequence, constructions of combat stress reactions (CSR) that have prevailed through time are many and varied. Even to this day, model building takes place on the shifting sands of diagnostic controversies about acute stress disorder, acute and chronic post-traumatic stress disorder, and other trauma-related reactions. It is not by any means only a matter of academic interest that the DSM-IV (American Psychiatric Association [APA], 1994) and ICD-10 (World Health Organization [WHO], 1994) respectively construe these as anxiety disorders and adjustment disorders. Consistent with formulation, the former promotes fear reduction models of early intervention, whereas the latter places emphasis on practical help and support to improve the quality of survivors' recovery environment or day-to-day coping. Through the 1980s and to the present, more flexible etiological perspectives have engendered altogether different coexistent models of early intervention apposite a range of survivor populations within the military, emergency services, or civilian settings.

As illustrated by the stages of forgetting and remembering, each phase in the dynamic cycles of transition is underpinned by different models of early intervention. Whether the general model in vogue is of denial and not wanting to know (a no-need model) or one that gives heed to compelling impulses to promptly mobilize intensive care for large numbers of survivors (an acute-dependency model) or a more reflective, evidence-based approach (a learning-from-survivor model), each position emerges through selective perceptions of lessons learned from accrued experience. As for the selective bias that operates in these evaluations of evidence, a crucial intervening consideration is that of prevailing psychosocial and military-political imperatives.

Model Building from Modest Beginnings

Awareness of personal, psychological, medical, peer, and social factors in the genesis of CSR reactions stretches back at least to Grecian times. In fourth century BCE, Herodotus and Socrates spoke of soldiers' intensely felt conflict between the imperatives to fight in order to win battles and understandable inclinations to flee to preserve life (Herodotus, 1998, book 7, para. 230, book 9, para. 71). The model they discussed was based on conflict resolution and consideration of how to resolve soldiers' personal dilemmas. While survival and comfort may have exerted a pull in one direction, soldiers would also be warned about dire humiliations suffered through public derision, abuse, and disgrace heaped upon troops who withdraw from battle. In contrast, victorious soldiers as well as those killed in battle would be honored and commemorated through elaborate rituals sometimes involving whole nations. Thus, this model recognized remedial influences exerted by the very same individuals, groups, or communities in whose service soldiers had risked their lives.

The position of mercenaries has always been an interesting one in that their contractual arrangements differ from those of serving personnel drawn from a particular society. Militias, societies, or nations served by mercenaries owe nothing beyond an agreed fee. Inspiration to loyal service has historically been mediated through the mercenary privilege of looting and pillage after victorious battle. But those who were injured had to fend for themselves without mercy. Those killed were buried in anonymous mass graves. For surviving mercenaries there is, to this day, no model of early intervention other than survival through self-care and moving on to further military assignments for new paymasters. More by default than planning, the model by which surviving mercenaries adjusted to acute battle reactions is still to find distractions and live by high-risk strategies likely to retraumatize.

In 1678 the German doctor Johannes Hofer described a condition or state of nostalgia in soldiers characterized by a psychological reaction of longing for the past. Historically it is interesting to note that the military context that gave rise to this new diagnosis was a determination to curtail negative influences exerted by some soldiers upon the morale and operational readiness of front-line units. Typically, those given the new diagnosis were soldiers who had the greatest battle exposure and had been most affected by their experiences. It is interesting to note that although the model of early intervention for this condition was never made explicit, the imperative was to deliver acute services that removed obstacles to combat efficiency. Nostalgia is the predecessor of multitudes of terms since used to describe CSR.

Mass mobilization of citizens for military service became standard practice during the late eighteenth century at about the time of the French Revolution. A new ideology of citizenship engendered a new perspective on disasters and atrocities. The emergent model was based on a view that adverse incidents were rooted in social injustice.

Survivor needs should therefore, at least in principle, be addressed through measures rooted in recognition of a community's collective responsibility to help those affected. During this pivotal phase of history, a social welfare model of early intervention found a natural place within a context of citizen empowerment and the democratic election of accountable public servants.

During the American Civil War (1861–1865), soldiers reported syndromes of responses now recognized as psychosomatic reactions associated with feelings of extreme fear and panic. Drawing upon the then-emergent science of medicine, symptoms were interpreted as indicative of cardiovascular disorder. So it came to be that palpitations, cardiac pain, rapid pulse, and respiratory problems were attributed to overstimulation of special nerve centers at the base of the heart. Battlefield observations defined syndromes of reactions variously labeled "irritable heart," "soldiers' heart," and "Da Costa's syndrome" (Da Costa, 1871; Myers, 1870). In keeping with principles of medicine as practiced at the time, early intervention involved drug treatments to eliminate symptoms of underlying physical or organic pathology.

Hand in hand with increased dependence on medical model treatments for psychological and psychosomatic reactions came demands for specialist medical services at or near front lines. These developments mediated a radical shift in the premises used to develop models of trauma aftercare. As psychosocial perspectives waned, an era of medical-model domination became ascendant. In some quarters this persists to the present and offers simple treatment prescriptions. Disappointingly, these models ignore previously recognized remedial effects of peer support, self-help, and social support. In some respects, this pure medical model set back the cause of securing effective acute help for soldiers with CSR.

An elaboration of the medical model of early intervention occurred at the beginning of the twentieth century when the Russian Army made its first attempts to provide CSR treatments at or very near front lines. This innovation had become a practical necessity spearheaded by problems associated with very long evacuation lines. Although not derived from a recognition of clinical need, a serendipitous merger occurred between strategic imperatives and soldiers' welfare. While still rooted in orthodox medical models of early intervention, military authorities incorporated considerations of military culture and operational imperatives to pioneer some of the now widely accepted principles of forward psychiatry.

When World War I (WWI) erupted, there was scant recognition of the importance of emotions in the etiology CSR then labeled "shell shock" (Myers, 1915). This was consistent with the medical model applied at the time which perpetrated the orthodoxy of constructing suffering as impaired physical health. Remarkably, its etiology failed to take account of the horrors, trauma, and losses associated with high-fatality trench warfare. To have done otherwise would have transformed a seductively simple CSR model into one that was inconveniently complex. Such denials of the obvious mirrored high level of media censorship about what was going on in battle zones paired with a general refusal to know the intolerable. No wonder the war poets met with disbelief and accusations of being unpatriotic.

From July to December 1916, more than 16 000 cases of shell shock were recorded among British battle casualties. Eventually this condition proved to be the third most frequent cause for discharge from the British Army in WWI. At such magnitudes the numbers of affected soldiers must have stretched services beyond their reasonable limits. In part this indicates significant limitations in any model of acute care that disregards emotions and psychosocial processes. But also, it suggests that some horrors and traumas have devastating psychological and health consequences that cannot be addressed by early intervention alone. In such instances, acute care might do well to limit its aspirations to not aggravate personal crises or do more damage.

Further impetus for model change arose from observations of troops at or near front lines. Some soldiers appeared to be particularly vulnerable to CSR. Shell shock also proved highly contagious within particular combat units. That the extent of symptom formation proved to be independent of the degree of exposure to explosions and physical shock was not readily reconcilable with the prevailing model of acute disorder. Notwithstanding such findings, models of early intervention continued to draw heavily on physical causes and reliance on physical treatments for symptom elimination. Remedies were sought by evacuating operationally compromised troops away from the trenches to hospitals for acute treatment followed by medical discharge, demobilization on health grounds, and dispatch home.

Clinical research conducted in nonmilitary settings at about this time generated evidence that eventually prompted a revision of models for early intervention for troops and, for the first time, civilians as well. Particularly notable are Hesnard's (1914) studies of emotional changes in survivors of explosions. In time these helped foster a better understanding of CSR among front-line troops but did not significantly change acute service provision.

Away from the relentless pressures of acute treatment, Myers (1915) concluded the "shell shock" diagnosis was clinically unhelpful. His view was shared by others who in 1917 renamed the condition "effort syndrome" (Merskey, 1991). This represented, at least conceptually, a long overdue shift of emphasis away from physical causes as determinants of early intervention models. Rest and relaxation started to be recognized as core components in effective care. Later studies of English, French, and German soldiers concluded that at least 80% of "shell shock" cases had an emotional etiology (Weisaeth, 2003).

Myers (1940) also played an important role in rejecting presumed connections between various acute trauma-related syndromes and "organic molecular commotion in the brain." He did so by drawing attention to close similarities between soldiers' presentations and those of civilian patients diagnosed with hysteria. The latter condition was known to have a psychological etiology. It can be seen that as social circumstances started to change with public opinion being better informed about the horrors of war, so a view of etiology took hold in which turbulent emotions and conflicts were greater recognition. Consequently, new early intervention models were required. All the same, organic etiologies enjoyed continued favor within European military services even after their presumed physical basis had proven to be largely incorrect.

A wind of change, which still blows, is revealed in the formative influences of Salmon (1917). He took the view that "war neurosis" arose in soldiers caught in a dilemma of dynamic tensions. Resonant with views held in ancient Greece, conflicts arose from tensions between combat situations perceived as intolerable and the reassurances conferred by a more bearable neurotic condition. Salmon pointed out that although acute intervention models involving evacuation or discharge secured survival, they also removed soldiers from peer support in established groups with common objectives and goal. Emergent social network models of early intervention shifted the focus of care toward practical help, guidance, and support provided at or near front lines. For those requiring intensive health care, backup was available in advanced neurological hospitals some miles from front lines or at base hospitals. This model of care is a core aspect of Salmon's principles of forward psychiatry. Novelty of approach is discernible in the notion that effective early intervention could be achieved through peer and professional support, reassurance, and rest paired with an explicit expectation that traumatized soldiers would return to their front-line units (Myers, 1940; War Office Committee, 1922). Occupational therapy was introduced for the first time (Brock, 1918) along with a growing acceptance that troops take an active role in their own recovery as well as that of their peers.

Use of this model within advanced posts of the French Army's medical service neuropsychiatric department as early as 1916 is reported to have achieved a return to front line duties in up to 90% of psychiatric casualties. Interestingly, the outcome criterion was resumed fighting which accommodated the operational imperative of maintaining battle capability.

Greater recognition of psychological processes in the etiology of traumatic stress reactions contributed to a focus on conflicts and dilemmas within individual soldiers rather than traumatic stresses inherent in combat. From this arose a personal conflict resolution model of early intervention. It stipulated that the manner in which soldiers resolved seemingly irreconcilable aspects of their predicament was a function of their personalities. If outcomes of soldiers' attempts to resolve their conflicts were unhelpful for the war effort or in some ways deemed misguided by those in authority, the cause of dysfunction was presumed to be individual weakness (Eder, 1916; Shepard, 2000).

Within this model, CSR was construed as arising from a failure of personal will. As such it had moved from being medical to moralist and judgmental, which in turn inspired interventions involving disciplinary measures such as brutal administrations of painful electric shock (Moran, 1945). Other cruel techniques were used with the explicit aim of inflicting suffering to change the balance of conflicted considerations that made soldiers incapable of fighting. Early intervention followed principles similar to those used by officers faced with insubordination. Refusal to follow orders to return to the battlefield would engender ever more painful punishments until compliance was achieved. Thus the shift to psychosocial models of early intervention created a climate of care in which trauma survivors were not only held personally responsible for their disabling state but also considered to be the cause of their personal failures to benefit from acute clinical interventions (Weisaeth, 2003).

Models of Early Intervention during World War II

Lessons learned about early intervention had largely been forgotten or set aside by 1939. Once again, ill-considered interventions such as evacuation from battle line positions became common practice. Consequently, benefits to affected soldiers were compromised and military forces suffered severe losses of manpower (Ahrenfeldt, 1958;

Stouffer *et al.*, 1949). The notion of CSR casualties being to blame for their condition retained considerable currency within military medical services during this period and has persisted in some quarters to the recent past (Jones, 1994).

Etiological formulations of traumatic stress responses during the initial stages of WWII tended to assume personal weakness in the face of conflicts involving fear of death or being maimed, on the one hand, and loyalty to fellow soldiers, on the other. This perspective had its roots in WWI models sustained by a misinterpretation of published evidence and accumulated experience. One line of investigation had shown that nervous disposition, lack of experience, higher age, and being a reservist increase risks of developing CSR. While some of the worst excesses of WWI-era models were not replicated, moralist and judgmental models engendered clinical approaches involving dismissive judgments of sufferers as malingerers, hysterics, or simulators not deserving of any compassion. A more dispassionate interpretation of this same evidence led others to advocate preventive selection, preventive measures, and better officer leadership training to reduce casualty rates. Early intervention in the form of prevention came into its own during WWII thus making space for psycho-educational models of practice. As the war progressed, preventive measures became better coordinated through the selection of personnel, better training for active military service, teaching officers about group cohesion, developing good leadership practices, and taking care to strengthen motivation and improve morale (Arenfeldt, 1958).

A significant etiological shift occurred about this time. Evoked reactions culminating in CSRs were linked to exhaustion rather than debilitating psychic conflicts. By implication, the core problem addressed by early intervention became tiredness for which a natural remedy is rest in secure settings. The emergent model incorporated principles of ***brevity*** and ***immediacy*** of early interventions. Focus on psychological reactions occupied ***centrality*** of clinical concerns with an ***expectancy*** to rejoin their combat units. This care was provided acutely in close proximity to front lines and was characterized by ***simplicity*** of approach (the BICEPS model of early intervention for treating soldiers: brevity, immediacy, centrality, expediency, and proximity; Freeman, Moore, & Freeman, 2009).

Reviews of wartime studies gauging return rates to front-line service showed success in up to 70% of cases referred for CSR. In conjunction with these psychosocial models, other healthcare professionals took an active role in trauma care (Shepard, 2000). With the successes of this approach, an optimistic view took hold that traumatic stress reaction were typically temporary and transitory psychological states.

WWII was instrumental in promoting models of early intervention that increasingly recognized social and environmental influences in both the genesis and resolution reactions evoked by trauma or critical incidents. In the process of so doing, better informed and more humane links were drawn between similarities in the presentations of troops with CSR and civilian patients diagnosed with neurotic conditions not necessarily associated with overwhelming experience. This was so for psychosomatic disorders and conversion syndromes irrespective of patient group. In this way, some early intervention strategies and subsequent ongoing care for functional disorders remained rooted in the notion of resolving personal conflicts whether experienced in battle or not. Eissler (1986) claimed this approach was effective with positive prognoses.

Models of Early Intervention for More Recent Wars

With the onset of the Korean War (1950–1953), CSR management failed to take account of experiences accrued during earlier conflicts. On this occasion the situation was promptly remedied in order to maintain troop numbers and front-line operational readiness. This was achieved using a BICEPS rest model with stricter-than-normal Salmon principles. In consequence, evacuation rates were as low as 35 per 1000 soldiers in active duty (Glass, 1954).

By the late 1950s accrued expertise had again been thrown to the wind as US military involvement in Vietnam increased. At its start, spurious CSR statistics and low evacuation rates were used to claim excellent morale. But the manner in which this war was conducted differed from that of previous and subsequent conflicts. With its painful progress and many sobering setbacks came a recognition of both the scope and limitations of early intervention as a function of particular circumstances prevailing during military campaigns.

US strategic and operational weakness was created by failures to screen and select troops about to be dispatched overseas. Combat readiness was further compromised by mobilizing troops for 12-month rotations. Front-line conditions increasingly mirrored and instigated broad-based sociopolitical developments stateside. For instance, an increasingly radical and articulate human rights movement drew attention to the fact that disproportionate numbers of ethnic minority soldiers were called up for front-line service. Soldiers and civilians started questioning the merits of this war, and front-line protests were expressed as ordinary soldiers openly defied the authority of officers. Crises of morale and motivation caused a collapse of coherent military purpose in full view of the world's media.

Within the broader moral, ethical, political, military, and racial crises embodied in this war, acute care could not resolve behavioral and disciplinary problems, manage drug abuse, moderate soldier and civilian opposition to the war, or change the perception of its lack of purpose and meaning. It should not go unnoticed that no significant innovations in CSR care took place during the Vietnam War era. If there were any early interventions to speak of, they consisted largely of outspoken opposition to political, social, and military complacency. Never before or since has the model of early intervention been as explicitly political as it was during this time and up to the mid-1970s.

Weisaeth (2001) undertook a comprehensive review of early intervention for soldiers during recent wars, covering the period of armed civil conflict in Northern Ireland that started during the late 1970s, the Falklands War in 1982, and subsequent wars in the Balkans. Significant steps have been taken to interrupt recurrent cycles of gaining experience in the delivery of early intervention that is later forgotten, along with a denial of soldiers' acute-care needs. To this end formal structures are in place to monitor quality of provision (Solomon & Benbenishti, 1986). The BICEPS model of early intervention has been simplified to be synchronous with the nature of modern warfare. It draws extensively on principles of proximity, immediacy, and expectancy (PIE). This is the model applied for NATO forces (Mehlum, 2003).

A 1993 review of early intervention for United Nations peacekeeping operations and troops mobilized for the First Gulf War (1990–1991) drew attention to the diversity of

CSRs evoked by modern warfare. A case in point is "UN Soldiers' Stress Syndrome," a significant risk factor for which are situations in which soldiers are scared of their own anger. This tends to arise when rules of engagement do not allow action to be taken to stop atrocities, injustices, and suffering. Evoked syndromes of reactions are associated with relatively high morbidity (Weisæth, Aarhaug, & Mehlum, 1993; Weisæth & Sund, 1982). In contrast to traditional guidelines for CSR, UN soldiers experiencing problems with self-control are typically evacuated from front-line duties and sometimes returned to their countries of origin.

In keeping with a growing appreciation of diversity in CSRs, the First Gulf War is clinically notable for generating a new diagnosis of Gulf War syndrome. Its complexities have set the stage for a broader recognition that recent wars are fought on many different fronts simultaneously and by different means. Past conceptualizations of CSRs may therefore not apply for more recent or future wars. Models of early intervention deemed effective in past conflicts may no longer be so. Furthermore, clinical observations made in the early aftermath of recent combat exposures point to presentations closely related to medically and psychologically unexplained syndromes such as chronic fatigue syndromes or multiple chemical sensitivities rather than the more typical post-traumatic stress syndromes (Wessely, 2001).

It is a matter of historical record that these more recent observations about the diversity of CSRs and presumed effectiveness of acute intervention for preventing later trauma-related sequelae are echoed in post-war experiences of many veterans of earlier conflicts. Contrary to expectations, many have found that the presumed benefits of acute- and intermediate-phase interventions were more a promise than reality. Possible reasons are that assumptions of etiological similarity between reactions evoked by different trauma in diverse survivor groups are incorrect. Such a formulation fails to consider differences that arise through long-term exposures to intense battle trauma and the typically less dramatic responses evoked in the wake of civilian trauma. Actual life-threatening trauma engenders qualitatively different psychological responses and processes compared to those mediated by dynamic conflicts unfolding within a person's imagination (Gersons, 2003).

Evolving Models for a Rapidly Changing World

The role and status attained by early intervention both military and civilian settings are some of the positive legacies of the Vietnam War. First inspirations were drawn from the long-term plight and problems experienced by some of its veterans. Given their suffering and postwar adjustment difficulties, pertinent questions were soon asked about the prospect of preventing such functional disorders by early intervention initiatives. Thus it came to be that unsubstantiated claims were made for certain forms of early intervention being effective means of preventing post-traumatic stress disorder (PTSD) and related conditions evoked within military, emergency services, and civilian settings (Michell, 1983).

Consideration of which models of acute intervention informed these claims of effective prevention gives rise to much confusion. For instance, Mitchell's (1983) assertions that crisis theory was the basis for critical incident stress debriefing and its numerous derivatives were summarily dismissed by Gersons (2003). A crisis of theory has effected a change toward more pragmatic and practically focused models of acute care. More than ever, interventions take account of survivors' actual needs established through negotiation, consultation, and accumulated experience. For instance, an information-sharing model of early intervention gives priority to explaining the nature of reactions evoked by trauma, their likely development over time, and when to seek professional help. To this end, a booklet or leaflet handed to survivors for their safe keeping might be used (Avery, 2003), as might trauma-related psycho-educational web sites. Gersons and Carlier (1993) adhered to a psycho-educational model in their approach to early intervention after major incidents in the Netherlands.

While past "product promotions" of early intervention claims a heritage derived from particular theories or etiological formulations, current approaches are primarily underpinned by a strong sense of pragmatism. Their objective is to deliver practical help in keeping with survivors' expressly stated wishes (Ørner and Schnyder, 2003) and facilitate psychosocial support with the aim of improving the quality of the recovery environment. Empirical support for this approach is provided by Brewin, Andrews, and Valentine (2000). In their review of the influence of post-incident variables for outcomes, not only did post-trauma social support emerge as the most powerful predictor of the course and development of trauma reactions, but also, on its own, it proved more powerfully predictive than all other variables combined.

As a diversity of early intervention models is being tolerated, so has a change in their presumed objectives and aims. Claims of clinical effectiveness have been substantially moderated in the light of systematic investigations of outcomes, a series of Cochrane Reviews (Rose & Bisson, 1998), and clinical guidelines such as those published by the National Institute for Health and Clinical Excellence (NICE). A more reasoned and rational aspiration to link practice to evidence has brought with it considerable moderations in claims about outcomes that are both realistic and realizable. For instance, claims of effects are no longer phrased in terms of "symptom elimination" or "preventing onset of PTSD." This is entirely consistent with recent more general developments within psychotraumatology. Firstly, that acute symptoms evoked by trauma have little predictive power of later adjustment difficulties (McFarlane, 2003). Secondly, that early reactions may have significant adaptive functions in improving chances of survival. If this is correct, it is ill-advised to intervene with the intention to eliminate reactions that are conducive to survival and improved longer term adjustment (Shalev, 2003; Shalev & Ursano, 2003). No longer preoccupied with symptoms, the focus of early intervention is to foster group cohesion and optimize benefits derived from using social support networks. Reflective practice has tempered impulsive overzealousness in responders and made sincere wishes to help resonant with the actual needs of recent trauma survivors.

Recently advocated approaches to early intervention, such as Support Post Trauma (SPoT) (Rick, O'Regan, & Kinder, 2006) and Trauma Risk Management (TRiM) (Greenberg *et al.*, 2005), derive their models of care directly from the notion of psychological first aid that was first advocated by Raphael (1986). No single model prevails but they have some common elements. Acute interventions are phased according to survivors' needs and incorporate elements of disparate theories and models. In these ways they foster a climate in which tolerant pluralism can set roots. With this changed ethos has come greater tolerance of coexistence and cohabitation so that flexibility of approach is no longer anathema.

Once again, and truth told, when historical perspectives are under consideration, recent transformations of models for acute care mirror ongoing social, political, and military trends. Above all else, more humanitarian approaches have emerged from worldviews that are increasingly global paired with respect for evidence published in the public domain. "Product champions" know they are subject to the balances of public scrutiny and must generate evidence of effectiveness or risk being dismissed.

References

Ahrenfeldt, R. H. (1958). *Psychiatry in the British army in the Second World War.* New York: Columbia University Press.

American Psychiatric Association (1994). *Diagnostic and statistical manual of mental disorders* (4th ed.) (DSM-IV). Washington, DC: APA.

Brewin, C. R., Andrews, B., & Valentine, J. D. (2000). Meta-analysis of risk factors for posttraumatic stress disorder in trauma-exposed adults. *Journal of Consulting and Clinical Psychology, 68,* 748–766.

Brock, A. J. (1918). The re-education of the adult: The neurasthenic in war and peace. *Sociological Review, 10,* 25–40.

Da Costa, J. M. (1871). On irritable heart: A clinical study of a form of functional cardiac disorder and its consequences. *American Journal of the Medical Sciences, 61,* 17–52.

Eder, D. (1916). The psychopathology of the war neurosis. *Lancet, 2,* 264–268.

Eissler, K. R. (1986). *Freud as an expert witness: The discussion of war neuroses between Freud and Wagner-Jauregg.* Madison, CT: International University Press.

Freeman, S. M., Moore, B. A., & Freeman A. (2009). *Living and surviving in harm's way.* New York: Routledge.

Gersons, B. P. R. (2003). Historical background: Social psychiatry and crisis theory. In R. J. Ørner & U. Schnyder (Eds.), *Reconstructing early intervention after trauma* (pp. 14–24). Oxford: Oxford University Press.

Gersons, B. P. R., & Carlier, I. V. E. (1993). Plane crash crisis intervention: A preliminary report from the Bijlmermeer Amsterdam. *Journal of Crisis Intervention and Suicide Prevention, 14,* 109–116.

Glass, A. J. (1954). Psychotherapy in the combat zone. *American Journal of Psychiatry, 110,* 727.

Greenberg, N., Cawkill, P., March, C., and Sharpley, J. (2005). How to TRiM away at post traumatic stress reactions: Traumatic risk management – now and the future. *Journal of the Royal Naval Medical Services, 91,* 26–31.

Herodotus. (1998). *The histories.* Oxford: Oxford University Press.

Hesnard, A. (1914). Les troubles nerveux et psychiques consécutifs aux catastrophes navales. *Revue de Psychiatrie, 18,* 139–151.

Jones, F. D. (1994). From combat to community psychiatry. In R. Zajtchuk (Ed.), *Textbook of military medicine: Part 1. Military psychiatry: Preparing in peace for war* (pp. 227–238). Washington, DC: Department of the Army.

McFarlane, A. C. (2003). Early reactions to traumatic events: The diversity of diagnostic formulations. In R. J. Ørner & U. Schnyder (Eds.), *Reconstructing early intervention after trauma* (pp. 45–57). Oxford: Oxford University Press.

Merskey, H. (1991). Shell shock. In G. E. Berrios & H. L. Freeman (Eds.), *150 years of British psychiatry 1841–1991* (pp. 245–267). London: Royal College of Psychiatrists.

Mitchell, J. (1983). When disaster strikes: The critical incident stress debriefing process. *Journal of Emergency Medical Services, 8*, 36–39.

Moran, C. E. (1945). *The anatomy of courage.* London: Constable.

Myers, A. B. R. (1870). *On the etiology and prevalence of disease of the heart among soldiers.* London: Churchill.

Myers, C. S. (1915). A contribution to the study of shell shock. *Lancet, 1*, 316–320.

Myers, C. S. (1940). *Shell shock in France 1914–18.* Cambridge: Cambridge University Press.

Ørner, R. J., & Schnyder, U. (2003). Progress made towards reconstructing early intervention after trauma: Emergent themes. In R. J. Ørner & U. Schnyder (Eds.), *Reconstructing early intervention after trauma* (pp. 249–266). Oxford: Oxford University Press.

Raphael, B. (1986). *When disaster strikes: A handbook for the caring professionals.* Boston: Unwin Hyman.

Rick, J., O'Regan, S., & Kinder, A. (2006). Early intervention following trauma: A controlled longitudinal study at Royal Mail Group. *Report 435*, November. Brighton, UK: Institute for Employment Studies.

Rose, S., & Bisson, J. (1998). Brief early interventions following trauma: A systematic review of the literature. *Journal of Traumatic Stress, 11*, 697–710.

Salmon, T. W. (1917). The care and treatment of mental diseases and war neuroses ("shell shock") in the British army. *Mental Hygiene, 1*, 509–547.

Shalev, A. Y. (2003). Psychobiological perspectives on early reactions to traumatic events. In R. J. Ørner & U. Schnyder (Eds.), *Reconstructing early intervention after trauma* (pp. 57–65). Oxford: Oxford University Press.

Shalev, A. Y., & Ursano, R. J. (2003). Mapping the multidimensional picture of acute responses to traumatic stress. In R. J. Ørner & U. Schnyder (Eds.), *Reconstructing early intervention after trauma* (pp. 118–130). Oxford: Oxford University Press.

Shepard, B. (2000). *A war of nerves: soldiers and psychiatrists.* Jonathan Cape, London.

Solomon, Z., & Benbenishti, R. (1986). The role of proximity, immediacy and expectancy in frontline treatment of combat stress reaction among Israeli CSR casualties. *American Journal of Psychiatry, 143*(5), 613–617.

Stouffer, S. A., Lumsdaine, A. A., Harper, M., Williams, R. M., Brewster Smith, M., Irving, L., *et al.* (1949). *The American soldier: Vol. 2 Combat and its aftermath.* US Studies in Social Psychology in World War II. Princeton, NJ: Princeton University Press.

War Office Committee of Enquiry into Shell Shock. (1922). *Report of the committee, Cmd 1737.* London: HMSO.

Weisæth, L., Aarhaug, P., & Mehlum, L. (1993). *The Unifil study. Report – Part I. Results and recommendations.* Oslo: Headquarter Defence Command Norway, The Joint Medical Service.

Weisæth, L., & Sund, A. (1982). Psychiatric problems. *International Review of the Army, Navy and Air Force Medical Services, 55*, 109–116.

Wessely, S. (2001). Psychological injury: Fact and fiction. In A. Braidwood (Ed.), *Psychological injury: Understanding and supporting* (pp. 33–44). London: Department of Social Security, The Stationery Office.

World Health Organization (WHO). (1994). *The ICD-10 classification of mental and behavioural disorders: clinical descriptions and diagnostic guidelines.* Geneva: Author.

2

Evidence-based Trauma Management for Organizations: Developments and Prospects

Jo Rick and Rob B. Briner

This chapter looks at how organizational responses to workplace trauma have changed over the first decade of the twenty-first century. It describes evidence-based management approaches and their increasing importance in this context as well as considers some of the more recent developments in organizational approaches to trauma management.

Background

During the 1980s and 1990s, there had been growing awareness of the risks that workplace trauma posed to employees across a range of organizational settings beyond those traditionally perceived as high risk. Increasingly, organizations sought to develop and implement trauma management practices. At the time, the predominant organizational response to trauma was that of *psychological debriefing*, also known as *critical incident stress debriefing*. Normally in the form of a structured single-session intervention, delivered to all people involved in an incident (regardless of individual symptoms) soon after exposure to a traumatic event (within 24–48 hours), psychological debriefing aimed to "reduce the incidence, duration, and severity of, or impairment from, traumatic stress" (Everly and Mitchell, 1999).

The development and use of psychological debriefing techniques in occupational settings that first emerged in the 1980s are described elsewhere in this volume. By the late 1990s, the approach had been adapted for use across a variety of organizational settings. The first systematic review pulling together the best available research evidence on the effects of psychological debriefing had just been published (Wessely, Rose, & Bisson, 1998), as had our own research on managing workplace trauma for the UK Health and Safety Executive (Rick, Perryman, *et al.*, 1998; Rick, Young, & Guppy, 1998). Debate about the safety of psychological debriefing and the limitations of the research evidence was intense. The originators of the technique claimed that

International Handbook of Workplace Trauma Support, First Edition. Edited by Rick Hughes, Andrew Kinder, and Cary L. Cooper.

psychological debriefing would mitigate the harmful effects of traumatic stress and accelerate normal recovery processes. Yet the research evidence available at the time simply did not support these assertions. On the contrary, detailed examination of the most robust evaluations of psychological debriefing showed that, at best, it had no effect on symptoms. More worrying was the finding that, at worst, psychological debriefing could be harmful for some individuals post trauma.

The first systematic review of research evidence (Wessely *et al.*, 1998) concluded:

> There is no evidence that single session individual psychological debriefing is a useful treatment for the prevention of post traumatic stress disorder after traumatic incidents.

In other words, the review found that psychological debriefing was not effective as a treatment – it did not reduce trauma symptoms or prevent the onset of post-traumatic stress disorder (PTSD).

At the same time, there was a lack of evidence or guidance about what interventions, if any, were effective in an organizational setting (i.e., there was nothing to help organizations manage the aftermath of a workplace trauma or know what appropriate support to offer employees). The challenge for organizations, then, was what to do for the best given the limited evidence available?

Back in 2000, we presented a paper at the British Psychological Society's annual Occupational Psychology conference (Rick and Briner, 2000) that attempted to answer just this question. The thinking in that paper formed the basis of an article that subsequently appeared in the journal *Counselling at Work* (Rick and Briner, 2004).

Our original paper attempted to tread a line between the best available evidence on practice and the clear needs of organizations and individuals – to marry the research evidence on clinical effectiveness with organizational and individual need for guidance. Our approach was to highlight the treatment and management strands within the debate and to focus on what was valued within organizations.

Psychological debriefing was typically offered to all employees exposed to a traumatic event regardless of their symptoms or responses. We argued that generally "treating" people who are reacting normally to a situation is not considered good practice. However, sensitive and supportive management of traumatic incidents is of high importance to employees and employers alike.

Our proposed approach for organizations attempted to disentangle the clinical treatment elements of psychological debriefing from what could be considered good management practice. We argued that debriefing appeared to offer a fusion between a clinical practice of dubious benefit and highly valued management processes. The act of introducing debriefing to organizations had necessitated the development of management policies, structures, and procedures for the support of employees within those organizations post trauma. We proposed that removing the problematic elements of debriefing – the intense re-exposure of traumatized employees to the incident they have just experienced (one of the defining characteristics of the psychological debriefing intervention) – but maintaining the management structures and systems could be a way of providing effective support to employees without clinical risk.

At the time there was little research evidence to support this approach; rather, it was theorizing on the basis of clinical evidence, management practices, and existing psychological knowledge about the relationships between individual employees and their organizations.

Since then there has been growing recognition of the importance of using evidence to underpin practice. The concept of evidence-based approaches has permeated many sectors. Understanding what "evidence" means, what constitutes evidence, and whose evidence counts assists organizations in drawing on the best available evidence to make informed decisions about effective trauma management approaches. This chapter next considers developments in evidence-based practice, then describes developments in organizational trauma management.

Evidence-based Practice – From Medicine to Management

Who would deny the importance of using the best available evidence, including evidence from scientific research, to make decisions, particularly decisions that have important consequences? Although few deny this, it has increasingly become apparent that, across a range of disciplines and professions, practitioners are not using evidence as effectively as they might. This has given rise to a clearer definition of evidence-based practice, which is now apparent in many fields including medicine, professions allied to medicine, education, government policy making, criminology, and, perhaps most recently, management. It is worth noting that the subject of this book and this chapter – trauma management in the workplace – encompasses one of the first fields to embrace evidence-based practice, medicine, and one of the last, management.

What is Evidence-based Management, and Where Did It Come From?

There are many reasons why practitioners might not have or use the best available evidence in making decisions: peer-reviewed journals are not accessible to many practitioners; even if accessible, the research articles may be incomprehensible; there are few practice-relevant reviews of evidence; evidence learned in training may become rapidly outdated; practitioners may be under a lot of pressure to make decisions quickly, which is felt to preclude a detailed consideration of evidence; practice may be driven more by fad and fashion than evidence; local politics rather than evidence may determine how decisions are made; and so on.

In the case of medicine, a key event which focused attention on the limited use of evidence was an editorial published in the *British Medical Journal* (Smith, 1991, p. 798) in which it was claimed that "only about 15% of medical interventions are supported by solid scientific evidence." This acted as a catalyst for many subsequent activities all of which aimed in different ways to enhance and expand the use of evidence in medicine. An early and widely used definition was provided by Sackett *et al.* (1997, pp. 2–3), who described evidence-based medicine as "integrating individual clinical expertise with the best available external clinical evidence from systematic research" in making decisions about patient care.

While there are, of course, many differences between management and medicine, the earliest writings on evidence-based management (Pfeffer & Sutton, 2006; Rousseau, 2006) drew explicitly on models of evidence-based practice that had been developed in medicine. One of the most recent definitions of evidence-based management likewise draws on definitions from evidence-based medicine (Briner, Denyer, & Rousseau, 2009, p. 19):

> Evidence-based management is about making decisions through the conscientious, explicit, and judicious use of four sources of information: practitioner expertise and judgment, evidence from the local context, a critical evaluation of the best available research evidence, and the perspectives of those people who might be affected by the decision.

One of the most widespread misconceptions of evidence-based management is that it involves *only* the use of certain forms of scientific evidence. However, as this definition makes explicit, research evidence is just one of four sources of evidence or information that are used to make decisions. We will now discuss some other aspects of this definition.

The four sources of information or evidence

The first source, *practitioner expertise and judgment*, refers to any relevant knowledge and expertise that practitioners already have which they can bring to bear on the decision. It is likely that in many practitioner contexts, this is the source of information that is relied on most heavily. But, as is the case for any form of evidence, some of it may be valid and highly relevant and some of it may not be and, as we discuss below, this is why all forms of information need to be used in a critical way.

Evidence from the local context is any evidence that is specific to and gathered from that context including evidence from inside the organization such as HR data, performance metrics, absence data, organizational history, recent events, changes in the local economy or social context, and so on. This seems particularly important in management as many organizations appear keen to copy what other apparently successful organizations do, yet all interventions need to take account of the local context as what is effective in one setting may not be in another.

The third area of information is *a critical evaluation of the best available research evidence.* Searching for, identifying, collecting together, and critically analyzing a body of evidence relevant to the decision can be a complex and time-consuming process. For this reason, various methods of systematic review and research synthesis have evolved to help make the process more effective. Systematic reviews are essentially reviews on existing research, which start with a clear question or problem. They determine at the start what types of evidence will in principle be relevant and of sufficient quality given the question. They then search systematically through potentially thousands of published studies in order to identify *all* the available evidence. Finally, the data from all these studies are pulled together so we can build up a comprehensive picture of what is known and, equally important, what is *not* known about that problem or question.

Without such reviews, focused on practice problems or questions, it is challenging for individual practitioners or organizations to make best use of this source of information. Indeed, as discussed in this chapter, it was a systematic review (Wessely *et al.*, 1998) of the effects of psychological debriefing which first helped to establish clearly our knowledge about this intervention.

The fourth and last area of evidence that is incorporated into an evidence-based management decision is *the perspectives of those people who might be affected by the decision*. This is for both ethical and practical reasons. If employees feel a particular course of action is particularly harmful or beneficial, then we may want to take this into account for ethical reasons. If employees or managers believe an intervention is unlikely to work for particular reasons, taking account of these views may mean that a better decision will be made.

The conscientious, explicit, and judicious use of evidence

One of the challenges of evidence-based practice in any field, including evidence-based management, is how evidence from these different sources of information can actually be used to help inform a decision. First, the use of evidence needs to be *conscientious*. This means putting effort in to gather the evidence and being persistent in trying to get as much relevant information as possible from each of these sources. Second is the idea that we need to be *explicit* in our use of evidence. This means finding ways to record, detail, and communicate the evidence that is being gathered so that all those involved can see what evidence is there. Last, and perhaps most important, is that evidence needs to be used in a judicious way. How valid and relevant is the evidence we have? We may have a lot of evidence, but after close examination we may decide it is of very poor quality and therefore should not be used to influence the decision.

Evidence-based Management in Practice

If evidence-based management is such a good idea, why aren't we all doing it already? Of course, organizations and managers always use some form of evidence to some extent when making decisions. What is less clear is whether this evidence is gathered from these four sources and if is used in a conscientious and critical way. Some of the barriers to evidence-based practice have been discussed in this chapter. But how can we facilitate the use of evidence-based management approaches? There are many possible ways of doing this and, as the first *Handbook of Evidence-Based Management* (Rousseau, 2012) shows, these entail changes in the way we educate managers and practitioners, changes to the structure and content of continuing professional development, changes in the way researchers review and understand the evidence that already exists, and changes in the way organizations think about and make decisions.

At its core, evidence-based management is about being more open, honest, and critical about what we know and what we don't know and using that evidence and the gaps in that evidence to inform organizational decisions. Our discussion next of trauma management in the workplace provides an excellent example of both the challenges of and need for evidence-based approaches.

Developing Evidence-based Trauma Management for Organizations: What Emerged from the Early Debate on Psychological Debriefing?

What emerged a decade ago was a striking contrast between the conclusions drawn from different sources of evidence. Critical evaluations of the best available research evidence demonstrated the ineffectiveness and possibly harmful consequences of psychological debriefing; practitioner expertise and judgment amongst those delivering debriefing were broadly positive, and commitment to the technique remained strong. Additionally there was research as well as anecdotal evidence that employees (those who might be affected by organizational policy on debriefing) reported finding the trauma management process helpful. McNalley, writing in 2004, summarized the situation as follows:

> Psychological debriefing–the most widely used method–has undergone increasing empirical scrutiny, and the results have been disappointing. Although the majority of debriefed survivors describe the experience as helpful, there is no convincing evidence that debriefing reduces the incidence of PTSD, and some controlled studies suggest that it may impede natural recovery from trauma. Most studies show that individuals who receive debriefing fare no better than those who do not receive debriefing. Methodological limitations have complicated interpretation of the data, and an intense controversy has developed regarding how best to help people in the immediate wake of trauma.

Local context (in this case, the organizational settings in which debriefing was used) differed considerably from the settings in which the research had been conducted and although there was research evidence about the clinical impact of psychological debriefing, there was little recognition of the other powerful drivers of individual and organizational behavior. In addition to the issues raised about evidence and research, a fundamental force for the continued use of debriefing was the individual and organizational need to respond. Raphael *et al.*, writing in the *British Medical Journal* in 1995, eloquently describe these organizational and individual needs (regardless of legal or health requirements) that are the natural response of staff wanting to support colleagues and to see their organization react appropriately:

> Why is debriefing so successful as a social movement and so believed in as an ideology, given that there have been no adequate demonstrations of beneficial effects . . .? Debriefing may be perceived so positively because it meets many needs: the need of those not directly affected to overcome their sense of helplessness and the guilt of surviving, to make restitution, and to experience and master vicariously the traumatic encounter with death; the needs of those directly affected to speak of what has happened, understand it, and gain control; and the symbolic need for workers and management to assist those who suffer and . . . show concern.

Raphael and colleagues went on to suggest, correctly, that debriefing would remain in practice, at least for a time, as it met some real and symbolic needs: psychological debriefing remained a popular organizational response to workplace trauma in some settings.

To some extent the situation reflected the primacy of personal experience, local needs, and anecdotal evidence over critical reviews of research findings. A first-hand account of how psychological debriefing was experienced as helpful, or the absolute commitment of individual debriefers and their belief in the efficacy of the services they provided, often made far more compelling arguments for action than academic journal articles.

However, the preference of one form of evidence over another does have inherent dangers. Research evidence showed that regardless of individual recipient and provider opinion on psychological debriefing, it did not reduce trauma symptoms in exposed populations in general and it could be harmful for some. Its continued widespread use could:

- Provide a largely neutral clinical impact, taking resources away from other possibly more clinically effective interventions.
- Put some vulnerable people at risk of increased trauma symptoms.
- Militate against the development and evaluation of other approaches including "more individualised and longer term programmes focusing on recovery and rehabilitation for those who have been traumatised" (Raphael *et al.*, 1995).

What Needed to Change?

Ten years ago, there were barriers to the implementation of evidence-based practice in organizational trauma management. In particular, several aspects of the evidence available did not meet organizational needs:

- Research evidence about early intervention following trauma was limited to a few studies and focused almost exclusively on one intervention – debriefing.
- The evidence was mainly concerned with the clinical effectiveness of psychological debriefing (i.e., in reducing symptom levels or preventing PTSD) and not with evidence about recipients' perceptions of its utility or helpfulness.
- The research focused on individuals. Although it was acknowledged that workers across a range of occupational groups and in different workplace settings can experience events that can lead to psychological trauma and PTSD, there was no research on employees or organizations.
- There is a profound sense of duty, both legal and moral, for organizations and individuals within them to respond appropriately, yet there was little or no guidance for employees and employers as to how to deal with traumatic situations at work.

There was a need for research evidence about other types of trauma management intervention, considering wider outcomes, located in organizational settings. There was also the need for an approach which could integrate evidence from different sources, particularly where that evidence appeared contradictory.

Building evidence takes time, and for many organizations there was an urgent need for immediate practical guidance about how to respond to traumatic incidents at work. It was this need that our original work sought to address. Hobfoll and colleagues (2007) identify a similar lack of guidance about mass trauma intervention more generally and propose a similar solution:

> [T]o date, no evidence-based consensus has been reached supporting a clear set of recommendations for intervention during the immediate and the mid-term post mass trauma phases. . . .
>
> This has left the field without an evidence-based framework for post-disaster psychosocial intervention. This gap in the field has led to a search for an evidence-informed framework for post-disaster psychosocial intervention. One solution to the lack of direct research evidence for such interventions has been to both extrapolate from related fields of research to create evidence-informed practices and to attempt to gain consensus from researchers and practitioners.

In line with the challenges posed by a lack of specific forms of evidence and the need to take into account a range of perspectives and needs highlighted in this chapter, where robust evidence does not exist to guide practice, we have to draw on proven theory, a range of other evidence, and guidance or practice from other related fields (see Chapter 21 for a fuller description of this approach).

What New Evidence Has Emerged?

First and foremost, evidence has continued to accrue in relation to the impact of psychological debriefing. This evidence does nothing to dispel the concerns first raised by Wessely *et al.* in 1998. On the contrary, an update of the original Wessely review (Rose *et al.*, 2002) serves to reinforce those early messages about the ineffectiveness of psychological debriefing in reducing trauma symptom levels and its possible harmful consequences for some individuals. In addition, the UK National Institute for Health and Clinical Excellence (NICE) – the body charged with assessing evidence and providing guidance on treatments and procedures in the UK National Health Service (NHS) – conducted its own review of psychological debriefing (NICE, 2005). This review concluded that for individuals who have experienced a traumatic event, brief, single-session interventions (debriefing) that focus on the traumatic incident should not be offered.

There have been important developments in other areas. The biggest change perhaps comes in the form of what evidence is now available about what clinicians and organizations *can do* to support individuals in the aftermath of a trauma. Both specific evaluations of interventions and general advice about managing in the aftermath of a psychological trauma have emerged over the last few years.

The NICE guidance identifies a number of responses that could be incorporated into or help guide organizational planning and management of traumatic workplace events. In line with current thinking on the importance of different evidence sources, the NICE guidance is based on three different forms of evidence: evidence from scientific research, evidence from committee reports or opinions and/or experiences of respected authorities, and recommended good practice based on the clinical experience of the NICE guideline development group itself.

More General Guidance about How and When to Intervene

Practical and social support are recognized as having an important role in the recovery of individuals from PTSD, particularly in the immediate aftermath of the trauma. Clearly, where a traumatic incident occurs at work the organization's policies and procedures will determine the immediate response and how the needs of those individuals involved are taken in to consideration. This underlines the importance of organizational responses in the immediate post-trauma period.

The NICE guidelines emphasize the following:

- Provision of practical, emotional, and social support is important immediately following exposure to a traumatic incident.
- Consider watchful waiting when symptoms are mild and have been present for less than 4 weeks after the trauma.
- Arrange a follow-up contact within 1 month.

For individuals at high risk of developing PTSD following a major disaster, NICE also recommends that consideration should be given (by those responsible for co-ordination of the disaster plan) to the routine use of a brief screening instrument for PTSD at 1 month after the disaster.

Finally, the NICE guidance recommends that healthcare professionals should:

- Identify the need for appropriate information about the range of emotional responses that may develop.
- Provide practical advice on how to access appropriate services for these problems.

The way in which such activities can be incorporated in to organizational planning is discussed in detail in Chapter 13.

Specific Interventions for Trauma Symptoms

In the medium and longer term, the NICE guidance focuses on the provision of clinical treatment. Two forms of intervention for the treatment of PTSD have gained prominence: eye movement desensitization and reprocessing (EMDR), and trauma-focused cognitive-behavioral therapy (TF-CBT). Where organizations provide occupational health or staff welfare services, they should consider including these specific interventions (or referral to specialist providers) as part of their trauma management procedures.

The NICE guidance also specifies that providers should:

- Ensure that psychological treatment is regular and continuous (usually at least once a week) and is delivered by the same person.
- Ensure that trauma-focused psychological treatment is offered regardless of the time elapsed since the trauma.

- Do not routinely offer non-trauma-focused interventions (such as relaxation or nondirective therapy) that do not address traumatic memories.

Organization-specific Guidance

The NICE guidance is designed predominantly for use with patients in a health service context. Whilst much of it can help guide organizational practice, it is still concerned primarily with the treatment of symptoms. Other new evidence has started to emerge that more directly addresses organizational and employee needs. This evidence highlights interventions that essentially promote good management practices following workplace trauma.

Trauma Risk Management (TRiM) in the Royal Navy

One such approach, Trauma Risk Management (TRiM), has been developed in the Royal Navy. TRiM is described as

> a proactive, post traumatic peer group delivered management strategy that aims to keep employees of hierarchical organizations functioning after traumatic events, to provide support and education to those who require it and to identify those with difficulties that require more specialist input.
>
> (Greenberg *et al.*, 2011)

The TRiM approach involves selecting serving personnel in junior management positions (with no medical training). These personnel are then trained as TRiM practitioners and embedded within units.

The training is described as wide ranging, covering subjects such as

- The psychological aspects of incident site management,
- how to plan for personnel's psychological needs after an event,
- how to conduct a semi-structured risk assessment interview,
- how to conduct basic psycho-educational briefings, and
- how and when to liaise with managers and medical/welfare staff.

The aim is that personnel trained in the TRiM system will be able to "manage the psychological aftermath of a whole incident" (http://www.kcl.ac.uk/kcmhr/research/trim/index.html).

Anecdotal evidence suggests that the TRiM approach is well accepted within the service. However, rigorous testing of TRiM has also been undertaken by a group based at Kings College London and the Institute of Psychiatry in the United Kingdom. Prior to the TRiM training program being put in place, Greenberg and his colleagues (2008; see also Frappell-Cooke *et al.*, 2010) measured psychological health and occupational functioning on 12 Royal Navy warships. Six of these were then randomly allocated to the TRiM training condition. Following implementation of the training, measures of psychological health and occupational functioning were again taken on all 12 ships between 12 and 18 months later.

The results of the study found that the TRiM approach, as suggested from the anecdotal evidence, was acceptable to those who might benefit from it. Importantly the study also established that there was no evidence of any harm to personnel as a result of taking part in the TRiM process. From an organizational perspective, the group also found that organizational functioning was better on ships that had the TRiM system in operation, suggesting that the approach may benefit organizational efficiency.

Organizational approaches to trauma management

A second study to provide evidence in this area over the last few years is that of the Support Post Trauma (SPoT) system at Royal Mail Group (RMG). The RMG experience is described in detail in Chapter 13, but is worth mentioning here to highlight its similarity, despite a very different organizational setting, to the TRiM approach.

Both TRiM and SPoT:

- Train serving personnel with no medical background to provide support in the aftermath of a trauma.
- Focus primarily on managing the situation rather than claim to reduce symptoms or prevent the onset of PTSD.
- Embody the principles highlighted by the NICE evidence-based guidance of;
  - watchful waiting
  - providing social, emotional, and practical support to the individuals affected; and
  - doing no harm.

Conclusions

Since we first wrote about how organizations should manage traumatic incidents at work, there have been considerable changes in the nature and content of the evidence available to help organizations shape and develop their responses to workplace trauma. Overall, the evidence is becoming more rounded, and whilst there is still a need for further research, the debate is no longer polarized around one form of intervention (psychological debriefing).

What's stayed the same?

Research evidence on psychological debriefing has increased in volume, but the message has not changed:

> Psychological debriefing is either equivalent to, or worse than, control or educational interventions in preventing or reducing the severity of PTSD, depression, anxiety and general psychological morbidity. There is some suggestion that it may increase the risk of PTSD and depression. The routine use of single session debriefing given to non selected trauma victims is not supported. No evidence has been found that this procedure is effective.
>
> (Rose *et al.*, 2002)

> For individuals who have experienced a traumatic event, the systematic provision to that individual alone of brief, single-session interventions (often referred to as debriefing) that focus on the traumatic incident should not be routine practice when delivering services.
> (NICE, 2005)

Although psychological debriefing still has its proponents, other forms of response have come to the fore, backed by evidence of clinical effectiveness or evidence that they carry no risk of harm to recipients.

What's changed?

Broadly speaking there have been changes in three important areas:

- Evidence on the effectiveness of specific clinical interventions for the treatment of symptoms – in particular, trauma-focused CBT and EMDR
- Evidence on overall responses to traumatic situations – identifying what is needed at different stages of the aftermath, what type of intervention is appropriate, for whom, and when.
- Evidence specific to organizations on good trauma management practices.

The result of growing evidence from research and practice over the last 10 years is that organizations are now in a much better position to plan and implement management practices that will assist and support their employees in dealing with potentially traumatizing experiences at work. Chapter 21 looks in detail at how this might be done. Future research will usefully establish the evidence in relation to specific or developing organizational practices and explore how they can be further refined.

References

Briner, R. B., Denyer, D., & Rousseau, D. M. (2009). Evidence-based management: Concept clean-up time? *Academy of Management Perspectives, 23*, 19–32.

Everly, G. S., Jr., & Mitchell, J. T. (1999). *Critical Incident Stress Management: A new era and standard of care in crisis intervention* (2nd ed.). Ellicott City, MD: Chevron Publishing.

Frappell-Cooke, W., Gulina, M., Green, K., Hacker Hughes, J., & Greenberg, N. (2010). Does Trauma Risk Management reduce psychological distress in deployed troops? *Occupational Medicine, 60*(8), 645–650.

Greenberg, N., Langston, V., Iversen, A. C., & Wessely, S. (2011). The acceptability of Trauma Risk Management within the UK Armed Forces. *Occupational Medicine.* doi: 10.1093/occmed/kqr022.

Greenberg, N., Langston, V., & Jones, N. (2008). Trauma Risk Management (TRiM) in the UK Armed Forces. *Journal of the Royal Army Medical Corps, 154*(2), 123–126.

Hobfoll, S. E., Watson, P., Bell, C. C., Bryant, R. A., Brymer, M. J., Friedman, M. J., *et al.* (2007). Five essential elements of immediate and mid-term mass trauma intervention: Empirical evidence. *Psychiatry, 70*, 283–315.

McNalley, R. J. (2004). Psychological debriefing does not prevent posttraumatic stress disorder. *Psychiatric Times, 21*(4). Retrieved from http://www.psychiatrictimes.com/ptsd/content/article/10168/54486

National Institute for Health and Clinical Excellence (NICE). (2005, March 23). *CG26 Post-traumatic stress disorder (PTSD): NICE guideline.* London: Author.

Pfeffer, J., & Sutton, R. I. (2006). Evidence-based management. *Harvard Business Review, 84* (1), 62–74.

Raphael, B., Meldrum, L., & McFarlane, A. C. (1995). Does debriefing after psychological trauma work? *British Journal of Medicine, 310*, 1479.

Rick, J., & Briner, R. (2000). Trauma management vs stress debriefing: what should responsible organisations do? Paper presented at the British Psychological Society Occupational Psychology Conference, Brighton, UK.

Rick, J., & Briner, R. (2004, Autumn). Trauma management vs stress debriefing: What should responsible organisations do? *Counselling at Work.* Retrieved from http://www.bacpworkplace.org.uk/journal_pdf/acw_autumn04_a.pdf

Rick, J., Perryman, S., Young, K., Guppy, A., & Hillage, J. (1998, May). *Workplace trauma and its management: A review of the literature.* HSE Contract Research Report 170/98. London: Health and Safety Executive.

Rick, J., Young, K., & Guppy, A. (1998, November). *From accidents to assaults: How organisational responses to traumatic incidents can prevent post-traumatic stress disorder (PTSD) in the workplace.* HSE Contract Research Report 195/98. London: Health and Safety Executive.

Rose, S. C., Bisson, J., Churchill, R., & Wessely, S. (2002). Psychological debriefing for preventing post traumatic stress disorder (PTSD). *Cochrane Database of Systematic Reviews*, (2), art. no. CD000560. doi: 10.1002/14651858.CD000560.

Rousseau, D. M. (2006). Is there such a thing as evidence-based management? *Academy of Management Review, 31*, 256–269.

Rousseau, D. M. (2012, March). *Handbook of evidence-based management: Companies, classrooms and research.* New York: Oxford University Press.

Sackett, D. L., Richardson, W. S., Rosenburg, W., & Haynes, R. B. (1997). *Evidence-based medicine: How to practice and teach EBM.* London: Churchill Livingstone.

Smith, R. (1991). Where is the wisdom...? The poverty of medical evidence. *British Medical Journal, 303*, 798–799.

Wessely, S., Rose, S., & Bisson, S. (1998). A systematic review of brief psychological interventions (debriefing) for the treatment of immediate trauma related systems and the prevention of post-traumatic stress disorder. *The Cochrane Library*, (4).

3

Large-scale Trauma: Institutionalizing Pre- and Post-trauma Prevention, Intervention, and Treatment

Joyce A. Adkins and Bryan M. Davidson

The science and practice of prevention and intervention related to trauma have been significantly influenced by the experience of the military around the globe, both in the past and in the present. The Global War on Terror represents one of the longest continuous periods of armed conflict involving multiple countries since World War II. Unlike previous conflicts, lessons learned about stress and trauma have been implemented during this period, rather than waiting until after the war to research, analyze, and implement new practices. While neither the science nor the practice has reached optimal levels of effectiveness, numerous insights and best practices can be gleaned for further research and practical advances. Through better understanding the state of the art in the general military and veteran community, the practice in the community at large can also benefit.

The effects of war are many, affecting the whole person to include physical, psychological, social, spiritual, and family health and functioning, while also affecting the organizations and communities in which these individuals and their families live and to which they return post combat. While there are readily identifiable physical and environmental threats to health and well-being, the more abstruse threats are related to psychosocial stress and trauma. Both the environmental variables and the psychological effects associated with those exposures require careful delineation in order to design and implement effective prevention, intervention, and treatment strategies. There can be a tendency to use either a broad-brush generalized approach or an isolated, fragmented reaction to each new situation without taking the time to fully understand the complex dynamics involved, especially when those interventions are perceived as urgently needed such as in a time of war. Unfortunately, that approach can serve to create more instability and can potentially lead to unintended negative consequences.

An overarching strategy to confront stress and trauma in times of war can be guided by the general principles outlined for preventive stress management in organizations (Quick *et al.*, 1997) as well as for the prevention of occupationally related psychological

International Handbook of Workplace Trauma Support, First Edition. Edited by Rick Hughes, Andrew Kinder, and Cary L. Cooper.

disorders (Sauter, Murphy, & Hurrell, 1990). Although the threats to health associated with stress and trauma are not entirely or even primarily psychological, the psychosocial threats to health and the psychological concerns and conditions that result have been the most challenging to address in the profession of arms. There has been considerable research in the area of work-related psychological conditions, but strategies for prevention and intervention can be categorized into four broad categories:

1. *Create and maintain fit and healthy personnel.* Prevent or mitigate adverse effects on individuals associated with risk factors through building a robust system of accurate and effective information, education, and training along with developing positive skills and capabilities for individual-level prevention.
2. *Engineer protection and organizational health promotion factors.* Protect workers from environmental threats or risk factors through modification of the work environment. Focus on the organization, reduce the threats to health posed by the occupational environment, and foster the development of health-engendering factors into the occupational context to provide a protective function for all personnel from a population health perspective.
3. *Provide an enriched, comprehensive continuum of psychological health services* ranging from early intervention to a complete system of treatment, rehabilitation, and return and reintegration to work to counter the potentially negative effects of adverse events.
4. *Conduct ongoing surveillance* of psychological disorders and risk factors as well as routine program evaluation to ensure data-based decision making.

These broad domains correspond to the pillars of the US military's Force Health Protection and Readiness program, which are: Fit and Healthy Individuals, Prevention and Protection, and Full Continuum of Care.

Force health protection is a common term in the military community but often tends to be related to traditional environmental health hazards, such as contaminated water, vector-borne illness, and physical injuries associated with weapons and war-related physical activities. However, force health protection for psychological injuries is also a vital component to protecting the health of the force.

All three pillars of Force Health Protection are embedded in a system of *health surveillance* which supports and refines program implementation and change. In addition, the programs and policies included in Force Health Protection strategies are further enhanced through targeting interventions to particular demographic variables and through the implementation of specific interventions based on temporal variables associated with the circumstances surrounding an individual's occupational, family, and community environment.

Stress and Trauma Exposure: Mapping the Landscape of Military Stress and Trauma in the Context of Combat

The broad area of stress and health has a long, complex history. It is commonly understood that the terminology associated with stress has become vague and imprecise, with informal use of the term creating misinformation in the general public as well as in the professional community. The introduction of issues related to trauma can serve to compound the problem. The relationship between stress and trauma is also unclear. Is there a continuous or discontinuous relationship between stress and trauma? Is trauma just an extreme stress exposure, or is it in an entirely different category of events in the human experience? Resolution of these questions is beyond the scope of this discussion. However, it warrants an explicit statement that stress and trauma are both quantitatively and qualitatively different, albeit closely related, categories of exposure. Occupational stress is an exposure variable associated with the structure and function of work. It is not an individual-level disease state nor does it describe a diagnosis or medical condition. Traumatic stress is also an exposure or experience and is not a label for a disease or disorder. While continuous exposure to unremitting stress certainly can affect the individual's vulnerability to the effects of trauma, they are not the same exposure, which may warrant varying methods for prevention and intervention depending on the nature, frequency, and duration of exposure to each. Occupational stress and especially prolonged occupational stress exposure are risk factors in that they reduce the capabilities of individuals to respond to the experience of traumatic stress.

Military occupational stress

The context of war represents a specific occupational environment for military personnel. The profession of arms brings with it an organizational culture as well as a set of duty requirements which in the civilian world would be considered a workplace or occupational setting. However, military members serve their country rather than engage in traditional work. That service involves a more comprehensive commitment than civilian jobs. A military member serves 24 hours a day, 7 days a week. There are no limits on duty hours or hazardous work environments because of the very nature of national defense. The concepts of occupational stress and preventive stress management continue to apply, but they are magnified by additional military-specific aspects of the total context of never being truly off-duty. Related, there are numerous buffers and protective factors engineered into the military culture that off-set some of the additional risk factors. Those protective factors in the military environment include additional social support that comes from the camaraderie and membership in the military community. In addition, the meaningfulness and purpose of mission that come from national service are also significant buffers to stress. With some notable caveats, however, the science associated with the rich research in civilian occupational stress is relevant to the prevention of threats to psychological health associated with military occupational stress and the potential trauma of combat.

All the elements of occupational stress are experienced by military personnel. However, there are external constraints involved in changing the environment that may not be present in the civilian occupational environment. The upper limit in terms of the number of personnel in the military is set by the US Congress, so when duty requirements increase, such as during time of war, personnel cannot be equally or quickly increased. When personnel deploy, the requirements to maintain home-station operations continue. Therefore the workload and the length of duty day increase proportionate to the decrease in personnel generated by the deployment of sizable portions of personnel to the war zone. Concepts frequently associated with civilian work–life balance can become untenable. These constant, unremitting, low-level stressors generate fatigue in individuals and reduce organizational resilience awaiting the onset of stressors of increasing intensity.

Differential effects of stress

For purposes of better understanding the context of military stress, the general term of *stress* can be divided into several types or categories: acute, chronic, cumulative, residual, and traumatic. Each can have an effect on health and well-being as well as the functional behavior of both the individual and the group or organization in which those individuals live and work.

Acute stress represents a transaction between the individual and the environment in which a threat, pressure, or demand is perceived. In a combat environment, the response by an individual to an acute-stress situation actually facilitates facing potential threats and can improve reaction and action when danger is present. However, no human being is capable of being in a state of constant hyper-alertness for indeterminate periods of time. Human physiology was intended to respond to emergent situations then to recover to a state of homeostasis. *Chronic stress* conditions result when the threat situation and threat assessment on the part of the individual persist without interruption and recovery is not possible. In the face of chronic, unremitting stressful conditions, even persistent low-intensity pressure, the same responses that assist in an acute emergent situation now become counterproductive, not from an individual failing but because the nature of human capability cannot withstand such constant demands. While there are physical and cognitive methods by which individuals can increase their capacity to withstand a state of constant threat, all humans will eventually develop fatigue in situations of prolonged duration of exposure which will degrade human performance and negatively impact health.

Cumulative stress results in situations in which there are multiple, frequent, high-intensity threat situations or curtailed periods of relief from chronic stress situations that fail to remit for a sufficient period of time to allow for recovery. Again, fatigue, degraded performance, counterproductive behavior, and ill health can result from constant and cumulative exposure to stressful situations. The countermeasures for situations of chronic and cumulative stress involve first modifying the situation whenever possible to reduce the intensity, duration, or frequency of exposure. Correspondingly, increasing the capabilities of individuals to withstand the threat situation or to change their appraisal of the situation, to manage their physiological reaction, or to find cognitive methods to facilitate relief from the situation can reduce the effects on health and well-being while also preserving performance integrity.

Residual stress can be used to refer to those internal demands and pressures that each individual experiences as a result of unresolved issues and concerns, commonly referred to as *emotional baggage*. Residual stress varies widely among individuals. These thoughts and emotions most often remain below the state of overt conscious thought, but they still require individual resources for management. While residual stress is generally believed to be limited to situations in childhood and/or early adulthood, significant emotional events, such as traumatic events that occur at any time in the life span, can add to the residual stress experienced by an individual. In this way, baseline PTSD symptom levels resulting from unresolved past trauma can represent one source of residual stress; and recent research has demonstrated that predeployment PTSD symptoms contributed an additional 20% of the variance for post-deployment PTSD scores (Vasterling *et al.*, 2010).

It is possible to reduce this stress burden through resolving the issues, but the effort involved is often so uncomfortable that individuals fail to take that action alone. Psychotherapy, psychological counseling, spiritual counseling, or even social support groups and self-help strategies such as journaling can all reduce residual stress. However, during times of combat, opportunities are limited for individuals to process experiences, especially those involving combat trauma. Combat operations are high-paced and frequently encountered leaving thoughts, perceptions, and emotions surrounding those events largely unprocessed and simply added to the burden of residual stress, reducing the resources available to the individual to manage additional stressful situations without degradation in performance or health. A note of caution is advised in combat-related emotional processing during the actual period of combat; the use of psychotherapy techniques to process past or residual stress or to divert attention to introspective thought has been discouraged during combat. Techniques that distract or divert the war fighter can have detrimental effects on combat operations. Combat stress techniques focus, instead, on bolstering coping strategies and individual-level defenses against thoughts or emotions that may interfere with combat focus or situational awareness during times of active threat in combat scenarios. Applying more introspective psychological techniques during combat can be life-threatening. Use of supportive techniques, peer support, and cognitive reframing can be more effective and lead to increased resilience during post-combat operations.

All these sources of stress tend to accumulate and can surpass an individual's stress management resources. When few resources are available and a traumatic event occurs, individuals who might otherwise manage the situation effectively can be negatively impacted by the event. In this way, reducing the overall stress burden can reduce the adverse effects of trauma by increasing resources available to the individual to manage whatever situations arise.

Operational and deployment stress

With deployments come additional stress hazards associated with the deployment itself and compounded by combat operations which bring potential for significant trauma. *Operational and deployment stress* is associated with maintaining operations in a deployed location, and is separate from potential trauma associated with the experience of combat.

Operational deployments begin long before the individual departs home station. Considerable effort is applied toward training and preparation for the mission as well as preparation of the family for the absence of the military member for an extended period of time. This preparation can begin to create a condition of fatigue. The trip itself is typically long and uncomfortable. When the individual arrives at the deployed location, it is a similar process to anyone who relocates to a foreign country, with a different culture, and a different living environment. In addition, the living conditions are typically not comparable to home life; they are frequently uncomfortable and unpleasant, and meeting basic daily needs can become complicated. Group-living conditions are common, reducing the availability of privacy, with little individual control over personal space. The hours of duty are long, with little availability of leisure time or activities. Off-duty time is limited, if available at all. Individual deployers often remark that a deployment is a marathon, not a sprint. Every day is much the same routine with long hours and little relief from operational requirements or harsh environmental conditions. Military personnel are generally required to adapt instantly to the new living and working environment while also launching immediately into a high-threat combat environment.

The intensity of operations is also high. The war of insurgency associated with what has been called a war on terrorism is quite different from previous global conflicts in that the threat of violence surrounds military and civilian environments. The insurgents are not readily identifiable. With the frequent use of improvised explosive devices, travel of all kinds (and especially vehicular travel) is hazardous. These circumstances lead to a continually heightened state of alert, with urgent and emergent conditions perceived as being present at all times. Deployers refer to this state of high alert as "the juice" or a continuous jolt of stress-related hormones. When time comes for sleep or relaxation, it can be difficult to turn off "the juice", making sleep less than restful. When the deployers return to home station, "the juice" often continues. Moreover, this hyper-alert state can become habitual such that individuals miss the exhilaration when they return to a steady routine at home, leading to increase in risk-taking behavior or use of alcohol or drugs to replicate the feeling that they experienced in the deployed setting.

Social support systems are different. Emotional support, typically obtained from family members and close friends, is not easily accessed. There are numerous forms of communication, either telephone or internet based. However, electronic communication is not always satisfying. The advantage over previous wars is the ease of immediate communication with family. However, this communication is a double-edged sword. Problems that arise at home are reported to deployed personnel, but they have little ability to positively impact problems or circumstances. The knowledge of problems without the control to remedy them can serve to increase perceived stress. For units that train and deploy together, social support can be a strong buffer to stress. However, for those who deploy as individuals, new social networks must be created in the deployed environment before that protective factor is realized.

Combat and traumatic stress

Trauma is associated with a situation in which the individual feels helpless, hopeless, or horrified by an event that potentially affects life or limb of either self or others. In a combat situation, those events certainly can occur for troops whose mission involves the conduct of armed combat. However, trauma, similar to stress, is an individually perceived experience. It can be problematic when people outside the situation tend to rate the potential trauma of a situation based on their own perceptual lens.

The experiences in combat can be, but are not always, considered traumatic situations for military personnel. Military personnel are exceptionally well trained. The extensive, realistic training of military personnel builds a sense of self-efficacy and competence with the goal that they will not feel helpless when confronted with the violence of combat. Reliance on sufficient numbers of combatants along with appropriate and familiar equipment reduces the potential for feelings of hopelessness. Soldiers are not helpless in the face of combat, and they have hope that they will receive back-up from their fellow war-fighters armed with the best technology and equipment available. Therefore, the most effective deterrent to a traumatic response in combat is effective training for the duties at hand along with adequate numbers of personnel armed with reliable weapons and equipment. Individuals removed from the circumstances may inappropriately assume that military personnel respond with the same emotional response they would have in the same situation. However, the experience and perception of the situation are very different for trained military personnel and the general public. Policies or programs based on presumed experiences without the depth of understanding of the full situation experienced by actual personnel can be less than optimal and may not have the intended results.

Of course, no amount of training or preparation can overcome all potential trauma. The experience of witnessing the death of a friend or colleague, or the dismemberment of human beings, results in an emotional response for everyone. Feelings of grief, loss, and horror will be experienced in actual combat. In addition, the experience of causing the death of another human being, even enemy combatants who are threatening your life, holds great potential for traumatic responses. Emotional preparation through education and understanding of what could come about in combat can assist; but building a strong, resilient base of personal resources will be critical to assisting military personnel involved in combat arms.

Two buffers to stress can be useful in this situation: perception of control and social support. Perception of control comes through training, practice, and knowledge. Physiological control strategies that assist military members in understanding and modifying their physiological arousal, such as heart rate and breathing, can increase perceptions of control over individual responses to traumatic situations. Increased social support strategies to assist fellow soldiers in processing emotionally laden experiences can also serve as a strong buffer to the effects of combat stress situations.

However, the frequency and intensity of these situations can curtail the time and social resources available to adequately implement personal control strategies or process experiences with friends and colleagues. In addition, if chronic and cumulative stress hazards have reduced the coping resources of the individual, then the effects of combat can make a more significant and/or lasting impact on the health and well-being of the individuals involved. Reducing the overall burden of stress can build resilience in the face of trauma by increasing resources and capabilities to manage threats to psychological health that emerge. Prolonged and persistent exposure to unremitting nontraumatic stress depletes resources and reduces the ability of individuals to flex or adapt in the face of severe adversity.

It is important to note, however, that a small proportion of military personnel who deploy to a combat region are actually involved in combat situations. Those military personnel who deploy but are not directly involved in the violence of combat are not typically at any higher risk for traumatic sequelae, such as post-traumatic stress disorder (PTSD), than military personnel who do not deploy. They are subject to the impact of general stress exposure associated with the operational environment, but symptoms associated with traumatic exposure are not any more likely than those at home station.

Data related to the experience and effects of trauma

In the Armed Forces, or any population of interest, the attempt to quantify the impact of occupational stress exposure or potentially traumatic events is challenging. The constructs of the underlying occupational stress, traumatic events, resulting sequelae, and successful recovery are all dynamic both in definition and in progression over time. Attempts to assess these variables at the group or population level tend to compound the difficulties. Interpreting data collected in this area requires clarification of underlying definitions and research methodology in order to draw reliable conclusions.

For example, the media have reported rates of PTSD for those returning from deployment as high as 12% or more based on cross-sectional survey research (e.g., Hoge & Castro, 2006; Hoge *et al.*, 2004; Miliken, Auchterlonie, & Hoge, 2007). Given the authoritative backing of peer-reviewed journals, it is extremely difficult to argue against the data. Closer review of all the available data, however, can present a somewhat different picture. Statistics related to combat stress and the associated effects emanate from the methodology of data collection. The US Department of Defense conducts medical surveillance across many disease states or health burdens. In a review of selected mental health disorders of active duty US Armed Forces personnel from 2007 to 2010, the rates of PTSD were markedly lower than the rates reported in the popular media (Armed Forces Health Surveillance Center, 2010). Of the different military branches, the Army was consistently highest, with an average rate of 3% of the population affected with new cases of PTSD, and the overall military population stood at less than 2%. The primary reason for the discrepancy is data collection methodology. The lower rates were derived from actual clinical diagnoses whereas the higher rates were driven by cross-sectional self-report surveys or screening instruments administered to combat personnel and then extrapolated to all deployed personnel.

When research links the rates of PTSD with deployment, the implication is that all deployers experience an equal threat for trauma exposure and subsequent PTSD development. However, deployments alone do not precipitate an episode of PTSD; rather, it is exposure to combat as well as baseline levels of PTSD symptoms that appear to be associated with risk for trauma-related disorders (Vasterling *et al.*, 2010). From 2002 to 2009, there were 3.3 million deployments to Iraq or Afghanistan representing 2 million individuals deploying once and 800 000 deploying multiple times (Tan, 2009); however, only a fraction of those were actually involved in combat operations. From March 2003 through April 2011, 5966 members died from hostile and nonhostile causes in Iraq or Afghanistan and 43 000 were wounded in action (US Department of Defense Manpower Data Center [DMDC], 2011). Deployers in general have different exposures to stress and trauma situations. Extrapolation across these different populations leads to erroneous results. PTSD-like symptoms were more prevalent among those with increased frequency of direct combat exposures (Hoge *et al.*, 2004) and, as would be expected, among those with severe combat-related physical injury (Grieger *et al.*, 2006). Similarly, exposure to lower levels of combat-related stress can actually result in reports of positive mental health changes and a reduction in predeployment PTSD symptom severity (Schnurr, Rosenberg, & Friedman, 1993; Vasterling *et al.*, 2010).

Frequency and duration of operational stress exposure are relevant variables. Duration of deployment varies with personnel deploying anywhere from 4 months upward to over 24 months. Research is scant on the health effects of varying lengths of deployment and the optimal amount of recovery time between deployments proportionate to the length. However, frequency of exposure to armed combat tends to increase the probability of mental health consequences. Hoge and colleagues (2004) found a nearly linear relationship between mental health symptoms and frequency of direct combat exposure. In a group of veterans presenting for medical treatment, Seal and colleagues (2009) also found that greater combat exposure was related to greater probability of presentation of PTSD symptoms. Insomuch as duration of deployment also increased exposure to direct combat, then deployment duration is also a risk factor for psychological consequences.

Results from the Millennium Cohort Study, a prospective study that follows 140 000 military members across more than 20 years to determine the health effects of military service, similarly found that new-onset, self-reported PTSD symptoms were identified in from 7.6% to 8.7% of deployers who reported combat exposures, from 1.4% to 2.1% of deployers who did not report combat exposures, and from 2.3% to 3.0% of nondeployers (Smith *et al.*, 2008). Health surveillance data confirm that military members who never deploy are more likely to experience PTSD as well as other mental health conditions than members who do deploy. While it may, at first thought, seem counterintuitive, those members with existing mental health conditions are likely to have medical restrictions that prohibit their deployment. Only the healthiest and most resilient service members deploy and are more likely to demonstrate characteristics which serve a protective function allowing resilience in the face of the adversity of combat. Studies have confirmed that military service members who screen in the lowest 15% of mental and/or physical health prior to combat exposure are more vulnerable to developing post-deployment PTSD, suggesting that fit and healthy personnel demonstrate more resilience even in the face of trauma (LeardMann *et al.*, 2009). Thus, while all research methods can have utility, the differences in findings are likely to drive very different assumptions about the degree and nature of the problem present.

Unfortunately, the repercussions of overstating the incidence of PTSD may have an iatrogenic impact upon veterans. While unemployment rates in the United States in 2010 hovered around 10% (US Bureau of Labor Statistics, 2011), the unemployment rate for veterans was as high as 21% (Associated Press, 2010), and some believe that the reason for such discrepancy is the perception that employers who hire veterans automatically take on additional risk because veterans are somehow damaged by their military experience. Educational health messages, such as those to inform the public that PTSD is not an automatic response for military members who deploy and that PTSD is a treatable condition, fail to generate the same level of publicity, leaving both veterans and employers with the perception that PTSD renders large numbers of veterans significantly disabled for life. In addition, there has been a significant increase in the focus on brief assessments and screening for PTSD and other mental health conditions and concerns post-deployment, with little research on the potential negative impact of false-positive results.

Similarly, erroneous inferences are possible in association with the rates of PTSD reported over time in our military community. Over the past decade, considerable efforts have been applied to reducing barriers to mental health care, both in specialty clinics and through the addition of behavioral health providers into primary care settings. Efforts to decrease stigma associated with mental health care have also received increased attention. It is difficult to say if any increase in care seeking or diagnosis of mental health conditions is a result of an increased incidence of disease, an increased number of mental health professionals, or a decrease in stigma.

In terms of treatment, it is important to note that PTSD is a treatable condition. Clinical research has found that approximately 50% of cases will remit spontaneously, with complete recovery occurring within three months (American Psychiatric Association, 1994). For individuals who continue to experience symptoms, evidence-based treatments, which include prolonged exposure therapy (Foa, Keane, & Friedman, 2000) and cognitive processing therapy (Monson *et al.*, 2006; Resick, Monson, & Chard, 2007), have been found to be successful in remediating symptoms. The diagnosis of PTSD is, therefore, not a life sentence to misery. If public statements are made suggesting that PTSD is unresponsive to treatment, or if disability ratings issued by military or veteran agencies are predetermined without the benefit of treatment, those actions can work at odds with the efficacy of treatment as well as the return and reintegration of PTSD patients into productive society. Expectations of successful recovery can and should be used to assist in encouraging individuals to seek and remain in treatment, while also bolstering the efficacy of the treatment received.

Psychological research in general has been traditionally more adept at finding or confirming potential problems than in reporting trends that may be within normal limits. While alarming results may garner more resources for an issue, this approach can result in a misalignment of limited resources, leaving significant gaps in services. While PTSD is a valid concern and warrants the presence of effective treatment options, it must be considered within the broader context of the variety of responses to trauma, including normal, healthy recovery. In addition, diverting resources to the identification and treatment of PTSD disproportionate to the scope of the problem reduces resources for health concerns associated with the wear and tear of occupational and operational stress such as depression, anxiety disorders, or substance abuse as well as chronic health conditions such as fatigue, pain, and sleep disturbance. Lifestyle health concerns, such as tobacco use and obesity, are also growing concerns post deployment. Without the recognition that they represent a significant health burden associated with deployment, resources can be misaligned toward sole treatment of trauma-spectrum disorders.

To more effectively prepare for and respond to the threat of trauma at the large-scale, organizational level, it is critical to avoid over-focusing on clinical diagnosis and formal treatment for trauma taken in isolation. The information presented in this section is intended to encourage policy and program personnel to consider the entire spectrum from primary prevention through clinical care, resulting in improved asset allocation. In no way should this discussion be interpreted as suggesting that PTSD is not a serious consequence of war, or as trying to minimize the potentially devastating effect it can have for those who experience this condition, either acutely or chronically. It is intended to broaden the scope of concern to other potential conditions rather than focusing disproportionately on one diagnostic category. In addition, prevention and protection measures must necessarily focus on generalized stress and not exclusively on trauma in order to have the desired scope of effect for all deployed personnel and their families. Given the environment of stress and trauma involved in the military, especially during periods of armed conflict, policies and programs must necessarily be multidimensional following the Force Health Protection model.

Promoting Community, Organizational, and Individual Health

Force Health Protection programs have been developed and implemented enterprise-wide specifically focused on the broad categories of stress and trauma along with potential military-related psychological conditions and concerns. Combat stress is a recognized concern in the military, so programs and policies have been in place throughout history. As the Global War on Terror progressed, the existing programs and policies were subjected to review and revision to keep pace with the best available science as well as to tailor efforts to the particular conditions presented in the current period of conflict.

Fit and healthy force

Prior to entering military service, potential members must meet the physical and mental health standards of their chosen military service. These overall military standards include additional rigorous requirements for particularly high-risk career fields such as aviation or special operations. From this select baseline, preparedness for operational challenges is enhanced by continuous, realistic specialized training in each career field. Detailed, rigorous training within an organizational unit engenders resilience, creates confidence, and fosters esprit de corps that ultimately serve as protective factors against the multiple stressors encountered in the course of military and combat operations. The organizational culture, military values, traditions, effective leadership, peer and commander support, competence that comes from training, and confidence in military equipment and weaponry provide a foundation of overall well-being on which community and healthcare support rests.

From the beginning of basic military training, there are systematic processes in place to transform individuals from a variety of backgrounds into strong members of a cohesive team. Physical strength and endurance are built through regular exercise. Confidence in individual military skills becomes ingrained through classroom and field training along with behavioral repetition or drills. Marching with precision becomes a hallmark metaphor for interdependence that combines individual skill with mutual cooperation. These foundations are important in filling a reservoir of strength from which to draw in times of need and assist the individual in feeling competent and capable rather than hopeless when confronted with the life-threatening experiences of war. This combined confidence in individual skills along with the ability to effectively relate to others is heavily built upon a foundation laid in the first 18 years of life, which reinforces the need for strict accession criteria and liberal use of early separation from the military upon finding a mismatch between occupational demand and an individual's ability to meet that demand (Christeson, Taggart, & Messner-Zidell, 2009). These early force-shaping measures are significant methods for ensuring the maintenance of future health under the duress of potentially traumatic events.

Protecting the force continues at the first duty assignment and throughout a career. Systems have been put in place for honing military technical skills as well as deployment training and readiness. Since 2001, additional emphasis has been given to the psychosocial factors relevant to team cohesion and stress management. The traditional rigor applied to physical health is being carried over to mental, social, and spiritual domains. Comprehensive Soldier Fitness (CSF), the US Army's model for building multidimensional fitness, has received significant attention in the psychology community for its potential application to other organizations (see Cornum, Matthews, & Seligman, 2011). Just as a soldier has personal protective equipment (PPE) such as body armor and a helmet to protect his physical integrity, CSF and other resiliency-building models attempt to reduce the individual vulnerabilities and build up more effective coping strategies – a sort of PPE for cognitive, emotional, and social integrity. Strength-based tools from the field of positive psychology, including those that promote optimism, self-efficacy, flexibility, faith, and effective problem solving, can be taught and learned (Reivich, Seligman, & McBride, 2011). This program not only builds individual skills, but also reduces environmental threats that may be implicit in a unit, such as stigmatizing help-seeking behavior as a sign of weakness rather than a part of the process of growth and a return to full health.

Prevention efforts such as CSF are intentionally nonspecific and attempt to lay a general foundation of stress management skills. Other prevention programs are more targeted and focus on pre-exposure preparation for those units in career fields where exposure to combat or disturbing events is more likely. The goal is to provide interactive-style education about normal reactions to abnormal events, the expectation of a full recovery, the maintenance of healthy behaviors, buddy support, and appropriate help-seeking behavior.

Community and organizational health

Comprehensive support systems are available to military personnel and their family members. These include a wide range of community and family support, family advocacy, legal, educational, and chaplain services, as well as health and fitness programs and facilities. These support programs facilitate prevention, education, and counseling services for the broad array of everyday life stressors to improve and maintain the overall level of health and quality of life of military families. Every effort is made to reduce the background stress of everyday life before, during, and after deployment and to encourage overall positive health states.

Worksite wellness programs. Broad scope health and fitness programs constitute key elements of community wellness programs at every military installation and in theaters of operations. These programs routinely include stress management training tailored for military members and families depending on the nature of stress hazards for the location, career field, and mission of the individual installations and services.

Community and family support. Peer and professional support services are institutionalized across the military community. Family and community support centers, family readiness groups, and family ombudsman programs provide a broad network of support to families.

Chaplains. Multifaith chaplains are an integral part of the military community. They provide family counseling and care for the spiritual needs of the community. A chaplain is assigned to every military unit and provides a ministry of presence that includes getting to know the needs of the unit. They deploy with units to maintain that presence. Although chaplains do not provide medical treatment, they offer confidential counseling and often facilitate access to other avenues of care. They also provide much of the return and reunion educational content for the deployment cycle support program and are an important part of suicide prevention efforts.

Organizational health. Each military service has begun a process of integrating behavioral health specialists into operational units. These consultants provide essential feedback to leadership in maintaining the health of their unit through organizational interventions and through optimizing the health and well-being of their people. Though known by various names, these specialists not only monitor the pulse of the organization through their daily integration into the activities of the unit, but also provide coaching, education, and individual-level consultation and referral both in garrison and through combat stress teams in the deployed environment. While these programs vary in their level of maturation, future standardization and institutionalization hold promise for building sustainable organizational health across the military services.

Early intervention and secondary prevention programs

Employee assistance programs. The US Department of Defense provides a system-wide, confidential, nonmedical counseling and work–life program that can be accessed through telephone, the internet, and email as well as through family and personal counseling in local communities across the country. Counseling and support services are provided for virtually every area of life's problems for all active duty and reserve component personnel and their families. Face-to-face counseling is provided at no cost. The purpose of this counseling is to promote early identification and intervention for issues before they reach clinical significance.

Self-help programs. A number of web-based programs are available to the military community related to stress and coping. These programs range from mental health self-assessment programs, online educational programs, and virtual and interactive programs (e.g., http://www.afterdeployment.org). Use of social media for social support and behavior change has also been popular with younger military personnel and families.

Resilience and strength-based prevention programs. Through advances in the field of positive psychology, the military services have adopted training to enhance performance by leveraging strength-based skills (Cornum *et al.*, 2011). Rather than waiting until people need clinical care, training in cognitive flexibility, behavioral relaxation, and value-centered decision making is designed to sufficiently bolster the ability for members to face stress and trauma effectively without a significant deterioration in functioning or health.

Comprehensive deployment-related health assessment, education, and training programs. Deployment-limiting and disqualifying medical conditions are continuously assessed and identified at multiple points across the deployment cycle. A comprehensive, deployment-related series of assessment and educational procedures have been established to ensure that service members who deploy are in good health and that post-deployment health conditions and concerns are identified and addressed early.

Education is a mandatory requirement of each deployment-related health assessment process. A robust continuum of outreach, awareness, and education has been established to ensure that service members and their families are aware of potential signs and symptoms to watch for as well as the healthcare avenues available to treat these issues. Each individual is different. There will never be an exact right moment in time to assess each and every individual. Arming them with education provides them with tools they need to assess their ongoing situation and seek care when it is appropriate for them. Cross-functional education and training from chaplains, family programs, and mental health functions are provided across the deployment cycle, with specific emphasis on post-deployment reunion, reintegration, and readjustment.

Pre-deployment education and training programs assist military members and families in understanding and preparing for deployments. Return and reunion programs and marriage enrichment programs assist families in understanding the dynamics of separation as well as relationship-related stress often associated both with separation and with combat exposure. In addition, the military requires annual training in suicide prevention for all service members. Suicide data collection and reporting have been standardized to assist in prevention and intervention efforts. Suicide prevention programs have also been implemented in middle schools and high schools with high concentration of military children to include an evidence-based prevention program designed to educate children, teachers, and parents about depression and self-injurious behavior that has proven effective in lowering distress and identifying children at risk for mental health conditions and concerns in the school system.

Veterans outreach and transition programs. The military services have formed unprecedented partnerships with veteran agencies to provide outreach, prevention, and transitional care to our separating and retiring military members as well as to our reserve component members who have been released from active duty subsequent to deployment. The agencies collaborate to provide outreach and education on mental health concerns and benefits and provide information about benefits and services available to veterans. Coalitions formed across the country form a strong community-based safety net to veterans returning to their community after service in a combat environment.

Combat operational stress management program

Throughout the current conflict, the appreciation for mental health assets has risen dramatically, enabling more personnel to receive support tailored to their situation and condition, ranging from basic support through comprehensive interdisciplinary mental health treatment and rehabilitation. During combat operations, combat stress teams provide consultation to leadership in prevention and health promotion strategies specific to combat operations. They provide rest and restoration services for war fighters who experience adverse responses to the rigors of war. They also provide a full range of counseling and mental health treatment appropriate to the theater of war and coordinate medical evacuation when conditions warrant a higher level of care or removal from the intensity of combat. Psychological health experts are also called upon for command consultation when units have been potentially affected by combat events, working collaboratively with unit leadership to decide the most appropriate response to best serve the personnel as well as the mission.

When someone is exposed to a potentially traumatic event, such as in combat, it is common for many to expect a pathological reaction, so a primary role for both military leadership and mental health professionals is to normalize what are initially troublesome reactions. These actions are considered pre-interventional and do not represent clinical care. Simply providing safety and comfort, meeting basic needs, and listening can be effective components in disaster response. While numerous models exist for psychological first aid (National Center for PTSD, 2006), basic elements of support are the core components of all of them. Formal treatment is neither expected nor preferred at this stage.

In recent years, Critical Incident Stress Management (Everly & Mitchell, 1999) procedures in response to traumatic events, including stress debriefings and defusing, have come under increasing scrutiny. The military community historically used Critical Incident Stress Management methods liberally both in a deployed environment and in-garrison. However, with research calling into question the effectiveness of this approach (Devilly, Gist, & Cotton, 2006; Tuckey, 2007), debriefing use has been significantly curtailed. Nevertheless, research within the US Department of Defense has supported time-driven debriefings within the repertoire of care when it is properly facilitated for a voluntary cohesive team (Adler *et al.*, 2007, 2009).

Comprehensive continuum of care

The Department of Defense's health care program for active duty and retired members of the uniformed services, their families, and survivors serves over 9 million beneficiaries (including approximately 1.4 million active duty service members) through a worldwide network of 70 military in-patient facilities, more than 800 military medical and dental clinics, and the department's private sector health plan business partners.

Behavioral health in primary care. Behavioral health providers are being integrated into primary care settings to assist with behavioral components of physical health concerns as well as to assist primary care providers to identify and appropriately treat and/or refer mental health conditions and concerns. This integration helps to reduce the stigma of seeking mental health care for those who may be reluctant or not quite ready to attend specialty care. Behavioral health providers are also being integrated into less traditional primary care portals such as internal medicine, women's health, and pediatric clinics.

Mental health specialty care. Mental health care is available to military personnel and their families in military treatment facilities through civilian mental health providers through the military health insurance program. Reserve component members who have served in combat can receive care through community-based counseling centers, called Vet Centers, and through the Department of Veterans Affairs Medical Centers after return from operational deployments.

Telehealth and technology. Reserve component military members and their families have proven to be a particularly difficult-to-reach population. They do not serve full-time in the military, and the families are not necessarily located in close proximity to a military or federal government facility. To better serve these widely dispersed members of the military community, technology has been leveraged and telehealth methods of both prevention and treatment have been implemented. These efforts are in their infancy, but hold great promise to provide a full continuum of care around the world.

Conclusion

The terrorist attacks that occurred on September 11, 2001, had an impact that was felt around the world. These events, coupled with the subsequent prolonged periods of armed conflict in Afghanistan and Iraq, have brought into public and professional focus the impact of trauma and prolonged exposure to high-stress environments on individuals, organizations, and communities. It is significant to note that historically many advances in the field of psychology have occurred during times of war. It is unfortunately the case that more concern and more resources are allocated to these critical issues when large numbers of people are visibly affected. Regardless of the reasoning, however, the concentrated attention has led to expanded clarification of the processes associated with trauma as well as occupational stress related to combat and deployments along with development of prevention, intervention, and treatment plans and programs to mitigate the impact of war on the military population and communities in which they live. Much remains to be learned, but many of the best practices gleaned from and continuously applied and concurrently studied during the decade following 9/11 have served to shape program and policy development as well as to shape future research and practice efforts both in the military and in the civilian community.

References

Adler, A. B., Bliese, P. D., McGurk, D., & Hoge, C. W. (2009). Battlemind debriefing and battlemind training as early interventions with soldiers returning from Iraq: Randomization by platoon. *Journal of Consulting and Clinical Psychology, 77*(5), 928–940.

Adler, A. B., Castro, C., & McGurk, D. (2007). *Battlemind psychological debriefings.* Washington, DC: Walter Reed Army Institute of Research.

American Psychiatric Association. (1994). *Diagnostic and statistical manual of mental disorders* (4th ed.) (DSM-IV). Washington, DC: Author.

Armed Forces Health Surveillance Center. (2010, April). Selected mental health disorders among active component members, U.S. Armed Forces, 2007–2010. *Medical Surveillance Monthly Report (MSMR), 17*(11), 2–5.

Associated Press. (2010, March 13). Government finds 21.1 percent unemployment rate for young veterans of Iraq, Afghanistan wars. *Minneapolis Star-Tribune.* Retrieved from http://www.startribune.com/lifestyle/87466772.html

Christeson, W., Taggart, A. D., & Messner-Zidell, S. (2009). *Ready, willing and unable to serve: 75 percent of young adults cannot join the military.* Washington, DC: Mission: Readiness, Military Leaders for Kids.

Cornum, R., Matthews, M. D., & Seligman, M. E. P. (2011). Comprehensive soldier fitness: Building resilience in a challenging institutional context. *American Psychologist, 66*(1), 4–9.

Devilly, G. J., Gist, R., & Cotton, P. (2006). Ready! Fire! Aim! The status of psychological debriefing and therapeutic interventions: In the work place and after disasters. *Review of General Psychology, 10*(4), 318–345.

Everly, G. S., Jr., & Mitchell, J. T. (1999). *Critical Incident Stress Management: A new era and standard of care in crisis intervention.* Ellicott City, MD: Chevron.

Foa, E. B., Keane, T. M., & Friedman, M. J. (2000). Introduction. In E. B. Foa, T. M. Keane, & M. J. Friedman (Eds.), *Effective treatments for PTSD: Practice guidelines from the International Society for Traumatic Stress Studies* (pp. 1–17). New York: Guilford.

Grieger, T. A., Cozza, S. J., Ursano, R. J., Hoge, C., Martinez, P. E., Engel, C. C., & Wain, H. J. (2006). Posttraumatic stress disorder and depression in battle-injured soldiers. *American Journal of Psychiatry, 163,* 1777–1783.

Hoge, C. W., & Castro, C. A. (2006). Post-traumatic stress disorder in UK and US forces deployed to Iraq. *Lancet, 368,* 837.

Hoge, C. W., Castro, C. A., Messer, S.C., McGurk, D., Cotting, D. I., & Koffman, R. L. (2004). Combat duty in Iraq and Afghanistan, mental health problems, and barriers to care. *New England Journal of Medicine, 351,* 13–22.

LeardMann, C. A., Smith, T. C., Smith, B., Wells, T. S., & Ryan, M. A. K., for the Millennium Cohort Study Team. (2009). Baseline self-reported functional health predicts vulnerability to posttraumatic stress disorder following combat deployment: prospective US military cohort study. *British Medical Journal, 338,* 1273.

Miliken, C. S., Auchterlonie, J. L., & Hoge, C. W. (2007). Longitudinal assessment of mental health problems among active and reserve component soldiers returning from the Iraq war. *Journal of the American Medical Association, 298,* 2141–2148.

Monson, C. M., Schnurr, P. P., Resick, P. A., Friedman, M. J., Young-Xu, Y., & Stevens, S. P. (2006). Cognitive processing therapy for veterans with military-related posttraumatic stress disorder. *Journal of Consulting and Clinical Psychology, 74,* 898–907.

National Center for PTSD, National Center for Traumatic Stress Network. (2006). *Psychological first aid: Field operations guide* (2nd ed.). Retrieved from http://ncptsd.va.gov/ncmain/ncdocs/manuals/nc_manual_psyfirstaid.html

Quick, J. C., Quick, J. D., Nelson, D. L., & Hurrell, J. J. (1997). *Preventive stress management in organizations.* Washington, DC: American Psychological Association.

Reivich, K. J., Seligman, M. E. P., & McBride, S. (2011). Master resilience training in the U.S. Army. *American Psychologist, 66*(1), 25–34.

Resick, P. A., Monson, C. M., & Chard, K. M. (2007, January). *Cognitive processing therapy (veteran/military version): Therapist's manual.* Washington, DC: Department of Veterans' Affairs.

Sauter, S. L., Murphy, L. R., & Hurrell, J. J. (1990). Prevention of work-related psychological disorders: A national strategy proposed by the National Institute of Occupational Safety and Health (NIOSH). *American Psychologist*, *45*, 1146–1158.

Schnurr, P. P., Rosenberg, S. D., & Friedman, M. J. (1993). Change in MMPI scores from college to adulthood as a function of military service. *Journal of Abnormal Psychology*, *102*, 288–296.

Seal, K. H., Metzler, T. J., Gima, K. S., Bertenthal, D., Maguen, S., & Marmar, C. R. (2009). Trends and risk factors for mental health diagnoses among Iraq and Afghanistan veterans using Department of Veterans Affairs health care, 2002–2008. *American Journal of Public Health*, *99*, 1651–1658.

Smith, T. C., Ryan, M. A. K., Wingard, D. L., Slymen, D. J., Sallis, J. F., Kritz-Silverstein, D., Millennium Cohort Study Team. (2008). New onset and persistent symptoms of post-traumatic stress disorder self reported after deployment and combat exposures: Prospective population based US military cohort study. *British Medical Journal*, *336*, 366–371.

Tan, M.(2009, December 20). A million soldiers deployed since 9/11. *Army Times*, 12. Retrieved from http://www.armytimes.com/news/2009/12/army_deployments_121809w/

Tuckey, M. R. (2007). Issues in the debriefing debate for the emergency services: Moving research outcomes forward. *Clinical Psychology: Science and Practice*, *14*, 106–116.

Vasterling, J. J., Proctor, S. P., Friedman, M. J., Hoge, C. W., Heeren, T., King, L. A., *et al.* (2010). PTSD symptom increases in Iraq-deployed soldiers: Comparison with nondeployed soldiers and associations with baseline symptoms, deployment experiences, and post-deployment stress. *Journal of Traumatic Stress*, *23*, 41–51.

US, Bureau of Labor, Statistics. (2011). Retrieved from http://www.tradingeconomics.com/united-states/unemployment-rate.

US Department of Defense Manpower Data Center (DMDC). (2011). Personnel and procurement reports and data files: Military casualty information. Retrieved from http://siadapp.dmdc.osd.mil/personnel/CASUALTY/castop.htm

4

Commonalities and New Directions in Post-trauma Support Interventions: From Pathology to the Promotion of Post-traumatic Growth

Stephen Regel and Atle Dyregrov

Introduction

The field of workplace trauma support, especially in relation to the provision of Critical Incident Stress Management (CISM) strategies and processes – especially that of Critical Incident Stress Debriefing (CISD), also widely known as Psychological Debriefing (PD) – has been fraught with controversy and confusion throughout the 1990s and 2000s (in both CISD/PD, the structure of the meeting is basically the same, but for the names of the last three phases); however, for the sake of clarity and the purposes of this chapter, the term *Psychological Debriefing* (PD) will be used throughout.

This has been driven not only by confusion over terminology, but also by a plethora of literature, which has continually revisited "old ground" but added little to clarify, inform, and develop the field of peer support group or individual crisis intervention. This chapter will aim to address a number of issues in order to lend clarity to the debate by beginning with (1) providing a brief overview of the development of PD and the direction of research in the field; (2) examining the current trend to re-invent the wheel by renaming CISM and PD interventions; (3) deconstructing some of the myths surrounding PD, especially that of retraumatization through the provision of PD; and (4) examining and illustrating how the literature has added to the lack of clarity and confusion. The chapter will also go on to examine some key issues in the provision of PD within the context of post-trauma support, such as training in PD, the role of assessment prior to initiating peer support group crisis intervention meetings, and what the NICE Guidelines for PTSD (National Institute for Health and Clinical

International Handbook of Workplace Trauma Support, First Edition. Edited by Rick Hughes, Andrew Kinder, and Cary L. Cooper.

Excellence [NICE], 2005) *really* recommend with regard to early interventions. We will then move on to argue for the use of CISM, PD, and other workplace trauma support interventions within the context of social and organizational support given that there is significant evidence to indicate that social support is seen as a significant protective factor following exposure to traumatic events (Joseph, 2003) and the lack of social support as a major risk factor (Brewin *et al.*, 2000). Finally we will argue for a more sophisticated approach to research and practice in the field of workplace trauma support by suggesting a new paradigm for understanding the rationale for providing such support. This will be through addressing issues related to measuring the effectiveness of interventions within the context of post-traumatic growth rather than the presence or absence of pathology, an issue, which has dominated the literature with regard to assessing the effectiveness of early interventions. The promotion of psychological well-being (PWB) and resilience will also be addressed.

The Debate So Far – Deconstructing Myths about Group Peer Support and PD

The provision of CISM and PD as originally conceived by Mitchell (1983, 1988) and Dyregrov (1989) has been utilized amongst high-risk groups in a wide variety of contexts and settings throughout the world for over two decades. It was widely used within the United Kingdom, especially during the 1980s which has often been referred to as a "decade of disasters," beginning with the fire at Bradford City football ground in 1985 with the loss of 56 lives, through to the sinking of the *Marchioness* pleasure boat on the river Thames in London in 1989 with 51 fatalities. In between, of course, there were some very high-profile disasters, the most significant of these being the terrorist attack on Pan Am Flight 103, now known as the Lockerbie bombing, named after the small town in Scotland where most of the wreckage came down resulting in a total loss of life of 270 people, including 11 from the town. In between the dates above in the United Kingdom, there were a number of other transportation, industrial, and sporting disasters resulting in many deaths and injuries.

CISM, PD, and other support interventions were utilized in the United Kingdom by some of the emergency services involved in those disasters, but the published literature is scant with a few exceptions (Hodgkinson & Stewart, 1991). In those relatively early days of post-trauma support, there was of course literature from the United States (Mitchell, 1988) Scandinavia (Dyregrov, 1989), and Australia (Raphael, 1990) describing the use and possible benefits of post-trauma support in the workplace, with emergency service workers and subsequently such interventions came to be practiced in the United Kingdom.

However, it was not long before there were doubts expressed about the effectiveness of early interventions, especially PD, as it had become an increasingly popular intervention. PD gradually became the focus of academic scrutiny (Bisson & Deahl, 1994). The Bisson and Deahl editorial in the *British Journal of Psychiatry* was entitled "Psychological Debriefing and the Prevention of Post Traumatic Stress – More Research Is Needed" and sparked the beginning of a debate where, to paraphrase Churchill, never in the field of crisis intervention has so much been written by so few to the confusion of so many. Though, of course, the "few" have now become many and papers continue to be published with the same message, which is generally that psychological debriefing increases distress and therefore should cease (Raphael, Meldrum, & McFarlane, 1995; Rose, Bisson, & Wessley, 2003). However, Bisson and Deahl asked some very valid and pertinent questions, noting that there were methodological problems in the debriefing studies they reviewed. These were:

- The studies were not prospective.
- They had small sample sizes.
- There was an absence of a control group.
- There were varying degrees of trauma.
- Other confounding variables were ignored.
- Low response rates.
- Sampling bias.
- Lack of uniformity of CISD/PD.
- The timing variance.
- Questionnaire versus interview results.

Firstly, the most notable of the shortcomings noted was that of the lack of uniformity of the provision of PD, which will be addressed in more detail in the discussion on training in PD. Secondly, the methodological shortcomings noted above, whilst valid, did not appear to have had much impact when randomized control trials (RCTs) were chosen for inclusion in the chapter in the NICE Guidelines (NICE, 2005) on "Early Interventions for PTSD." Surprisingly, the studies chosen for inclusion reflected many of the methodological shortcomings highlighted in this section by Bisson and Deahl.

The studies chosen for inclusion in the NICE Guidelines were seven RCTs of individual psychological debriefing: Bisson *et al.* (1997), Conlon *et al.* (1999), Hobbs, Mayou, and Harrison (1996), Lee *et al.* (1996), Mayou *et al.* (2002), and Rose *et al.* (1999), $n = 629$. Studies involved individuals who had experienced a range of traumatic events including road traffic incidents, assaults, miscarriages, fires, and unspecified other incidents. Psychological debriefing was delivered between 10 hours and 31 days after the incident, with duration of 30–120 minutes. Five studies were of individual treatment only. One study included some debriefing of groups of 2–5 PTSD sufferers and another included family members in some debriefing sessions. All debriefing interventions were single sessions and included education about traumatic stress, expression of emotions, and planning for the future. In view of the above, it is hardly surprising that a number of myths and misconceptions have arisen about the provision of PD as there is no sense of logic or cohesion about the research or within the literature. The research has relied heavily upon RCTs. In attempting to satisfy the rigorous methodological criteria demanded of Level I evidence, many RCTs become detached from clinical reality, losing validity and rendering the findings clinically meaningless. Level I RCTs are not the sine qua non of evidence-based medicine. PD challenges the hegemony of RCTs, lending credibility to observational studies and more qualitative-orientated research. Deahl *et al.* (2001) have suggested that future trials of debriefing should employ a wider range of outcome measures than hitherto and assess social and occupational function, personality, substance misuse, and other factors.

Myth 1: all the research indicates it can be harmful and it has negative effects

A frequent comment or observation made by many mental health professionals, when asked what their opinions are of CISM and PD, is usually to respond that these interventions (1) are ineffective and (2) make people worse, and that (3) *all* the research says it is harmful. However, often when pressed further, they can rarely articulate the reasons for these claims.

A fairly straightforward review of the evidence indicates quite clearly that there are only two studies that suggest (in the context of the methodology and outcomes used) negative effects. In the study on burn trauma (Bisson *et al.*, 1997), the authors acknowledged that the vagaries of randomization meant that all the subjects with the highest levels of subjective life threat, previous psychological morbidity, and previous psychological treatment, all factors predictive of poor psychological outcome, were in the intervention group. In addition, the authors described PD as involving "intense imaginal exposure to a traumatic incident" (p. 80). Imaginal exposure is a psychological technique, which is often used with trauma survivors as part of a trauma-focused cognitive-behavioural therapy (CBT) package for PTSD. As a technique it involves the patient reliving the traumatic experience as if it were actually occurring again, describing their experience in the first person and in the present tense. It is a demanding and anxiety-provoking procedure, which is conducted only after careful assessment and consent of the patient within the confines of an established therapeutic relationship. PD does not entail "imaginal reliving" of the event, merely a brief overview of the individual's (and group's) perspective and experience of what happened, their thoughts and reactions in the context of that exposure, followed by education, advice, and guidance about common reactions, the course of such reactions, suggestions for future coping, and "signposting" to appropriate support and assistance if any signs of initial distress persist, increase, or become increasingly problematic.

The study involving road accident victims (Hobbs *et al.*, 1996) also had significant limitations that were not described in the original paper as submitted by the *British Medical Journal*. The authors in a later book chapter, describing the study in detail, acknowledged that: "the interventions were undertaken instead by the research assistant. The intervention therefore immediately followed the screening interview, with which it became merged to some degree, and interviewer 'blindness' was inevitably compromised" (Hobbs & Adshead, 1997, p. 167).

Furthermore, the two studies just mentioned failed to achieve equivalent group membership at pretest (debriefed groups had more severe injuries in both studies). These differences may well have influenced post-intervention outcomes. Moreover, the deterioration in the psychopathology of the debriefed group in the road accident study, although statistically significant, was so slight as to be clinically irrelevant. Therefore, the two *most quoted studies* that cast doubts on the efficacy of debriefing are methodologically flawed and thus cannot be seen as representative of research in this field.

In addition, recent group randomized trials (Adler *et al.*, 2008) and early intervention studies (Adler *et al.*, 2009), conducted with military populations found that PD was well received, was not found to be harmful, and appeared to have positive outcomes relevant to a military organization. The Adler *et al.* study (2008) found that one of the "excellent features" (p. 262) of PD was "its sensitivity to, and engagement in, work cultures and its emphasis on peer processes" (p. 262).

Myth 2: psychological debriefing is a form of therapy or counseling

It is almost impossible to find any critique or discussion about early interventions for trauma that mention counseling and PD in the same context. Many of those who strongly support the notion of intervening early after trauma (including the authors of this chapter) have consistently made the distinction between interventions designed for high-risk groups such as those in the emergency services and the military, such as CISM and PD, and trauma-exposed populations in a civilian context (Busuttil, Turnbull, & Neal, 1995; Dyregrov, 2001, 2003; Regel, 2007, 2010; Regel, Dyregrov, & Joseph, 2007; Richards, 1994). The present authors have also highlighted and clarified these distinctions further by describing the implications of recent research on early interventions with individuals and families (Dyregrov & Regel, 2011). The importance of disentangling the distinction between the origins of CISM and PD and the populations it was originally designed to be used with, and the subsequent inappropriate application of its use, cannot be over-emphasized.

However, the trend for erroneously describing PD as counseling or treatment was given the most emphasis and impetus by the *Cochrane Review*. The Cochrane report on PD (Rose, Bisson, & Wessley, 2003) has been interpreted as providing evidence that PD could have negative effects on people because it concluded, "There is no current evidence that psychological debriefing is a useful treatment for prevention of post traumatic stress disorder after traumatic incidents. Compulsory psychological debriefing for victims of trauma should cease" (p. 10). This conclusion was arrived at not only through the inclusion of a number of poorly designed early intervention studies, but also what was clearly a complete misunderstanding and misconception of the aims and of objectives of CISM and PD by describing the intervention as a "treatment." In addition, PD was rarely, if ever, compulsory. Furthermore, given that PD as an intervention is used between 72 hours and 14 days post event, even the most inexperienced clinician would question its use as a "treatment" for PTSD as the condition cannot be formally diagnosed until one month has passed.

Myth 3: the aim of CISM, PD, and early intervention is to prevent the development of PTSD

When CISM and PD were first conceived and introduced, Mitchell (1983) suggested that the PD component would prevent the development of PTSD. This notion has, of course, not only inevitably given rise to many of the misconceptions about the aims of CISM and PD but also influenced the focus and direction of the plethora of literature (i.e., the reviews and meta-analyses generated by the debate, the early research, and perhaps most significantly what constitutes early interventions for trauma). However, the aims and rationale of providing early interventions for high-risk groups (and other populations) have moved on significantly since the development of the model.

CISM in its current form (despite what it is called or described as) is designed to provide peer support, education, and monitoring of individuals and groups of emergency workers, such as fire-fighters and police officers, exposed to potentially traumatic experiences through the course of their work. Therefore there is an acknowledgment that emergency service workers may present differently from other trauma survivors, where a single traumatic event is the primary focus of their fears. The armed forces are also more commonly involved in peacekeeping or humanitarian duties and, as a result, they can be exposed to considerable human suffering, with no immediate threat to themselves and in this respect are increasingly similar to emergency workers. Typical risk scenarios in emergency workers include:

- Repeated experience of a variety of traumatic incidents which entail varying degrees of a sense of personal threat often combined with the witnessing of harm or death to others, rather than after a single incident
- An incident where the individual makes some personal identification with a victim or event
- Repeated intense exposures over a period of time leading to accumulated risk
- Major terrorist incidents, or disasters with multiple loss of life, especially those including children

CISM programs within these organizations comprise a number of elements, which include:

- *Pre-crisis education*: within and for individuals and groups within organizations
- *Assessment*: of the nature, potential, and actual impact of the incident on individuals and groups involved
- *Defusing*: a brief peer group support meeting within the first 12–24 hours
- *Psychological Debriefing (PD)*: the structured group meeting held 72 hours to 14 days post incident
- *Specialist follow-up*: for ongoing psychological therapy support if necessary; usually provided "in house" by organizations or outsourced where necessary

Myth 4: it was and still is compulsory in many organizations

One of the reasons that PD was seen to be unhelpful was the perception that it was a mandatory process. This myth has been promulgated, again, by a clear lack of understanding of the history, development, and contextual use of PD as a crisis intervention process (Wessley, 2003). It has never been compulsory for many of the organizations that use PD as part of a CISM protocol. Both authors of this chapter have conducted regular and frequent training on CISM and PD with many of the emergency services in the United Kingdom, including agencies abroad (e.g., the International Committee of the Red Cross [ICRC], the United Nations High Commissioner for Refugees [UNHCR], and UNICEF), and are not aware of it being a compulsory procedure. This also includes two major police forces in the United

Kingdom, the Greater Manchester Police (GMP), and the Police Service of Northern Ireland (PSNI). However, it *is* compulsory within the context of Trauma Risk Management (TRiM), the CISM process in current vogue, as described above, within the Royal Marines, the British Foreign and Commonwealth Office, the BBC, and a few emergency service providers in the United Kingdom.

Myth 5: many clinical practice guidelines recommend it should be discontinued or ceased

The focus here has been the NICE Guidelines and the chapter on "Early Interventions for PTSD" (NICE, 2005). This is a misleading title as the chapter covers a variety of treatment options for PTSD; however, to include a brief crisis intervention strategy designed as a form of peer support to be carried out within the first month following exposure to a potentially traumatic event, before PTSD can be diagnosed, only serves to further confuse any reasoned discussion surrounding early interventions for trauma. On scrutiny of the NICE guidelines on early interventions, there is a cautionary note, which states:

> given the evidence that there is unlikely to be a clinically important effect on subsequent PTSD, *we do not recommend that systematic, brief, single session interventions that focus on the traumatic incident are provided individually to everyone who has been exposed to such an incident.* (p. 84, emphasis added)

This has never ever been the case within statutory bodies, especially in the United Kingdom, without a consensual agreement of the parties affected. In the experience of one of the authors (SR), there were numerous Occupational Health and Welfare Departments in the emergency services in the United Kingdom who felt completely at odds at what they believed was good practice in providing support for individuals affected by incidents in the workplace. This recommendation was incongruous with their reality of having to see many individuals who were affected, but who were not part of a group (e.g., the police officer confronting a life-threatening situation alone). They would often question their normal good practice of an individual assessment, followed by appropriate interventions and follow-up, and often needed reassurance that they were not contravening good practice. In reality, individual psychological debriefing is almost never conducted with every individual who has been exposed to a traumatic incident unless there is a thorough needs assessment of the incident, the impact on the individual or group affected, and a range of other factors, which will depend on the context and circumstances.

The NICE guidelines also cite the British Psychological Society (BPS) report, *Psychological Debriefing* (BPS, 2002), which states that: "The provision of psychological debriefing as a community support and cohesion strategy rather than a treatment intervention to prevent PTSD is beyond the scope of this guidance" (p. 82). Finally, the NICE clinical practice recommendations within the guidelines indicate that:

> All health and social care workers should be aware of the psychological impact of traumatic incidents and their immediate post-incident care of survivors and offer practical, social and

> emotional support to those involved. Support and guidance is likely to cover reassurance about immediate distress, information about the likely course of symptoms, and practical and emotional support in the first month after the incident. (p. 85)

What the guidelines fail to do is to describe how this should be done, and therefore this has been left open to interpretation.

Commonalities in CISM Interventions

CISM has been described as a comprehensive, systematic, and integrated multicomponent crisis intervention package. It enables individuals and groups to receive assessment of need, practical support, and follow-up following exposure to traumatic events in the workplace, facilitating the early detection and treatment (where appropriate) of post-trauma reactions (Everly, 1989).

Over the past decade, there have been a number of attempts to reinvent CISM, especially after the publication of the studies described above, the *Cochrane Review*, and the NICE Guidelines. Around the late 1980s, the London Metropolitan Police, and many other police forces in the United Kingdom, utilized CISM including PD, but adapted the technique from Mitchell and Dyregrov, which became known as the *Three-Stage model*, comprising:

- Facts (equating to the “introduction and facts” stage of psychological debriefing)
- Feelings (equating to the “thoughts and reactions” stage of psychological debriefing)
- Future (equating to the “normalization, future planning and coping, and disengagement stages” of psychological debriefing)

This adaptation of the technique was in use for a number of years, but a current trend is to retrain many peer support teams (e.g., in the emergency services in CISM and PD), combining all the main aspects of CISM and utilizing one of the author’s (AD) approach to training in PD, which is the most “hands-on” aspects of CISM for peer supporters. However, in view of the controversy generated by selected studies, the *Cochrane Review*, and the NICE guidelines, there has been a trend for some organizations to develop new CISM models. As a result, CISM still continues to be used by a number of organizations but described and “packaged” in different ways; for example, critical incident processing (CIPR) (Galliano, 2002), which contains all the CISM elements recommended and described above; and emotional decompression (Kinchin, 2007), a relatively recent addition to the modification of the PD aspect of CISM. It is a hybrid of the three- and seven-stage psychological debriefing techniques, but with the stages renamed. There are almost no differences with the basic PD structure.

There are two further additions to CISM developments. The most popular addition to the CISM literature and practice is Trauma Risk Management (TRiM). The TRiM model was developed by Norman Jones and Peter Roberts, two experienced mental health professionals working in the British military psychiatric services, specifically for the British Royal Marines (Jones & Roberts, 1998). TRiM is described as a post-traumatic management strategy based on peer group assessment for hierarchical organizations (Jones, Roberts, & Greenberg, 2003). TRiM is CISM in everything but name, despite efforts to refute this in the literature (Greenberg *et al.*, 2008). It also contains all the CISM components as advocated by Mitchell and Everly (2003), such as assessment, planning, risk assessment, support and monitoring, follow-up, and early referral for specialized therapeutic intervention if required. Most importantly, it utilizes the three-stage technique of psychological debriefing (Facts, Feelings, and Future) for individuals and groups 72 hours post incident. This aspect of TRiM is conducted using a Before, During, and After (BDA) grid, and the authors acknowledge its adherence to the three-stage model developed by the London Metropolitan Police and adapted directly from the universally familiar Mitchell seven-stage debriefing model. Even a cursory inspection of the BDA grid demonstrates all the questions asked in the Mitchell model (e.g., "What happened during the event? Who was present? How do you think it happened?" And, crucially, "Describe the emotions that were experienced."). The authors state clearly that "It builds on the positive aspects of psychological debriefing, whilst aiming to avoid deep emotional exploration related to the traumatic event" (Jones, Roberts, & Greenberg, 2003, p. 474). This raises an issue that goes to heart of the debate and the controversy surrounding debriefing, one that has also attained the status of an urban myth and permeated all the literature, most specifically that PD involves a "deep exploration of the emotional aspects of the traumatic event and it is this aspect which makes people worse." It is worth remembering that the *only* evidence of recipients of PD being made worse was in the two studies most frequently cited as described above, and there was minimal training offered to the debriefers (an issue we shall address in this chapter).

Another recent development within the US military has been the use of "Battlemind Debriefing" (Adler *et al.*, 2009). This utilizes an overall CISM structure, and the staff conducting the interventions were trained in the seven-stage Mitchell model, but received further training for the Battlemind intervention. The Battlemind model merely makes adaptations to the Mitchell framework. The debriefings in this method were also mandated, but there was a difference in the group sizes (i.e., platoons ranging from 18 to 25 individuals or large groups ranging from 126 to 225). Again, as in TRiM, there is an introduction phase, an event phase, a thought and reaction phase where participants were asked to "describe the worst part of their experience" (p. 933), a symptom phase, and then a teaching phase. In the final phase, participants were provided a positive framework for thinking about the deployment and mental health resources identified. As can be seen from the descriptions of TRiM and Battlemind interventions, both draw heavily on the original CISM and PD interventions but have tailored them slightly differently to suit a very structured military context; also, both are mandatory.

On close inspection, therefore, models that purport to offer different solutions to post-trauma support can all be seen to involve PD and CISM under new acronyms. This also means that many organizations are offering similar post-incident support systems but modifying the process to suit their organizational context and needs (e.g., TRiM and Battlemind, which are well suited to a military context but perhaps less suited to civilian settings).

Training in CISM and PD

The issue of training, particularly in relation to PD, has never been significantly addressed in the literature. However, the research suggests that debriefing tends to be beneficial only when led by a trained, experienced debriefer (Arendt & Elklit, 2001; Dyregrov, 1999; Mitchell & Everly, 1995, 2003; Regel, 2007). Mitchell and Everly (1995) recommend mental health professionals to become familiar with the emergency services "culture" before providing any service to those personnel, so that participants can relate to them. Therefore, a cultural knowledge of the organization is vital if the credibility of the facilitator is to be established, which is why many organizations, particularly the emergency services and the military, advocate the use of peer supporters.

The major factor, that may also have contributed to the negative outcomes achieved in the two most oft-cited studies, suggest that a lack of training for those facilitating the interventions was a significant confounding variable. This was recognized by the NICE guidelines in the chapter on Early Interventions, "The training and qualifications of the debriefers was not comprehensively described in any of the studies" (p. 83). It is also later stated in the Clinical Summary on Early interventions for PTSD, "there is a paucity of methodologically sound early intervention studies, containing detailed descriptions of training and fidelity checks on interventions used" (p. 84).

The British Psychological Society's *Report on Psychological Debriefing* (BPS, 2002) highlighted the need for ongoing training, assessment of competence, regular updating, and supervision of practice in the chapter on "Training, Supervision and the Assessment of Competence" (see Hughes, 2002, pp. 25–30). The main BPS recommendations with regard to training were as follows:

- There should be a comprehensive understanding for the use of CISD/PD.
- Specialist training should be undertaken.
- There should be a learning of a series of competencies.
- There should be assessment of facilitators' competencies.
- There should be regular updating and professional development.
- There should be regular practice and reflection on practice.
- There should be supervision of practice.
- People selected for training should have a background in mental health or counseling.

Both authors would agree with all of the above, but for the last of the recommendations (i.e., that only those with a background in mental health or counseling should facilitate PD). From our extensive experience of conducting training courses for debriefers, we have found that a mental health background is no guarantee for making good debriefers, and many peer-support personnel (e.g., police and fire fighters) have shown remarkably good skills for understanding and facilitating crisis groups. How well debriefers are trained in the structure and process of a debriefing will determine their ability to lead the group. It is also often evident from our extensive experience of training in this field that mental health practitioners can struggle with the structure of a PD and frequently slip into a "counseling" mode, tending to focus on one individual rather than consider and contain the needs of the whole group. In addition, training should be conducted in small groups of 12–14 people, with ample opportunity for practice and video feedback of practice, something both authors have advocated and taught over the past two decades.

A training model on CISM and PD developed by one of the authors (SR) and utilized within the Centre for Trauma, Resilience and Growth (CTRG) has been to conduct a 5-day training program for a maximum of 14 participants, mainly peer supporters from the same organization. This comprises one-and-a-half days of theory; for example, common responses to traumatic events, the course of these reactions, assessment of risk and vulnerability factors, assessment of incidents and predebriefing preparation, current research and developments in the field of early interventions, CISM, and PD. The main theoretical influence on the facilitation and other essential aspects of PD taught is Dyregrov's (1997) model, which places more emphasis on process than the Mitchell model. The former has been developed within a European context and may reflect a different tradition for groups and structure than in the United States.

This is followed by a live, videotaped demonstration of the PD process. The rest of the course is dedicated to the practice of PD whereby each participant has to practice the facilitation of a carefully scripted, organizationally specific PD role-play, lasting approximately 45–50 minutes. Throughout the training, close attention is paid to questioning styles, using cognitively framed questions. Emphasis is also placed on the process and the educational aspect of crisis intervention. Role-plays are videotaped and assessed using a Competency Checklist developed by the CTRG to facilitate individual, detailed feedback to the participants. Every participant is given the opportunity to have three perspectives of the PD process throughout the training: that of facilitator, group member, and observer. The role-plays are always facilitated in pairs, and participants are given a DVD of their practice for reflection and use as an aide memoire for the future. There is always detailed discussion and feedback after each role-play. Emphasis is placed on maintaining the structure and process, including all the practical aspects of attention to environment, the nature and culture of the group, microcommunication, and the management of potentially difficult disclosures and situations that may occur in the PD process (see Dyregrov, 1997). The theoretical and practice content are currently being carefully modified to reflect the emphasis on the promotion of psychological well-being, resilience, and growth.

The CTRG training also promotes safe practice through the careful assessment of incidents that may require group crisis interventions or peer support. The assessment of incidents has been paid little attention in the literature with a few exceptions (Dyregrov, 1999, 2003). Every organization will differ in their practice for initiating and conducting PD group meetings. In many organizations, the coordination of the peer support process will be different according to protocols, procedures, needs, specific circumstances, and issues such as risk and security considerations. In addition, every group and incident will be different. In order to optimize the effectiveness of the meeting and integrity of the group process, there needs to be a thorough assessment of the incident, which needs to take into account a variety of factors. Again, these will also differ between organizations and agencies.

The assessment needs to be conducted by the facilitators in cooperation with CISM co-coordinators and the department responsible for co-coordinating and managing the peer support interventions and process. This should involve a discussion surrounding the points below as appropriate or relevant. The facilitators need to be aware of some or all of the areas listed below. The list is not intended to be inclusive, and there may indeed specific factors or considerations that affect the context and idiosyncratic circumstances, including the workplace environment. Some areas for consideration in the assessment process prior to providing a group crisis intervention such as PD should cover:

- Careful review of the incident – is there a need?
- Planning, logistical, and environmental considerations
- The impact – is it on only an individual or a group?
- If it is a group, the nature and context of each individual's involvement (e.g., who should be present at the meeting? Is the group homogeneous or heterogeneous?)
- The scale of incident, the involvement of others (e.g., where appropriate military, peacekeepers or warring factions, and media), and other complicating factors
- The nature and culture of the group: size, culture, constituency, gender, culture, cohesiveness, and potential for conflicts and anger within group
- The risk assessment – of individuals or group where appropriate, on others affected or attending
- Coordination and monitoring of follow-up and support arrangements
- Organizational considerations
- Lessons learned (individual, operational, and organizational)

Careful consideration of the above and other relevant factors will ensure a more effective meeting (Dyregrov, 1997). PD if conducted without the consideration of these factors will lead to an undesirable outcome for both participants and facilitators and potentially promulgate the myth of PD "doing harm." In addition, there is adherence to the recommendations of international guidelines as suggested by NICE, the BPS, and the recent report by the Australian Centre for Posttraumatic Mental Health (2011) on peer support in the workplace. This recent and important peer support project aimed to achieve an international consensus of expert opinion on a range of issues in peer support. These peer support guidelines were developed using the Delphi methodology, which recognizes the value of experts' opinions, experience, and intuition when full scientific knowledge is lacking (Linstone and Turoff, 1975). This study aims to inform the practice of peer support internationally on the basis of the best available advice from experts and practitioners in the field. Ninety-two participants, who were experts and practitioners, were involved in the study. Their roles included peer supporter, peer support co-coordinator, academic/researcher, trainer/educator, manager/administrator, policy maker, and clinician. Eight key domains of recommendations emerged from the project findings.

A starting point for these recommendations is the consensus view that all high-risk occupations and settings should have a well-planned, integrated, and tailored peer support program for their current employees as well as, for a limited time, once employment with the organization ceases. The report recognizes that each context, however, is different and suggests that the recommendations should not be interpreted rigidly but, rather, should be implemented as appropriate to the specific context of the program. This is particularly important since there is currently an absence of objective empirical evidence for the effectiveness of peer support in improving psychosocial outcomes. Indeed, the authors strongly support the establishment of properly designed and controlled research trials to inform our understanding of the effectiveness of these models. However, two key recommendations in the context of this section of the chapter is that on training and accreditation. The recommendation is that peer supporters should: (1) be trained in basic skills to fulfill their role (such as listening skills, psychological first aid, and information about referral options), (2) meet specific standards in that training before commencing their role, and (3) participate in on-going training, supervision, review, and accreditation.

Future Directions in Post-trauma Support Interventions – from Pathology to Post-traumatic Growth

In the context of post-trauma support in the workplace, CISM and interventions such as PD have been the focus of much criticism over the past decade, with much of the negative judgments arising from the misunderstanding and misapplication of both the CISM and PD interventions. The process of using a crisis intervention framework, which allows individuals and groups to use a structured discussion to understand and contextualize their experience, has been lost in paradigmatic posturing. Much criticism of the model is based on the idea that it purported to prevent the development of PTSD, even though there is little evidence that any intervention can prevent the development of PTSD and all robust and comprehensive post-incident support interventions are primarily concerned with minimizing the impact of such events and putting appropriate measures in place to optimize recovery and promote well-being (Barber & Lawrence, 2004). Despite the numerous claims that CISM and PD are "harmful," there is ample evidence (as indicated above) that no organizations have "banned" the use of such interventions, but merely repackaged the CISM and PD process to suit their own organizational needs. Whilst this is perfectly understandable and acceptable, it has erroneously created a certain mythology about what is "new" or more evidence based, though even slightly modified versions of CISM, such as TRiM, have not been shown to be more effective or yield more positive results (Greenberg *et al.*, 2010).

A major theoretical shift has occurred in the past decade with regard to our understanding of the way individuals understand, make sense of, and process traumatic experiences. It is especially relevant to see this shift within the context of early interventions for trauma and our understanding of many of the issues discussed within the chapters of this book. Previously, in the context of trauma, psychology has been largely concerned with the darker side of human experience. Since the emergence of positive psychology over a decade ago, this imbalance has been corrected. One of the strengths of this approach is the emphasis on the integration of negative and positive aspects of the experience. A topic that best characterizes this integrative position and has begun to attract interest from clinicians is that of post-traumatic growth – the study of how people change in positive ways in their struggle with adversity. It is now well established that stressful and traumatic events may serve as a trigger toward personal growth and positive change (Joseph & Linley, 2006). The positive changes that have been observed following trauma and adversity have been variously labeled as *adversarial growth* (Linley & Joseph, 2004a) and *benefit finding* (Affleck & Tennen, 1996; Tennen & Affleck, 2002).

Post-traumatic growth is a wide-ranging concept, still in development, but to date three broad domains of positive change have been noted throughout the literature (Tedeschi & Calhoun, 1996). Firstly, relationships are enhanced in some way. For example, people describe that they come to value their friends and family more, and feel an increased sense of compassion for others and a longing for more intimate relationships. Secondly, people change their views of themselves in some way; for example, they have a greater sense of personal resiliency, wisdom, and strength, perhaps coupled with a greater acceptance of vulnerabilities and limitations. Thirdly, people describe changes in their philosophy of life, for example finding a fresh appreciation for each new day and revaluating their understanding of what really matters in life, thus developing new priorities.

Tedeschi and Calhoun's model was developed in light of theoretical work in the post-traumatic stress literature pointing to the importance of appraisal processes (Janoff-Bulman, 1992), their own and others' extensive empirical work in growth (Tedeschi *et al.*, 1998), and their clinical experience (Calhoun & Tedeschi, 1999). The functional-descriptive model they propose discusses how traumatic events serve as seismic challenges to the pre-trauma schema, by shattering prior goals, beliefs, and ways of managing emotional distress. When these schemas are shattered in this way, this shattering leads to ruminative activity, as people try to make sense of what has happened and to deal with their emotional reactions to the trauma. In the initial stages, this ruminative activity is more automatic than deliberate (consistent with the re-experiencing and avoidance symptom clusters within PTSD). Although this automatic ruminative activity may often be distressing, it is indicative of cognitive activity that is directed at rebuilding the pre-trauma schema. This ruminative process is influenced by social support networks that provide sources of comfort and relief, as well as being influenced by new coping behaviors and the options that are available for the construction of new, post-trauma schemas. Successful coping at this stage facilitates disengagement from goals that are now unreachable, and beliefs that are no longer tenable in the post-trauma environment, together with decreased emotional distress. As successful coping aids adaptation, the initial ruminative activity that was characterized by its automatic nature shifts toward a more effortful ruminative activity. This effortful ruminative activity is characterized by narrative development, part of which may be the search for meaning. Interacting with this process is the experience and self-identification of adversarial growth. Importantly, although this shift toward more effortful ruminative activity represents growthful adaptation, it does not exclude the possibility of some enduring distress from the trauma, but at a lower level than was experienced in the immediate aftermath.

It is therefore time for a paradigm shift from a disease model and the language of disease that focuses on PTSD and the reduction of psychopathology (which is often difficult to identify at an early stage) to that of the promotion of mental health, psychological well-being, resilience, and growth. For example, one suggestion for a new conceptualization of post-traumatic growth proposed by Joseph and Linley (2008) was to reframe post-traumatic growth as an increase in psychological well-being

(PWB) as opposed to subjective well-being (SWB) as defined by Ryff's (1989) conceptualization of PWB as consisting of the six interrelated domains of autonomy, environmental mastery, positive relations with others, openness to personal growth, purpose in life, and self-acceptance.

Health promotion is defined by the World Health Organization (WHO) as "the process of enabling people to increase control over and to improve their health" (WHO, 1998, p. 1). Others have defined health promotion as a series of actions or "activities that increase the levels of health and well being and actualize or maximize the health and potential of individuals, families, groups, communities and society" (Murray *et al.*, 2006, p. 44).

Currently measures of change focus on the reduction of pathology or the elimination of symptoms, and this has been reflected in the literature on the effectiveness of CISM and PD through the use of measures and questionnaires which reflect these changes and the absence or presence of PTSD symptoms. Building on the theoretical notions discussed here within the context of early interventions for trauma and interventions such as PD, another way of measuring or conceptualizing change would be to use measures such as the recently developed Psychological Well-being Post-traumatic Changes Questionnaire (PWB-PTCQ) (Joseph & Regel, 2010). This questionnaire was developed to assess posttraumatic growth as representing an increase in PWB. The psychometric properties are good (Joseph *et al.*, 2011). Across three samples, evidence was provided that the PWB-PTCQ possesses a single-factor structure (invariant across clinical and general populations), high internal consistency, six-month stability, convergent validity with existing measures of post-traumatic growth, concurrent validity with personality and coping measures, predictive validity of change in well-being over time, discriminant validity with social desirability, prediction of clinical caseness, and, most importantly, incremental validity over and above existing measures of post-traumatic growth as a predictor of SWB.

These developments in our conceptualization of how human beings deal with the experience of traumatic situations further build on our understanding of resilience. The recent literature recognizes that, for many people, a traumatic event catalyzes internal resources of competence, coping, and resilience (Bonanno, 2004; Mitchell & Mitchell, 2006). Flach (1990) discussed the concept of resilience in relation to surviving a traumatic event by defining it as:

> Psychobiological resilience is the efficient blending of psychological, biological and environmental elements that permits human beings … to transit episodes of chaos necessarily associated with significant periods of stress and change successfully. (Flach, 1990, p. 40)

Antonovsky (1996) defined categories of resources available to people that determine whether stress becomes pathogenic or salutogenically strengthening. These resources include material resources, knowledge and intelligence, ego strength, mastery of flexible, rational and far-sighted coping strategies, social supports, commitment to one's social group, cultural stability, a stable system of values and beliefs derived from one's philosophy or religion, a preventative health orientation, and genetic or constitutional strengths (Sullivan, 1989, p. 337). All of these factors were mentioned by fire fighters in a recent study (Blaney, 2009).

In view of the evidence and emerging trends described here, there needs to be a significant shift from pathogenesis to the promotion of health and well-being. There is growing evidence that this philosophical shift is taking place (Blaney, 2009; Mitchell & Mitchell, 2006). CISM and PD are evolving beyond the medical model into one of promoting mental health and well-being, enhancing natural resilience, and promoting post-traumatic growth. This therefore requires new perspectives and paradigms of service delivery, outcome measures (such as the PWB-TCQ, described in this section), and robust qualitative research methodologies. Furthermore, there is ample evidence that CISM (including PD) is still widely practiced in a wide variety of organizations and agencies, both in the United Kingdom and abroad (Regel, 2007).

The relatively recent publication of a case study, encompassing all the elements of what is essentially psychological debriefing and follow-up, appears to have given legitimacy to its use as an intervention, although the technique is not described as such in the report. It is worthy of note that the authors of this study come from previously opposing sides of the debate regarding the use of psychological debriefing, and the publication can therefore be seen as some sort of rapprochement regarding the use of early interventions following exposure to traumatic experiences (Bisson *et al.*, 2007). In addition, the debate and discussion surrounding early interventions for trauma need to be reasoned, well informed, and linear, rather than continuing the cyclical repetition of all the myths and misconceptions which have dogged researchers and practice over the past decade.

At the time of writing, one of the authors (AD) is actively involved in the organization of crisis support being offered to individuals, families, communities, and first responders after the tragedy involving a mass shooting on the island of Utoeya, Norway, in July 2011. To deny the survivors and the bereaved access to experienced professional support, which has been carefully considered, because of methodologically flawed "evidence" would be unethical to say the least. This calls into sharp focus the distinctions that need to be made with regard to the diverse nature of early interventions following exposure to traumatic events, whether they are in the workplace or in other contexts. The confusion over terminology and practice needs to be finally laid to rest through rational consensus. Early interventions for trauma need to be moved on from their previous narrow confines and focus on pathological responses to the more sophisticated conceptual framework of mental health promotion, psychological well-being, and post-traumatic growth.

References

Adler, A. B., Bliese, P. D., McGurk, D., Hoge, C. W., & Castro, C. A. (2009). Battlemind debriefing and battlemind training as early interventions with soldiers returning from Iraq: Randomization by platoon. *Journal of Counselling and Clinical Psychology, 77*(5), 928–940.

Adler, A. B., Litz, B. T., Castro, C. A., Suvak, M., Thomas, J., Burrell, L., *et al.* (2008). A group randomized trial of critical incident stress debriefing provided to U.S. peacekeepers. *Journal of Traumatic Stress, 21*(3), 253–263.

Affleck, G., & Tennen, H. (1996). Construing benefits from adversity: Adaptational significance and dispositional underpinnings. *Journal of Personality, 64*, 899–922.

Antonovsky, A. (1996). The salutogenic model as a theory to guide health promotion. *Health Promotion International, 11*(1), 11–18.

Arendt, M., & Elklit, A. (2001). Effectiveness of psychological debriefing. *Acta Psychiatrica. Scandanavica, 104*, 1–15.

Australian Centre for Posttraumatic Mental Health. (2011). Development of guidelines on peer support using the Delphi methodology. Retrieved from http://www.acpmh.unimelb.edu.au.

Bisson, J. I., Brayne, M., Ochberg, F. M., & Everly G. S. (2007). Early psychosocial intervention following traumatic events. *American Journal of Psychiatry, 164*, 1016–1019.

Bisson, J., & Deahl, M. (1994). Psychological debriefing and prevention of post-traumatic stress. *British Journal of Psychiatry, 165*, 717–720.

Bisson, J. I., Jenkins, P. L., Alexander, J., *et al.* (1997). Randomised controlled trial of psychological debriefing for victims of acute burn trauma. *British Journal of Psychiatry, 171*, 78–81.

Blaney, L. S. (2009). Beyond "knee jerk" reaction: CISM as a health promotion construct. *Irish Journal of Psychology, 30*(1–2), 37–57.

Bonanno, G. (2004). Loss, trauma and human resilience: Have we underestimated the human capacity to thrive after aversive events? *American Psychologist, 59*, 20–28.

Brewin, C. R., Andrews, B., & Valentine, J. D. (2000). Meta-analysis of risk factors for posttraumatic stress disorder in trauma-exposed adults. *Journal of Consulting and Clinical Psychology, 68*(5), 748–766.

British Psychological Society (BPS). (2002). *Psychological debriefing: Report by British Psychological Society Professional Affairs Working Party.* London: British Psychological Society.

Busuttil, W., Turnbull, G. J., & Neal, L. A. (1995). Incorporating psychological debriefing techniques within a brief group psychotherapy programme for the treatment of post-traumatic stress disorder. *British Journal of Psychiatry, 167*, 495–502.

Calhoun, L. G., & Tedeschi, R. G. (1999). *Facilitating post-traumatic growth: A clinician's guide.* Mahwah, NJ: Lawrence Erlbaum.

Conlon, L., Fahy, T. J., & Conroy, R. (1999). PTSD in ambulant RTA victims: A randomized controlled trial of debriefing. *Journal of Psychosomatic Research, 46*(1), 37–44.

Deahl, M. (2001). Evaluating psychological debriefing: Are we measuring the right outcomes? *Journal of Traumatic Stress, 14*(3), 527–529.

Dyregrov, A. (1989). Caring for helpers in disaster situations: Psychological debriefing. *Disaster Management, 2*(1), 25–30.

Dyregrov, A. (1997). The process in psychological debriefings. *Journal of Traumatic Stress, 10*(4), 589–605.

Dyregrov, A. (1999). Helpful and hurtful aspects of psychological debriefing groups. *International Journal of Emergency Mental Health, 3*, 175–181.

Dyregrov, A. (2003). *Psychological debriefing: A leader's guide for small group crisis intervention.* Ellicott City, MD: Chevron.

Dyregrov, A., & Regel, S. (2011, August). Early Interventions following exposure to traumatic events: Implications for practice from recent research. *Journal of Loss and Trauma.*

Everly, G. S., & Mitchell, J. T. (2000). The debriefing "controversy" and crisis intervention: A review of lexical and substantive issues. *International Journal of Emergency Mental Health, 2*(4), 211–225.

Flach, F. (1990). The resilience hypothesis and post traumatic stress disorder. In M. E. Wolfe & A. D. Mosnaim (Eds.), *Posttraumatic stress disorder: Etiology, phenomenology and treatment.* Washington, DC: American Psychiatric Press.

Galliano, S. (2002, July). Critical incident processing. *Counselling and Psychotherapy Journal*, 18–20.

Greenberg, N., Langston, V., Everitt, B., Iverson, A., Fear, N. T., Jones, N., *et al.* (2010). Cluster randomized controlled trial to determine the efficacy of Trauma Risk Management (TRiM) in a military population. *Journal of Traumatic Stress*, *23*(4), 430–436.

Greenberg, N., Langston, V., & Jones, N. (2008). Trauma Risk Management (TRiM) in the UK Armed Forces. *Journal of Army Medical Corps*, *154*(2), 123–126.

Hobbs, M., Mayou, R., & Harrison, B. (1996). A randomised controlled trial of psychological debriefing for victims of road traffic accidents. *British Medical Journal*, *313*, 1438–1439.

Hobbs, M., & Adshead, G. (1997). Preventive psychological intervention for road crash survivors. In M. Mitchell (Ed.), *The aftermath of road accidents* (pp. 159–253). London: Routledge.

Hodgkinson, P., & Stewart, M. (1991). *Coping with catastrophe: A handbook of disaster management*. London: Routledge.

Hughes, O. (2002). Training, supervision and assessment of competence. In British Psychological Society (Eds.), *Psychological debriefing: Report by British Psychological Society Professional Affairs Working Party*. London: British Psychological Society.

Janoff-Bulman, R. (1992). *Shattered assumptions: Toward a new psychology of trauma*. New York: Free Press.

Jones, N., & Roberts, P. (1998). *Risk management following psychological trauma: A guide for RMC Combat Stress Practitioners*. London: HQRM.

Jones, N., Roberts, P., & Greenberg, N. (2003). Peer group risk assessment: A post-traumatic management strategy for hierarchical organizations. *Occupational Medicine*, *53*, 469–475.

Joseph, S. (2003). Social support and mental health following trauma. In W. Yule (Ed.), *Post-traumatic stress disorders: Concepts and therapy* (3rd ed.). Chichester: Wiley.

Joseph, S., & Linley, P. A. (2006). Growth following adversity: Theoretical perspectives and implications for clinical practice. *Clinical Psychology Review*, *26*, 1041–1053.

Joseph, S., Maltby, J., Wood, A., Stockton, H., Hunt, N., & Regel, S. (2011, August). Psychological Well-Being – Post Traumatic Changes Questionnaire: Reliability and validity. *Psychological Trauma: Theory, Research, Policy and Practice*.

Joseph, S., & Regel, S. (2010). The Psychological Well-Being – Post Traumatic Changes Questionnaire. In S. Regel & S. Joseph (Eds.), *Post traumatic stress: The facts*. Oxford: Oxford University Press.

Kinchin, D. A. (2007). *Guide to psychological debriefing: Managing emotional decompression and post traumatic stress disorder*. London: Jessica Kingsley.

Laurence, L., & Barber, G. (2004). Debriefing in the fire service. *Counselling at Work*, *11*, 11–13.

Lee, C., Slade, P., & Lygo, V. (1996). The influence of psychological debriefing on emotional adaptation in women following early miscarriage: A preliminary study. *British Journal of Medical Psychology*, *69*, 47–58.

Linley, P., & Joseph, S. (2004). Positive change following trauma and adversity: A review. *Journal of Traumatic Stress*, *17*(1), 11–21.

Linstone, H. A., & Turoff, M (1975). *The Delphi Method: Techniques and applications*. Reading, MA: Addison-Wesley.

Mayou, R. A., Ehlers, A., & Bryant, B. (2002). Posttraumatic stress disorder after motor vehicle accidents: 3 year follow-up of a prospective longitudinal study. *Behaviour Research and Therapy*, *40*, 665–675.

Mitchell, J. T. (1983). When disaster strikes: The Critical Incident Stress Debriefing process. *Journal of Emergency Medical Services*, 36–39.

Mitchell, J. T. (1988a). Development and functions of a Critical Incident Stress Debriefing Team. *Journal of Emergency Medical Services*, 43–46.

Mitchell, J. T. (1988b). The history, status and future of Critical Incident Stress Debriefings. *Journal of Emergency Medical Services*, 47–51.

Mitchell, J. T., & Everly, G. S. (1995). Critical Incident Stress Debriefing (CISD) and the prevention of work-related traumatic stress among high risk occupational groups. *Psychotraumatology – Key Papers and Core Concepts*, 267–280.

Mitchell, J. T., & Everly, G. S. (2001). *Critical Incident Stress Debriefing: An operations manual for the prevention of traumatic stress among emergency services and disaster workers* (3rd ed.). Ellicott City, MD: Chevron.

Mitchell, J. T., & Everly, G. S. (2003). Critical Incident Stress Management and Critical Incident Stress Debriefings: Evolutions, effects and outcomes. In B. Raphael & J. P. Wilson (Eds.), *Psychological debriefing: Theory, practice and evidence* (2nd ed., pp. 71–90). Cambridge: Cambridge University Press.

Mitchell, S., & Mitchell, J. (2006). Caplan, community and CISM. *Journal of Emergency Mental Health*, *8*, 8–14.

Murray, R., Zentner, J., Pangman, V., & Pangan, C. (2006). *Health promotion strategies throughout the lifespan*. Toronto: Pearson Education.

National Institute for Health and Clinical Excellence (NICE). (2005). *The management of PTSD in adults and children in primary and secondary care*. National Clinical Practice Guideline 26. Wiltshire, UK: Gaskell Press.

Raphael, B. (1990). *When disaster strikes – a handbook for the caring professions*. London: Unwin Hyman.

Raphael, B., Meldrum, L., & McFarlane, A. C. (1995). Does debriefing after psychological trauma work? *British Medical Journal*, *310*, 1479–1480.

Regel, S. (2007). Post trauma support in the workplace: The current status and practice of Critical Incident Stress Management (CISM) and Psychological Debriefing (PD) within organisations in the United Kingdom. *Occupational Medicine*, *57*, 411–416.

Regel, S. (2010). Does psychological debriefing work? *Healthcare Counselling and Psychotherapy Journal*, *10*(2), 14–18.

Regel, S., Dyregrov, A., & Joseph, S. (2007). Psychological debriefing in cross-cultural contexts: Ten implications for practice. *International Journal of Emergency Mental Health*, *9*(1), 37–45.

Richards, D. (1994). Traumatic stress at work: A public health model. *British Journal of Guidance & Counselling*, *22*(1), 51–64.

Rose, S., Bisson, J., & Wessely, S. (2004). Psychological debriefing for preventing post traumatic stress disorder (PTSD) (Cochrane review). *Cochrane Library*, issue 3.

Rose, S., Brewin, C. R., & Andrews, B. (1999). A randomized controlled trial of individual psychological debriefing for victims of violent crime. *Psychological Medicine*, *29*, 793–799.

Ryff, C. D. (1989). Happiness is everything or is it? Explorations on the meaning of psychological well-being. *Journal of Personality and Social Psychology*, *57*, 1069–1081.

Sullivan, G. (1989). Evaluating Antonovsky's salutogenic model for its adaptability to nursing. *Journal of Advanced Nursing*, *14*, 336–342.

Tedeschi, R. G., & Calhoun, L. G. (1996). The post-traumatic growth inventory: Measuring the positive legacy of trauma. *Journal of Traumatic Stress*, *9*, 455–471.

Tedeschi, R. G., Park, C. L., & Calhoun, L. G. (1998). Post-traumatic growth: Conceptual issues. In R. G. Tedeschi, C. L. Park, & L. G. Calhoun (Eds.), *Post-traumatic growth: Positive changes in the aftermath of crisis* (pp. 1–22). Mahwah, NJ: Lawrence Erlbaum.
Tennen, H., & Affleck, G. (2002). Benefit-finding and benefit-reminding. In C. R. Snyder & S. Lopez (Eds.), *Handbook of positive psychology* (pp. 584–597). New York: Oxford University Press.
Wessely S., & Deahl, M. (2003). In debate: Psychological debriefing is a waste of time. *British Journal of Psychiatry*, *183*, 12–14.
World Health Organisation. (1998). *The WHO health promotion glossary*. Geneva: Author.

Part B

The Legal and Business Imperatives to Manage Trauma Effectively

5

The Trauma Impact on Organizations: Causes, Consequences, and Remedies

Ronald J. Burke

Cantor Fitzgerald: A Case Study Following the Events of 9/11

Cantor Fitzgerald, an international brokerage firm with about 1000 employees, lost 659 employees when a hijacked plane struck the World Trade Center (WTC) on September 11, 2001. Some employees jumped to their death to avoid fires. Not one employee working above where the plane hit made it out alive. Cantor Fitzgerald was effectively destroyed. Howard Lutnick, CEO of Cantor Fitzgerald, asked a college friend to come to New York and record ongoing events which resulted in a best-selling book (Barbash, 2003). Cantor Fitzgerald was on the verge of collapse following the attack. Lutnick, predictably, was in shock. Family members and friends were anxious to find out whether their loved ones were alive. Cantor Fitzgerald's employees were a closely knit group. Some surviving Cantor Fitzgerald employees were injured (mostly burns) and in hospital; some of these employees later died in hospital from their burns. Their London office was still functioning, however.

Within a day or two, Cantor Fitzgerald decided to continue. Wall Street was to shut down for a few days. Cantor Fitzgerald had about 150 of its NYC employees alive and accounted for. Most wanted to work. Cantor Fitzgerald approached its London office to see what help they might be able to provide.

Lutnick spent the first week going to funerals, establishing a nearby disaster recovery site in New Jersey, going to hospitals, setting up a relief fund for spouses and children, dealing with the media, and with continuing the business. All employees worked long hours with little sleep under intense pressure. Within a few days, Cantor Fitzgerald began functioning. It hired new traders and obtained office furniture, computers, and phones. Their staff became even more cohesive and engaged.

Lutnick concluded there needed to be a meeting place for families and obtained use of the Pierre Hotel, which was operated by another friend. Cantor Fitzgerald provided tables, chairs, phones, and counselors. Family members were anxious to find out what

International Handbook of Workplace Trauma Support, First Edition. Edited by Rick Hughes, Andrew Kinder, and Cary L. Cooper.

had happened to other family members working for the firm. Dozens of grief counselors and clergy were available. Bulletin boards were created to post information on loved ones. Information was provided to families about insurance, death certificates, benefits, salaries, and how to file a missing persons report. Some family members were angry at Cantor Fitzgerald and Lutnick. They had expected answers when none were available.

Other brokerage houses that were not affected by 9/11 were still very competitive and not sympathetic to or supportive of Cantor Fitzgerald. Cantor Fitzgerald also opened an office in nearby Darien, Connecticut, as it worked its way back to more complete functioning. On October 1, Cantor Fitzgerald held a ceremony for family and friends in Central Park at which Rudy Giuliani made some remarks. Lutnick continued to be criticized by some family members for not doing enough.

Cantor Fitzgerald is now fully functioning. Barbash (2003) identified some steps that seemed to be important in its recovery. These included first finding out who was alive and dealing with the emotions of the families and their situation, deciding what to do next, only focusing on those parts of the business that had some chance of succeeding, having the staff sometimes function as grief counselors, reconnecting the phone lines, benefiting from help provided by some of its suppliers and customers, attending funerals and hospitals, and eventually stopping the payment of salaries to dead employees because of the strain this placed on survival (an action that, again, was criticized by families) but providing access to insurance, death certificates, and family counseling.

The bulk of the research and writing on the effects of traumatic events in the workplace and in organizations has focused on individual employees, bystanders, and families. These include the effects of terrorist attacks on survivors or innocent bystanders, police officers involved in shooting incidents, fire fighters involved in putting out blazes, assisting survivors, and recovering bodies, survivors of mine and oil rig disasters, members of emergency response teams dealing with the aftermath of organizational trauma, and military veterans upon their return home. Almost no research and writing have focused on the effects of organizational trauma on the organization itself.

Organizational Trauma: A Definition

What do we mean by *organizational trauma*? Organizational trauma is a set of potential organizational responses to internal or external acts or events (Hopper, 2010). These acts of events are "caused" by one or more individuals or by natural forces that result in psychological distress, property damage, injury, or death to one or more employees of an organization. These acts of events can affect one or more organizations simultaneously. *Organizational trauma* is used interchangeably with *trauma effects in organizations.*

Examples would include the following:

- Organizational restructuring and downsizing resulting in new job assignments or job losses (Noer, 1993).
- Organizational mergers producing high levels of uncertainty and ambiguity.
- Murders in the workplace by a disgruntled employee or non-employee (Denenberg & Denenberg, 2010; Lester, 2011).

- A spate of employee suicides as have occurred in Foxconn, a Chinese company, and France Telecom, a French company, over the past 2–3 years.
- Acts of terrorism such as occurred in organizations located in the World Trade Towers on 9/11 (Burke & Cooper, 2008).
- Disasters occurring in mines, on oil rigs, in nuclear submarines, in nuclear power plants, and in communities by errors or accidents in organizations (e.g., the Union Carbide plant in Bhopal, India, in 1984).
- Natural disasters caused by earthquakes or hurricanes that affect workplaces and communities.
- Worker injuries or deaths (e.g., the collapse of scaffolds holding window washers on skyscrapers).

The nature and magnitude of these traumatic events also can vary, though all have powerful effects on employees, their families, and the organization. Some traumatic events or acts are small and limited to one workplace or organization (e.g., disgruntled employee killing coworkers, or a cluster of employee suicides in Foxconn). Other traumatic events or acts are large and span several workplaces (e.g., terrorist attacks on the WTC, Oklahoma City bombing, anthrax attacks on US government buildings, mine explosions, gas explosions, and oil rig fires).

A traumatic event or disaster is any destructive event or act that disrupts the functioning of an organization. Acts or events that damage the workplace also damage the identity of individuals; work is an important life role for many. Naturally occurring disasters include wildfires, floods, hurricanes, tornadoes, and earthquakes. Man-made disasters, intentional or unintentional, include fires, explosions, hazardous chemical release, and acts of violence including terrorism. Technological disasters include nuclear power plant releases, chemical plant explosions, oil spills, and equipment failures.

Why Do Catastrophic Events Occur in Organizations?

Roberts and Martelli (2011) observe that organizations today have a greater capacity to create catastrophic accidents. Organizations are larger, more complex, more interconnected, more technologically sophisticated, and under more pressure to generate profits to shareholders. These factors will increase the number of man-made disasters. Mistakes, errors, and disasters are inevitable in organizations. And errors today can have catastrophic effects.

Shrivastava (1987) identified three causes of the Bhopal tragedy: human, organizational, and technical.

- *Human.* low employee morale, labor–management conflicts
- *Organizational.* low importance of the Bhopal plant to Union Carbide, frequent changes in plant top management
- *Technical.* tightly coupled nature of the technology

Consider the 2010 BP oil rig disaster in the Gulf of Mexico. Experts typically respond to the problems of technological failure by applying more technology. They encourage the use of "more reliable blowout preventers" and "mandatory relief wells" instead of examining the deeper issue of whether we should even undertake oil extraction in such high-risk situations by reducing our dependence on oil, or the important role played by human and organizational factors.

Organizational Responses to Trauma

Suicides in the workplace

Foxconn, a Chinese company that had 10 employee suicides in the first five months of 2010, decided to raise employee salaries and informed their customers that they would increase their prices as a consequence. Foxconn hired counselors, was planning to bring in monks, and had set up a suicide helpline. Foxconn workers toiled under harsh work conditions and long work hours, lived in crowded dormitories, and endured heavy enforcement of discipline on the assembly lines and fines for minor work infractions. In addition, Foxconn decided it would no longer pay compensation to families of suicide victims (US$14 600) to discourage further suicides. Foxconn concluded that some individuals committed suicide to give their families money.

France Telecom had more than 20 workers commit suicide over an 18-month period in 2008–2009. Suicide notes sometimes identified job stress and job dissatisfaction as causes. France Telecom was privatized from a government organization to a private sector organization. The unions blamed restructuring cuts, extreme pressure, bullying, and poor management as major problems. Some employees had to make major job changes, and other employees were terminated in the transition. The employing organization acted to address workplace stress and provide better medical support to employees.

Deaths at work due to accidents

South African organizations now make increased use of trauma management initiatives such as employee assistance programs (Maiden & Terblanche, 2007). For example, the suicide of a senior plant maintenance official working in a chemical plant was followed by individual counseling with 76 employees; six other employees sought out private therapy. A South African chemical plant had an industrial accident when a tank exploded, killing several people and injuring many more (Maiden & Terblance, 2007). Surviving employees feared that another explosion might occur. The company hired employee assistance program counselors who conducted debriefing sessions with 852 employees.

Organizations obviously need to inform the individual's family. Typically there is some anger directed at the organization. Counseling is often made available to the family. The organization must also participate in a formal accident investigation. Some organizations train workers to provide immediate care help to all surviving accident victims and bystanders.

Witnessing deaths in the workplace

Shalev (2004) noted that some individuals, not under threat themselves but having seen body parts or bodies burned beyond recognition, experienced more long-term distress than survivors of an attack.

Toronto has a subway system, and from 1998 to 2007 there were 100 unsuccessful and 150 successful suicide attempts by individuals throwing themselves under the subway train. Both unsuccessful and successful suicides are sometimes viewed by other passengers and/or subway operators. Viewers of suicide indicate that these images remain with them for some time. One subway operator said he witnessed three suicides over a four-month period. The Toronto Transit Commission (TTC) works with a Toronto hospital and Health Center to provide a support structure for train operators who suffer post-traumatic stress disorder (PTSD).

There are some jobs in which incumbents face concerns about their own death (e.g., soldiers, miners, police officers, power plant operators, and fire fighters) – death awareness – because they work in dangerous jobs. Some employees work in jobs that expose them to the deaths of others (e.g., doctors, nurses, paramedics, funeral directors, grief counselors, and rescue workers). Grant and Wade-Benzoni (2009) propose a four-stage model of death awareness at work that includes situational triggers (mortality triggers such as mortality exposure, death awareness anxiety, and reflection), individual and work context contingencies (coping behaviors, individual work orientation, and job design factors), and work behaviors (stress-related withdrawal behaviors such as absenteeism, tardiness, and turnover, and cooperative behaviors such as helping others and changing to a service job). Managers in organizations where safety and physical dangers are high (e.g., mining, oil rigs, and construction) might encourage death reflection to increase adherence to safety practices.

Impact on individuals and organizations

Walter, Hall, and Hobfoll (2008) use Hobfoll's conservation of resources (COR) theory to explain the influence of events such as mass casualties on organizations. COR theory states that individuals and groups try to get, keep, and protect things that they most value (resources). Stress results from the actual loss of resources, threats of such loss, or failure to get resources after much trying. Individuals and groups must also use resources to keep resources, stop their losses, recover from their losses, or get more resources. Individuals and groups lacking resources are also more vulnerable to losing resources and less able to gain resources. Traumatic events and their demands (physical, social, and psychological) cause individuals and groups to quickly lose resources (e.g., fear, anxiety and concerns for others, damage to their workplace).

Organizational Responsibility and Liability

Organizations are responsible for creating and maintaining safe work environments. This involves reducing or eliminating potentially harmful effects of known causes (e.g., unsafe work behaviors). It also involves employee training in safe job behaviors, providing and encouraging (ensuring) the use of appropriate equipment, rewarding safe behaviors, and abiding by established safety regulations (Clarke, 2010; Kowalski-Trakofler *et al.*, 2011). Organizations face legal and financial liability when they fail to abide by prevailing regulations. A few individuals in management positions with the now-defunct Union Carbide company in the Bhopal disaster were recently fined a relatively small amount and sentenced to jail, again for a relatively small amount of time, many years after the tragedy.

Organizations shoulder some responsibility to minimize risk exposure. They may be legally liable if they fail to do all they could to protect their employees. One might also argue that organizations have a moral obligation to consider level of risk (see Regel, 2007). McFarlane and Bryant (2007) write that organizations need to anticipate possible traumatic exposure and to develop clear policies and procedures on how this will be addressed.

Lindahl (2008) reports a decision of the Virginia "Supreme Court that awarded a firefighter/paramedic compensation when he developed chronic, disabling PTSD after responding to a fire."

Financial Costs of Organizational Disasters and Trauma

A study of the after-effects of 9/11, conducted two months after the attacks, estimated the cost of the primary damage to the New York and New Jersey area to be $25–30 billion. About half of this amount resulted from damage to physical property and costs of necessary clean-up. The other half involved loss of life. The secondary cost to business was estimated at $2.7–4.8 billion, associated with reductions in both sales and new manufacturing orders. There were also costs associated with loss of foreign investment, reduced travel and tourism, and job loss in most affected industries (e.g., airlines, hotel and restaurants, and tourism).

In the United States in the fall of 2001, anthrax was used as a weapon of terror (Cole, 2009; Day, 2003; Pastel, 2008). Anthrax was sent through the mail and resulted 22 cases of deliberate contamination. These acts occurred in three metropolitan areas and affected the lives of a large number of people. These anthrax attacks shut down some businesses (particularly government offices) and placed another strain on individuals. Governments spent over $250 million in the last half of 2001 and early 2002 investigating real and fake anthrax attacks.

Injured and survivors suffering anxiety or PTSD who cannot return to work are typically covered by worker's compensation in most countries. These costs are borne indirectly or directly by the organizations in which "injured" workers were employed, by various levels of government, and eventually by the taxpayers.

Psychological and Physical Health Costs of Organizational Disaster and Trauma

Shirom *et al.* (2008), in a study of a large sample of Israeli workers, reported that fear of a potential terrorist attack was associated with greater health difficulties. This supports, for example, the conclusions that survivors of a mining or oil rig disaster are more likely to be in greater psychological distress when they return to their previous jobs.

The explosion, fire, and eventual sinking of the British Petroleum (BP)–run oil rig in the Gulf of Mexico in late April 2010 had several impacts on BP and their employees. These include the following:

- Eleven deaths.
- Loss of this well's oil production.
- The costs of cleaning up the oil spill and resulting damage to surrounding beaches and wetlands, and payments to businesses affected by the oil spill, will be billions of dollars.
- Damage to BP's reputation and image.
- A drop in the value of BP's share price: the value of BP had dropped by about half in early June 2010.
- Increasing government surveillance of future exploration for off-shore oil that will require greater preventative measures of potential disasters (more redundancy in systems).
- A moratorium on further development of oil rigs, costing BP time and lost investments and loss of jobs to oil rig workers.
- Increased government and BP responses to inspect current and future oil rigs in the Gulf and elsewhere.
- Greater scrutiny of the government agencies responsible for allowing off-shore oil exploration and drilling. The US Minerals and Management Services (MMS), the agency responsible for these activities, has come under criticism for its lax approach to regulation and monitoring, and its head, Elizabeth Birnbaum, was fired by the Obama administration. MMS was responsible for monitoring and regulating offshore drilling. MMS has come under fire for supposedly lax environmental oversight for some period of time and having too close a relationship with the organizations they were monitoring (e.g., employees of MMS and oil firms vacationing together).
- The spectacle of Halliburton (builders of the oil rig), BP (operators of the oil rig), and Transocean (owners of the oil rig) pointing fingers at and blaming each other in hearings before the US government on who was responsible for the oil rig disaster.
- There has been the suggestion that the oil rig crew was short-handed and overworked.
- Criminal charges will possibly be filed against BP.

Post-traumatic stress disorder

PTSD is the most common reaction to workplace trauma. Characteristics of PTSD (American Psychiatric Association, 2000) are re-experiencing the phenomenon (e.g., recurrent distressing dreams), avoidance and numbing (e.g., avoidance of thoughts, feelings, and conversations), increased arousal (e.g., irritability or outbursts of anger), and changes in functioning (e.g., the level of one's current well-being)

To meet these criteria, individuals have "to be exposed to a traumatic event that involves actual or threatened death or serious injury, or a threat to the physical integrity of self and others" (Bisson, 2007, p. 3).

PTSD has been treated by using psychotherapy, medication, critical incident stress debriefing, and action research, and by providing social support, strengthening coping resources, undertaking system-wide interventions including mental health evaluations of employees, providing access to mental health professionals, monitoring hazards, developing minimum standards for safety and health training for all staff, using safety equipment, and providing on-site training and medical care during the early phases of a disaster (McFarlane & Bryant, 2007).

Organizational Challenges During Traumatic Events

Dealing with emotional trauma

Emotional trauma in organizations limits individual performance and effectiveness. Organizational development (OD) processes can influence the healing of emotional trauma (de Klerk, 2007). Survivors and witnesses of those directly involved are also affected by organizational trauma. Trauma lasts well beyond the trauma event itself. Emotional wounds must be opened and cleaned before they can heal. Leaders have a key role in facilitating healing. Important steps in the healing process involve acknowledging the existence of trauma, providing a safe place, bringing the trauma into awareness, and allowing the expression of and dealing with the emotions.

Walter *et al.* (2008) offer five principles for dealing with individuals following a disastrous act or event. These include proving a sense of safety, providing calming, providing a sense of self- and collective efficacy, promoting connectedness, and promoting hope. They indicate why these principles are helpful and identify specific examples of managerial intervention supporting each of the five principles.

Maintaining work motivation following trauma

Singh (2008) discusses the unique challenges of maintaining work motivation during "trying times." He collected data from the New York Presbyterian Hospital (NYPH), located close to the WTC. NYPH had lost some of its employees and spouses, partners, and family members of its employees, but it had a disaster plan in place and had to respond to the needs of its own workforce and the incoming needs of those affected directly by the events of 9/11.

NYPH's disaster plan went into effect within minutes of the first plane hitting the WTC. NYPH's Emergency Medical Services sent vehicles, technicians, and paramedics to the WTC. NYPH treated over 125 victims of these terrorist attacks.

NYPH staff reported immediate psychological and emotional effects as they worried about family and friends. Hospital social workers set up counseling centers for staff and visitors and established additional phone lines for families calling the hospital.

NYPH's employees continued to work effectively, against greater challenges (e.g., many walked to work). This was attributed to solid leadership by trained women and men; the nature of the organization and profession emphasized care and devotion to their profession. Employees also felt a sense of patriotism – this was an attack on the United States – and inspiration from Mayor Rudy Giuliani. Many employees just wanted to help. Finally, employees developed a greater respect for their coworkers and their jobs. Key factors in maintaining work motivation then were patriotism, high levels of professionalism, involved management, and leadership best characterized by "compassionate autocracy." Hospital leadership was sensitive and caring for staff but took immediate control of the situation and what needed to be done.

Dealing with the wider community

When a large organization is involved, developing an understanding of organizational trauma becomes complex as witnessed in the case study of Cantor Fitzgerald. Several parties are involved: organizational leaders and their representatives, employees, family and friends of employees, lawyers, the media, and various government agencies. Organizations experiencing traumatic events need to have a plan and strategy in place to both provide information and deal with the anger of family, friends, and the local community, media reporting, and various levels of governments (Blatt, 2001–2002).

Emergency Preparedness

Gershon and her colleagues (2008) review the role of workplace emergency preparedness to protect the health and safety of workers and employing organizations.

Emergency plans need to include how to report emergencies to senior management, how to evacuate the workplace, how to shut down important equipment and/or procedures, and how to account for all employees after an evacuation.

Following 9/11, a new rule passed in NYC in 2006 required all high-rise buildings designated as office space to have an approved emergency action plan that must be implemented in these buildings.

The US Occupational Safety and Health Administration includes standards of an emergency preparedness plan (see Gershon *et al.*, 2008, p. 246). These include a written emergency action plan, employee training on the plan including when employees are first hired, establishment of a chain of command and the designation of an emergency response team coordinator, emergency communication systems including alarm systems, emergency escape procedures and emergency escape routes, assigning rescue and medical duties and training those employees who are to perform them (emergency response teams), and clearly marked exits and exit routes which discharge directly to a street or other open place, among others.

Increased building security has also been introduced following 9/11 and the Oklahoma City bombing of 1995. This involves the use of individual searches, physical barriers, fences, gates, guard stations, and video surveillance equipment. A decision was made following the Oklahoma City bombing of the federal Alfred P. Murrah Building to not have day care centers in some or many government buildings as a result of the deaths of children in a day care center in this building.

Benefits Resulting from Traumatic Organizational Events

Are there any benefits that result from such examples of organizational trauma? Strangely enough, there may be some. Consider the following:

- More government regulations of work conditions that might be exposing individuals and organizations to undue risk.
- Safer equipment (e.g., in mines).
- Greater morale and cohesion among organizational employees following a traumatic event or act.
- The development of disaster preparation, response, and continuity plans.
- The development of greater resilience in the face of potentially traumatic events or acts.
- Organizational diagnosis and intervention to identify causes of traumatic events (e.g., suicides in both Foxconn and France Telecom).
- Development of trauma counseling services within organizations.
- Development of communication strategies and programs to minimize damage to an organization's reputation and image.
- Development of back-up or redundant resources and systems to maintain an organization's functioning.
- Rehearsal for potential organizational traumatic events (e.g., building evacuation routes, and emergency medical training of staff).
- Leaning from a disaster event in order to reduce the likelihood of the same event occurring again.

Madsen (2009) notes that investigations that almost always follow industrial, transportation, and mining disasters are carried out and prevent future tragedies. Using data from US coal mining disasters, he found that mines learn to prevent future disasters by both direct and vicarious experiences with disasters. Others (Baum & Dahlin, 2007; David, Maude-Griffin, & Rothwell, 1996; Hood, 2004) have found similar learning benefits from train crashes, Bhopal, and the Three Mile Island nuclear accident in 1979.

Traumatized versus Resilient Organizations

As noted above, organizational trauma caused by catastrophic acts of events can exist at the individual, group, and organizational levels. And resilience can also exist at the individual, group, and organizational levels.

Resilience has been defined as "an ability to sustain a shock without completely deteriorating, that is most conceptions of resilience include some idea of adapting or 'bouncing back' from a disruption" (Kendra & Wachtendorf, 2003, p. 41). Resilience is appropriately seen as an art as well as the application of scientific knowledge and techniques (Sutcliffe & Vogus, 2003)

Gittell *et al.* (2006, 2008) examined the effects of 9/11 on the airline industry. The US airline industry suffered huge losses in the aftermath of 9/11. Flights were cut and airline employees were laid off. But some airlines coped better than others. Airlines that fared better were characterized by having positive employee relationships, more viable business models, and greater financial resources, and made less use of layoffs as a result; these were all sources of resilience.

Dutton, Quinn, and Pasick (2002a, 2002b) chronicle the responses of Reuters America (RA) to the 9/11 attack. RA had a data center in the WTC and four at other NYC sites, one in Washington, DC, and one in the Sears Tower in Chicago. Some RA employees were out of the NYC site at conferences of business. Dutton and her colleagues conducted interviews with 30 RA employees between October 19 and November 8, 2001, to capture RA's responses. They developed three categories of response.

1. Turning chaos into order by establishing a command center an hour and a half following the attacks, giving priority to finding their employees, and continuing to serve their clients, with leaders developing an approach for dealing with the situation. Clear priorities were established: employees first, then customers, then the business itself.
2. Searching for employees and finding all but the six who died during the attacks, taking care of families by offering support and resources, providing counseling help, and holding large town hall meetings using teleconferencing to allow all of their employees to take part.
3. Technical recovery by contacting customers to see how RA might be of help and quickly replacing the data system lost in the WTC using systems located in their other facilities.

RA found that these responses mirrored their culture. RA took pride in their commitment to their people. RA became more flexible and innovative in dealing with the tragedy. Interpersonal relationships became more open and trusting. RA was seen as more humane, attaching greater importance to employees. RA became more results oriented and less process or rule oriented.

Kelly and Stark (2002) interviewed managers and employees in financial services firms having offices in the WTC and nearby buildings whose firms began trading less than a week following the attacks. They found that strong personal relationships among employees, the use of teams, and self-organization in the form of nonhierarchical relations sped up the recovery process.

Kendra and Wachtendorf (2003) studied the rebuilding of the Emergency Operations Center (EOC) following its destruction in the WTC attacks to better understand resilience. EOC continued to function despite the destruction of its physical location. EOC called on the resources of NYC and other jurisdictions. Employee relations and assigned roles existing before 9/11, and organizational response patterns, contributed to EOC resilience.

Buenza and Stark (2003, 2006) found that successful recovery in the financial services firms they studied was a function of planning and spontaneity, redundancy, and self-organization, associated with organizations having nonbureaucratic and nonhierarchical structures.

Argenti (2002) drew five lessons based on interviews with managers about their experiences and responses to 9/11 that apply to any organizations facing a crisis that affects employee motivation and confidence. These were:

1. Get on the scene: managers must be visible, show that dealing with staff anxiety and distress is their top priority, lead, and provide a sense of calm, reason, and humanity.
2. Communicate using channels carefully and creatively.
3. Focusing on the business, the work, and employees' needs to engage both the head and the heart.
4. Develop a plan including contingency work sites and important lines of communication.
5. Innovate and be flexible, but in ways consistent with the organization's guiding values and culture.

Sheffi (2005) examined ways that organizations might recover from major trauma. He suggests a two-pronged strategy: lowering vulnerabilities (increased dependence, interdependence, globalization, and uncertainties) and increasing resilience (increasing resources to rebound quickly). Resilience involves establishing redundancy in systems, developing back-up systems, allowing quick movement of people information and products from those parts of the organization that are still functioning, and creating a "can-do" culture that can get the job done under any circumstances.

Sheffi (2005) notes the following aspects of cultures of resilient organizations: an emphasis on results, teamwork and collaboration, informal networks and personal relationships, leadership at all levels, employees who are engaged in their work, lots of communication, and an emphasis in innovation and flexibility, among others.

The Importance of Leadership

Frohman (2006) and Frohman and Howard (2008) describe how an organization in Israel, headed by Frohman, continued working during the scud missile attacks of the Persian Gulf War of 1990–1991. They offer three lessons learned from this experience: focus on long-term survival and long-term consequences, do the opposite of what people expect, and trust your instincts.

Former NYC Mayor Rudy Giuliani received high marks for his role immediately following the 9/11 attacks. Giuliani responded immediately, was highly visible for some time after, offered support, comfort, and hope, and seemed to be in command of the situation (Giuliani & Kirson, 2007).

Planning and Preparing for Disasters

Planning and preparing for disasters have been described by some (e.g., McConnell & Drennan, 2006) as "mission impossible." Disasters and crises are low-probability events placing huge demands on scarce resources. But others (e.g., Bazerman &

Watkins, 2004; Watkins & Bazerman, 2003) suggest that there are signs and signals that likely portend disasters. Watkins and Bazerman identify three barriers to identifying what they term "predictable surprises": psychological, organizational, and political. Psychological barriers include cognitive biases (e.g., seeing things as being better than they are), organizational barriers involve silos and fragmented information (e.g., limiting the understanding of potential threats), and political barriers involve power differences in the organization in terms of whose voices are really heard.

They offer a three-stage process for improving an organization's ability to predict surprises: recognition, prioritization, and mobilization.

- *Recognition.* the organization scans the environment for emerging threats. Managers need to pay attention here.
- *Prioritization.* did the organization assign priority to threats having the highest costs? Serious threats need to be identified.
- *Mobilization.* did the organization adopt precautionary measures to address the risks involved? Managers need to respond appropriately and effectively.

Conclusions

Natural events and man-made acts can and will cause trauma to organizations and within organizations. Severe organizational trauma can lead to the death of organizations (e.g., Enron and Arthur Andersen), cause damage to the organization's reputation (e.g., Toyota and BP), result in employee and community deaths (e.g., Bhopal), have a negative impact on employee well-being, diminish employee motivation, and threaten the very survival of the organization. But the evidence indicates that both individuals and organizations can be or are resilient.

Events causing organizational trauma will always occur (Perrow, 1981, 1984; Starbuck & Farjoun, 2005), hopefully less frequently and with less damage to individuals and organizations. Following a disastrous event, organizations can proceed on one of two paths: toward recovery or toward decline and disintegration. Individual and organizational health before a disastrous act or event is a strong predictor of embarking on the path to recovery. Leadership and organizational culture are central in this regard (Vivian & Hormann, 2002; Weick, 2003).

Most writing to date has emphasized the final stage of a disaster – dealing with trauma – rather than on prevention. We know a considerable amount on response shortcomings and the ways in which individuals behave in disasters. Progress is being made also in the training of disaster mental health professionals. But applying these resources to prevention, as well as disaster response, is sorely needed.

Apparently, individuals' competence and health make the difference in both preparedness and responsiveness. Healthy individuals and healthy organizations can perform heroically in the face of tragedy.

Acknowledgment

Preparation of this chapter was supported in part by York University.

References

American Psychiatric Association. (2000). *Diagnostic and statistical manual of mental disorders* (4th ed., text rev.). Washington, DC: Author.

Argenti, P. (2002, December). Crisis communication lessons from 9/11. *Harvard Business Review*, 103–109.

Barbash, T. (2003). *On top of the world: Cantor Fitzgerald, Howard Lutnick and 9/11*. New York: HarperCollins.

Baum, J. A. P., & Dahlin, K. B. (2007). Aspiration performance and railroads' patterns of learning from train wrecks and crashes. *Organizational Science*, *18*, 368–385.

Bazerman, M., & Watkins, M. D. (2004). *Predictable surprises: The disasters you should have seen coming and how to prevent them*. Boston: Harvard Business School Press.

Bisson, J. L. (2007). Post traumatic stress disorder. *British Medical Journal*, *334*, 789–793.

Blatt, F. (2001–2002, Summer). The Lassing mine disaster: A retrospective. *Australian Journal of Emergency Management*, 35–43.

Buenza, D., & Stark, D. (2003). The organization of responsiveness: Innovation and recovery in the trading rooms of Lower Manhattan. *Socio-Economic Review*, *1*, 135–166.

Buenza, D., & Stark, D. (2006). *Trading sites: Destroyed, revealed, restored*. New York: Social Science Research Council.

Burke, R. J., & Cooper, C. L. (2008). *International terrorism and threats to security: Managerial and organizational challenges*. Cheltenham, UK: Edward Elgar.

Clarke, S. (2010). Managing the risk of workplace accidents. In R. J. Burke, S. Clarke, & C. L. Cooper (eds.), *Occupational health and safety: Psychological and behavioral challenges*. Surrey, UK: Gower.

Cole, L. A. (2009). *The anthrax letters: A bioterrorism expert investigates the attack that shocked America*. New York: Skyhorse.

David, P. A., Maude-Griffin, R., & Rothwell, G. (1996). Learning by accident? Reductions in the risk of unplanned outages in U.S. nuclear power plants after Three Mile Island. *Journal of Risk and Uncertainty*, *13*, 175–198.

Day, T. G. (2003). The Autumn 2001 anthrax attack on the United States Postal Service: The consequences and responses. *Journal of Contingencies and Crisis Management*, *11*, 110–117.

de Klerk, M. (2007). Healing emotional trauma in organizations: An O. D. framework and case study. *Organizational Development Journal*, *25*, 49–56.

Denenberg, R. V., & Denenberg, T. S. (2010). Workplace violence: The American experience. In R. J. Burke & C. L. Cooper (Eds.), *Risky business: Psychological, physical and financial costs of high risk behavior in organizations* (pp. 375–402). Surrey: Gower.

Dutton, J., Quinn, R., & Pasick, R. (2002a). *The heart of Reuters: Part A*. Ann Arbor: Center for Positive Organizational Scholarship, Ross School of Business, University of Michigan.

Dutton, J., Quinn, R., & Pasick, R. (2002b). *The heart of Reuters: Part B*. Ann Arbor: Center for Positive Organizational Scholarship, Ross School of Business, University of Michigan.

Frohman, D. (2006, December). Leadership under fire. *Harvard Business Review*, 1124–1131.

Frohman, D., & Howard, R. (2008). *Leadership the hard way*. San Francisco: Jossey-Bass.

Gershon, R. R. M., Qureshi, K. A., Barocas, B., Pearson, J., & Dopson, S. A. (2008). Worksite emergency preparedness lessons from the World Trade Center Evacuation Study. In R. J. Burke & C. L. Cooper (Eds.), *International terrorism and threats to security: Managerial and organizational challenges* (pp. 232–266). Cheltenham, UK: Edward Elgar.

Gittell, J. H., Cameron, K. S., Lim, S., & Rivas, V. (2006). Relationships, layoffs, and organizational resilience: Airline industry responses to September 11. *Journal of Applied Behavioral Science, 42*, 300–329.

Gittell, J. H., Cameron, K., Lim, S., & Rivas, V. (2008). Airline industry responses to September 11th. In R. J. Burke & C. L. cooper (Eds.), *International terrorism and threats to security: Managerial and organizational challenges* (pp. 267–290). Cheltenham, UK: Edward Elgar.

Giuliani, R., & Kirson, K. (2007). *Leadership*. New York: Hyperion.

Grant, A. M., & Wade-Benzoni, K. A. (2009). The hot and cool of death awareness at work: Mortality cues, agency and self-protective and pro-social motivations. *Academy of Management Review, 34*, 600–622.

Hood, E. (2004). Lessons learned? Chemical plant safety since Bhopal. *Environmental Health Perspectives, 112*, 353–359.

Hopper, E. (2010). *Trauma and organizations*. London: Karnac Books.

Kelly, J., & Stark, D. (2002). Crisis, recovery, innovation: Learning from 9/11. *Environment and Planning A, 34*, 1523–1533.

Kendra, J. M., & Wachtendorf, T. (2003). Elements of resilience after the World Trade Center disaster: Reconstructing New York City's Emergency Operations Center. *Disasters, 27*, 37–53.

Kowalski-Trakofler, K., Vaught, C., McWilliams, L. J., & Reisman, D. R. (2011). Psychological and behavioral aspects of occupational safety and health in the U. S. coal mining industry. In R. J. Burke & C. L. Cooper (Eds.), *Occupational health and safety: Psychological and behavioral challenges* (pp. 197–222). Surrey, UK: Gower.

Lester, D. (2011). Violence in the workplace. In R. J. Burke & C. L. Cooper (Eds.), *Occupational health and safety: Psychological and behavioral challenges* (pp. 179–196). Surrey, UK: Gower.

Lindahl, M. W. (2008). A new development in PTSD and the law: The case of *Fairfax County v. Mottram*. *Journal of Traumatic Stress, 17*, 543–546.

Madsen, P. M. (2000). These lives will not be lost in vain: Organizational learning from disasters in US coalmining. *Organization Science, 22*, 861–875.

Maiden, R. P., & Terblanche, L. (2007, June). Managing the trauma of community violence and workplace accidents in South Africa. openUP. Retrieved from http://137.215.9.22/bitstream/handle/2263/3316/Maiden_Managing(2006).pdf?sequence=1.

McConnell, A., & Drennan, L. (2006). Mission impossible? Planning and preparing for crisis. *Journal of Contingencies and Crisis Management, 14*, 59–70.

McFarlane, A. C., & Bryant, R. A. (2007). Post-traumatic stress disorder in occupational settings: Anticipating and managing the risk. *Occupational Medicine, 57*, 404–410.

Noer, D. (1993). *Healing the wounds: Overcoming the trauma of layoffs and revitalizing downsized organizations*. San Francisco: Jossey-Bass.

Pastel, R. H. (2008). Psychosocial impacts of bioterrorism and stress in the wake of 9/11. In R. J. Burke & C. L. Cooper (Eds.), *International terrorism and threats to security: Managerial and organizational challenges* (pp. 10–125). Cheltenham, UK: Edward Elgar.

Perrow, C. (1981). Normal accidents at Three Mile Island. *Society, 18*, 17–26.

Perrow, C. (1984). *Normal accidents*. New York: Basic Books.

Regel, S. (2007). Post-trauma support in the workplace: The current status and practice of critical incident stress management (CISM) and psychological debriefing (PD) within organizations in the UK. *Occupational Medicine, 57*, 411–416.

Roberts, K., & Martelli, P. (2011). A variegated approach to occupational safety. In R. J. Burke & C. L. Cooper (Eds.), *Occupational health and safety: Psychological and behavioral challenges* (pp. 323–342). Surrey, UK: Gower.

Shalev, Y. (2004). Further lessons from 9/11: Does stress equal trauma? *Psychiatry, 67*, 174–176.

Sheffi, Y. (2005). *The resilient enterprise.* Cambridge, MA: MIT Press.

Shirom, A., Toker, S., Berliner, S., Shapira, I., & Melamed, S. (2008). Fear of terror and health: A study of apparently healthy employees. In R. J. Burke & C. L. Cooper (Eds.), *International terrorism and threats to security: Managerial and organizational challenges* (pp. 126–152). Cheltenham, UK: Edward Elgar.

Shrivastava, P. (1987). *Bhopal: Anatomy of a crisis.* Cambridge, MA: Ballinger.

Singh, P. (2008). Maintaining work motivation during trying times. In R. J. Burke & C. L. Cooper (Eds.), *International terrorism and threats to security: Managerial and organizational challenges* (pp. 219–231). Cheltenham, UK: Edward Elgar.

Starbuck, W. H., & Farjoun, M. (2005). *Organization at the limit: Lessons from the Columbia disaster.* Malden, MA: Blackwell.

Sutcliffe, K. M., & Vogus, T. (2003). Organizing for resilience. In K. S. Cameron, J. E. Duitton, & R. E Quinn (Eds.), *Positive organizational scholarship: Foundations for a new discipline* (pp. 94–110). San Francisco: Berrett-Koehler.

Vivian, P., & Hormann, S. (2002). Trauma and healing in organizations. *OD Practitioner, 34*, 37–42.

Walter, K. H., Hall, B. J., & Hobfoll, S. E. (2008). Not business as usual: The psychological impact of terrorism and mass casualty on business and organizational behavior. In R. J. Burke & C. L. Cooper (Eds.), *International terrorism and threats to security: Managerial and organizational challenges* (pp. 81–104). Cheltenham, UK: Edward Elgar.

Watkins, M. D., & Bazerman, M. H. (2003, March). Predicable surprises: The disasters you should have seen coming. *Harvard Business Review*, 72–80.

Weick, K. E. (2003). Positive organization and organizational tragedy. In K. S. Cameron, J. E. Dutton, & R. W. Quinn (Eds.), *Positive organizational scholarship: Foundations of a new discipline* (pp. 66–80). San Francisco: Berrett-Koehler.

Weick, K. E., & Sutcliffe, K. M. (2001). *Managing the unexpected: Assuring high performance in an age of complexity.* San Francisco: Jossey-Bass.

6

ASSIST: A Model for Supporting Staff in Secure Healthcare Settings after Traumatic Events That Is Expanding into Other European Territories

Annette Greenwood, Carol Rooney, and Vittoria Ardino

Introduction

This chapter explores the impact of psychological trauma on health professionals and their organization from two different traditions. The UK context presents a more established culture and police in terms of trauma services, whereas Italy offered trauma services mainly to abused and neglected children overlooking the need of comprehensive trauma centers and the need of trauma support to professionals. The chapter presents a conceptual and applied framework of understanding to assist and support health professionals in their response to a work-related trauma. Furthermore, the chapter discusses the sustainability and transferability of a model across different contexts, such as the trauma center funded by the Italian Red Cross in Milan, Italy.

Individuals working in caring professions are among the occupational groups identified as being at high risk of work stress (Smith *et al.*, 2000). Evidence suggests that employees of the UK National Health Service (NHS) or other similar private settings reported greater occupational stress and minor psychiatric disturbance than did other job groups in the United Kingdom (Wall *et al.*, 1997). The task of working with people who present serious mental health problems with propensity to violence can be challenging, and at times harmful and traumatic. Such issues are particularly pertinent to managers and staff working in secure mental health environments, where there is a higher likelihood of verbal abuse and physical assault. According to the NHS Security

International Handbook of Workplace Trauma Support, First Edition. Edited by Rick Hughes, Andrew Kinder, and Cary L. Cooper.

Management Service, in 2004–2005 there were 43 301 reported physical assaults on NHS staff in mental health and learning disability environments, which they say is considerably higher than in other areas of patient care.

Trauma literature has widely explored the psychological outcomes of staff being exposed to violent clients. However, there is limited empirical evidence on the benefits of a support framework for staff who have experienced a work-related traumatic event and potentially post-traumatic reactions. This chapter evaluates the importance of psychological support for staff working in secure mental health hospitals and illustrates a model of psychological first aid, ASSIST, implemented at St. Andrew's Healthcare, a charity that provides secure mental health services to a range of patients who have difficult and challenging mental health problems in the United Kingdom. The chapter aims also to describe the experience of staff working with violent clients, highlighting the differences between working with such or other client groups. Furthermore, this chapter addresses what factors are related and contribute to the development of post-traumatic reactions in this professional group. Finally, the chapter presents how the ASSIST model can be transferred and implemented in a different European context and with professionals working with different client groups.

Trauma within Organizations

Traumatic events can impact people and organizations affecting psychological well-being. Providing appropriate and timely support for health professionals (the staff) and business (the organization) is considered as a key protective factor to prevent serious outcomes following traumatic exposures. The challenge that is a thread throughout this chapter starts in the 1990s with an incident in the NHS and goes through to the present day to the current work at St. Andrew's Healthcare. In particular, the challenge is to mediate the potential conflict between staff needs and those of organizational management.

The work at St. Andrew's stems back to early developments of intervention for supporting staff exposed to trauma and of debriefing services present in the UK NHS. In the United Kingdom, there were major incidents, such as the *Piper Alpha* oil rig disaster, the sinking of the *Herald of Free Enterprise* passenger ferry, Hillsborough stadium, and the Bradford football disasters that increased awareness of the importance of appropriate psychological support for individuals involved. The NHS policy had also recognized the importance of supporting professionals who provided help in the occasion of such incidents on the premise that they could have been potentially traumatized. As a result, several professionals working with victims of such incidents had been trained according to a support model known as *psychological debriefing*, which was theorized by Mitchell (1993).

Psychological well-being was a developing field within the NHS, and by the year 2000 all NHS Trusts were directed to offer a staff counseling service for staff. For example, the Royal College of Nursing (RCN) commissioned a guidance document entitled "Staff Counselling in the Health Service: A Guide for Employers and Managers"; Greenwood (2002) went some way toward helping ensure that services met a quality standard.

Greenwood's early work in the early 1990s whilst employed at a large acute NHS Trusts was to develop and to provide psychological models of support for staff and the organization following a major incident. For example, one of the first major incidents in the United Kingdom in the NHS was the Beverly Allitt incident at the Grantham Hospital in Lincolnshire. This incident has had a profound impact on NHS policy. Beverly Allitt systematically harmed and murdered children whilst employed as a nurse on a children's ward. She injected the children with insulin and then, when they collapsed, took on the role of the nurse trying to save the child's life. This act had a profound impact upon the parents, the healthcare staff, and the public within the United Kingdom. The hospital had to continue providing healthcare to the general population and, at the same time, involved professionals required support to prevent a disruption of patient care. After the initial shock response to the trauma of the Allitt incident, staff received ongoing support for five years. This period was very difficult for the staff and the organization, which were under intense scrutiny from the public and the media, as they tried to recover from the legacy of the Allitt incident. Known in the UK media as the "Angel of Death," Beverly Allitt challenged how nurses were viewed in the United Kingdom. Developing and providing psychological support services that meet the needs of the staff and the organization were at times conflicting. The Clothier Report (1994) had a significant impact upon how confidentiality and recruitment of healthcare staff would be implemented within the United Kingdom.

Occupational health departments and healthcare managers needed to assess staff for resilience and to screen for mental health problems as the NHS worried about employing another healthcare professional who might harm or kill patients. This incident posed implications for both the staff, who found it was not safe to declare mental health problem difficulties, and for the managers, who worried about who they were recruiting and the possible consequences for the organization and themselves. Many managers were left fearful, on the one hand not discriminating against individuals with mental health problems and on the other hand needing to be confident that the healthcare professional was safe to practice.

Staff Support in Secure Mental Health Services

Work with patients with mental health issues implies many demands in terms of staff resilience and coping. In the 1990s the NHS implemented secure mental health services including the new specialty of forensic nursing.

Several authors have attempted to identify the essential skills required to work in secure mental health environments (Kitchener *et al.*, 1992). Staff working in secure mental health environments are exposed to challenging behavior, and occasionally frightening experiences and interpersonal violence. Within the United Kingdom, medium-secure forensic mental health services were developed and expanded rapidly in the 1990s caring for patients who have enduring serious mental health problems and a history of convictions and/or challenging behavior. A scoping exercise carried out by the United Kingdom Central Council (UKCC), the national professional body for nursing in the United Kingdom, into the nature of nursing in secure environments highlighted that the patients' mental disorder and offending patterns pose intense demands upon nurses as they are required to maintain empathic relationships while maintaining an awareness of the need to focus on risk (UKCC, 1999). Killian and Clark (1996) identified complex interpersonal issues associated with nursing in secure environments, observing that most nurses have to balance their dual role of providing clinical care and of being a "custodial" figure. Kitchener *et al.* (1992) identified that conflict and ambiguity associated with forensic nurses' therapeutic and controlling roles were common problems. Mason (2002) also identified role tensions as a significant factor in nursing practice in secure care.

Several studies recommended that in secure services, given the nature of the patient groups, services should provide some support to staff; nevertheless, there was very little guidance to implement such a service (Coffey and Coleman, 2001; Kirby *et al.*, 1995; Mason, 2002).

Local county-wide emergency response structures grew from a focus on practical assistance, to include a debriefing and counseling service that was also available locally to mental health service staff in the aftermath of a serious incident. Staff was trained in critical incident stress debriefing using the Mitchell model (Mitchell & Everly, 1993). However, there was no policy to assist managers in making informed decisions about the modalities for requesting external support for arranging formal debriefing, and for explaining support options to their staff. The result was a confused picture which returned ineffective debriefs, disappointed staff who were frustrated in their expectations, and occasional psychological harm caused by misguided attempts to provide support. Consequently, staff felt unsupported by the organization with a consequent effect on their well-being, their views about the organization, and ultimately their motivation to continue working.

Psychological Consequences

Nursing in secure environments requires the ability to manage risks and to foster resilience and coping abilities to face traumatizing situations. Staff may experience two different types of traumatic exposure depending on whether they are the primary or secondary victim of an incident occurring in their work environment.

Post-traumatic stress disorder (PTSD) is a long-lasting anxiety response following a traumatic or catastrophic event and consists of various symptoms in the following groups: persistent re-experience of the traumatic event (e.g., including images, thoughts or perceptions, and recurrent distressing dreams), persistent avoidance of trauma-associated stimuli (e.g., thoughts, feelings, or conversations; efforts to avoid activities, places, or people that arouse recollections of the trauma; and inability to recall an important aspect of the trauma), and persistent symptoms of increased arousal such as difficulty falling or staying asleep (American Psychiatric Association, 1994). Isolated PTSD symptoms have been reported implicitly or explicitly by numerous authors.

Direct exposure

Workplace violence is a major cause of direct trauma exposure for who work with secure-unit patients. The likely typologies of violence toward staff involve:

Type 1: intrusive violence. Criminal intent by strangers, terrorist acts, mental illness– or drug-related aggression, and protest violence.
Type 2: consumer-related violence. Consumer, client, or patient (and family) violence against staff; vicarious trauma to staff; and staff violence to clients or consumers.
Type 3: relationship violence. Staff-on-staff violence and bullying, and domestic violence at work.
Type 4: organizational violence. Organizational violence against staff; and organizational violence against consumers, clients, or patients (Merchant & Lundell, 2001; Peek-Asa, Runyan, & Zwerling, 2001).

Violence and harm against staff lead to different effects: bio-physiological, cognitive, emotional, and social effects.

Bio-Physiological effects. Anxiety or fear is the most frequently reported bio-physiological effect. Anxiety may occur in a generalized form (Lanza, 1983; Ryan & Poster, 1989; Whittington & Wykes, 1992) and fear may relate, for example, to the workplace (Bin Abdullah *et al.*, 2000; Arnetz & Arnetz, 2001) or patients (Fernandes *et al.*, 2002). Fear can be differentiated as fear of the perpetrator or of other patients (Hauck, 1993; Lanza, 1983; Whittington & Wykes, 1992), fear for oneself or one's family (Fry *et al.*, 2002), fear of permanent side effects of the assault or of becoming physically dependent on others (Lanza, 1983), fear of retaliation toward the aggressor (Lanza, 1983; Ryan & Poster, 1989), or fear of coworkers or of the future (Hauck, 1993).

Guilt or self-blame or shame is also a prominent reaction to aggression reported in a majority of the studies. Some nurses feel guilty for not handling the situation in a more appropriate way (Hauck, 1993).

Patient aggression may lead in a minority of cases to a fully established PTSD or isolated symptoms of this. Caldwell (1992) reports that 137 of 224 (56.1%) of traumatized clinical staff in psychiatry contracted some PTSD symptoms but only 22 (9.8%) suffered the full clinical PTSD.

Cognitive effects. Various threats to personal integrity or pride are reported, with some victims perceiving themselves as disrespected, unappreciated, violated, robbed of their rights (Gates, Fitzwater, & Meyer, 1999), humiliated (Lanza *et al.*, 1991; Hauck, 1993), compromised (Chambers, 1998), or intimidated, harassed, and threatened (Fry *et al.*, 2002). Others perceive themselves as being at the mercy of the perpetrator (Hauck, 1993), whilst yet others experience disbelief that the assault occurred (Adams & Whittington, 1995; Chambers, 1998; Hislop & Melby, 2003; May & Grubbs, 2002; Ryan & Poster, 1989). Denial or rationalization of the assault (Lanza, 1983) or disbelief at being involved in the incident (Chambers, 1998) may also occur. Some incidents can lead to a radical transformation of the conception of the world, with some victims stating that nothing will ever be the same again (Hauck, 1993), that the event has a disruptive meaning (Flannery, Hanson, & Penk, 1995), or that the world has become less predictable (Gates *et al.*, 1999; Hauck, 1993) or threatening (Hauck, 1993).

Emotional effects. Emotional reactions constitute the greatest variety of symptoms, with anger being the most frequently reported. Anger may be directed toward the nurses themselves, superiors (Hauck, 1993), or the institution (Chambers, 1998; Hauck, 1993). The range of percentages of nurses experiencing anger is greatest in the emergency services, with rates from 14% (Fernandes *et al.*, 2002) to 68.6% (Abdullah *et al.*, 2000). Two groups of authors report higher anger rates following verbal aggression than physical aggression (Fernandes *et al.*, 2002; O'Connell *et al.*, 2000). Feelings of guilt and self-blame are sometimes reinforced by superiors placing the blame for assaults on the victims (Hauck, 1993). Guilt sometimes occurs in conjunction with shame and may lead to impairments in self-confidence (Hauck, 1993).

Social effects. Assaults can affect (Gates *et al.*, 1999; Suserud, Blomquist, & Johansson, 2002) or undermine the nurse–patient relationship and lead to behaviors such as less eagerness to spend time with residents, less willingness to answer residents' call lights (Gates *et al.*, 1999), avoiding patients (Flannery *et al.*, 1995; Levin, Hewitt, & Misner, 1998), or adopting a passive role (Chambers, 1998). Some nurses report becoming callous toward patients (Levin *et al.*, 1998).

Nurses' perceptions of their job competency and security at, or satisfaction with, the workplace may be affected. Some of the assaulted nurses question the normality of a job in which workers are assaulted (Hauck, 1993), or even consider changing jobs (Bin Abdullah *et al.*, 2000; Lanza, 1983). Some actually change ward or employer (O'Connell *et al.*, 2000). Many assault victims feel insecure at work (Bin Abdullah *et al.*, 2000; Hauck, 1993; May and Grubbs, 2002; Vincent *et al.*, 2000), more vulnerable (Fry *et al.*, 2002; Lanza, 1983), or less in control (Lanza *et al.*, 1991).

Patient aggression and assault can lead to real or perceived impairments in professional performance, leading nurses to doubt the quality of their work (Bin Abdullah *et al.*, 2000) and their competency (Whittington & Wykes, 1992), or perceive themselves as having failed.

Indirect exposure

According to the *Diagnostic and Statistical Manual of Mental Disorders* (APA, 1994), exposure to trauma incorporates experiencing, witnessing, and learning about a traumatic event, implying that an individual does not have to directly experience the event to eventually develop post-traumatic symptoms of PTSD; such an occurrence is more appropriately termed *secondary traumatization* (Bride, Radey, & Figley, 2007). Secondary traumatization was first recognized in the literature following emergency services research in the 1970s whereby rescue workers were observed to display PTSD symptoms after attending to victims (Iliffe & Steed, 2000). This catalyzed the conception of four main conceptual frameworks of secondary trauma: vicarious traumatization (McCann and Pearlman, 2001), burnout (Freudenberger and Robbins, 1979), secondary traumatization, and compassion fatigue (Figley, 1995). Although these four concepts differ according to their respective schools of thought, they all fundamentally describe the negative effects of indirect exposure to another's trauma.

McCann and Pearlman (1990) first presented the term *vicarious traumatization* (VT) to describe the effect of counseling victims of sexual abuse on the counselor. Based on the *constructionist self-development theory* (whereby individuals are said to construct and construe their own realities), the cumulative exposure to another's trauma is predicted to negatively alter a clinician's cognitions about the world, the self, and others (Saakvitne, Tennen, and Affleck, 1998). The term *burnout* describes the symptoms of emotional exhaustion that are the result of job strain, lowered feelings of achievement and accomplishment, and the destruction of idealism (Anderson, 2000; Maslach, 1982). *Burnout* is said to occur when working with any client group and is often viewed as the end result of compassion fatigue (Adams, Figley, & Boscarino, 2008).

Compassion fatigue (CF) can occur through caring for any client group, although literature has mainly focused on work with trauma victims. According to Figley (1995), as carers are fundamentally guided by altruism and compassion for humanity, direct exposure to a client's suffering may cause CF whereby the caregiver experiences a reduced capacity to be empathic. Aside from reduced empathy, compassion fatigue incorporates a variety of physical (e.g., serious illness) and psychological (e.g., anxiety, and intrusive nightmares) symptoms and also various social problems (e.g., relational problems) all indicative of the PTSD symptomatology (Valent, 2002), except that the trauma exposure is indirect.

A few studies have investigated CF in relation to professionals dealing with other client groups such as domestic violence counselors (Iliffe & Steed, 2000), substance abuse counselors (Fahy, 2007), telephone counselors (Dunkley & Whelan, 2006), and even CF experienced by cancer care providers (Najjar *et al.*, 2006).

Of particular concern is the limited research attention dedicated to the potential experience of CF in secure mental health staff. Janik (1995) described such staff as "sin-eaters" in that they are subjected to the poisonous and caustic images, memories, and experiences of offenders on a daily basis. Thus staff are said to amass these toxic mental images that may conflict with their assumptions about the world, and it is this accumulation of disturbing mental material that may lead to CF and even burnout (Janik, 1995). There have been a number of studies (Abendroth & Flannery, 2006; Frank & Adkinson, 2007; Frank & Karioth, 2006; Jenkins & Elliot, 2004; Maytum, Heiman, & Garwick, 2004) investigating compassion fatigue amongst nurses. However, there has been limited research attention on nurses working in acute mental health settings (Jenkins & Elliot, 2004) and as yet CF research has not extended to forensic nurses. The existing CF studies on nurses have found that nurses do experience CF and furthermore that their vulnerability to CF is attributable to the caring and empathic nature of nursing (Najjar *et al.*, 2009).

Role ambiguity has also been highlighted as a contributor to job stress for secure unit staff (Alexander-Rodriguez, 1983; Gulotta, 1987; Johnson, 1995). Moreover, the conflicting aims of healthcare staff, to be caring versus disciplinarian, can be a source of much tension between the two disciplines that are required to work cohesively. This dissonance is termed "double-agentry" (Johnson, 1995) and is said to negatively impact upon staff job satisfaction and as a result act as a stressor.

However, there are several protective factors that have been found to mitigate staff from feelings of stress and emotional exhaustion. These include autonomy and control (Flanagan & Flanagan, 2002; Sprang, Clark, & Whitt-Woosley, 2007), emotional support from coworkers (Jenkins & Elliot, 2004), social support (Melchior *et al.*, 1997), compassionate self-care and monitoring empathy (Larson & Bush, 2006), increasing job satisfaction (Musa & Hamid, 2008), and sufficient supervision (Adams *et al.*, 2008).

Trauma Support Models in Two European Countries

The next two sections of the chapter describes trauma response services for staff that are provided in two European counties, the United Kingdom and Italy. An exploration of the different cultural approaches is discussed. One of the significant results of the author's work in this area is how a psychological first aid model developed in the United Kingdom can be transferable and the skills and learning can enable a cross-cultural development of trauma support services. The new project is to be funded by the Italian Red Cross for a trauma center in Milan, Italy.

The Experience of St. Andrew's Healthcare, United Kingdom

St. Andrew's Healthcare is a charity that provides secure mental health services to a range of patients who have difficult and challenging mental health problems. There is a strong culture of caring for patients who cannot be cared for in other NHS or independent provider units, because of their uniquely challenging needs and specialist care required.

The focus of patient care is on assessment, treatment and recovery, and the development of coping skills. However, with a population of patients who require secure care, there will almost inevitably be times when staff are subject to verbal abuse and physical assault, and experience psychological trauma.

At St. Andrew's Healthcare, there was recognition that a structure was needed to provide a range of supportive and restorative interventions for staff. This thinking culminated in the development of the post of trauma advisor, with a remit to review policies and develop structures for staff support.

Developing a Model of Trauma Support That Benefits the People and the Business

Not all trauma services are based on a critical incident management model particularly when the incidents are an ongoing issue for the organization. It is good business sense for some organizations to plan services that will be central in educating and preparing its staff for possible traumatic events and ways of managing them. An example of how providing trauma support has benefitted the business was the appointment of a trauma advisor to a secure mental health hospital charity group. This role was central to the development, design, and delivery of the trauma response service provided. The post holder needed to have experience of managing and clinically delivering services to traumatized staff, a working experience of a secure mental health hospital environment, and an understanding of organization cultures. The organization wanted to establish a service for staff, not in reaction to some major incident or event, but to develop a range of services, policies, and strategies to support staff well-being within the organizational environment where violent threats and assaults did occur. The hospital has a national profile for providing care to patients with difficult and challenging behaviors and identified that supporting staff was at the core of ensuring and maintaining high standards of care for the patients.

The new service was set up within the clinical risk management team and worked closely with the health and safety team, occupational health and human resources, and the hospital directors. A brief scoping and development plan that included feedback from the people (the staff) and the organization (the business) formed the baseline of the service model. This inclusion of both ensured success and worked toward enhancing the support for staff, with many reported feeling more valued by the organization, and the organization in turn found that its profile with the monitoring external agencies such as the Care Quality Commission and the Health and Safety Council paid dividends. According to Rick, O'Regan, and Kinder (2006), "The way individual employees perceive the support offered by their organization can develop effective trauma management practices." The ASSIST trauma response service was the primary-level service for staff, and a secondary level was provided by the commissioning of an external employee assistance program (EAP) that had the skills to support traumatized staff and was willing to take time to understand the organizational context of secure mental health hospitals. The service was provided by ATOS Healthcare, and staff could access this service day or night every day of the year.

The model takes account of the National Institute of Health and Clinical Excellence (NICE) guidelines for the treatment of PTSD (2006) and the work of Boorman (2009) whose recent review of NHS occupational health services suggests that early contact and support for staff on sick leave ensure they have an earlier return to work. Indeed, Boorman's review builds upon the notion suggested by Black (2007) that "work does improve an individual's mental health." However, for the staff who have been violently assaulted at work there is a need for a robust framework of support and organizational policies and standards that can direct and help them and their managers to enable a sustainable return.

All organizations want staff to be present, motivated, and effective, with the resources to cope with the demands of the job. A structure for the support and recovery of staff who are working in challenging conditions should be seen as a positive investment in these ways, which will ultimately impact positively on patient care:

- Improved motivation and morale; loyalty.
- Improved staff well-being.
- Reduced sickness absence.
- Faster recovery and return to work.
- Decrease in injury claims.
- Increase in staff feeling valued.
- To reduce turnover of staff – the cost of training staff to work in secure mental health is a valuable resource the organization does not want to lose.

The St. Andrews ASSIST Trauma Response Model of Psychological Support

The model developed for the staff seriously assaulted or traumatized at work is based on a psychological first aid response which provides in the first session an assessment, support, normalization, and follow-up sessions to monitor and support an early return to work when fit. As Guy and Guy noted, "It is essential to note that PTSD and trauma symptoms are not suitable for counselling; indeed talking therapies will embed the trauma further and possibly vicariously traumatise the counsellor" (2009, p. 19). There has been much debate regarding the effectiveness of debriefing sessions following a traumatic incident, and indeed NICE (2006) suggest that single-session debriefs should be avoided. This is in contrast to the suggestion by many working in this area that a brief intervention is helpful. Tehrani, for example, argues that:

> The law attempts to protect employees by placing a duty of care on the employer. In order to meet this duty the organization has to put in place a number of policies and procedures including ensuring that there are adequate risk assessments, safe practices and post trauma support for employees that come involved in traumatic events. (2004, p. 62)

The NICE guidance on the treatment of PTSD is aimed at the general public and does not discuss how support and treatment for employees should be developed and delivered. It is to be noted that whilst the staff using the service had signs and symptoms of post-traumatic stress (PTS), none presented with PTSD. Therefore, the NICE guidance has limitations for some populations. Indeed Rick *et al.* (2006) raised an important question, "What can organisations do for employees post trauma that is both safe and effective?" Indeed, Regal and Joseph (2010) noted that "in the NICE guidelines, that none of the studies that showed negative effects contained any descriptions of the training given to those carrying out the interventions," but rather it is well accepted by those receiving the intervention that it is helpful.

The psychological first aid model (ASSIST) employed at St. Andrew's Healthcare, in response to staff who have been seriously assaulted or traumatized, has at its core a corporate policy and procedure to enable staff and the managers to use and to promote an organizational culture where getting and asking for support is seen as normal practice. Early on in the development of this model of support, staff reported, "It happens anyway, but we need something." This starting point became the focus of an academic paper presented at an international conference (Greenwood & Rooney, 2010) because the staff wanted something that valued them, was a place that they could access, and was confidential. An important factor for the staff was that anyone providing psychological support needed to understand the organizational culture and context of working on a locked ward and secure mental health. Some of the staff requested group debriefs and saw them as a vehicle for teams to express collectively their anxieties regarding incidents at work. Feedback from those attending these sessions was that it helped team dynamics which could be dealt with in a safe environment. They felt valued by their manager and the organization for providing the time and space they needed when discussing the impact of working with patients with violent and threatening behaviors, even if this was due to an underlining serious mental health problem.

A senior nurse who had worked for over 14 years stated,

> I remember feeling why me and then feeling why I am tearful and unable to cope, I always thought that I was a strong person and could cope with anything so why was I in such a state. I also began to question my belief in my job role, what had I said or done wrong for this to happen to me, was I in the right job, could I go back to the unit, where this patient remained. . . . I seemed to have lost me, and was afraid I would not return.

This example highlights the notion that trauma can impact anyone, even if they have years of experience in their job. The impact of psychological trauma is real, and developing a model that supports the staff and the organization needs to be founded on a commitment by the organization to have policies and procedures in place. These need to inform both the staff and the managers on how to manage violence and traumatic incidents that injure staff in the workplace. Most organizations in healthcare have policies for any process or event that could occur within the hospital. The St Andrew's Healthcare psychological first aid model is embedded within the core policy structure, and the policy has a step-by-step action plan for managers and staff to follow. In this way it becomes normal practice after a violent or traumatic incident for staff to be referred for a psychological first session with the trauma advisor. The impact for the organization is that an organizational culture is developed where asking for help and offering support are part of the practice of all managers within the hospitals. Organizational learning is of paramount importance to the success of any service. Statistical feedback of key findings and themes enables practices and clinical risk assessments to be reviewed, and learning and changes in practice to be view as positive steps. By closing the loop and feedback, the organizational culture is changed by where staff no longer state, "It happens anyway." All identifying personal information is removed to ensure the confidentiality of the individual staff accessing the service.

The ASSIST Model

The ASSIST model is not a treatment for PSTD; rather, it is a psychological first aid intervention to provide a quick, accessible, structured support that promotes psychological well-being.

The model is based on a brief psychological first aid debrief model and is employed on a 1 + 2 session basis. Staff are seen within 4–5 days of the incident or event. The attendance at the sessions is voluntary, but an organizational policy outlines the procedures for managers and staff following an incident or event at work. The model's design has been influenced by other models of psychological first aid. Examples include SPOT, developed by the Post Office, and TRIM the MOD model and the Psychological Debrief (PD) approach developed by Regel and Joseph (2010) at the Centre for Trauma, Resilience and Growth, in Nottinghamshire. Importantly, the ASSIST model includes assessment of psychological awareness and function post trauma. Research suggests that the traumatized brain shuts down during the exposure to trauma, with the amygdala, the part of the brain responsible for rapid processing of emotionally loaded stimuli that leads to an emotional response, being affected. Staff have reported that often their responses post trauma can include the loss of the ability to modulate. Being unable to regulate your emotional response on a secure ward environment could be a risk to the staff member and their team. Two responses that seem to be retained are the fight-or-flight response. An example at times of post-event stress could be "I'm not here" or "It's happening again," and the person's emotional response over-rides the normal cognitive function.

Staff who have been assaulted typically say "I'm OK"; indeed, many survivors of trauma report the same. If the amydgala does shut off and, in the short term, processes changes to deal with the demands of the trauma, one question that arises is how can we be sure that it is switched on again? Holmes *et al.* (2011) and Kennerley and Kischka (2011) have found evidence that suggest all trauma intervention should consider the neurological response to trauma, which is a growing area of research. Working in a secure environment, staff rely upon their developed "relational security" to promote safety for the patients and staff. Indeed, relational security is a key component of all staff working at St. Andrew's. A brief function assessment that focuses on enabling the individual to monitor their affect and develop strategies has been developed. Emotional regulation helps enable the staff to understand their own responses to emotional triggers associated with the trauma and to take back some control as a coping strategy. And, most importantly, it ensures that they are safe to return to working in a secure environment.

ASSIST

- Assessment of the individual or groups needs context and background. A function checklist is employed to assess psychological awareness and response to possible psychological triggers.
- Structured session provides the individual with an understanding of the impact of psychological trauma and information of normal signs and symptoms and possible reactions.
- Strategies to help cope with the psychological impact of the signs and symptoms of trauma are discussed.
- Information regarding ongoing support – 24/7 helpline.
- Sign posting and referrals to specialist help and treatment for staff who have continued frequency of signs and symptoms or who report life difficulties related to trauma reactions.
- The model takes account of the individual's natural resilience and ability for psychological growth and recovery.

Offering support to staff after an event that has happened at work makes good business sense. Staff have reported that they feel more valued by the organization, and being given the time to attend sessions has been one of the keys to their recovery. They report that having an understanding of the process and impact of trauma helps to stop fears that they are developing mental health problems that will not go away.

The Experience of Trauma Support in Italy

Italy has a long tradition in child protection services with a few excellent trauma centers for abused and neglected children. Likewise, the Italian NHS and some private centers have been committed to develop services for female victims of violence. However, Italian trauma care provision has overlooked the need for developing a comprehensive trauma center dedicated to the interdisciplinary treatment and study of trauma and its psychological reactions. In 2010, the Italian Society for Traumatic Stress Studies was established with the aim of coordinating trauma research and clinical services in the Italian context. Furthermore, most available services responding to traumatic events, such as mass disasters, motor incidents, and interpersonal violence, have not embedded into their organizations specialized support for professionals.

The 2009 earthquake in Abruzzo in Italy represents a good example of the confused picture in terms of support to psychologists, and health professionals who intervened on site to provide care to victims. The psychosocial intervention in Abruzzo followed a guideline – promoted by the government – which implied specific steps that professionals had to follow to intervene in mass disasters. The primary goal was to guarantee survival and psychological well-being through an activation of personal and community resources. Secondary goals implied intervention monitoring directed to victims, their families, and rescue workers. Interestingly enough, the recommendations do not provide any specific model for supporting post-trauma response in nurses and psychologists on site.

Thanks to an agreement with the Italian Red Cross, the first Italian comprehensive trauma center was recently funded in Milan, northern Italy, with the aim of offering an out-patient specialty mental health service to the community with an explicit commitment to ensure trauma support and care to disadvantaged, under-served, and severely impacted youth and adults. The mission of the trauma center is to "support individuals, families and communities that have been impacted by trauma and adversity to re-establish a sense of safety and predictability in the world, and to provide them with state-of-the-art therapeutic care as they reclaim, rebuild and renew their lives." Another relevant aim of the center is to provide psychological support to respond to an eventual traumatic exposure according to St. Andrew's model. Finally, the center will be actively engaged in clinical research dedicated to advancing the field of traumatic stress in regard to understanding and treatment of disorders and conditions associated with traumatic exposure. Ongoing research will also test the St. Andrew's model in terms of preventing PTSD symptoms in staff and volunteers.

The trauma center in Milan is explicitly committed to provide psychological aid to all staff and volunteers who provide care to victims. For this reason, in the second phase of its implementation the St. Andrew's psychological first trauma model will be transferred to the center. The center, therefore, intends to respond to the need for implementing more advanced models to support staff exposed to trauma and to overcome the cultural resistance of health professionals to seek help and support when threatened by post-traumatic reactions. The ASSIST model will be presented to stakeholders and to the scientific board of the center to ensure that the organization will endorse all steps involved in the implementation of the psychological aid for staff.

Psychologists working in the center will be trained according to the procedures previously described so that they will ensure support to volunteers involved in the clinical work. An external psychotherapist will be trained to support psychologists working in the center. Volunteers will be offered voluntary sessions fortnightly unless a major incident happens. In case of a major incident, volunteers will be invited to a session within 4–5 days of the incident or event.

Future Developments and Challenges

In the United Kingdom at St. Andrew's Healthcare, a further development is the rolling out of a trauma awareness training program for all managers. The training has at its heart the ASSIST trauma response policy which outlines a step-by-step approach for managing trauma incidents at work.

This chapter is a starting point, and illustrates the possibility of transferring psychological models of support across European countries. The authors acknowledge that cultural differences can be a barrier, but working together the basic principles of psychological first aid are transferable. In the United Kingdom the benefits to the staff at St. Andrew's Healthcare, a hospital with a reputation for caring for challenging and violent patients, is paying dividends to their psychological well-being at work. The organization has benefitted by being more informed and prepared in supporting the staff and ensuring the continuing care for the patients who used its services. The ASSIST model of trauma support does inform the organization (the business) and the staff (the people), and it is seen as an ongoing process and not a quick fix. The business of secure mental health hospitals by its nature means that the patient will be difficult and have the potential to harm. The model offered at St. Andrews gives a clear message to the staff that the organization takes seriously the need for the support for staff and helps challenges the organizational view that "It happens anyway."

Finally the most noteworthy finding from working together on this chapter is that the ASSIST psychological first aid trauma response model which was designed and developed in the United Kingdom will be employed to train the new recruits at the International Red Cross center in Milan, Italy.

Future research will follow this development of the cross-cultural exchange of psychological trauma support and knowledge.

References

Abendroth, M., & Flannery, J. (2006). Predicting the risk of compassion fatigue: A study of hospice nurses. *Journal of Hospice and Palliative Nursing*, *8*, 346–356.

Adams, J., & Whittington, R. (1995). Verbal aggression to psychiatric staff: Traumatic stressor or part of the job? *Psychiatric Care*, *2*, 171–174.

Adams, R. E., Figley, C. R., & Boscarino, J. A. (2008). The compassion fatigue scale: Its use with social workers following urban disaster. *Research on Social Work Practice*, *18*(3), 238–250.

Alexander-Rodriguez, T. (1983). Prison health: A role for professional nursing. *Nursing Outlook*, *31*(2), 115–118.

American Psychiatric Association. (1994). *Diagnostic and statistical manual of mental disorders* (4th ed.). Washington, DC: American Psychiatric Association.

Arnetz, J. E., & Arnetz, B. B. (2001). Violence towards health care staff and possible effects on the quality of patient care. *Social Science and Medicine*, *52*, 417–427.

Bin Abdullah, A. M., Khim, L. Y. L., Wah, L. C., Bee, O. G., & Pushpam, S. (2000). A study of violence towards nursing staff in the emergency department. *Singapore Nursing Journal*, *27*, 30–37.

Black, C. (2007, March 17). *Working for a healthier tomorrow*. London: TSO. Retrieved from http://www.dwp.gov.uk/docs/hwwb-working-for-a-healthier-tomorrow.pdf

Boorman, S. (2009). *NHS health and wellbeing: The Boorman review*. London: National Health Service.

Bride, B. E., Radey, M., & Figley, C .R. (2007). Measuring compassion fatigue. *Clinical Social Work Journal*, *35*, 155–163.

Caldwell, M. F. (1992). Incidence of PTSD among staff victims of patient violence. *Hospital and Community Psychiatry: A Journal of the American Psychiatric Association*, *43*(8), 838–839.

Chambers, N. (1998). "We have to put up with it – don't we?" The experience of being the registered nurse on duty, managing a violent incident involving an elderly patient: A phenomenological study. *Journal of Advanced Nursing*, *27*, 429–436.

Clothier Report. (1994). *Independent inquiry relating to the deaths and injuries on the children's ward at Grantham and Kesteven General Hospital*. London: HMSO.

Coffey, M., & Coleman, M. (2001). The relationship between support and stress in forensic community mental health nursing. *Journal of Advanced Nursing*, *34*(3), 397–407.

Fahy, A. (2007). The unbearable fatigue of compassion: Notes from a substance abuse counselor who dreams of working at Starbucks. *Clinical Social Work Journal*, *35*(3), 199–205.

Fernandes, C. M., Raboud, J. M., Christenson, J. M., Bouthillette, F., Bullock, L., Ouellet, L., & Moore, C. F. (2002). The effect of an education program on violence in the emergency department. *Annals of Emergency Medicine*, *39*, 47–55.

Figley, C. R. (1995). Compassion fatigue as secondary traumatic stress: An overview. In C. Figley (Ed.), *Compassion fatigue: Coping with secondary traumatic stress disorder* (pp. 1–20). New York: Brunner/Mazel.

Flanagan, N. A., & Flanagan, T. J. (2002). An analysis of the relationship between job satisfaction and job stress in correctional nurses. *Research in Nursing and Health*, *25*(4), 282–294.

Flannery, R., Hanson, M., & Penk, W. (1995). Patients' threats: Expanded definition of assault. *General Hospital Psychiatry*, *17*, 451–453.

Frank, D., & Adkinson, L. (2007). A developmental perspective on risk for compassion fatigue in middle-aged nurses caring for hurricane victims in Florida. *Holistic Nursing Practice*, *21*(2), 55–62.

Frank, D. I., & Karioth, S. P. (2006). Measuring compassion fatigue in public health nurses providing assistance to hurricane victims. *Southern Online Journal of Nursing Research*, *7*(4), 1–13.

Freudenberger, H. J., & Robbins, A. (1979). The hazards of being a psychoanalyst. *Psychoanalytic Review*, *66*, 275–296.

Fry, A. J., O'Riordan, D., Turner, M., & Mills, K. L. (2002). Survey of aggressive incidents experienced by community mental health staff. *International Journal of Mental Health Nursing*, *11*, 112–120.

Gates, D. M., Fitzwater, E., & Meyer, U. (1999). Violence against caregivers in nursing omes: expected, tolerated, and accepted. *Journal of Gerontological Nursing*, *4*, 12–22.

Greenwood, A. (2002). *Counselling for staff in health service settings: A guide for employers and managers*. London: Royal College of Nursing Press.

Greenwood, A., & Rooney, C. (2010, October 27). It happens anyway: A trauma response model for dealing with the consequences of violence and aggression in low and medium secure mental health hospitals. Paper presented at the 2nd International Conference on Violence in the Health Sector, Amsterdam.

Gulotta, K. C. (1987). Factors affecting nursing practice in a correctional health care setting. *Journal of Prison and Jail Health, 6*, 3–22.

Guy, K., & Guy, G. (2009, Spring). Psychological trauma: Keith Guy and Nicola Guy introduce the rewind trauma intervention model. *Counselling at Work: BACP Journal.* Retrieved from http://www.mentalhealthacademy.com.au/journal_archive/bacpwp098.pdf

Hauck, M. (1993). *Die Wut bleibt – Gewalt von Patienten gegenüber Pflegenden* [The anger remains – patient violence towards nurses]. Aarau: Kaderschule für die Krankenpflege.

Hislop, E., & Melby, V. (2003). The lived experience of violence in accident and emergency. *Accident and Emergency Nursing, 11*, 5–11.

Holmes, E. A., James, E. L., Blackwell, S. E., & Hales, S. (2011). They flash upon that inward eye. *The Psychologist*, 340–343.

Iliffe, G., & Steed, L. G. (2000). Exploring the counselor's experience of working with perpetrators and survivors of domestic violence. *Journal of Interpersonal Violence, 15*, 393–412.

Janik, J. (1995). Overwhelmed corrections workers can seek therapy. *Corrections Today, 57*(7), 162–163.

Jenkins, R., & Elliott, P. (2004). Stressors, burnout and social support: nurses in acute mental health settings. *Journal of Advanced Nursing, 48*, 622–631.

Johnson, C. L. (1995). Cultural diversity in the late-life family. In R. Blieszner & V. H. Bedford (Eds.), *Handbook of aging and the family* (pp. 307–331). Westport, CT: Greenwood.

Kennerley, H., & Kischka, U. (2011, February 26–27). Two-day workshop on brain, emotion and cognitive therapy, University of Hong Kong.

Killian, M., & Clark, N. (1996). The multidisciplinary team: The nurse. In C. Cordess & M. Cox (Eds.), *Forensic psychotherapy: Crime, psychodynamics and the offender patient.* London: Jessica Kingsley.

Kitchener, D., Davidson, C., & Burnard, P. (1996). Integrated care pathways: Effective tools for continuous evaluation of clinical practice. *Journal of Evaluation of Clinical Practice, 2*(1), 65–69.

Lanza, M. L. (1983). The reactions of nursing staff to physical assault by a patient. *Hospital Community Psychiatry, 34*, 44–47.

Lanza, M., Kayne, H., Hicks, C., & Milner, J. (1991). Nursing staff characteristics related to patient assault. *Issues in Mental Health Nursing, 12*, 253–265.

Larson, D. G., & Bush, N. J. (2006). Stress management for oncology nurses: Finding a healing balance. In R. M. Carroll-Johnson, L. M. Gorman, & N. J. Bush (Eds.), *Psychosocial nursing care along the cancer continuum* (2nd ed., pp. 587–601). Pittsburgh, PA: Oncology Nursing Society.

Levin, P. F., Hewitt, J. B., & Misner, S. T. (1998). Insights of nurses about assault in hospital-based emergency departments. *Image Journal of Nursing Scholarship, 30*, 249–254.

Mason, T. (2002). Forensic psychiatric nursing: A literature review and thematic analysis of role tensions. *Journal of Psychiatric and Mental Health Nursing, 9*, 511–520.

May, D. D., & Grubbs, L. M. (2002). The extent, nature, and precipitating factors of nurse assault among three groups of registered nurses in a regional medical center. *Journal of Emergency Nursing, 28*, 11–17.

Maytum, J. C., Heiman, M. B., & Garwick, A. W. (2004). Compassion fatigue and burnout in nurses who work with children with chronic conditions and their families. *Journal of Pediatric Health Care, 18*, 171–179.

McCann, I. L., & Pearlman, L. A. (1990). Vicarious traumatization: A framework for understanding the psychological effects of working with victims. *Journal of Traumatic Stress, 3*(1), 131–149.

Melchior, M. E. W., Bours, C. J. J. W., Schmitz, P., & Wittich, Y. (1997). Burnout in psychiatric nursing: A meta-analysis of related variables. *Journal of Psychiatric and Mental Health Nursing, 4*, 193–201.

Merchant, J. A., & Lundell, J. A. (2001). Workplace violence intervention research workshop, April 5–7, 2000, Washington, DC: Background, rationale, and summary. *American Journal of Preventive Medicine, 20*, 135–140.

Mitchell, J. T. (1990). Law enforcement applications of critical incident stress teams. In J. T. Reese, J. M. Horn, & C. Dunning (Eds.), *Critical incidents in policing* (pp. 201–212). Washington DC: Government Printing Office.

Mitchell, J. T., & Everly, G. S. (1993). *Critical incident stress debriefing: An operations manual for the prevention of traumatic stress among emergency services and disaster workers.* Ellicott City, MD: Chevron.

Musa, S. A., & Hamid, A. A. R. M. (2008). Psychological problems among aid workers operating in Darfur. *Social Behavior and Personality, 36*(3), 407–416.

Najjar, N., Davis, L. W., Beck-Coon, K., & Doebbeling, C. C. (2009). Compassion fatigue: A review of the research to date and relevance to cancer-care providers. *Journal of Health Psychology, 14*, 267–277.

National Institute for Clinical Excellence (NICE). (2006). *Post-traumatic stress disorder: The management of PTSD in adults and children in primary and secondary care.* Clinical Guidelines 26. London: NHS.

O'Connell, B., Young, J., Brooks, J., Hutchings, J., & Lofthouse, J. (2000). Nurses' perceptions of the nature and frequency of aggression in general ward settings and high dependency areas. *Journal of Clinical Nursing, 9*, 602–610.

Peek-Asa, C., Runyan, C. W., & Zwerling, C. (2001). The role of surveillance and evaluation research in the reduction of violence against workers. *American Journal of Preventive Medicine, 20*, 141–148.

Regel, S., & Joseph, S. (2010). *Post traumatic stress.* Oxford: Oxford University Press.

Rick, J., O'Regan, S., & Kinder, A. (2006). *Early intervention following trauma: A controlled longitudinal study at the Royal Mail.* London: British Occupational Health Research Foundation.

Ryan, J. A., & Poster, E. C. (1989). The assaulted nurse: Short-term and long-term responses. *Archives of Psychiatric Nursing, 3*, 323–331.

Saakvitne, K. W., Tennen, H., & Affleck, G. (1998). Exploring thriving in the context of clinical trauma theory: Constructivist self-development theory. *Journal of Social Issues, 54*, 279–299.

Smith, A., Brice, C., Collins, A., Mathews, V., & McNamara, R. (2000). *The scale of occupational stress: A further analysis of the impact of demographic factors and type of job.* HSE contract research report no: 311/2000. Sudbury, UK: HSE Books.

Sprang, G., Clark, J. J., & Whitt-Woosley, A. (2007). Compassion fatigue, compassion satisfaction, and burnout: Factors impacting a professional's quality of life. *Journal of Loss and Trauma, 12*, 259–280.

Suserud, B. O., Blomquist, M., & Johansson, I. (2002). Experiences of threats and violence in the Swedish ambulance service. *Accident and Emergency Nursing, 10*(3), 127–135.

Tehrani, N. (2004). *Workplace trauma: Concepts, assessment, and Intervention.* Hove, UK: Brunner-Routledge.

United Kingdom Central Council (UKCC) (1999). *Nursing in secure environments: A scoping study conducted by the Faculty of Health, University of Central Lancashire.* London: Author.

Valent, P. (2002). Diagnosis and treatment of helper stresses, traumas, and illness. In C. R. Figley (Ed.), *Treating compassion fatigue* (pp. 17–38). New York: Brunner-Routledge.
Wall, T. D., Bolden, R. I., & Borrill, C. S. (1997). Minor psychiatric disorder in NHS trust staff: Occupational and gender differences. *British Journal of Psychiatry*, *171*, 519–523.
Whittington, R., & Wykes, T. (1992). Staff strain and social support in a psychiatric hospital following assault by a patient. *Journal of Advanced Nursing*, *17*, 480–486.

7

SAV-T First: Managing Workplace Violence

Kate Calnan, E. K. Kelloway, and Kathryne E. Dupré

Research on workplace violence has proliferated in recent years (Kelloway, Barling, & Hurrell, 2006), and sufficient data have now accumulated to dispel many of the myths surrounding this issue (Barling, Dupré, & Kelloway, 2009). Enduring among these myths, however, is the suggestion that workplace violence is so rare as to not warrant serious concern.

Defined as the physical or psychological behaviors intended to cause employees *physical* harm or threat (LeBlanc & Kelloway, 2002; Schat & Kelloway, 2005), workplace violence is costly to a wide array of stakeholders and a continuous threat to both employees and organizations. Although less predominant that nonphysical acts of aggression (Baron & Neuman, 1998), as many as 1 in 20 employees have reported being physically assaulted while at work (National Center on Addiction and Substance Abuse, 2000). In their analysis of a national probability sample of employed Americans, Schat, Frone, & Kelloway (2005) reported that 6% of the workforce had experienced an act of workplace violence in the preceding 12 months. The aim of this chapter is to explore the current understanding of workplace violence and examine the intervention strategies available to employers for reducing the occurrence and impact of work related violence.

Defining Workplace Aggression and Violence

The terms *workplace aggression* and *workplace violence* are often used interchangeably. They are, however, two empirically distinct constructs (LeBlanc & Kelloway, 2002). Unlike workplace aggression, which encompasses a broader array of harmful behavior, definitions of workplace violence tend to focus more on the physical aspects of harm (LeBlanc & Kelloway, 2002). That is, workplace aggression is physical or verbal in nature and involves intended *psychological* or *physical* harm or threat of harm toward others in a work-related context (Barling *et al.*, 2009). Workplace violence, on the other hand, is a subset of aggression and is typically defined as physical or psychological behaviors intended to cause or threaten *physical* harm (LeBlanc & Kelloway, 2002; Schat & Kelloway, 2005). Additionally, a variety of labels such as

International Handbook of Workplace Trauma Support, First Edition. Edited by Rick Hughes, Andrew Kinder, and Cary L. Cooper.

bullying (Rayner & Cooper, 2006), *workplace harassment* (Bowling & Beehr, 2006; Rospenda, 2002), and *dysfunctional and deviant workplace behavior* (Martinko, Gundlach, & Douglas, 2002) have been used to describe relatively similar and overlapping behaviors, some of which can be classified as workplace aggression or violence (Barling *et al.*, 2009).

Operationalizations of workplace violence differ in the severity of the behaviors captured. That is, workplace violence can include a range of physical threats and acts (Barling, 1996). For example, pushing and shoving are among less severe behaviors that constitute workplace violence, whereas assault and murder would be considered the most extreme (Barling, 1996). The latter are infrequent (Schat *et al.*, 2005) and occur in highly unique situations (Barling *et al.*, 2009).

The Prevalence of Workplace Violence

In terms of prevalence, workplace violence again differs from nonviolent acts of workplace aggression (Baron & Neuman, 1996, 1998; Baron, Neuman, & Geddes, 1999). This is in line with societal and organizational tolerance norms (Aquino & Thau, 2009). Organizations and individuals alike typically do not tolerate physical aggression, while, on the other hand, there is a higher degree of tolerance, perhaps because of the seemingly less adverse consequences, for nonphysical aggression. Research suggests, however, that the potential psychological outcomes associated with being the target of a less extreme versus a more extreme act of aggression are not necessarily dissimilar (see, e.g., O'Leary & Jouriles, 1994). Nonetheless, given tolerance levels, it would posit that less physically violent behaviors would take place more frequently. Data support this proposition: Compared to workplace violence, acts of aggression are much more prevalent with up to 70% of employees reporting having experienced nonphysical acts of workplace aggression (Einarsen & Raknes, 1997). Similarly, while they obtained a 6% prevalence rate for workplace violence, Schat *et al.* (2005) reported that 41% of their national probability sample had experienced acts of workplace aggression.

Moreover, it is widely understood that it is possible for workplace aggression to exist without the inclusion of workplace violence, but not the other way around (Baron & Neuman, 1996). That is, it appears that workplace aggression is often a precursor to workplace violence (see Barling, 1996; Dupré & Barling, 2006; Glomb, 2002), thus making the incidence of nonviolent aggressive acts (e.g., verbal abuse and psychological aggression in the form of yelling, cursing, and name calling; Barling, 1996) much more prevalent in the workforce (Einarsen & Raknes, 1997).

Sources of Workplace Violence

Before examining various prevention strategies, it is important to first examine the different types of workplace violence. It is common to distinguish among four types (i.e., sources) of workplace violence, differentiated by the source or perpetrator of the violence (California Occupational Safety and Health Administration [Cal/OSHA], 1995). Type

I workplace violence encompasses criminal acts committed by an individual(s) with no prior relationship to the organization. In this context, the particular organization is the target of a crime such as a robbery (Braverman, 1999). Type II workplace violence involves a physical and/or psychological attack from a client or customer (Braverman, 1999). In this context, the perpetrator has a relationship with the organization. Together Type I and Type II (i.e., violence committed by organizational outsiders) are the most common types of workplace violence (Barling *et al.*, 2009; LeBlanc & Kelloway, 2002). Specifically, employees are four times more likely to experience violence from an outsider (i.e., member of the public; Schat *et al.*, 2005) than an organizational insider (e.g., supervisor or coworker).

The third classification of workplace violence involves former or current employees (Braverman, 1999) committing violent acts toward supervisors, coworkers, subordinates, and other staff. These violent acts are typically reactions to some type of trigger within the organization such as perceived injustice, abusive supervision, or role stressors (Barling, 1996; Dupré & Barling, 2006; Greenberg & Barling, 1999). Prevention techniques for this type of violence are usually targeted at the reduction or removal of such triggers. Although Type III violence attracts the most attention (Kelloway *et al.*, 2006), it is infrequent in comparison to the other types. Indeed, it is not uncommon for researchers to report no experience with Type III violence within a sample (e.g., LeBlanc & Kelloway, 2002).

Finally, Type IV workplace violence is referred to as intimate partner (or family and friend) violence. Although this type of violence originates from the home, there can be substantial spillover into the workplace (Barling *et al.*, 2009). Specifically, this fourth type of violence is thought to account for 1–3% of workplace violence incidents (Durhart, 2001). For the majority of employees their place of work functions as a social address, and therefore, partners from the past or present, friends, and family have a way of finding the individual on any given day (Barling *et al.*, 2009). Intimate partner violence typically involves work disruption (Pollack, Austin, & Grisso, 2010), stalking, and on-the-job harassment (Swanberg, Logan, & Marke, 2006), and can have several damaging effects for organizations such as tardiness and absenteeism, property damage, and production setbacks (Barling *et al.*, 2009).

Intervention Strategies

The occupational health literature identifies three types of workplace interventions: primary, secondary, and tertiary (Schat & Kelloway, 2006). Primary interventions are concerned with the prevention of workplace violence by eliminating risk factors in the workplace. Secondary interventions assume workplace violence has occurred or is imminent and aims to reduce its negative outcomes (Schat & Kelloway, 2006). Lastly, tertiary interventions assume workplace violence has occurred and that physical or psychological harm has been done (Schat & Kelloway, 2006). The focus, therefore, is to reduce the negative impact through counseling or therapeutic techniques (Cooper & Cartwright, 1997). In the following sections we will examine the role of each type of intervention and how it is best applied to the prevention, reduction, and treatment of workplace violence.

Preventing workplace violence

The most commonly recommended intervention to prevent workplace violence is for organizations to develop policies and practices geared toward preventing workplace violence. Typically such policies would articulate a no-tolerance policy and define what is unacceptable (Wassell, 2009). Written policies should include what constitutes acceptable and unacceptable behaviors in the workplace (Smith, 2000). Policies should also specify a requirement to treat coworkers and customers with respect. Encouraging positive behaviors, as opposed to simply prohibiting undesirable behaviors, will impact cultural norms as well as help to set tolerance levels and expectations. It is typically recommended that policies must also offer procedural guidance for reporting and dealing with acts of workplace violence (Smith, 2000). These procedures should be effectively communicated to employees, and reporting should be encouraged. Employees, in no way, should be made to feel that they would be blamed for doing something to instigate any form of aggressive behavior in others, even that stemming from customer dissatisfaction. Moreover, it is important to make sure resources are in place to help employees when violence is reported.

Although we see no harm in developing and implementing policies designed to prevent workplace violence, we note that such policies are still comparatively rare with only 30% of US organizations having a formal policy pertaining to workplace violence (US Department of Labor, 2006). Perhaps more importantly, workplace violence policies are aimed primarily at employees of a given organization. This is significant given that, as previously stated, organizational outsiders perpetrate the vast majority of acts of workplace violence. Individuals committing a criminal act (e.g., robbery) are unlikely to be dissuaded by the existence of an organizational policy. Although organizational clients may be more attentive to policies, it is easy to imagine individuals under care (e.g., individuals with dementia, or prisoners in custody) who may not be aware of, or respectful of, organizational policies. Considering the data on the frequency and sources of workplace violence, we conclude that policies alone are unlikely to be an effective deterrent of workplace violence.

Other organizational policies and practices designed to prevent workplace violence are neither grounded in experience nor supported by research. Thus, organizations may be advised to implement screening or selection policies designed to "weed out" employees with violent tendencies. Unfortunately, the screening criteria are often so broad as to result in numerous false positives and in the exclusion of many nonviolent individuals. Moreover organizations implementing such strategies run a real risk of running afoul of human rights or fair employment standards (see, e.g., Day & Catano, 2006). Finally, we reiterate: employees do not perpetrate most acts of workplace violence. Relying on strategies that control the risk of violence from employees does not address the actual nature of the workplace violence problem.

We suggest that a more effective means of preventing workplace violence is to understand and manage the risk factors associated with the perpetration of violent acts. In particular, we propose a model of risk assessment (the SAV-T model) that focuses on the identification and management of both occupational/task and imminent risk factors. We suggest that the conduct of such risk assessments is the most effective way to prevent violence in the workplace.

To date, research has established key contributors to workplace violence. Dealing with the public, for example, puts employees at the greatest risk for experiencing workplace violence as either criminal targets (as in the first type of workplace violence) or the focus of dissatisfied clients (as in the second type of workplace violence; Braverman, 1999; Kelloway, Catano, & Day, 2011). At-risk industries, therefore, are those that offer some type of public service. Examples include healthcare, transportation, education, corrections, retail services, and hospitality (Kelloway *et al.*, 2011; LeBlanc & Kelloway, 2002).

We summarize the literature on occupational risk factors with the acronym SAV-T:

1. *Scheduling.* Working alone, at night, and on weekends has been shown to increase the risk of workplace violence (LeBlanc & Kelloway, 2002). Unsafe scheduling is most commonly linked to criminal acts (i.e., Type I violence).
2. *Authority.* Having decision-making power or authority over others, supervising others, or being in a position to deny services can increase the likelihood of workplace violence (e.g., Hearnden, 1988). This risk factor typically leads to client- or customer-based (Type II) violence but may also result in aspects of Type III violence (i.e., subordinate to supervisor).
3. *Valuables.* Working with or around valuables such as money, alcohol, prescription medication, and jewelry can put employees at an increased risk of Type I violence (LeBlanc & Kelloway, 2002; Schat & Kelloway, 2006). That is, individuals who work with valuables are more likely to experience criminal acts of violence such as robbery.
4. *Taking care of others.* Providing physical and emotional care also increases employees' risk of experiencing violence. This job characteristic is particularly associated with experiences of Type II (i.e., patient- and client-related) violence. Additionally, taking care of others may require travel to client homes. In these situations the workplace is no longer a stable location making the risk for violence greater still (see for example, Barling, Rogers & Kelloway, 2001).

The occupational risks identified above can be addressed by altering aspects of the work environment (Schat & Kelloway, 2006) to provide enhanced protection for employees and to deter potential acts of violence. Target hardening is one approach to be considered. It involves altering the environment in a way that makes it more difficult for employees to be harmed. Installing security devices such as surveillance cameras, protective glass dividers, and curved mirrors and employing security personnel can make acts of criminal violence increasingly difficult and less enticing (Schat & Kelloway, 2006). Additionally, increasing visibility further helps to deter violent acts (Wassell, 2009). Removing sight barriers, situating cash registers in front of windows, and having effective lighting can protect against robbery or other acts of violence (e.g., LeBlanc, Dupré, & Barling, 2006). Such interventions are essential for occupations that meet increased risks of public and criminal violence and are considered to be the best intervention technique for the latter (Wassell, 2009). Overall, interventions that alter aspects of the work environment are an effective approach to workplace violence prevention and have demonstrated reductions in criminal acts of workplace violence by 30–84% (Wassell, 2009). Further, research indicates that the more protective measures put in place, the greater the protection for employees (Wassell, 2009).

Imminent risk

Although striving to prevent workplace violence by managing and eliminating occupational risk factors is important, it is difficult to address some of the factors identified as salient predictors of workplace violence. For example, one risk factor is taking care of others – but this risk is also the definition of health care. Moreover, "target-hardening" strategies (e.g., the provision of barriers or body armor) would be unpalatable in a healthcare or social service environment. Therefore, the assessment of risk must move beyond the task level to identify and mitigate the risk of violence in particular situations.

Although we believe that organizations need a primary function on preventing workplace violence, we also recognize that employees will be exposed to risks and that it may be impossible to achieve "zero" violence in some occupations. In such occupations, the focus shifts from trying to eliminate workplace violence toward thinking about how better to equip employees to handle violence when it occurs. That is, we must focus on the recognition and assessment of imminent risk.

Imminent risk is the short-term risk of violence occurring in any given situation. This type of risk is based on the notion of the assault cycle. The assault cycle is a model identifying the way in which a situation can escalate from aggressive behavior to an act of violence. Specifically, the assault cycle is composed of five phases. The first phase is a baseline phase or standard behavior. The second phase refers to an escalation period whereby an individual gets increasingly agitated by a triggering event(s). The third phase is the act of violence, while the fourth and fifth phases represent de-escalation and a return to the baseline behavior. This progression illustrates two main points: specifically, that aggressive behaviors can escalate into physical violence, and that this progression can be seen and anticipated. When the risk of violence is imminent for any given occupation, it is crucial for employees to be able to recognize the warning signs of escalation.

All too often, individuals misinterpret or ignore these warning signs and end up being the victim of an attack. When warning signs go unnoticed or are misinterpreted, it is often the result of a lack of training on risk assessment and awareness (Lieber, 2007). In order to reduce incidents of violence by preventing escalation, individual behavior must be modified so that recognition cues are easily identifiable and appropriate responses are implemented.

Behavioral interventions are designed to prevent violence by altering individual behavior (Schat & Kelloway, 2006). Training initiatives are an example of a behavioral intervention. The majority of training initiatives can be classified across two dimensions: prevention focused (i.e., primary intervention) versus consequence focused (i.e., secondary intervention) and target directed versus assailant directed (Schat & Kelloway, 2006). *Prevention-focused training* is a primary intervention technique aimed at teaching employees how they can avoid acts of workplace aggression. *Consequence-focused training*, on the other hand, is a secondary intervention technique. The focus of this approach is to teach employees how to respond if they have experienced workplace aggression. The majority of workplace violence training is done from a preventative standpoint (Schat & Kelloway, 2006); however, in several occupations (e.g., healthcare) secondary interventions are also employed and warranted (Schat & Kelloway, 2003). Determining which type of training is best suited to a particular organization requires a needs analysis (Arthur *et al.*, 2003). That is, organizational and personal factors should be examined to determine who should be trained and what type of training (i.e., prevention versus consequence focused) would be most relevant and of benefit to the organization and its employees.

Behavioral intervention training can include a wide array of techniques and approaches, all geared toward the reduction of violence. In the retail industry, for example, it is recommended that staff be trained in conflict resolution tactics and customer service. Training interventions specific to risk identification and assessment, however, are less prevalent and in need of continued development.

To date, behavioral training interventions have received a great deal of empirical support. Knowledge and awareness training, in particular, can help employees recognize imminent risk and respond appropriately. Violence risk training has also demonstrated improvements in employee confidence and perceptions of preparedness for future violent encounters (Beech & Leather, 2006; Ishak & Christensen, 2002; Schat & Kelloway, 2003). Specifically, a study examining training effectiveness in hospital staff indicated those who had received training experienced higher levels of perceived control. This, in turn, translated to an increase in well-being and a reduction in fear of future violence (Schat & Kelloway, 2003). When examining occurrence rates pre- and post-training, however, results are mixed. Frequently, few if any reductions are noted (Wassell, 2009). This suggests that although training can have a positive effect on individual perceptions of violence it does little to reduce the occurrence of violence, yet training initiatives still strive to achieve such outcomes (Wassell, 2009).

Considering the benefits (e.g., increased sense of control) of employee-focused training interventions, as well as the need for risk awareness and assessment training, we extend the SAV-T model to identify the signs of escalation and to propose strategies for intervention in the situation. In this context, the SAV-T acronym represents escalation cues, that is, *swearing*, *agitation*, *volume*, and *threat*. By itself swearing or agitation may not indicate impending violence; however, when two or more criteria are coupled together it becomes indicative of escalation and may lead to violence. As part of this component of the SAV-T intervention, employees are taught to set boundaries early on. That is, at the first sign of agitation or swearing, employees should make it clear that aggressive behavior will not be tolerated. Consequences for unacceptable behavior should be communicated. Overall, boundaries should be simple, reasonable, and clear, and individuals must be willing to enforce them if the need should arise (Caraulia & Steiger, 1997). Additionally, as escalation (e.g., swearing, agitation, and volume) develops employees can engage in several other de-escalation techniques (Caraulia & Steiger, 1997). Empathic listening can be a useful approach to de-escalating a risky situation. Empathic listening involves acknowledging what the aggressor is saying by restating their arguments, refraining from making judgments, providing undivided attention, and focusing on feelings (Caraulia & Steiger, 1997). Moreover, defusing the situation is another alternative. Diffusing involves using humor (when appropriate) and distractions, making requests, and providing suggestions for next steps and remedial action.

Continuous monitoring of the situation is crucial. If escalation persists, employees should call for security or assistance. Lastly, we identify threat as a particular risk factor. Although not all threats lead to actual violence, it is also clear that actual incidents of violence are typically preceded by threat (Braverman, 1999). Therefore, we highlight the role of threats suggesting that if a threat is made employees should focus on escaping the situation. Specifically, they should cease any attempts at de-escalation and remove themselves from the situation entirely.

When faced with a potentially violent situation, it is important for risk assessments to be immediate and simple (Caraulia & Steiger, 1997; Schat & Kelloway, 2005). Complex assessments or judgments based on past knowledge of the aggressor are time dependent and ultimately ineffective. Moreover, to increase the likelihood of transfer to a high stress situation, that is, the probability that an individual will remember and use the SAV-T model, simplicity is a must (Schat & Kelloway, 2005).

It is important to emphasize that a thorough assessment of the interventions aimed at managing and minimizing workplace violence is lacking (see also Runyan, Zakocs, & Zwerling, 2000). In general, the evaluation of organizational interventions aimed at managing employee stress, health, and well-being has proven challenging for both researchers and practitioners (see, for example, Cox *et al.*, 2007; Randall, Griffiths, & Cox, 2005). As discussed by Cox *et al.* (2007), an approach to the evaluation of workplace violence interventions could be to focus on traditional research methods. However, given the complexities associated with organizations, as well as the limitations associated with traditional research methods in the real world, such an approach is challenging (see Cox *et al.*, 2007). As such, Cox *et al.* (2007) offer a broad and innovative framework for the evaluation of organizational interventions for work stress. Likewise, in order to advance workplace violence knowledge, researchers and practitioners must ensure that they continue to develop and refine their approaches to the evaluation of workplace violence interventions, and this is equally true of the SAV-T model.

Post-event interventions

Despite the many preventative measures, workplace violence may still occur and it is necessary for organizations to put in place strategies that help workers recover after experiencing negative health or behavioral effects of workplace violence. Tertiary or post-event interventions are designed to do just that. Specifically, the purpose of taking such an approach is to help the worker cope, recover, and eventually return to work. Post-event intervention techniques are particularly relevant for healthcare settings and other high-risk occupations where workplace violence is imminent. Such interventions come in many forms. Most commonly, employee assistance programs (EAPs), critical incident debriefing, and psychological first aid strategies are employed by organizations.

Employee Assistance Programs

Employee assistance programs were initially developed in the 1970s as a means of dealing with alcoholism (Pollack *et al.*, 2010). At that time, EAPs allowed organizations to support their employees, while maintaining a workforce and reducing turnover. Since their initial inception, EAPs have grown substantially in the support services they offer. EAPs are now designed to help employees with a wide breadth of work-related and non-work-related problems (Barrett, 2011) that, if left untreated, may have detrimental effects for job performance and well-being (Pollack *et al.*, 2010). Additionally, support services offered by EAPs are confidential, providing workers with secure opportunities to seek help without fear of retaliation or discrimination (Barrett, 2011; Pollack *et al.*, 2010). Moreover, many benefits and services offered by EAPs are extended to employees' families. This can help reduce financial burdens and facilitate work–life demands.

Most commonly EAPs are independent services that are contracted for hire. Contracted EAPs typically include 24-hour help lines, counseling services, and referrals for more specified services (Barrett, 2011; Pollack *et al.*, 2010). They may also include training, case monitoring, follow-up services, wellness programs, and HR-related consultation (Pollack *et al.*, 2010; Smith, 2000). Although comparatively rare, some organizations offer internal EAP services. Internal EAPs primarily consist of the same services as contracted EAPs with the exception of phone-based helplines.

EAPs have grown to be a widely implemented employee resource tool. As of 2008, 58% of all American employers were providing their employees with EAP services (Pollack *et al.*, 2010). From an individual perspective, EAPs provide an excellent resource for helping workers cope with experiences of workplace violence. Many organizations report using EAPs as their main source of support for intimate partner violence (i.e., Type IV) (Pollack *et al.*, 2010).

From an organizational perspective, when workplace violence occurs, managers similarly turn to EAPs for assistance. Under these circumstances, however, many EAPs tend to be ill equipped to handle such issues (Braverman, 1999). That is, many EAPs are unaware how to assess risks for workplace violence and cope with threats (Braverman, 1999). Instead, it is suggested that EAPs should play an important role in violence prevention as opposed to treatment. Specifically, EAPs would be better suited to gather information, recommend specialists, aid in preliminary decision making regarding specific incidents, and lend support for general program planning (Braverman, 1999).

Although their main objective is to assist employees with personal and/or work-related issues, EAPs also have utility for management and human resources departments (Smith, 2000). Specifically, EAPs may be a helpful resource for identifying organizational issues (Barrett, 2011) as well as developing company-wide interventions (Nobrega *et al.*, 2010). One study in particular assessed EAP personnel on the prevention and management of job stress at an *organizational* level. Findings demonstrated high degrees of knowledge and awareness pertaining to health outcomes resulting from stress, as well as a variety of techniques for dealing with job stress in the workplace (Azaroff *et al.*, 2010). Thus, it seems EAPs have the ability to contribute to individual as well as organizational health and well-being. EAPs, however, are rarely utilized to their full potential. When interviewed, contracted and internal EAP personnel identified a lack of involvement in organizational initiatives (Nobrega *et al.*, 2010). Several barriers were said to directly lead to their exclusion. Namely, an inability to access company management and (for contracted EAPs) the sense of contract vulnerability were among the most notable reasons for this lack of involvement (Nobrega *et al.*, 2010). Internal EAPs may have an advantage in overcoming these particular barriers in that they are exempt from funding and contract threats, and are likely to be more ingrained in the organization. In this regard, internal EAPs should be able to form the required organizational alliances needed to effect workplace interventions (e.g., primary interventions; Nobrega *et al.*, 2010). This ability is strengthened when organizations have an existing agenda to improve workplace wellness (Nobrega *et al.*, 2010).

A confidentiality breach is of additional concern when considering the role of EAPs in the development of organizational interventions. That is, because it is crucial for EAPs to maintain employee confidentiality, it may be difficult to determine the appropriate content that EAPs need to share with management in order to aid organizational initiatives (Braverman, 1999). This, along with potential adverse employee reactions, may partially explain limited organizational usage of EAPs for company-wide interventions.

When considering the potential effectiveness of an EAP, it is important to be cognizant of the differences among EAPs in terms of their relationship with the organization (i.e., internal versus contracted services), the comprehensiveness of the program, and the delivery approaches offered (Barrett, 2011). Although similar in the services they offer, internal and contracted EAPs differ considerably in terms of their capacity to effect workplace interventions and the extent to which employees favor them. As briefly discussed, internal EAPs may have a slight advantage in effecting organizational, as opposed to individual, interventions. Unlike contracted EAP personnel, internal EAP personnel are engrained in the organization and likely have more access to resources enabling them to secure the positive relationships and support needed to facilitate such interventions. Internal EAPs, however, tend to be less favored by employees due to concerns over confidentiality, retaliation, and perceived support (Harlow, 1998). That is, employees are much less likely to use internal EAP services for fear that their concerns will not be kept confidential or be used against them in a way that may threaten their job (Harlow, 1998). Additionally, employees are less likely to use internal EAPs if they perceive little organizational or peer support for the program (Harlow, 1998).

Furthermore, EAPs can vary extensively in terms of the services they offer and to whom within the organization they are offered to (Barrett, 2011). The comprehensiveness of a program, as well as the number of people who have access to it, will directly affect its cost (Barrett, 2011). Generally, the more comprehensive the EAP is, and the greater the access to the EAP, the higher the cost. Moreover, EAPs vary in terms of delivery. That is, specific elements of the program may differ in terms of what approaches are made available to employees. For example, counseling services may be offered over the phone or face-to-face. Recently, web-based services have increased in popularity (Barrett, 2011). The Internet has provided a unique opportunity for counseling via email, chat services, and discussion forums. This trend can only be expected to continue with EAP-related application software likely to appear shortly (Barrett, 2011).

Psychological Debriefing

Commonly employed as a group-level intervention, debriefing was initially used on an individual basis and formally referred to as *crisis intervention approaches* (Everly, Flannery, & Mitchell, 2000). Psychological debriefing typically takes place within a short time frame following an event and is frequently carried out in a single session (Emmerik *et al.*, 2002). The primary goals of psychological debriefing are to reduce immediate distress and prevent further psychological trauma (Caraulia & Steiger, 1997; Emmerik *et al.*, 2002). Psychological debriefing is seen as a favorable approach because it provides an opportunity for employees to discuss their reactions to a violent event (Adler *et al.*, 2008; Caraulia & Steiger, 1997). Three types of psychological debriefing are identified in the literature: *critical incident stress debriefing* (CISD), *process debriefing*, and the *Raphael model*. The latter two are variations of the first and differ only slightly in structure and content; therefore, most of the following discussion pertains to CISD in particular.

The effectiveness of CISD is debated; however, the majority of research suggests it has little to no effect as a post-event intervention (Ruzek *et al.*, 2007). For example, a meta-analytic review comparing natural recovery processes to CISD found that individuals who participated in CISD did not show any improvements beyond those who did not participate (Emmerik *et al.*, 2002). Interestingly, employees who have experienced it (Emmerik *et al.*, 2002; Litz *et al.*, 2002) favor CISD. Despite its favorability across samples, CISD is consistently ineffective in reducing negative post-event outcomes (e.g., post-traumatic stress; Adler *et al.*, 2008; Emmerik *et al.*, 2002).

Furthermore, results of several studies indicate that CISD may actually have a negative effect (Adler *et al.*, 2008; Bisson, 2003; Mayou, Ehlers, & Hobbs, 2000). One study in particular found that when CISD was compared to two other early psychological interventions, certain negative outcomes (e.g., alcohol abuse) actually increased as a result (Adler *et al.*, 2008). For the most part CISD is not seen as harmful, but overall research supports the notion that CISD is not an effective intervention approach.

Several suggestions as to why CISD has limited effects have been proposed. Specifically, the length of treatment may be too brief for positive effects to develop (Emmerik *et al.*, 2002; Ruzek *et al.*, 2007). Moreover, CISD may unintentionally deter employees from exploring additional treatment options (Ruzek *et al.*, 2007). From a design perspective, many studies are flawed with methodological issues calling into question the validity of their results (Ruzek *et al.*, 2007). Lastly, early psychological interventions of a different nature may be incorrectly labeled a psychological debriefing in the literature potentially making overall conclusions regarding this technique less valid (Emmerik *et al.*, 2002).

Psychological First Aid

Psychological first aid (PFA) is designed to "provide individuals with skills they can use in responding to psychological consequences of disasters [i.e. workplace violence] in their own lives, as well as in the lives of their family, friends, and neighbours" (Everly *et al.*, 2006, p. 131). PFA encompasses eight "core actions" that incorporate empirically supported intervention practices for dealing with violence as well as other physical and psychological traumas. The goals of PFA are to improve perceived safety, offer support and emotional assistance, disseminate information, and provide access to additional services that may be required (see Ruzek *et al.*, 2007).

PFA is gaining popularity as a good choice for post-crisis intervention (Everly *et al.*, 2006). When severe incidents such as violence occur in the workplace, individuals typically experience some type of stress reaction and often desire some form of psychological support (Everly *et al.*, 2006). PFA is a useful tool for such situations as it encompasses a variety of psychological and resource supports. Similarly to EAPs, PFA interventions are designed to help individuals establish contact with a variety of support systems to ensure their well-being and continued care. Linkages with informal support systems (e.g., family, friends, and coworkers) as well as formal support systems (e.g., community programs, hospitals, and faith-related resources) have been noted (Everly *et al.*, 2006).

Unlike interventions that are individually targeted, PFA can be administered in a group setting. This has the potential to offer many unique benefits. Specifically, group-level interventions can help build structure, reestablish trust, decrease perceptions of isolation, and most notably validate individual experiences (Ulman, 2004).

Despite its utility, Ruzek *et al.* (2007) caution that "PFA is designed to fit into a large, more comprehensive disaster/trauma response intervention whose components collectively address a broad spectrum of disaster-related problems and associated needs" (p. 30). Thus, PFA may not be appropriate or sufficient if used on its own.

When it comes to the delivery of PFA, it is important to have trained personnel guide the intervention (Ruzek *et al.*, 2007). Currently, attempts to increase PFAs availability are being made. Field manuals are being developed and used to train individuals on how to appropriately administer PFA (Everly *et al.*, 2006). School personnel, medical professionals, and clergy are among those currently being trained (Ruzek *et al.*, 2007).

Cognitive-Behavioral Therapy

Cognitive-behavioral therapy (CBT) can take different approaches but intends to reduce negative outcomes associated with experiencing a trauma. CBT may be more or less cognitively or behaviorally oriented. More specifically, CBT includes several cognitively oriented approaches such as education, anxiety management training, and cognitive restructuring (i.e., a technique used to help individuals change the way they think about and perceive the negative event; Ruzek *et al.*, 2007). In vivo exposure is a behaviorally oriented CBT approach that helps individuals learn to cope with the traumatic event in the context in which it initially occurred. Lastly, imaginable exposure therapy is both cognitively and behaviorally oriented, and focuses on helping individuals cope by having them imagine the event in a safe and comfortable environment (Joseph & Gray, 2008).

There is a great deal of empirical support for CBT's effects in preventing negative long-term outcomes (Ruzek *et al.*, 2007). Studies have shown that CBT successfully prevents PTSD and depression more than education and support on their own

(Ruzek *et al.*, 2007). Moreover, longitudinal data suggest that CBT's positive effects can last up to four years after the initial exposure (Bryant, Moulds, & Nixon, 2003). More research is needed, however, to establish CBT's utility for workplace violence trauma. Typically, CBT is most effective when there is some sort of physical or environmental destruction. Workplace violence is slightly different in this regard as it is often individually targeted and rarely results in external or property damage (Ruzek *et al.*, 2007). Moreover, workplace violence is more directly personal than an environmental event, making it potentially more challenging to overcome.

Conclusion

Although less prevalent that nonphysical acts of aggression, workplace violence remains a serious concern, especially for those employed in the public service sector where the risk of violence is the greatest. Environmental and organizational interventions can directly reduce the occurrence of violence and help prevent its future occurrence. Behavioral interventions offer useful approaches to employee-focused training. As well, behavioral interventions help make the shift from focusing on elimination toward a focus on better equipping employees to deal with an imminent risk of workplace violence. Lastly, several post-event interventions such as EAPs and PFA provide beneficial treatment options for employees physically and psychologically harmed by workplace violence, while others, namely CISD, should be implemented with caution.

References

Adler, A. B., Litz, B. T., Castro, C. A., Suvak, M., Thomas, J. L., Burrell, L., *et al.* (2008). A group randomized trial of critical incident stress debriefing provided to U. S. peacekeepers. *Journal of Traumatic Stress, 21*, 253–263.

Aquino, K., & Thau, S. (2009). Workplace victimization: Aggression from the target's perspective. *Annual Review Psychology, 60*, 717–741.

Arthur, W., Jr., Bennet, W., Eden, P. S., & Bell, S. T. (2003). Effectiveness of training in organizations: A meta-analysis of design and evaluation features. *Journal of Applied Psychology, 88*, 234–245.

Azaroff, L. S., Champagne, N. J., Nobrega, S., Shetty, K., & Punnett, L. (2010). Getting to know you: Occupational health researchers investigate employee assistance professionals' approaches to workplace stress. *Journal of Workplace Behavioural Health, 25*, 296–319.

Barling, J. (1996). The prediction, experience, and consequences of workplace violence. In G. R. Vandenbos & E. Q. Bulatao (Eds.), *Violence on the job: Identifying risks and developing solutions* (pp. 29–49) Washington, DC: American Psychological Association.

Barling, J., Dupré, K., & Kelloway, E. K. (2009). Predicting workplace aggression and violence. *Annual Review Psychology, 60*, 671–692.

Barling, J., Rogers, G., & Kelloway, E. K. (2001). Behind closed doors: In-home workers' experience of sexual harassment and workplace violence. *Journal of Occupational Health Psychology, 6*, 255–269.

Baron, R. A., & Neuman, J. H. (1996). Workplace violence and workplace aggression: Evidence on their relative frequency and potential causes. *Aggressive Behaviour, 22*, 161–173.

Baron, R. A., & Neuman, J. H. (1998). Workplace aggression – the iceberg beneath the tip of workplace violence: Evidence on its forms, frequency, and targets. *Public Administration Quarterly*, *21*, 446–464.

Baron, R. A., Neuman, J. H., & Geddes, D. (1999). Social and personal determinants of workplace aggression: Evidence for the impact of perceived injustice and the type A behavior pattern. *Aggressive Behavior*, *25*, 281–296.

Barrett, S. (2011). Buyer's guide: Employee assistance programmes. *Employee Benefits*, 49–51.

Beech, B., & Leather, P. (2006). Workplace violence in the health care sector: A review of staff training and integration of training evaluation models. *Aggression and Violent Behavior*, *11*, 27–43.

Bisson, J. I. (2003). Single-session early psychological interventions following traumatic events. *Clinical Psychology Review*, *23*, 481–499.

Bowling, N. A., & Beehr, T. A. (2006). Workplace harassment from the victim's perspective: A theoretical model and meta-analysis. *Journal of Applied Psychology*, *91*, 998–1012.

Braverman, M. (1999). *Preventing workplace violence: A guide for employers and practitioners*. Thousand Oaks, CA: Sage.

Bryant, R. A., Moulds, M. L., & Nixon, R. V. (2003). Cognitive behaviour therapy of acute stress disorder: A four-year follow-up. *Behavioural Research and Therapy*, *41*, 489–494.

California Occupational Safety and Health Administration (Cal/OSHA). (1995). *Guidelines for workplace security*. Sacramento, CA: Author.

Caraulia, A. P., & Steiger, L. K. (1997). *Nonviolent crisis intervention: Learning to defuse explosive behavior*. Brookfield, WI: CPI.

Cooper, C. L., & Cartwright, S. (1997). An intervention strategy for workplace stress. *Journal of Psychosomatic Research*, *43*, 7–16.

Cox, T., Karanika, M., Griffiths, A., & Houdmont, J. (2007). Evaluating organisational-level work stress interventions: Beyond traditional methods. *Work and Stress*, *21*(4), 348–362.

Day, A. L., & Catano, V. M. (2006). Screening and selecting out violent employees. In E. K. Kelloway, J. Barling, & J. J. Hurrell (Eds.), *Handbook of workplace violence* (pp. 549–577). Thousand Oaks, CA: Sage.

Dupré, K. E., & Barling, J. (2006). Predicting and preventing supervisory workplace aggression. *Journal of Occupational Health Psychology*, *11*, 13–26.

Durhart, D. T. (2001). *Bureau of Justice Statistics special report: Violence in the workplace, 1993–1999*. NCJ 190076. Washington, DC: US Bureau of Justice Statistics.

Einarsen, S., & Raknes, B. (1997). Harassment in the workplace and the victimization of men. *Violence and Victims*, *12*, 247–263.

Emmerik, H., Kamphuis, M. C., Hulsbosch, A., & Emmelkamp, B. (2002). Threats of workplace violence and the buffering effect of social support. *Group and Organizational Management*, *32*, 152–175.

Everly, G. S., Flannery, R. B., & Mitchell, J. T. (2000). Critical incident stress management (CISM): A review of the literature. *Aggression and Violent Behaviour*, *5*, 23–40.

Everly, G. S., Phillips, S. B., Kane, D., & Feldman, D. (2006). Introduction to and overview of group psychological first aid. *Brief Treatment and Crisis Intervention*, *6*, 130–136.

Glomb, T. M. (2002). Workplace anger and aggression: Informing conceptual models with data from specific encounters. *Journal of Occupational Health Psychology*, *7*, 20–36.

Greenberg, L., & Barling, J. (1999). Predicting employee aggression against coworkers, subordinates and supervisors: The roles of person behaviors and perceived workplace factors. *Journal of Organizational Behaviour*, *20*, 897–913.

Harlow, K. C. (1998). Employee attitudes toward an internal employee assistance program. *Journal of Employment Counseling*, *35*, 141–150.

Hearnden, K. (1988). *Violence at work.* Industrial Safety Data File. London: United Trade Press.

Ishak, M., & Christensen, M., (2002). Achieving a better management for patients' aggressive behaviour: Evaluation of a training program. *Journal of Occupational Health and Safety Australia and New Zealand, 18,* 231–237.

Joseph, J. S., & Gray, M. J. (2008). Exposure therapy for posttraumatic stress disorder. *Journal of Behavior Analysis of Offender and Victim: Treatment and Prevention, 1,* 69–80.

Kelloway, E. K., Barling, J., & Hurrell, J.J. (2006). *Handbook of workplace violence.* Thousand Oaks, CA: Sage.

Kelloway, E. K., Catano, V., & Day, A. (2011). Counterproductive work behaviours. In *People and work in, Canada.,* Toronto, ON: Nelson.

LeBlanc, M. M., Dupré, K. E., & Barling, J. (2006). Predicting and preventing violence directed at employees by members of the public. In E. K. Kelloway, J. Barling, & J. Hurrell (Eds.), *Handbook of workplace violence* (pp. 261–280) Thousand Oaks, CA: Sage.

LeBlanc, M. M., & Kelloway, E. K. (2002). Predictors and outcomes of workplace violence and aggression. *Journal of Applied Psychology, 87,* 444–453.

Lieber, L. (2007). Workplace violence – what can employers do to prevent it? *Employment Relations Today, 34,* 91–100.

Litz, B. T., Gray, M. J., Bryant, R. A., & Adler, A. B. (2002). Early intervention for trauma: Current status and future directions. *Clinical Psychology: Science and Practice, 9,* 112–134.

Martinko, M. J., Gundlach, M. J., & Douglas, S. C. (2002). Toward and integrative theory of counterproductive workplace behaviour: A causal reasoning perspective. *International Journal of Selection and Assessment, 10,* 36–50.

Mayou, R., Ehlers, A., & Hobbs, M. (2000). Psychological debriefing for road traffic accident victims: Three year follow-up of a randomized controlled trial. *British Journal of Psychiatry, 176,* 589–593.

National Center on Addiction and Substance Abuse. (2000). *Report of the US Postal Service Commission on a Safe and Secure Workplace.* Washington, DC: Author.

Nobrega, S., Champagne, N., Azaroff, L. S., Shetty, K., & Punnett, L. (2010). Barriers to workplace stress interventions in employee assistance practice: EAP perspectives. *Journal of Workplace Behavioural Health, 25,* 282–295.

O'Leary, K. D., & Jouriles, E. N. (1994). Psychological abuse between adult partners: Prevalence and impact on partners and children. In L. L'Abate (Ed.), *Handbook of developmental and family psychology.* New York: Wiley.

Pollack, K. M., Austin, W., & Grisso, J. A. (2010). Employee assistance programs: A workplace resource to address intimate partner violence. *Journal of Women's Health, 19,* 729–733.

Randall, R., Griffiths, A., & Cox, T. (2005). Evaluating organizational stress-management interventions using adapted study designs. *European Journal of Work and Organisational Psychology, 14*(1), 23–41.

Rayner, C., & Cooper, C. L. (2006). Workplace bullying. In E. K. Kelloway, J. Barling, & J. J. Hurrell (Eds.), *Handbook of workplace violence* (pp. 121–146) Thousand Oaks, CA: Sage.

Rospenda, K. M. (2002). Workplace harassment, services utilization, and drinking outcomes. *Journal of Occupational Health Psychology, 7,* 141–155.

Runyan, C. W., Zakocs, R. C., & Zwerling, C. (2000). Administrative and behavioral intervention for workplace violence prevention. *American Journal of Preventive Medicine, 18* (4S), 116–127.

Ruzek, J. I., Brymer, M. J., Jacobs, A. K., Layne, C. M., Vernberg, E. M., & Watson, P. J. (2007). Psychological first aid. *Journal of Mental Health Counseling, 29,* 17–49.

Schat, A. C. H., Frone, M. R., & Kelloway, E. K. (2005). Prevalence of workplace aggression in the US workforce: Findings from a national study. In E. K. Kelloway, J. Barling, & J. J. Hurrell (Eds.), *Handbook of workplace violence* (pp. 47–89). Thousand Oaks, CA: Sage.

Schat, A. C. H., & Kelloway, E. K. (2003). Reducing the adverse consequences of workplace aggression and violence: The buffering effects of organizational support. *Journal of Occupational Health Psychology*, *8*, 110–122.

Schat, A. C. H., & Kelloway, E. K. (2005). Workplace violence. In J. Barling, E. K. Kelloway, & M. Frone (Eds.), *Handbook of work stress* (pp. 189–218). Thousand Oaks, CA: Sage.

Schat, A. C. H., & Kelloway, E. K. (2006). Training as a workplace aggression intervention strategy. In E. K. Kelloway, J. Barling, & J. J. Hurrell (Eds.), *Handbook of workplace violence* (pp. 579–605) Thousand Oaks, CA: Sage.

Smith, G. (2000). *Work rage: Identifying the problems, implement the solutions.* Toronto, ON: Harper Collins.

Swanberg, J. E., Logan, T. K., & Marke, C. (2006). The consequences of partner violence on employment in the workplace. In E. K. Kelloway, J. Barling, & J. J. Hurrell (Eds.), *Handbook of workplace violence* (pp. 351–380). Thousand Oaks, CA: Sage.

Ulman, K. H. (2004). Group interventions for treatment of trauma in adults. In B. J. Buchele & H. I. Spitz (Eds.), *Group interventions for treatment of psychological trauma.* New York: AGPA.

US Department of Labor. (2006). *Survey of workplace violence prevention*, 2005. USDL publication 06-1860. Washington, DC: Bureau of Labor Statistics.

Wassell, J. T. (2009). Workplace violence intervention effectiveness: A systematic literature review. *Safety Science*, *47*, 1049–1055.

8

The Occupational Implication of the Prolonged Effects of Repeated Exposure to Traumatic Stress

Alexander C. McFarlane

In occupational settings, the temporality of the exposure to a traumatic event and the onset of symptoms provide strong evidence for direct causality. This temporal relationship is often seen as central to approaches used in the workplace to minimize psychological injury. If the symptoms emerge immediately in the aftermath of an event, the causal role of that event logically argues for intervention. In contrast, if there is a significant temporal delay between the exposure and the onset of symptoms, the obligations to intervene and planning the strategies for the cumulative workplace stress are much more challenging. The existence of delayed-onset post-traumatic stress disorder (PTSD) has been controversial and a practical and theoretical challenging issue for psychiatry, particularly in occupational health settings and the related compensation claims management. From a psychopathological perspective, developing an etiological model to explain how there can be a prolonged delay between the exposure to an event and the onset of symptoms poses a major challenge.

Perhaps the area where this delayed emergence in symptoms has caused most controversy is in returning war veterans (Shephard, 2001). The longer the period between the end of the war and the emergence of symptoms, the greater the suspicion that arises about the causal role of combat exposure. For police, a similar suspicion exists with disability being seen to be the consequence of culture rather than traumatic exposure (Summerfield, 2011).

Understanding about the importance of the delayed effects of exposure to traumatic stress has recently increased because of the documented evidence about the important of these events not just to PTSD (de Boer *et al.*, 2011). Trauma exposures are accepted as important determinants of psychological and neurobiological vulnerability to depressive disorders (Heim & Nemeroff, 2001). This broader question also is an area of relevance to the prevention of work-related injuries. One of the key etiological theories that have been examined in the etiology of depression is the role of sensitization and kindling in the onset of the disorder (Post & Weiss, 1998). Whilst this model has focused on the importance of stressful events in the first episodes of illness and the decreasing role of exogenous events in subsequent episodes, the role of the burden of stress prior to the

International Handbook of Workplace Trauma Support, First Edition. Edited by Rick Hughes, Andrew Kinder, and Cary L. Cooper.

onset of the first episode of illness is also a related factor. These matters are relevant to the traumatic stress field because of the substantial level of comorbidity of PTSD and major depressive disorder and the need to better understand the mutuality of their etiology. Furthermore, the increasing body of evidence is suggesting the important role of traumatic events in the onset of major depressive disorder in the absence of PTSD (Laugharne *et al.*, 2010).

This chapter will examine the issue of the timing of the onset of PTSD following exposure to traumatic events and the emerging findings about the prevalence and nature of delayed-onset PTSD. Secondly, the contribution of the depression literature to these questions will be highlighted. The challenges presented by the complexity of these relationships will be dissected in terms of the issues that need to be addressed in the provision of occupational health care to first responders and to military personnel. The reactive nature of symptom management in the workplace means that important opportunities are lost due to an underestimation of the delayed and cumulative impact of traumatic stress.

Delayed-Onset PTSD

Delayed-onset PTSD was initially recognized in the DSM-III formulation published in 1980 (American Psychiatric Association (APA), 1980). DSM-IV specifically states that delayed-onset PTSD should be diagnosed if "the onset of symptoms is at least 6 months after the stressor" (APA, 2000, p. 468). From a theoretical point of view, this form of PTSD is likely to occur in individuals who have managed to contain their initial distress by adaptive means which are disrupted by subsequent stressors and/or the natural progression of the neurobiology has led to the manifestation of symptoms. A recent review about delayed-onset PTSD emphasized how much of the confusion about this construct has arisen from different definitions of delayed-onset PTSD (Andrews *et al.*, 2007). For example, different interpretations of the construct included an individual who has had subsyndromal symptoms who has subsequently crossed a threshold of clinical severity, as well as an individual who has been asymptomatic and then at some later point developed a disorder.

The existence of this delayed form of PTSD emphasizes how a traumatic experience can apparently lie relatively dormant within an individual only to become manifest at some future point. From the perspective of possible methods of prevention, the mechanism of how this subclinical state is triggered into a full-blown syndrome is important in terms of opportunities for intervention. This is particularly the case because subclinical symptoms would appear to leave an individual at risk of a progressive activation with exposure to further environmental stress or trauma exposure.

A meta-analysis of delayed-onset PTSD (Smid *et al.*, 2009) reviewed longitudinal studies with a mean duration of 25 months and the maximum range of 60 months. In the combined study population, 24.8% (95% CI = 22.6–27.2%) had delayed-onset PTSD. A regression analysis of this data showed that the proportion of individuals with delayed-onset PTSD was larger when the duration of follow-up was longer. They also found that traumatic exposures in military populations were association with a greater proportion of delayed-onset PTSD and with when the cumulative incidence was lower.

These observations would suggest there are important caveats on this paper. Andrews *et al.* (2007) further separated the rates identifying that 38.2% of military cases of PTSD and 15.3% of civilian cases were delayed onset, although a retrospective study by Frueh *et al.* (2009) in primary-care clinics for veterans did not support this conclusion. Andrews *et al.* (2007) also concluded that the very significant majority of delayed-onset cases of PTSD were preceded by prior symptoms but there was relatively little information available about the level of distress.

The probability of symptoms increasing across much longer time periods is not addressed by this literature. For example, a 20-year follow-up of a group of veterans with combat stress reactions and a group of veterans without stress reactions indicated that in the latter group, the total number of symptoms was greater 20 years after combat than at any of the previous assessment points. In other words, the degree of symptomatic distress even in those without PTSD does not seem to decrease with time. In those without combat stress reactions, 14% of individuals had PTSD in the first year in contrast to 26% at the 20-year assessment (Solomon & Mikulincer, 2006).

Smid *et al.* (2009) concluded that the risk of delayed-onset PTSD did not decrease between 9 and 25 months after the event, suggesting ongoing potential risk to individuals. They hypothesized that prodromal symptoms act to increase the allostatic load (McEwen, 2003) and hence the risk of developing PTSD. Importantly, they also found that the desire for compensation was unlikely to be a major factor in the onset of delayed-onset PTSD. Intervening life events were found to have been a significant risk factor for delayed-onset PTSD. Subsyndromal PTSD cases were also significantly more likely to go on to develop delayed-onset PTSD if they had met two or more PTSD criteria initially (Smid *et al.*, 2009).

In the occupational context, the evidence that has emerged means that this concept is no longer an issue of controversy, as there is now substantial epidemiological evidence demonstrating the validity of this construct. The challenge, however, emerges about the long tail of effect of exposure to traumatic events and how this should be managed to minimize the risk to the individual and the organization. Furthermore, symptoms that do not reach the threshold for disorder still may have long-term implications in terms of managing the further stress exposures of those workers deemed to be at risk. For this reason, the literature about subsyndromal PTSD is addressed.

The Issue of Subsyndromal PTSD

From an occupational perspective, the issue of the significance of patients who did not satisfy the full diagnostic criteria for PTSD presents a challenge. In workplace settings, individuals are seen as either healthy or sick/injured. The challenge posed by a dimensional view of health is challenging to say the least. However, identifying individuals at risk opens up opportunities for risk management and prevention, particularly by minimizing further exposures that confer risk.

Subsyndromal PTSD has been coined to describe the group of individuals who have a number of symptoms but do not satisfy the full DSM-IV criteria. A number of studies have highlighted that the avoidance symptoms generally are the set of criteria which form the barrier to the diagnosis of PTSD (Andrews, Slade, & Peters, 1999).

Interestingly, the ICD 10 criteria for PTSD (World Health Organization, 1992) do not require the presence of actual avoidance, rather referring to *preferred avoidance* of reminders. This difference reflects the lack of objective consensus about the optimal threshold for these phenomena.

The impact of subsyndromal PTSD has been examined systematically in several population studies, which suggest that the degree of disability is similar to that of PTSD. Furthermore partial PTSD has been associated with an increased risk of binge drinking of alcohol, potentially impeding recovery (Adams *et al.*, 2006). Furthermore, the pattern of health care utilization with partial PTSD more closely resembled that for individuals with full PTSD rather than those without symptoms (Gillock *et al.*, 2005). It has been argued that the associated disability warrants partial PTSD being accepted as a subcategory in any revision of the diagnostic criteria (Mylle & Maes, 2004). Another important issue is that patients with PTSD often have fluctuating symptoms and hence, at one point in time, may satisfy the full diagnostic criteria, but will not at another. The utility of subsyndromal PTSD is to capture such a longitudinal course of PTSD and would assist a better understanding of the nature of delayed-onset PTSD (Bryant & Harvey, 2002).

The relation between the acute symptoms and the emergence of PTSD is also an issue of considerable theoretical and clinical importance. It appears that the timing of the point of maximal intensity and the progressive reinforcement of the traumatic response are critical to the emergence of chronicity (Weisaeth, 1989). Foa (1997) noted that a poor response to treatment is predicted by delayed timing of the maximal traumatic response. She concluded that engagement with the traumatic memory is a critical dimension to the processing of these experiences and the lack of or delayed engagement means that the affect and intensity of the memory have been distanced. Anger and dissociation are two mechanisms that interfere with this process.

In the occupational setting, the findings about subsyndromal PTSD are important as they suggest that questioning whether a particular worker does or does not have PTSD is less important than identifying the accumulating risk of workplace exposures. The disability arising from subsyndromal PTSD is indicative that these individuals should not simply be dismissed as having a trivial or normal stress reaction. If assessed by an occupational physician or mental health professional, the clinical diagnosis that is often used for these individuals is an adjustment disorder, a diagnosis about which remarkably little literature exists despite its frequent use in clinical settings.

Other Post-traumatic Disorders?

In occupational settings, there has been considerable focus on PTSD because of the clear link that is established by the traumatic memories to a specific stressor. This interest has perhaps overemphasized the role of traumatic events in PTSD at the expense of minimizing their role in other psychiatric disorders. For example, a recent study of over 1000 traumatically injured patients identified that 12 months after the injury, the most common new psychiatric disorders were depression (9%), generalized anxiety disorder (9%), PTSD (6%), and agoraphobia (6%) (Bryant *et al.*, 2010). Hence PTSD is only one of a number of disorders that are related to traumatic exposures.

Similarly, a study of the 1991 Gulf War veterans found that in the 10-year post-deployment period, 18% of the veterans had developed an affective disorder in contrast to the 12% of the comparison group. Importantly, the comparison group had also had a number of deployments in peacekeeping and combat-related roles. In contrast, 5.4% of the Gulf War veterans had developed PTSD in contrast to 1.4% of the comparison population (Ikin *et al.*, 2004). A study of psychiatric disorders in the Canadian military demonstrated a similar relationship between combat and witnessing atrocities with depressive and anxiety disorders as well as PTSD (Sareen *et al.*, 2007). The relationship with PTSD, however, was substantially greater (witnessing atrocities PTSD AOR = 4.33 (2.8–6.7), major depression disorder AOR = 1.82 (1.3–2.5)).

There are no systematic reviews of the relationship between traumatic stress and major depressive disorder, despite the suggestion that the accumulated evidence is sufficient to warrant the conclusion that trauma plays a significant etiological role in the anxiety disorders other than PTSD and depressive disorders (Laugharne *et al.*, 2010). The only systematic examination of the evidence that links depression to traumatic events is within the "Statements of Principle" of the independent statutory body which determines grounds for Australian veterans' service entitlements, the Repatriation Medical Authority (2007). Using the standard of proof of a reasonable hypothesis, it was concluded that exposure to a major stressor was sufficient to be linked to a clinical onset of a depressive disorder if this had occurred within the previous 5 years. This included events such as experiencing a life-threatening event, being the subject of a serious physical assault or attack, or witnessing similar events. For events of a lesser magnitude, the window of effect is assessed to be only 1 year and this includes negative life events such as the breakdown of relationships, work difficulties, and so on (Repatriation Medical Authority, 2007).

The relationship between childhood trauma and major depressive disorder has been more systematically examined and reported in the literature from a neurobiological perspective (Heim & Nemeroff, 2001). This relationship had grown out of the increasing social awareness of the prevalence and high incidence of childhood maltreatment. The relationship between these exposures in early childhood and major depression is hypothesized to be mediated by changes in the hypothalamic pituitary adrenal (HPA) axis. It is now accepted from a substantial body of research that PTSD is highly comorbid with other conditions and not just in community samples (e.g., Breslau, 1998; Creamer *et al.*, 2001; Kessler *et al.*, 1995, 2005). However, the complexity of this relationship has not been given due attention.

Ginzburg, Ein-Dor, and Solomon (2009) have examined the patterns of comorbidity in a 20-year follow-up of a cohort of Israeli war veterans. This highlighted the high rates of triple comorbidity – namely, PTSD, anxiety, and depression – and found that this pattern was the rule rather than the exception. Using latent growth models, they were able to demonstrate that PTSD predicted the comorbid conditions rather than the reverse. The importance of these observations is that PTSD shares many common etiological mechanisms with the other anxiety disorders and major depressive disorder. To date, the literature has failed to focus on the commonalities of the etiological mechanisms, tending to treat the individual disorders as representative in differential etiological pathways. Logically, there must be substantial commonalities in the underlying etiological mechanisms for these disorders to coexist and the challenge therefore is to highlight what these shared etiological processes are.

In occupational settings, these are important questions because clinicians frequently fail to adequately assess the issue of comorbidity, making only a single diagnosis. Frequently, one clinician may focus on the depressive disorder in the worker, whereas another clinician might focus on the PTSD. This differential focus can lead to substantial disputes about etiology, prognosis, as well as the most appropriate course of action in regard to treatment. This situation is a recipe for poor clinical outcomes and disputes in the implementation of a rehabilitation program. The inadequacy of diagnosis accuracy, in standard clinical practice, was highlighted in a study of 1000 outpatients in an adult psychiatry department. The clinician's diagnosis was matched with a screening questionnaire in half the patients and a structured diagnostic interview in the other half. In the screening questionnaire group, clinicians diagnosed 7.2% of the patients with a PTSD, whereas an additional 18.6% were identified on the questionnaire. In the SCID-diagnosed groups, an additional 14.4% of the sample was found to have a PTSD above and beyond the cases identified by the clinicians (Zimmerman & Mattia, 1999). This lack of diagnostic accuracy of clinicians, even in a setting where they are aware that their assessments are being checked with a research protocol, has major implications for the inconsistencies of opinions in medico-legal settings.

The lack of systematic diagnostic agreement in occupational settings is one factor that has prevented a much-needed exploration of the shared etiology of depression and PTSD that is relevant to prognosis and the need to provide timely evidence-based care. One of the most destructive approaches in the management of a workplace injury is the use of competing expert reports. Achieving agreed standards of diagnostic assessments is in the interests of health professionals and injured workers alike.

The Relevance of Sensitization and Kindling to the Onset of Illness

Two critical constructs in understanding the etiology of delayed-onset PTSD are sensitization and kindling (McFarlane, 2010). Beginning with the observation of Kraepelin (1921), it is has been recognized that life events play a more significant role in the initial episodes of major depressive disorders in contrast to subsequent relapses. This has led to the development of the "kindling hypothesis" which can be specifically stated as "vulnerability to depressive relapse/recurrence is determined by the increased risk of particular negative patterns of information processing . . . increased reliance on these patterns of processing makes it easier for their future activation to be achieved on the basis of increasingly minimal cues" (Kendler *et al.*, 2001).

Sensitization, on the other hand, "refers to the observation that individuals who are repeatedly exposed to an environmental risk factor may develop progressively greater responses over time, finally resulting in a lasting change in response amplitude" (Collip *et al.*, 2008, p. 220). In essence, sensitization represents the progressively increasing response of an individual following the repeated exposure to a stressor (Bonne *et al.*, 2004; Eriksen & Ursin, 2004).

Sensitization provides a theoretical perspective for examining the risk of consequences of repeated exposure to major traumatic stresses prior to the onset of any disorder. Evidence demonstrates that repeated exposure to traumatic stresses in individuals increases the probability that they will suffer from PTSD with a further exposure (Bremner *et al.*, 1993; Post & Weiss, 1998; Yehuda *et al.*, 1995).

Copeland *et al.* (2007) investigated this question in a representative sample of 1420 children who were followed up annually through 16 years. They found that symptoms of PTSD were predicted by previous exposure to multiple traumas. This also demonstrated that the dose-dependent relationship between traumatic events and anxiety and depressive disorders was similar to that for trauma and post-traumatic stress symptoms. They found that this relationship was particularly marked for the onset of depressive disorders. In other words, the effects of trauma were not found to be symptom specific and across the age groups in this study, they found a strong relationship between trauma and the spectrum of anxiety, depressive, and disruptive behavior disorders they examined.

Breslau *et al.* (1999) in a longitudinal sample of 2181 young adults examined the risk of developing PTSD following an event. In particular, they identified that a history of any previous exposure to traumatic events predicted a greater risk of PTSD and that the probability of developing the condition was greater if the individual had had multiple previous exposures. These results were interpreted as being consistent with a sensitization hypothesis which was similar to the impact of childhood abuse as a predictor of PTSD in Vietnam veterans. In PTSD, the role of repeated exposure to traumatic events that initially cause only distress is proposed to cause a more severe and long-lasting disorder than PTSD that arises after a single exposure to a traumatic event (Post & Weiss, 1998).

Heim *et al.* (2008) have highlighted how this process of sensitization to symptoms arising from trauma exposure has been supported at a biological level. In a number of studies, a link between childhood trauma and sensitization of their neuro-endocrine stress response modified immune activation, glucocorticoid resistance, and reduced hippocampal volume has been identified. In particular, persistent sensitization of the stress response has been demonstrated as arising from exposure to childhood trauma, and this is associated with altered modulation of the HPA axis. These changes are in turn linked to symptoms of depression. Hence, there is a significant body of evidence demonstrating that repeated stress exposures prior to the onset of the first episode of a disorder increased the risk of PTSD and a range of other conditions, particularly depression.

An important question arises as to which individuals are vulnerable to the impact of further triggering and activation of their traumatic memories. One study of accident victims showed that those who went on to develop delayed-onset PTSD symptoms had significantly more symptoms 8 days after the trauma than those who did not develop PTSD. Importantly, the majority of the injury survivors in this study had low levels of symptoms in the acute setting, and these persisted over time. The initial symptoms which were particularly indicative of risk were intensity and frequency of the arousal and re-experiencing symptoms (O'Donnell *et al.*, 2007). These data demonstrate how subsyndromal PTSD can increase in severity due to the process of sensitization, particularly in the first year after a major traumatic incident.

In an occupational setting, this literature demonstrates a foreseeable risk of repeated exposures, particularly in an individual who is progressively recruiting symptoms which are yet to reach a level of clinical significance. In other words, further exposure of these individuals carries a significant attendant risk of worsening their symptom severity. As a consequence, a critical step in prevention of occupationally related mental disorders is to anticipate this risk and rotate individuals through high-stress roles, especially when they begin to develop disorders.

Sensitization and Kindling in Determining Future Episodes

Sensitization and kindling are also used to describe a second process. This pattern is related to a sensitization process to depressive states and predicts that with recurrent episodes of a major depressive disorder there will be a progressive diminution of the role of environmental stressors. For a long time it has been recognized that major depression has a complex etiology that involves both the role of stressful life events and genetic risk factors. In a recent meta-analysis (Stroud *et al.*, 2008), evidence was found that a first episode of major depression was more likely triggered by severe life stress and that the probability decreased that a recurrent episode would be triggered by stress. This finding was interpreted as meaning that milder events can trigger depressive episodes as the number of episodes of depression increases (stress desensitization). Ultimately this process can reach a point where the illness becomes relatively autonomous of stress in situations where the stressors are not required to trigger the onset or where what to an outside observer would be regarded as a stressor is absent (Patten, 2008).

Kendler *et al.* (2001) have concluded that kindling is particularly important in people who do not have a major genetic predisposition to depression. Furthermore, this kindling effect has been shown to disappear after approximately nine episodes, when the process becomes increasingly autonomous. This is in keeping with the kindling hypothesis that suggests that there is a threshold beyond which there can be no additional sensitization to the depressive state. This implies a transitioning to a pattern of increasingly autonomous illness, where an increasing number of episodes have been correlated with relative treatment refractiveness (Post & Weiss, 1998).

The commonality of these processes to other disorders like PTSD and obsessive-compulsive disorder has been emphasized (Post & Weiss, 1998). In the case of PTSD, once the disorder has developed, triggers play a particularly important role in reinforcing the intrusive memories and the associated psychophysiological activity. These memories become increasingly become more and more spontaneous as a consequence of a kindling-like progression. The support for the commonality of this mechanism also comes from the role of serotonin in the rate of development of amygdala kindling and the role of serotonin reuptake inhibitors in the treatment of all of these conditions.

Sensitization has also been hypothesized to play an important role via kindling-like mechanisms in pain syndromes. Given the significant interrelationship between PTSD and somatic symptoms, these shared mechanisms are of particular relevance in assessing the importance and contribution of stress exposure to conditions such as chronic back pain and fatigue-like syndromes (McFarlane, 2007; Post, 2002).

Sensitization and Kindling in Occupational Settings

These observations are of particular relevance to the issue of duty of care in managing repeated exposures which police officers and other emergency workers may face in the course of their work. This issue has often been misrepresented in the workplace due to incorrect beliefs about the apparent lack of impact of repeated stresses on workers (Adamou & Halem, 2003). An important legal precedent was reflected in the case of *Page v. Smith* (1996) where Lord Lloyd indicated that emergency service workers were "more hardened than non-professional ones and therefore it may be more difficult to foresee injury to them" (quoted in Adamou & Halem, 2003, p. 331), a position taken to limit the ability to access compensation by these groups. Clearly this judgement is not consistent with the sensitization literature which has provided substantive evidence of the accumulated risk that can occur for officers and military veterans through repeated exposures to traumatic events (McFarlane, 2009).

Importantly, in PTSD there are underlying neurobiological correlates for sensitization. Particularly in the domain of the HPA axis, this process of sensitization explains why individuals with PTSD become unusually reactive to stress which is manifest as exaggerated behavioral and biological responses to environmental challenge. These phenomena are clinically manifest as symptoms of hypervigilance, increased startle, and physiological arousal and distress on exposure to reminders. These need to be monitored prospectively in the occupational setting and if such a pattern of distress is increasing, the individual should be offered alternative duties that do not present the same risk of major trauma exposure.

Issue of Delay in Receiving Treatment

One of the core aims of early intervention in the workplace is to prevent the emergence of disability. Secondary prevention argues for the importance of early treatment and active management in the workplace before there is a substantial loss of function. The Australian Treatment guidelines for PTSD (Australian Centre for Posttraumatic Mental Health, 2007), which had a section dealing with emergency service personnel, recommended that "using the principles of secondary prevention, this [early treatment] minimises the development of a series of secondary patterns of adaptation that in themselves can present a significant disadvantage." While the logic supporting this approach is basic, the literature supporting it is less than complete. In particular, no studies that have considered the delay in implementation of treatment in PTSD (Duffey *et al.*, 2007; Friedman *et al.*, 2007; Gillespie *et al.*, 2002; Resick *et al.*, 2007) were specifically designed to answer this question.

The Resick *et al.* (2007) study compared CBT and cognitive reprocessing treatment and found that the duration of disorder did not influence the effectiveness. However, this conclusion does not negate the importance of early treatment. People who present late for treatment cannot be presumed to have the same disorder that has the same severity or course as those presenting to closer proximity to the trauma. Another study (Gillespie *et al.*, 2002), that was conducted following the Omagh bombing, found little differences in the outcome between those who presented early for care when compared with those presenting later. Again, those presenting in the immediate aftermath of bombing cannot be presumed to be the same as those presenting later. In particular, those presenting later may have a delayed-onset PTSD where the high levels of acute distress in those without a delayed onset may be indicative of differential risk factors. In contrast, the study of Duffy, Gillespie, & Clark (2007) of patients with PTSD in the context of terrorism in Northern Ireland did find that the longer the delay for presenting for treatment, the worse the outcome.

Hence, these studies do not provide a proper scientific investigation of this question about the delayed impact of treatment. The lack of investigation of this issue in PTSD in contrast to major depression is in fact surprising. Furthermore, there is a general consensus reflected in a number of recent publications about the gains that are obtained from the early intervention of psychiatric disorder (Beddington *et al.*, 2008; Herrman & Chopra, 2009; McGorry *et al.*, 2006).

Given the commonality of the underpinnings of the etiology in depression and PTSD, it is reasonable to extrapolate from the factors that have been identified as influencing treatment response.

1. *Duration of illness.* There are a number of studies which have examined the probability of recovery in individuals with major depression and the role of the impact of the duration of an episode. Kravitz *et al.* (2000) reviewed the evidence and in a study of a further sample identified that recovery from a major depressive episode was most strongly correlated with the length of the current episode. Similar findings have been identified in a number of other studies where the longer the illness length, the delay in the remission onset (Gormley, O'Leary, & Costello, 1999; O'Leary *et al.*, 2000; Scott *et al.*, 1992).
2. *Partial remission to treatment as a protector of relapse.* There is a substantial body of research that has examined the impact of partial remission following treatment in the course of a major depressive disorder. Pentor *et al.* (2003) followed up a population suffering from unipolar depression and identified that the relapse rate in patients with partial remission was 67%. This study emphasized the importance of complete remission as an issue required to decrease the rates of short-term relapse. This is an important finding in the context of the fact that about 80% of individuals have second episodes. These findings about major depressive episodes are pertinent particularly to individuals who have a comorbid major depressive disorder and PTSD but are also likely to be applicable to those without the secondary comorbidity.
3. *Shared neural circuitry.* The underlying circuitry neural regions which have been identified as being relevant to the etiology of PTSD are equally those involved in depression. For example, amygdala reactivity has been identified as a primary area of interest in PTSD, and this nucleus plays a central role in determining fear reactivity (Etkin & Wager, 2007; Shin & Liberzon, 2010). The amygdala has also been extensively involved in the investigation of depression (Drevets, Price, & Furey, 2008). For example, Ramel *et al.* (2007) highlighted how amygdala reactivity is an important issue in people with a history of depression in contrast to those without such a history. These results indicate how the amygdala plays a central role in modulating mood-congruent memory, particularly during the

induction of sad states of mind in individuals who are vulnerable to depression. Hence, a known risk for individuals with PTSD to have further exposure to environments that have traumatic triggers because of the obvious risks of activation of fear-related circuitry is similar to the risks for individuals with depression to activate the neural systems associated with vulnerability to negative emotions and the onset of depressive episodes (Hamilton & Gotlib, 2008; Leppänen, 2006).

4. Further evidence suggested the underlying biological mechanisms of how the duration of depression impacts cognitive functioning and disability. This effect is related to the sensitivity of the hippocampus to stress, a critical issue in PTSD. In depression, it has been found that the length of past depression impairs memory performance and that there is a significant toxic link between the burden of depression and cognition (Gorwood *et al.*, 2008).

Hence, this literature emphasizes that the similarity between the neurobiological underpinnings of PTSD and major depression would make it improbable that the literature that has demonstrated that the impact of partial treatment response and duration of illness in depression is not similar in PTSD. Hence, this raises important liability questions for employers in environments where there is a foreseeable risk of the workplace having a significant contribution to both PTSD and major depressive disorder. The Australian Defence Force has now been screening its members for a decade, and a similar practice is utilized in the US Armed Services (Wright *et al.*, 2002). In an environment where there is a foreseeable risk of these conditions such as the emergency services, an argument can be made why regular screening and assessment should be routine.

In particular, it has been argued elsewhere (McFarlane & Bryant, 2007) that there are special obligations upon employers to create health services that are aware of the diagnostic dilemmas and the propensity of individuals to present late for treatment. There are many cultural issues and issues of stigma that mean that it is unrealistic to expect that individuals will present with illness. These are important and controversial issues which are continuing to be litigated in court proceedings around the world. The limits placed on the responsibility of employers in this domain by legal jurisdiction will depend on the expert evidence available. Many research questions remain to be examined in this area of practice.

Emerging Issues

The relationship between workplace stress and cardiovascular disease has been an issue that has long provoked controversy. Interestingly, a range of studies has emerged in the last decade provide increasing evidence of a link between PTSD and hypertension. For example, the National Comorbidity Study concluded that PTSD had an independent relationship with hypertension, independent of depression, and this could possible account for the increased rates of cardiovascular disease associated with PTSD (Kibler *et al.*, 2009). In a recently completed two-phase epidemiological study of Australian Vietnam veterans, hypertension was found to be related to PTSD after controlling potential confounds (O'Toole & Catts, 2008). These emerging bodies of evidence require further validation, but are likely to propose new challenges in terms of employers' liability.

Further, a study from the US Department of Veterans' Affairs of normative aging (Kubzansky *et al.*, 2007) has been following a population since 1990. PTSD symptoms were found to have an association with an increased risk for coronary arterial disease in the absence of the full diagnosis, evidence that should be seen as requiring further examination. These findings need to be considered against the background of other work that has not found a relationship between military combat and coronary heart disease (Johnson *et al.*, 2010) in cohorts of World War II, Korean War, and Vietnam War veterans comparing them to noncombat veterans and community controls from the same population. The limitations of this protocol were that it did not examine the intervening role of PTSD and coronary heart disease and the study had limited power.

The possible significance of this relationship is also suggested by the fact that intervening risk factors such as hyperlipidaemia (David *et al.*, 2004; Kagan *et al.*, 1999; Karlovic *et al.*, 2004; Maia *et al.*, 2008; Perkonigg *et al.*, 2009; Scott *et al.*, 2008; Solter *et al.*, 2002; Vieweg *et al.*, 2007; Violanti *et al.*, 2006) have been associated with PTSD. Hence, there are a series of possible causal links now being suggested between PTSD and different cardiovascular risk factors as well as the end point of disease. Further issues that need to be clarified in the relationship between PTSD and the development of heart disease have recently been summarized (Kubzansky & Koenen, 2009).

In workers, compensation settings, the emerging literature about the relationship between cardiovascular disease, depression, and PTSD is likely to bring about a re-examination of the literature that previously has negated the role of stress in cardiovascular disease. Clearly this emerging relationship argues that employers with an interest in the health of their workforce need to consider the interrelationship between physical and mental health. These are inseparable dimensions that should be considered in unison.

Conclusion

One of the significant challenges in occupational health settings and workers compensation systems is the potentially delayed relationship between traumatic stress exposure and the emergence of psychopathology. The significant body of research which has been conducted in the last two decades has now provided valuable insights into the reality of delayed-onset PTSD and the fact that it represents the progressive accumulation of symptoms in individuals over time. Furthermore, intercurrent life stresses have been shown to play an important role impacting the trajectory of an individual's post-traumatic reactions.

These delayed effects represent significant challenges from a management perspective. To date, there has been remarkably little systematic work examining the cumulative burden of traumatic stress exposure in nonmilitary samples. The delayed risks to those who serve their nations are recognized through the substantial benefits and protections of veterans' affairs systems in many countries. Similar recognition and protection are likely to become an increasingly debated issue for police whose cumulative stress exposure to violence and death is often underestimated.

Central to understanding delayed-onset PTSD are the concepts of sensitization and kindling. These have been extensively studied in a range of other psychiatric disorders and highlight the commonality of the etiological mechanisms, particularly with major depressive disorder.

Depression has also been identified as a major consequence of exposure to traumatic events. Interestingly to date there has been little cross-fertilization between the etiological literature about depression and post-traumatic stress disorder. These two bodies of work demonstrate many similar mechanisms and, particularly in the area of neuroimaging, the circuitry involved in the two conditions has much in common. This would argue that many of the findings about the prognostic factors in depression could be readily applied to post-traumatic stress disorder in both research and litigation settings.

A particularly challenging area is the legal obligation of a working environment where there is a foreseeable risk of individuals developing PTSD. If it is accepted that the delay in providing treatment significantly worsens an individual's treatment response and ultimate prognosis, new obligations may emerge upon employers to screen populations and provide evidence-based treatments.

Finally, underpinning the etiology of depression and post-traumatic stress disorder, the concept of allostasis is critical (McEwen & Wingfield, 2003). PTSD is associated with a significant increase in allostatic load (McEwen & Stellar, 1993), and recently the importance of this process has been supported by increasing evidence about the causal association between PTSD and cardiovascular disease. This association is likely to lead to an increasing consideration of these matters in litigation settings.

Acknowledgments

This chapter is supported by National Health and Medical Research Council (NHMRC) program grant no. 300403. This chapter is a modified version of an earlier paper, McFarlane, A. C. (2010). The delayed and cumulative consequences of traumatic stress: Challenges and issues in compensation settings. *Psychology, Injury and Law*, *3*(2), 100–110.

References

Adamou, M. C., & Halem, A. S. (2003). PTSD and the law of psychiatric injury in England and Wales: finally coming closer? *Journal of the American Academy of Psychiatry and the Law*, *31*(3), 327–332.

Adams, R. E., Boscarino, J. A., & Galea, S. (2006). Alcohol use, mental health status and psychological well-being 2 years after theWorld Trade Center attacks in New York City. *American Journal of Drug and Alcohol Abuse*, *32*, 203–224. doi:10.1080/00952990500479522.

American Psychiatric Association. (1980). *Diagnostic and statistical manual of mental disorders* (3rd ed.). Washington, DC: Author.

American Psychiatric Association. (2000). *Diagnostic and statistical manual of mental disorders* (4th ed., text rev.). Washington, DC: Author.

Andrews, B., Brewin, C. R., Philpott, R., & Stewart, L. (2007). Delayed-onset posttraumatic stress disorder: A systematic review of the evidence. *American Journal of Psychiatry, 164*(9), 1319–1326.

Andrews, G., Slade, T., & Peters, L. (1999) Classification in psychiatry: ICD-10 versus DSM-IV. *British Journal of Psychiatry, 174*, 3–5.

Australian Centre for Posttraumatic Mental Health. (2007) *Australian guidelines for the treatment of adults with acute stress disorder and posttraumatic stress disorder.* Melbourne: Author.

Beddington, J., Cooper, C. L., Field, J., Goswami, U., Huppert, F. A., Jenkins, R., *et al.* (2008). The mental wealth of nations. *Nature, 455*(7216), 1057–1060.

Bonne, O., Grillon, C., Vythilingam, M., Neumeister, A., & Charney, D. S. (2004). Adaptive and maladaptive psychobiological responses to severe psychological stress: Implications for the discovery of novel pharmacotherapy. *Neuroscience and Biobehavioral Reviews, 28*(1), 65–94.

Bremner, J. D., Southwick, S. M., Johnson, D. R., Yehuda, R., & Charney, D. S. (1993). Childhood physical abuse and combat-related posttraumatic stress disorder in Vietnam veterans. *American Journal of Psychiatry, 150*, 235–239.

Breslau, N. (1998). Epidemiology of trauma and posttraumatic stress disorder. In R. Yehuda (Ed.), *Psychological trauma* (pp. 1–29). Washington DC: American Psychiatric Press.

Breslau, N., Chilcoat, H. D., Kessler, R. C., & Davis, G. C. (1999). Previous exposure to trauma and PTSD effects of subsequent trauma: Results from the Detroit Area Survey of Trauma. *American Journal of Psychiatry, 156*(6), 902–907.

Bryant, R. A., & Harvey, A. G. (2002). Delayed-onset posttraumatic stress disorder: A prospective evaluation. *Australian and New Zealand Journal of Psychiatry, 36*(2), 205–209.

Bryant, R. A., O'Donnell, M. L., Creamer, M., McFarlane, A. C., Clark, C. R., & Silove, D. (2010). The psychiatric sequelae of traumatic injury. *American Journal of Psychiatry, 167*, 312–320.

Collip, D., Myin-Germeys, I., & Van Os, J. (2008). Does the concept of "sensitization" provide a plausible mechanism for the putative link between the environment and schizophrenia? *Schizophrenia Bulletin, 34*(2), 220–225.

Copeland, W. E., Keeler, G., Angold, A., & Costello, E. J. (2007). Traumatic events and posttraumatic stress in childhood. *Archives of General Psychiatry, 64*(5), 577–584.

Creamer, M., Burgess, P. M., & McFarlane, A. C. (2001). Post traumatic stress disorder: Findings from the Australian National Survey of Mental Health and Wellbeing. *Psychological Medicine, 31*(7), 1237–1247.

David, D., Woodward, C., Esquenazi, J., & Mellman, T. A. (2004). Comparison of comorbid physical illnesses among veterans with PTSD and veterans with alcohol dependence. *Psychiatric Services, 55*(1), 82–85.

de Boer, J. C., Lok, A., Van't Verlaat, E., Duivenvoorden, H. J., Bakker, A. B., & Smit, B. J. (2011). Work-related critical incidents in hospital-based health care providers and the risk of post-traumatic stress symptoms, anxiety, and depression: A meta-analysis. *Social Science and Medicine, 73*, 316–326.

Drevets, W. C., Price, J. L., & Furey, M. L. (2008). Brain structural and functional abnormalities in mood disorders: Implications for neurocircuitry models of depression. *Brain Structure and Function, 213*(1–2), 93–118.

Duffy, M., Gillespie, K., & Clark, D. M. (2007). Post-traumatic stress disorder in the context of terrorism and other civil conflict in Northern Ireland: Randomised controlled trial. *British Medical Journal, 2, 334*(7604), 1147.

Eriksen, H. R., & Ursin, H. (2004). Subjective health complaints, sensitization, and sustained cognitive activation (stress). *Journal of Psychosomatic Research, 56*(4), 445–448.

Etkin, A., & Wager, T. D. (2007). Functional neuroimaging of anxiety: A meta-analysis of emotional processing in PTSD, social anxiety disorder, and specific phobia. *American Journal of Psychiatry, 164*(10), 1476–1488.

Frueh, B. C., Grubaugh, A. L., Yaeger, D. E., & Magruder, K. M. (2009) Delayed-onset posttraumatic stress disorder among war veterans in primary care clinics. *British Journal of Psychiatry, 194*, 515–520.

Gillespie, K., Duffy, M., Hackmann, A., & Clark, D. M. (2002). Community based cognitive therapy in the treatment of posttraumatic stress disorder following the Omagh bomb. *Behaviour Research and Therapy, 40*(4), 345–357.

Gillock, K. L., Zayfert, C., Hegel, M. T., & Ferguson, R. J. (2005). Posttraumatic stress disorder in primary care: prevalence and relationships with physical symptoms and medical utilization. *General Hospital Psychiatry, 27*, 392–399.

Ginzburg, K., Ein-Dor, T., & Solomon, Z. (2009, September 16). Comorbidity of posttraumatic stress disorder, anxiety and depression: A 20-year longitudinal study of war veterans. *Journal of Affective Disorders*. Epub ahead of print.

Gormley, N., O'Leary, D., & Costello, F. (1999). First admissions for depression: is the 'no-treatment interval' a critical predictor of time to remission? *Journal of Affective Disorders, 54*(1–2), 49–54.

Gorwood, P., Corruble, E., Falissard, B., & Goodwin, G. M. (2008). Toxic effects of depression on brain function: Impairment of delayed recall and the cumulative length of depressive disorder in a large sample of depressed outpatients. *American Journal of Psychiatry, 165*(6), 731–739.

Hamilton, J. P., & Gotlib, I. H. (2008). Neural substrates of increased memory sensitivity for negative stimuli in major depression. *Biological Psychiatry, 63*(12), 1155–1162.

Heim, C., & Nemeroff, C. B. (2001). The role of childhood trauma in the neurobiology of mood and anxiety disorders: Preclinical and clinical studies. *Biological Psychiatry, 49*(12), 1023–1039.

Heim, C., Newport, D. J., Mletzko, T., Miller, A. H., & Nemeroff, C. B. (2008). The link between childhood trauma and depression: Insights from HPA axis studies in humans. *Psychoneuroendocrinology, 33*, 693–710.

Herrman, H., & Chopra, P. (2009). Quality of life and neurotic disorders in general healthcare. *Current Opinion in Psychiatry, 22*(1), 61–68.

Ikin, J. F., Sim, M. R., Creamer, M. C., Forbes, A. B., McKenzie, D. P., Kelsall, H. L., *et al.* (2004). War-related psychological stressors and risk of psychological disorders in Australian veterans of the 1991 Gulf War. *British Journal of Psychiatry, 185*(2), 116–126.

Johnson, A. M., Rose, K. M., Elder, G. H., Chambless, L. E., Kaufman, J. S., & Heiss, G. (2010) Military combat and risk of coronary heart disease and ischaemic stroke in aging men: The Atherosclerosis Risk in Communities (ARIC) study. *Annals of Epidemiology, 20*, 143–150.

Kagan, B. L., Leskin, G., Haas, B., Wilkins, J., & Foy, D. (1999). Elevated lipid levels in Vietnam veterans with chronic posttraumatic stress disorder. *Biological Psychiatry, 45*(3), 274–277.

Karlovic, D., Buljan, D., Martinac, M., & Marcinko, D. (2004). Serum lipid concentrations in Croatian veterans with post-traumatic stress disorder, post-traumatic stress disorder comorbid with major depressive disorder, or major depressive disorder. *Journal of Korean Medical Science, 19*(3), 431–436.

Kendler, K. S., Thornton, L. M., & Gardner, C. O. (2001). Genetic risk, number of previous depressive episodes, and stressful life events in predicting onset of major depression. *American Journal of Psychiatry, 158*(4), 582–586.

Kessler, R. C., Berglund, P., Demler, O., Jin, R., Merikangas, K. R., & Walters, E. E. (2005). Lifetime prevalence and age-of-onset distributions of DSM-IV disorders in the National Comorbidity Survey Replication. *Archives of General Psychiatry, 62*(6), 593–602.

Kessler, R. C., Sonnega, A., & Bromet, E. (1995). Posttraumatic stress disorder in the National Comorbidity Survey. *Archives of General Psychiatry, 52*, 1048–1060.

Kibler, J. L., Joshi, K., & Ma, M. (2009). Hypertension in relation to posttraumatic stress disorder and depression in the US National Comorbidity Survey. *Behavioral Medicine, 34* (4), 125–132.

Kraepelin, E. (1921). *Manic-depressive insanity and paranoia* (Trans. R. M. Barclay & ed. G. M. Robertson). Edinburgh: E & S Livingstone.

Kravitz, H. M., Bloom, R. W., & Fawcett, J. (2000). Recovery from a recurrent major depressive episode. *Depression and Anxiety, 12*(1), 40–43.

Kubzansky, L. D., & Koenen, K. C. (2009). Is posttraumatic stress disorder related to development of heart disease? An update. *Cleveland Clinic Journal of Medicine, 76* (Suppl. 2), S60–S65.

Kubzansky, L. D., Koenen, K. C., Spiro, A., III, Vokonas, P. S., & Sparrow, D. (2007). Prospective study of posttraumatic stress disorder symptoms and coronary heart disease in the Normative Aging Study. *Archives of General Psychiatry, 64*(1), 109–116.

Laugharne, J., Lillee, A., & Janca, A. (2010). Role of psychological trauma in the cause and treatment of anxiety and depressive disorders. *Current Opinion in Psychiatry, 23*(1), 25–29.

Leppänen, J. M. (2006). Emotional information processing in mood disorders: a review of behavioral and neuroimaging findings. *Current Opinion in Psychiatry, 19*(1), 34–39.

Maia, D. B., Marmar, C. R., Mendlowicz, M. V., Metzler, T., Nobrega, A., Peres, M. C., *et al.* (2008). Abnormal serum lipid profile in Brazilian police officers with post-traumatic stress disorder. *Journal of Affective Disorders, 107*(1–3), 259–263.

McEwen, B. S. (2003). Mood disorders and allostatic load. *Biological Psychiatry, 54*(3), 200–207.

McEwen, B. S., & Stellar, E. (1993). Stress and the individual: Mechanisms leading to disease. *Archives of Internal Medicine, 153*(18), 2093–2101.

McEwen, B. S., & Wingfield, J. C. (2003). The concept of allostasis in biology and biomedicine. *Hormones and Behavior, 43*(1), 2–15.

McFarlane, A. C. (2007). Stress-related musculoskeletal pain. *Best Practice and Research Clinical Rheumatology, 21*(3), 549–565.

McFarlane, A. C. (2009) The duration of deployment and sensitization to stress. *Psychiatric Annals, 2, 39*(2), 81–88.

McFarlane, A. C. (2010). The long-term costs of traumatic stress: Intertwined physical and psychological consequences. *World Psychiatry, 9*(1), 3–10.

McFarlane, A. C., & Bryant, R. A. (2007) Post-traumatic stress disorder in occupational settings: Anticipating and managing the risk. *Occupational Medicine (London), 57*(6), 404–410.

McGorry, P. D., Hickie, I. B., Yung, A. R., Pantelis, C., & Jackson, H. J. (2006). Clinical staging of psychiatric disorders: A heuristic framework for choosing earlier, safer and more effective interventions. *Australian and New Zealand Journal of Psychiatry, 40*(8), 616–622.

Mylle, J., & Maes, M. (2004). Partial posttraumatic stress disorder revisited. *Journal of Affective Disorders, 78*, 37–48.

O'Donnell, M. L., Elliott, P., Lau, W., & Creamer, M. (2007). PTSD symptom trajectories: From early to chronic response. *Behaviour Research and Therapy, 45*(3), 601–606.

O'Leary, D., Costello, F., Gormley, N., & Webb, M. (2000). Remission onset and relapse in depression. An 18-month prospective study of course for 100 first admission patients. *Journal of Affective Disorders, 57*(1–3), 159–171.

O'Toole, B. I., & Catts, S. V. (2008). Trauma, PTSD, and physical health: An epidemiological study of Australian Vietnam veterans. *Journal of Psychosomatic Research*, *64*(1), 33–40.

Page v. Smith (1996). A.C. 155.

Patten, S. B. (2008). Sensitization: The sine qua non of the depressive disorders? *Medical Hypotheses*, *71*(6), 872–875.

Perkonigg, A., Owashi, T., Stein, M. B., Kirschbaum, C., & Wittchen, H. U. (2009). Posttraumatic stress disorder and obesity: Evidence for a risk association. *American Journal of Preventive Medicine*, *36*(1), 1–8.

Post, R. M. (2002). Stressful life events and previous episodes in the etiology of major depression in women: An evaluation of the "kindling" hypothesis. Do the epilepsies, pain syndromes, and affective disorders share common kindling-like mechanisms? *Epilepsy Research*, *50*(1–2), 203–219.

Post, R. M., & Weiss, S. R. (1998). Sensitisation and kindling phenomena in mood, anxiety and obsessive compulsive disorders: The role serotonergic mechanisms in illness progression. *Biological Psychiatry*, *44*, 193–206.

Ramel, W., Goldin, P. R., Eyler, L. T., Brown, G. G., Gotlib, I. H., & McQuaid, J. R. (2007). Amygdala reactivity and mood-congruent memory in individuals at risk for depressive relapse. *Biological Psychiatry*, *61*(2), 231–239.

Repatriation Medical Authority. (2007). Statement of principles concerning depressive disorders. No. 17. Retrieved from http://www.rma.gov.au/SOP/07/017.pdf

Resick, P. A., Nishith, P., Weaver, T. L., Astin, M. C., & Feuer, C. A. (2007). A comparison of cognitive-processing therapy with prolonged exposure and a waiting condition for the treatment of chronic posttraumatic stress disorder in female rape victims. *Journal of Clinical Psychiatry*, *68*(5), 711–720.

Sareen, J., Cox, B. J., Afifi, T. O., Stein, M. B., Belik, S. L., Meadows, G., *et al.* (2007). Combat and peacekeeping operations in relation to prevalence of mental disorders and perceived need for mental health care: findings from a large representative sample of military personnel. *Archives of General Psychiatry*, *64*(7), 843–852.

Scott, J., Eccleston, D., & Boys, R. (1992). Can we predict the persistence of depression? *British Journal of Psychiatry*, *161*, 633–637.

Scott, K. M., McGee, M. A., Wells, J. E., & Oakley Browne, M. A. (2008). Obesity and mental disorders in the adult general population. *Journal of Psychosomatic Research*, *64*(1), 97–105.

Shephard, B. (2001). *A war of nerves: Soldiers and psychiatrists in the twentieth century*. Cambridge, MA: Harvard University Press.

Shin, L. M., & Liberzon, I. (2010). The neurocircuitry of fear, stress, and anxiety disorders. *Neuropsychopharmacology*, *35*(1), 169–191.

Smid, G. E., Mooren, T. T., van der Mast, R. C., Gersons, B. P., & Kleber, R. J. (2009). Delayed posttraumatic stress disorder: Systematic review, meta-analysis, and meta-regression analysis of prospective studies. *Journal of Clinical Psychiatry*, *70*(11), 1572–1582.

Solomon, Z., & Mikulincer, M. (2006). Trajectories of PTSD: A 20-year longitudinal study. *American Journal of Psychiatry*, *163*(4), 659–666.

Solter, V., Thaller, V., Karlovic, D., & Crnkovic, D. (2002). Elevated serum lipids in veterans with combat-related chronic posttraumatic stress disorder. *Croatian Medical Journal*, *43*(6), 685–689.

Stroud, C. B., Davila, J., & Moyer, A. (2008). The relationship between stress and depression in first onsets versus recurrences: A meta-analytic review. *Journal of Abnormal Psychology*, *117*(1), 206–213.

Summerfield, D. (2011). Metropolitan Police blues: Protracted sickness absence, ill health retirement, and the occupational psychiatrist. *British Medical Journal*, *342*, d2127.

Vieweg, W. V., Julius, D. A., Bates, J., Quinn, J. F., III, Fernandez, A., Hasnain, M., *et al.* (2007). Posttraumatic stress disorder as a risk factor for obesity among male military veterans. *Acta Psychiatrica Scandinavica, 116*(6), 483–487.

Violanti, J. M., Fekedulegn, D., Hartley, T. A., Andrew, M. E., Charles, L. E., Mnatsakanova, A., *et al.* (2006). Police trauma and cardiovascular disease: Association between PTSD symptoms and metabolic syndrome. *International Journal of Emergency Mental Health, 8*(4), 227–237.

Weisaeth, L. (1989). Importance of high response rates in traumatic stress research. *Acta Psychiatrica Scandinavia Supplementum, 355*, 131–137.

World Health Organization. (1992). *The ICD-10 classification of mental and behavioural disorders: Clinical descriptions and diagnostic guidelines.* Geneva: Author.

Wright, K. M., Huffman, A. H., Adler, A. B., & Castro, C. A. (2002). Psychological screening program overview. *Military Medicine, 167*(10), 853–861.

Yehuda, R., Kahana, B., Schmeidler, J., Southwick, S. M., Wilson, S., & Giller, E. L. (1995). Impact of cumulative lifetime trauma and recent stress on current posttraumatic stress disorder symptoms in holocaust survivors. *American Journal of Psychiatry, 152*, 1815–1818.

Zimmerman, M., & Mattia, J. I. (1999). Is posttraumatic stress disorder underdiagnosed in routine clinical settings? *Journal of Nervous and Mental Disease, 187*(7), 420–428.

9

The Challenge for Effective Interventions in a Violent Society: Boundaries and Crossovers between Workplace and Community

Merle Friedman and Gerrit van Wyk

History of South Africa: A Legacy of Violence and Oppression

South Africa is unique in many ways, having emerged from the tyranny of apartheid into a new democratic dispensation. However, the legacy of apartheid continues to contaminate the possibility of healthy psychosocial development.

Understanding the present is possible only by setting it within the context of the past. The promise of the "Rainbow Nation" was created by a magnificent negotiation that illustrated that the impossible was possible. There were many moments when the leadership on both sides was challenged by issues that appeared insurmountable. There were other moments when right-wing radicals in the population threatened civil war. All of this was taking place within the background of ongoing violence that in fact may be described as a low-grade civil war, with many people dying on a daily basis. The elections of 1994 were the culmination of the struggle against an oppressive system that has been characterized as a crime against humanity, and managed to become the celebration of a miracle. It was also recognized that there would not be an easy transition into the new South Africa, despite the magic of the election and the peacefulness and camaraderie that infused the election process.

Today, it is common knowledge that post-apartheid, democratic South Africa is struggling with a range of social dynamics that seem to facilitate violence (Seedat *et al.*, 2009). Nationmaster (2010), a website comparing cross-nation performance using accepted statistics, ranks South Africa as the nation with the highest per capita rate of rape, and the second per capita rates for assault and murder. A review of the South African Police Service annual statistics reveals that certain neighborhoods contribute disproportionately to the national figures (South African Police Service, 2008–2009).

International Handbook of Workplace Trauma Support, First Edition. Edited by Rick Hughes, Andrew Kinder, and Cary L. Cooper.

These neighborhoods, so-called *townships*, are typically the products of the structural violence of Apartheid forced removals, and more recently of an unprecedented explosion of urbanization, leaving communities marginalized, dislocated, disintegrated, and under-resourced. In these townships political violence has been replaced by domestic abuse, sexual violence, and other violent crime (Hamber, 2000; Kaminer *et al.*, 2008; Seedat *et al.*, 2009). There are high levels of gang activity (Shields *et al.*, 2008), and studies have revealed that up to 80% of adolescents have been exposed to community, school, and/or domestic violence (Cluver, Fincham, & Seedat, 2009; Seedat *et al.*, 2004; Suliman *et al.*, 2005). Nearly 50% of school-age youth in Cape Town have witnessed murder (Shields *et al.*, 2008), while 33% have witnessed a family member being hurt or killed (Seedat *et al.*, 2004) and 31% have witnessed domestic violence (Suliman, 2005).

The antecedants to the violent crime in the country illustrate the difference between workplace trauma in South Africa and in many other developed countries in the world.

The following statement summarizes a useful perspective explaining the sociohistorical roots of violence:

> It has been argued that the legacy of apartheid has bequeathed to South Africa a "culture of violence." This has been rooted in the notion that violence in South Africa has become normative rather than deviant and it has come to be regarded as an appropriate means of resolving social, political and even domestic conflict. (Simpson, 1994)

Two streams of understanding emerge from Simpson's statement:

1. Structural violence was a direct result of the violence of the system of apartheid itself. This system denied the majority of South Africans opportunities for work and dignity.
2. Because violence was used to enforce apartheid, violence became accepted as the normative recourse to resolve problems for many people.

A large study commissioned by the South African government (Bruce, 2009) lends support to this perspective. This study outlines a number of precursers which characterize the high levels of violent crime that speak to the legacy of our past, of apartheid and colonialism: the impact of apartheid on families and the education system, the institutionalization of racism and racial domination, access to firearms, and impunity for criminals in the black township areas where police were solely involved in the rigorous enforcement of apartheid laws.

Despite the noble and courageous Truth and Reconciliation Commission (TRC) and its international success, its impact has not been all positive. There is a feeling among a large section of society that it did not go far enough. Some argue that perpetrators of violence under the previous regime have gained most by being given amnesty from prosecution, while victims gained very little. The financial compensation promised by the TRC to victims has not been forthcoming or adequate, and this has added to a sense of disillusionment and anger. Careful analysis of events post-TRC has demonstrated that a single series of events occurring on a national scale, however moving, cannot right the wrongs of successive generations (Friedman, 2000).

A further aspect which would add to the propensity to engage in violence is previous engagement in military action and its aftermath, in the now well-documented effects on military veterans (Engdahl *et al.*, 2003; Kulka *et al.*, 1990). Combatants on both sides of the freedom struggle, represented in veterans' organizations, frequently feel dispossessed, unrecognized, and sidelined and are unskilled or unemployable in a nonmilitary environment. In many cases this, in addition to their traumatic exposure, may be the basis of violent crime.

In summary, then, there are many possible results that emerge from the troubled history of South Africa that may explain the development of a society with a greater propensity for violence.

South Africa: The Current Context

Looking at South Africa since our new democratic dispensation in 1994 presents another set of complicated realities and issues. Riaan Malan (2009) cogently describes this complexity:

> [T]here is no such thing as a true story here. The facts might be correct, but the truth they embody is always a lie to someone else... we live in a country where mutually annihilating truths coexist entirely amicably.... The curse is that you can never, ever get it quite right. (p. ix)

If one then goes further and examines the current socio-economic and political features of the country, crime and violence become even more logical as the probable outcome. The socio-economic features include poverty, xenophobia, the development of a drug culture, joblessness, a very high rate of unemployment, a disappointment with government regarding poor service delivery and the lack of jobs, the rise of the new ultra-rich, and corruption that is becoming endemic.

The current situation in South Africa, clearly a result of our past, also has its own set of sociopolitical challenges that are related to high levels of crime. There has been and continues to be the need for the entire population to transform itself. The transformation required for South Africa is multifold. It involves the transformation of minds and hearts, the move away from racism of any kind and towards inclusiveness. It involves the transformation of sport, education, and other primary services. It requires the transformation of the economic, institutional, and organizational structures, as well as the physical transformation of towns and cities where apartheid was entrenched. It requires the upliftment from poverty, where South Africa already had an exceptionally high unemployment rate (25.3% in 2010). It is considered to be economically one of the most unevenly distributed countries in the world with a Gini coefficient of 65.[1]

Living and Working in a Violent Society

Within the many under-resourced communities in South Africa, the experience of trauma differs markedly from trauma experiences in economically developed countries where the mainstream trauma theories and interventions have been developed. Not only have people grown up within the violent contexts of these communities, but also they remain trapped within an environment of continuing threat and renewed violence. Hobfoll (2010) and others have shown, furthermore, that impoverished individuals are more trauma vulnerable and that they are less able to protect themselves against loss, or to recover from their losses.

Furthermore, in South Africa it is evident that those employees most exposed to violence in the workplace, such as armed robbery, assaults, hold-ups, and shootings, as well as industrial accidents, tend to be less skilled, less educated, and socio-economically disadvantaged. They tend to occupy frontline positions, such as cashiers, security guards, drivers, fuel station attendants, and shop assistants, and they also tend to come from under-resourced communities where they are exposed to ongoing violence. They find themselves in situations where they not only are more exposed to potentially traumatic events, but also lack resources to assist them in recovering from trauma.

The term *cross-over* refers to situations in which the violence crosses over the boundaries between inside and outside the workplace. Examples of such cross-overs are given in the next section.

Violence in the workplace which can result in violence outside of the workplace

Case example #1. A man approached one of the tellers in a bank and asked her to help him and his gang ensure a successful bank robbery. When the teller refused he told her that he knew where she stayed, how many children she has, and that they would be the ones to suffer.

Case example #2. An armed robbery took place in a restaurant. The robbery took place after the staff had left, when the managers were closing up for the night. The staff were incensed at the possibility of losing their Christmas bonuses and suspected management of engineering the disappearance of the cash. The chef was too terrified to cook with his staff wielding knives in the kitchen behind him, whilst the staff were too afraid to talk to the management about what their colleagues were saying as they feared repercussions when at home where staff members live in their neighborhoods.

Case example #3. A young woman in her early 20s resides in one of the many "dormitory townships" built 50 years ago to accommodate forcibly removed communities far removed from the main business districts. She shares a rundown apartment consisting of two rooms with seven family members. Domestic, alcohol, and drug abuse are rife. She is the only employed member of her family. The rest depend on the income she generates. She is employed as a cashier in a large convenience store. Her employment is on a temporary basis and involves shift work and long hours. Like most of the other cashiers in her store, she too is disgruntled and unhappy because they feel unfairly treated. Some of them have been held up more than once at gunpoint by robbers at their checkout desks, and they are constantly tense, anxious, and vigilant for a new attack. She often reports sick with a variety of somatic complaints. The attitude of management is that she can resign if she is unhappy, because there are many others queuing up for a job.

When those who are most vulnerable to community violence also commonly find themselves in an unsafe work environment, it exacerbates their already high levels of stress. It may, however, also be possible for some to find sanctuary and nurturance at work. In fact, it is entirely possible that a safe and supportive work environment could improve the ability of employees to cope with violence in their living environment. The following example illustrates this complex set of circumstances:

Case example #4. A housekeeper lives in a self-built shack in an outlying area wracked by gang violence. She leaves for work in the morning to travel a long distance with unsafe public transport where muggings are common. She is employed in a privileged and private-security estate, a total contrast to her own living circumstances. She has been traveling the same route for 10 years and has been housekeeping for the same family for many years. In some ways she feels part of the family and secure in her employment. Her employers have often assisted her in accessing resources and services she would otherwise not have been able to utilize. For her, the work situation is an escape from the constant sense of threat she experiences in her own home, and she often feels she would rather not go home. She experiences her employers as warm and caring, and she feels protected, validated, and reassured in her workplace.

The understanding of such cross-overs, from inside to ouside and from outside to inside the workplace, clearly changes the approach that must be taken when dealing with workplace violence in a place such as South Africa.

The Place of Early Intervention in a Violent Society

While much for the evidence for effective intervention has come from studies that were carried out on patients who walked into clinics some months or even years after a single traumatic event, the demands from the South African situation are entirely different. Here, people are living in extremely threatening environments, where simply getting to work every day can be running a gauntlet of traveling through ganglands, negotiating transport problems, and avoiding the threat of personal attack. For many, the workplace actually serves as what began to be termed an "island of safety" in their lives. When this environment is also breached, it often becomes the last straw which undermines their resilience.

Clinical psychologists with experience and expertise in the field of traumatic stress assumed the responsibility of devising interventions that were appropriate for the people to whom they were delivered. This began in 1989. The models of intervention that were developed became the basis for much early intervention within South Africa going forward. The essence of these interventions was using evidence from research and adapting it to meet the current local circumstances, and so the interventions have changed and developed in line with relevant literature.

There was a very strong need for competent care of staff, not only for their sake but also to support management who had neither the understanding nor the skills to deal with such a situation. In addition, there were of course a number of constraints in the provision of trauma services in the workplace. These constraints defined the way in which these services could be delivered. Firstly, there were very few mental health workers in South Africa in general, in relation to the population (Rock and Hamber, 1994). Also, very few of these psychologists knew anything about trauma and its intricacies. Most of them were centered in areas that did not make for easy access to large organizations and their branches.

Arising from these constraints, two models of intervention emerged. The early and very successful model was to train selected volunteers from within the organizations to do the work, and to supervise and upgrade their training on a regular basis. The other model was to train professionals who would be available to be called out when and if needed.

Nonprofessional counselors were both enthusiastic and competent in the work that they did. They were carefully selected and trained, and proved most effective. Not only did they provide an important service and inestimable goodwill within their organizations (Ortlepp and Friedman, 2001), but also they gained a great deal of personal meaning and status from their success and caring (Ortlepp and Friedman, 2001). Staff members in organizations were much more trusting of internal counselors who understood the organizational dynamics, showed great care and compassion, and committed themsleves because they were concerned for the workers. A great deal of care and attention was taken to mitigate the development of compassion fatigue or negative resilience (Friedman and Higson-Smith, 2003) and ensure effective supervision and ongoing training.

Despite the obvious success of the internal trauma systems that ran successfully for about 20 years, there were a number of factors that led to their demise. There was a drop in robberies aimed at large chains such as banks, who could afford to train and maintain internal systems. Armed and violent robberies, although still frequent, moved to sites that were not as well protected. Also, at this time the demands in business had begun to change.

As South Africa had successfully re-entered the international markets, the dynamics of operating on the global stage took precedence over national needs. The market demands to produce continuous growth year on year and the resultant pressure on organizations to reduce costs when no other sources of financial growth were available resulted in large-scale retrenchment with additional time and demand pressures on remaining staff. The internal volunteers who provided the core of the teams doing trauma care, and who were most often senior in their organizations, were no longer able to give of their time. This period coincided with the advent of employee assistance programs (EAP) which promised to take almost all "people care" issues off the hands of managers and into the realm of professional carers. Although this approach also made good economic sense, the results of EAP programs have been mixed.[2]

Models of Trauma Support and Intervention

The preferred approach to trauma support in the South African work place involves two elements, a pre-trauma program of trauma preparedness training, and a program of post-trauma support and intervention. Firstly, this approach endeavors to create a supportive work environment that is aware of, and sensitive and responsive to, the impact of violence on employees, particularly in those industries that are vulnerable to violence. Secondly, it also introduces psychological expertise in the immediate aftermath of incidents of violent trauma in the workplace to facilitate trauma recovery and intervene where necessary.

Trauma preparedness programs

Training in awareness and response of frontline staff: how to react during a hold-up and what to expect following the trauma. Given the frequency of workplace violence and the predicatable nature of this occupational hazard, it makes sense to prepare staff on how to react. Our experience has been that those who have had adequate preparation do not feel as helpless during the event and are much more resilient in the aftermath.

Training of management in trauma support. Management plays a primary role in the sense of well-being of their staff. To create a smoothly functioning and resilient system, management, who may not be directly exposed to the traumatic event, are empowered with particular information: what to do and what not to do, how they can be most helpful, and what to expect from their staff and themselves in the aftermath of a traumatic event. In fact, in our experience the response of management in the aftermath of a violent incident is a major factor in the ability of employees to deal with the impact of trauma.

In addition, a protocol for action is developed with each organization. It is attached to a board together with all other emergency responses and contains details of what to do with staff, how to secure the site for the police, and who to contact in management and for trauma counseling.

An example of a very useful protocol has been the provision of food and drink for the staff after an event. The theoretical basis for this is to interfere in the newly formed fear association with the workplace, and create an alternative comfort association using food and drink. The actions of ordering, setting up a table, and preparing food keeps people busy and active whilst waiting for the police. Staff respond very well to this, and the psychosocial impacts are beneficial.

Post-trauma response: two models

Research in early trauma intervention has focused primarily on the prevention of post-traumatic stress disorder (PTSD), and in this regard there is strong evidence that some of the early intervention models are ineffectual and possibly harmful (Bisson *et al.*, 2009). Consequently, in accordance with the ethical dictum "Above all, do no harm," the response of professional organizations and institutions has been the recommendation that all early intervention should cease.

Subsequently, in the course of the 2000s, early trauma intervention has been replaced by a hands-off approach. The so-called watchful waiting approach has been widely accepted and well documented: no active intervention, assessment only to recognize early risk factors or signs of developing PTSD, followed by treatment where necessary. An example of this approach was demonstrated in the aftermath of the London Tube bombings and well documented by Brewin and coworkers (2008).

It is widely accepted that debriefing and counseling may in fact "pathologize" an otherwise normal recovery (Bisson, 2007). However, Bisson has pointed out that while trying to not do too much, we may be doing too little (2007). This may be particularly true in work-related trauma where one finds a number of issues and considerations in the forefront that are somewhat different from trauma in other contexts. The objective of early intervention in work situations is not only to prevent PTSD. It is usually expected to also assist in:

- Limiting the development of avoidance patterns, absenteeism, resistance to the workplace, and staff turnover.
- Preventing breakdown of work relationships caused by hostility and anger.
- Rapidly restoring a sense of safety and normality in work environment.
- Facilitating positive peer support.
- Assisting management in dealing with the reactions of traumatized staff.

Also, on the part of management, and of trauma victims, it is often expected that early trauma response should facilitate and accelerate the recovery from typical traumatic stress reactions, such as fearfulness, reliving and rumination, irritability, and inability to concentrate, all of which affect performance and effeciency.

While the following two models have not been tested with any quantitative rigor, they have been developed and based on available evidence which highlight factors that play a roll in the recovery, or nonrecovery, from trauma. They have emerged from many years of working with trauma victims in the South African context, where there is a diversity of languages and cultures, and some disillusionment with the limitations of applying Eurocentric and Western findings to an African context.

Trauma support

At the core of the trauma support model is social support. It has been demonstrated comprehensively that lack of social support, or the existence of social stress, is one of the most critical factors in the etiology of PTSD (Andrews, Brewin, & Rose, 2003; Brewin, Andrews, & Valentine, 2000; Cluver *et al.*, 2009; Galea *et al.*, 2008; Guay, 2006; Norris *et al.*, 2002). Good social support mediates resilience in the face of repeated trauma (Hobfoll, 2010). Furthermore, of all the proven risk factors for PTSD, social support may be a prime factor in which intervention may prove preventative. Unlike other causative factors, such as a previous history of trauma, the nature of the traumatic event, and a history of other psychiatric disorders, the factor of social support is malleable and dynamic. It can be actively utilized in the pre-trauma situation, as well as the post-trauma situation, for instance by involving and activating spouses and other family members, friends, peers, and authority figures to improve the resilience of trauma victims. Social support is therefore a useful tool in assisting victims of violent trauma in a violent society.

In this model of trauma support, the general guideline is the five principles of early intervention in mass trauma as defined by an international panel of experts (Hobfoll *et al.*, 2007):

- Promote a sense of safety.
- Promote calming.
- Promote a sense of self-efficacy and collective efficacy.
- Promote connectedness.
- Promote hope.

Litz and Maguen (2007) make a valid point that early intervention is not necessarily only about prevention of PTSD and other trauma-related disorders. They propose a number of other goals, no less valid or useful, to guide early trauma support and intervention:

- Helping people to decrease, manage, or eliminate functional incapacities caused by trauma.
- Promoting and training individuals or groups to use positive coping strategies and healthy behaviors.
- Encouraging and assisting individuals or groups to develop, nurture, and take advantage of comforting, positive, and caring social supports.
- Targeting complicated bereavement or traumatic grief for special attention.
- Helping individuals cope with subsequent threat.

A typical trauma support process unfolds in three stages.

- In Stage One, the first few hours or up to two days following the incident, the focus is on providing direction and guidance in practical ways, structuring solutions to immediate problems (most importantly, the need for safety and protection), assessing (and, if necessary, bolstering) individuals' levels of social support, and responding empathically to the range of distressing emotions felt by the victims.
- These activities continue in Stage Two, which occurs after a few days and may last for two weeks. In addition, selected individuals are offered counseling or psychotherapy.
- Finally, in Stage Three, 2–4 weeks after the incident, we follow up and, reassess whether further interventions are needed at the individual or organizational level, and encourage organizations and individuals to consolidate their capacity for support in a resilient manner.

The trauma support staff act first as consultants or managers in the aftermath to trauma, rather than as counselors. They do not expect to deal exclusively with victims and they give attention to other important role players including work supervisors, work colleagues, and family members. Furthermore, it is recognized that different victims require different forms of help, and that different forms of help are appropriate at different times for the same individual. There is no prescribed recipe or standard procedure that can address the great variety of needs and demands following traumatic events.

A variety of strategies for normalizing psychological responses to trauma are part of any early trauma support, as well as guidelines for self-management, such as "Structure your time – keep occupied," "Reach out to others, and ask for support – do not try to be 'strong,'" "Do not make any big life decisions for a while," and "Be careful of drugs, alcohol, and medication to make things easier." These accord with similar guidelines published after the September 11, 2001, attacks in New York City and Washington, DC (Academy of Cognitive Therapy, 2002), and after the London bombings in July 2005 (Traumatic Stress Clinic, 2005), and support a balance between carrying on with life constructively and expressing and sharing one's emotional distress with supportive friends or colleagues in a manner that promotes reflection and processing of the implications of what has happened. In the words of Gist and Woodall (1999, p. 287): People are resilient; friends are important; conversation helps; time is a great healer; and look out for others while you look out for yourself.

A great deal of attention is given to social support, by identifying individuals who are vulnerable to isolation, and strengthening existing social support within peer groups at work and at home. Attempts are made to identify distressed individuals who are at risk because of factors that are complicating or obstructing the normal recovery, and to address these complicating factors through individual counseling or psychotherapy, or interventions in the family or workplace.

Furthermore, measures are taken to discourage survivors from moving into a "sick" role. Commonly, trauma victims are automatically given sick leave and medication following trauma, but there is little evidence that rest alone is a factor in recovery. Although medication can play a helpful role (Foa *et al.*, 1999), its provision can also undermine the individual's sense of efficacy in being able to rely on his or her own resources. In our experience, sick leave seems to contribute to greater absenteeism and does not contribute to an individual's sense of efficacy.

Four-Leg model

This intervention is based on a combination of what has emerged as the needs of individuals in the immediate aftermath of such an event, the results of the evidence-based research in the literature and the context of the South African working and community environment.

The four-leg model is an individual- rather than group-based intervention. An individual intervention was adopted when it became evident that group interventions could be counterproductive when dealing with personal responses to a traumatic event. This is true particularly in relation to workplace trauma where sensitivity to career-limiting information is paramount.

The context of this model is one of providing:

- a safe place for the individual,
- in which he or she can reassume a sense of mastery,
- appraise the situation in a more positive way,
- understand what has happened and what might happen emotionally and physically, and
- know how to deal with going home and living over the next days and weeks.

Before any counseling takes place, the counselors make sure that they have as much detail of the event and staff and management relations as is possible. When the counselors arrive, the police have already secured the site and carried out their police and forensic work, and staff may have been able to have some refreshment while waiting around.

All staff are invited for counseling irrespective of their level of involvement, as it is often difficult to ascertain the level of distress and impact of the experience in all people. As the effectiveness of counselors has become widespread and accepted, almost all members of staff usually come forward for some counseling.

As the aim of the intervention is to help the individual reassert a sense of mastery in his or her life, personal control in the session is of primary importance. Therefore, the procedure for the session is described, giving reasons for each activity, and permission is gained for each step of the way. In securing a safe place for the person when living in a violent society, the person's home circumstances, other levels of threat, and level of social support are addressed and emphasized in the sessions.

The first two legs of the model consititute the emotional work. It is only once the emotional aspects have been addressed that people can effectively attend to the educational details.

Brewin has described two separate neuropsychological processes involved in recovery from traumatic experiences (Brewin and Holmes, 2003). In short, they are:

1. Bringing under control vivid experiencing of trauma through flashbacks and nightmares, reaction of fear helpless and horror.
2. Conscious re-appraising of the event and its impact.

These are represented in the two forms of therapy that have emerged from evidence based research as successful: prolonged exposure and cognitive therapy. Aspects of these two approaches are reflected in the first two legs of the model.

First Leg: Tell the Story. The first leg of this model is about helping the individual to tell their story in detail, often multiple times, and, at times, facilitated. The aim here is to garner the effects of prolonged exposure. Telling the story to a supportive and concerned listener, who is mindful of the dangers of retraumatization, helps the person gain a sense of the detail of his or her experience, and mastery about what has happened both internally and externally. Brewin explicates the value of this approach in his dual representation theory. The aim is to aid the construction of detailed verbal memories by re-encoding the situationally accessed memory (SAM) into verbally accessed memory (VAM) (Brewin and Holmes, 2003).

Second Leg: Reframe the Guilt. The second leg of the model is devoted to reframing the often ubiquitous guilt. Negative self-referential thoughts and images are formed during and immediately after a negative event, and these are associated with prolonging negative mood. To work effectively with cognitive-behavioral therapy (CBT), changes must be effected by altering the processes that maintain these thoughts and images. Brewin suggests that this improves mood immediately and in the longer term also reduces the likelihood of problematic representations being activated in future (Brewin, 2006).

In this model the guilt is reframed with the individual very soon after the event and repeated over the course of the program. The reframed positive memory representation becomes the most prominent in the competition for retrieval (Brewin, 2006), and it is reinforced over time. We have referred to this process as *cognitive construction*, rather than what is commonly termed *cognitive reconstruction*. This is because it occurs very early on – within a few hours of the event, whilst the memories are still malleable, and before the consolidation of memories and cognitions around the trauma.

Current findings point to early cognitive interventions that prevent the consolidation of traumatic memories (Holmes *et al.*, 2009). These finding clearly add support to the use of reframing the guilt as an early intervention strategy. Working successfully though this leg most often produces a greater sense of relief and empowerment than anything else in the procedure.

Once this is successfully done, the person is now ready for the educational aspects of the intervention.

Third Leg: Normalize the Symptoms. This is carried out with the aid of a pamphlet, which is then given to the person to discuss with his or her family. If necessary, family members are counseled to assist them with their own vicarious responses and to help them understand and be most supportive and helpful to their family members.

Helping people to understand what may happen to them over the next few days or weeks removes the anxiety and concern over the usual early responses to traumatic stress and short-circuits the spiral of anxiety associated with their responses. Furthermore, seeing it documented reinforces the normality of the responses and also removes the concern about trying to remember, which is all but impossible at this time.

Fourth Leg: Go for Mastery. This aspect is aimed at assuring adequate social support; countering the avoidance response and the continuance of helplessness, and regaining a sense of mastery.

Helping the person ensure adequate social support by explaining its importance is emphasized at this time, as is facilitating his or her problem solving to ensure it. Also dealt with here are how to tell family members, understand their responses, and get their support.

In order to counter the sense of helplessness that is usually part of the traumatogenic experience, this issue is explained. The value of feeling "back in control" is emphasized, and its opposite highlighted to ensure that people are not booked off from work unless they are under serious threat to their health.

Finally the importance of regaining a sense of mastery and control by addressing issues of food; exercise and daily activities are also discussed. People respond very well to suggestions about eating healthily for energy, and avoiding the kinds of food that will retard their full engagement in life. Many have taken up exercise at this time and reported great benefits many years after the event.

This process is repeated, with variations depending on individual and situational properties another three times over the course of four weeks. At each point, counselors are at pains to carefully evaluate whether people need referrals to professionals. A team of professionals is on standby.

At this point (or earlier if necessary), referrals to professionals can be made or the sessions terminated.

Many thousands of people have experienced counseling with this intervention. It has been successful in terms of ease of training, supervision, and practice, with very low levels of PTSD reported in the long term where there is uncomplicated trauma exposure.

Conclusion

In conclusion, our aim in writing this chapter has been to share our many years of common and disparate experiences in an environment that is both politically sensitive and violent. We have shared a fairly detailed explication of both contexts and interventions in order to enable others who may live in challenging parts of the world to take our experience and develop their own context-appropriate interventions. Space has limited the detail to which we might have gone, but the golden rule for us has been to ensure that our interventions have stayed true to the developing researched literature while at the same time paying nuanced attention to the diverse population that we serve. We are constantly challenged by living and working in an environment in which both the helpers/therapists and victims are continuously under threat of violence, and where the boundaries and cross-overs are sometimes quite blurred. It often takes great courage to continue to work under such circumstances, and it is often difficult to know whether one is a helper or also a victim. We therefore pay tribute to all those nonprofessional and professionals who, on a daily basis, risk their mental health as well as their physical selves to help those in need.

Notes

1. Developed by an Italian statistician Corrado Gini in the 1910s, the Gini coefficient is commonly used to indicate income inequality in a society. The Gini coefficient is a number which has a value between zero and one. As the value of the coefficient rises, the higher the degree of income inequality in a society becomes (http://www.scribd.com/doc/328232/United-Nations-Gini-Coefficient).
2. As there have been no studies that adequately evaluate these interventions, this comment reflects perspectives of informed participants within diverse organizations. Criticisms leveled at the interventions have been that most EAP's deliver "one-off" interventions in groups. These interventions are often based on the critical incident stress debriefing model, the efficacy of which is highly questionable.

References

Academy of Cognitive Therapy. (2002). *Coping with traumatic events.* Philadelphia: Academy of Cognitive Therapy.

Andrews, B., Brewin, C., & Rose, S. (2003). Gender, social support, and PTSD in victims of violent crime. *Journal of Traumatic Stress, 16*, 421–427.

Bisson, J. (2007). Evidence based early intervention. Paper presented at the European Conference on Traumatic Stress, Opatija, Croatia.

Bisson, J., McFarlane, A., Rose, S., Ruzek, J., & Watson, P. (2009). Psychological debriefing for adults. In E. Foa, T. Keane, M. Friedman, & J. Cohen (Eds.), *Effective treatments for PTSD: Practice guidelines from the ISTSS* (pp. 83–105). New York: Guilford Press.

Brewin, C. R. (2006). Understanding cognitive behaviour therapy: A retrieval competition account. *Behaviour Research and Therapy, 44*, 765–784.

Brewin, C., Andrews, B., & Valentine, J. (2000). Meta-analysis of risk factors for posttraumatic stress disorder in trauma-exposed adults. *Journal of Consulting and Clinical Psychology, 68*, 748–766.

Brewin, C. R., & Holmes, E. A. (2003). Psychological theories of posttraumatic stress disorder. *Clinical Psychology Review, 23*, 339–376.

Brewin, C., Scragg, P., Robertson, M., Thompson, M., D'Ardenne, P., & Ehlers, A. (2008). Promoting mental health following the London bombings: A screen and treat approach. *Journal of Traumatic Stress, 21*, 3–8.

Bruce, D. (2009). *Why does South Africa have such high rates of violent crime? Supplement to the final report of the study on the violent nature of crime in South Africa*. Produced by the CSVR for the Justice, Crime Prevention and Security (JCPS) cluster. Johannesburg: Centre for the Study of Violence and Reconciliation.

Cluver, L., Fincham, D. S., & Seedat, S. (2009). Posttraumatic stress in AIDS-orphaned children exposed to high levels of trauma: The protective role of perceived social support. *Journal of Traumatic Stress, 22*(2), 106–112.

Engdahl, B. E., De Silva, P. L., Solomon, Z., & Somasundaram, D. J. (2003). Former combatants. In B. L. Green, M. J. Friedman, J. T. V. M. De Jong, S. D. Solomon, T. M. Keane, T. Martin, ... E. Frey-Wouters (Eds.), *Trauma interventions in war and peace: prevention, practice, and policy* (pp. 271–289). New York: Kluwer Academic/Plenum.

Foa, E. B., Davidson, J. R., & Frances, A. (1999). The expert consensus guideline series: Treatment of posttraumatic stress disorder: The Expert Consensus Panel for PTSD. *Journal of Clinical Psychiatry, 60* (Suppl. 16), 1–75.

Friedman, M. (2000). The Truth and Reconciliation Commission in South Africa as an attempt to heal a traumatized society In A. Y. Shalev, R. Yehuda, & A. C. McFarlane (Eds.), *International handbook of human response to trauma* (pp. 399–411). New York: Kluwer Academic/Plenum.

Friedman, M., & Higson-Smith, C. (2003). Building psychological resilience: learning from the South African Police Service. In D. Paton, J. M. Violanti, & L. M. Smith (Eds.), *Promoting capabilities to manage posttraumatic stress: perspectives on resilience* (pp. 103–118). Springfield, IL: Charles C. Thomas.

Galea, S., Tracy, M., Norris, F., & Coffey, S. F. (2008). Financial and social circumstances and the incidence and course of PTSD in Mississippi during the first two years after hurricane Katrina. *Journal of Traumatic Stress, 21*(4), 357–368.

Gist, R. & Woodall, S. J. (1999). There are no simple solutions to complex problems: The rise and fall of critical incident stress debriefing as a response to occupational stress in the fire service. In R. Gist & B. Lubin (Eds.), *Response to disaster: Psychosocial, community, and ecological approaches* (pp. 211–235). Brunner Mazel.

Guay, S. B. (2006). Exploring the links between posttraumatic stress disorder and social support: Processes and potential research avenues. *Journal of Traumatic Stress, 19*, 327–338.

Hamber, B. (2000). "Have no doubt it is fear in the land": An exploration of the continuing cycles of violence in South Africa. *Southern African Journal of Child and Adolescent Mental Health, 12*(1), 5–17.

Hobfoll, S. E. (2010). The impact of war terrorism and occupation on self and society. Paper presented at the International Conference on Cultural Psychiatry, Amsterdam.

Hobfoll, S. E., Watson, P., Bell, C. C., Bryant, R. A., Brymer, M. J., Friedman, M. J., *et al.* (2007). Five essential elements of immediate and mid-term mass trauma intervention: Empirical evidence. *Psychiatry, 70*(4), 283–315. Retrieved from http://www.psych.org/

Resources/DisasterPsychiatry/ResourcesfromOtherOrganizationsAgencies/ScientificLiterature/FiveEssentialElementsofImmediate.aspx

Holmes, E. A., James, E. L., Coode-Bate, T., & Deeprose, C. (2009). Can playing the computer game "Tetris" reduce the build-up of flashbacks for trauma? A proposal from cognitive science. *PLoS ONE*, *4*(1), e4153.

Kaminer, D., Grimsrud, A., Myer, L., Stein, D. J. & Williams, D. R. (2008). Risk for post-traumatic stress disorder associated with different forms of interpersonal violence in South Africa. *Social Science and Medicine*, *67*, 1589–1595.

Kulka, R. A., Schlenger, W. E., Fairbank, J. A., Hough, R. L., Jordan, B. K., Marmar, C. R., *et al.* (1990). In R. A. Kulka, W. E. Schlenger, J. A. Fairbank, R. L. Hough, B. K. Jordan, C. R. Marmar, *et al.* (Eds.), *Trauma and the Vietnam War generation: Report of findings from the National Vietnam Veterans Readustment Study* (pp. 265–275). New York: Brunner/Mazel.

Litz, B., & Maguen, S. (2007). Early intervention for trauma. In M. Friedman, T. Keane, & P. Resick (Eds.), *Handbook of PTSD: Science and practice* (pp. 206–229). New York: Guilford Press.

Malan, R. (2009). *Resident alien.* Johannesburg: Jonathan Ball.

Nationmaster. (2010). Crime statistics: Murders (per capita) (most recent) by country. Retrieved from http://www.nationmaster.com/graph/cri_mur_percap-crime-murders-per-capita

Norris, F., Friedman, M., Watson, P., Byrne, C., Diaz, E., & Kaniasty, K. (2002). 60,000 disaster victims speak. Part 1. An empirical review of the empirical literature, 1981–2001. *Psychiatry*, *65*, 207–239.

Ortlepp, K., & Friedman, M. (2001). The relationship between sense of coherence and indicators of secondary traumatic stress in non-professional trauma counsellors. *South African Journal of Psychology*, *31*(2), 38–45.

Rock, B., & Hamber, B. (1994). *Psychology in a future South Africa: The need for a National Psychology Development Programme.* Paper commissioned by the Professional Board of Psychology of the South African Medical and Dental Council. Zurich: CSVR Library. Retrieved from http://www.csvr.org.za/wits/papers/papbrbh.htm

Seedat, M., Van Niekerk, A., Jewkes, R., Suffla, S., & Ratele, K. (2009). Violence and injuries in South Africa: Prioritising an agenda for prevention. *The Lancet*, *374*(9694), 1011–1022.

Seedat, S., Nyamai, C., Njenga, F., Vythilingum, B., & Stein, D. J. (2004). Trauma exposure and post-traumatic stress symptoms in urban African schools. *British Journal of Psychiatry*, *184* (2), 169–175.

Shields, N., Nadasan, K., & Pierce, L. (2008). The effects of community violence on children in Cape Town, South Africa. *Child Abuse and Neglect*, *32*, 589–601.

Simpson, G. (1994, March 16–17). Business and Endemic Violence in South Africa: Surviving the disaster or managing the transition? Paper presented at the 6th South African Disaster Recovery Conference, Volkswagen Conference Centre, Midrand, South Africa.

South African Police Service. (2008–2009). *Strategic framework – annual report 2008/9.* Retrieved from http://www.saps.gov.za/saps_profile/strategic_framework/annual_report/2008_2009/2_crime_situation_sa.pdf

Suliman, S., Kaminer, D., Seedat, S., & Stein, D. J. (2005). Assessing post-traumatic stress disorder in South African adolescents: Using the child and adolescent trauma survey (CATS) as a screening tool. *Annals of General Psychiatry*, 4.

Traumatic Stress Clinic. (2005). *Coping with a major incident.* London: Camden and Islington Mental Health and Social Care Trust.

10

Adversity: Reconceptualizing the Post-trauma Response

Kevin Friery

Introduction

In its infancy, post-traumatic stress was largely perceived to be confined to people in the immediate vicinity of an event, who had been involved firsthand, and supportive and remedial interventions were similarly limited. In the twenty-first century, however, the world has changed so much that we may need to take an entirely different view and rethink the whole psychology and epidemiology of emotional trauma and the way we prepare and respond to it.

In the late eighteenth century in Great Britain, the introduction of mail coaches speeded up the rate at which information could travel between centers of population. A document could now travel from London to Edinburgh (a distance of some 400 miles) in 43 hours. A century later the French introduced optical communication by building towers with wooden signal boards – semaphore – and, under perfect lighting conditions, news could travel at 100 miles in 3 minutes. At 05:46 UTC on March 11, 2011, in Japan a massive earthquake struck; within seconds the news of this event, along with images and live commentary, was being communicated globally across distances of thousands of miles so that by 05:55 live voice reporting was available on most news stations. In the past 250 years, mass communication has changed out of all recognition and we now live in a world in which it is extremely difficult to restrict the news or to limit the speed at which it spreads.

Coupled with the speed with which news of events can now be communicated is the manner in which they are reported. One of the outstanding examples of this was the BBC News report from Michael Buerk (1984) on October 23, "Dawn. And as the sun breaks through the piercing chill of night on the plain outside Korem it lights up a biblical famine – now, in the 20th century. This place, say workers here, is the closest thing to hell on earth." In written form the words convey a story but, with the accompanying film footage, the story became a highly emotional multimedia experience introduced into people's living rooms in mid-evening, changing not only their state of knowledge about the famine in Ethiopia but also their emotional engagement

International Handbook of Workplace Trauma Support, First Edition. Edited by Rick Hughes, Andrew Kinder, and Cary L. Cooper.

with it. Film of dying infants, of corpses, and of humans in devastation had a major impact on the psyche of people who, before the advent of this type of reporting, might have retained more emotional detachment. Now those who remember the report might only hear the opening words, or read them, and re-experience their entire emotional response from decades ago.

The dividing line in mass communication between what is allowable and what is not has been challenged and become blurred; whereas in some countries news reporters are allowed to broadcast everything they see and hear, sparing no detail, other countries demand a more circumspect approach and limit the use of visual imagery and even of language. It means that people are exposed to far more opportunities to witness and become immersed in disturbing events elsewhere; perhaps war reporting is one of the most obvious examples of this issue.

This has also signaled a major change in social psychology.

There has always been philosophical and psychological discussion about the relationship between the individual and the wider social group; the relationship between "I" and "Thou," and between "Me" and "We," is explored not only in most psychotherapies but also in many education systems. The socialization of the individual can be seen as part of the core rationale for school-based education as opposed to private tuition. In this arena, the role of the media is significant. Mass communication through existing media had helped to create a society in which knowledge and information were shared and through which people might also be exhorted to act in communion. The individual, though, could still remain aloof and isolated, choosing to ignore communication or retain an intellectual distance between self and society. The growth of multimedia, with information flowing using different senses simultaneously, increased the feeling of the individual connected to a larger society and thereby increased the emotional connectivity. It has exponentially increased the rate at which information flows inward to the individual and it has led to the current situation in which someone watching a newscast in the United Kingdom can feel viscerally connected to events in Japan, for instance. It means that something happening thousands of miles away, to people whose existence had previously been unknown to us apart from in very general terms, can create within us a complete range of emotional states. Almost unacknowledged, we are now witnessing an effect that is almost the stuff of science fiction: the teleportation of emotion. In scientific communities the belief is that data can be teleported but life cannot; we are now becoming aware of the instantaneous global transmission of emotion.

In terms of post-traumatic stress response, we can see the effect of this sensitization to emotions in some of the requests for help.

This was borne out in London after the shooting of Jean Charles de Menezes in the Stockwell tube station in 2005 (BBC News, 2005). A number of organizations reached out for trauma support for their staff on that afternoon, even though they were not in the immediate vicinity or connected with the people or events. The very fact of being in the same city, hearing and seeing news reports, meant some people found the emotional load difficult to bear and feared that they would not be able to recover without help. The author was approached by one organization who said they had 500 people all suffering from traumatic shock as a result of the incident.

Gleick (1999), who wrote extensively about chaos theory, talks about the way in which communication has accelerated but we also now need to acknowledge that emotional transmission has kept pace with physical events. An emotional experience in China – adapting his "butterfly" quote – can have an emotional impact in the United States, but far quicker than ever before.

Given this state of affairs – and experience has taught us that such changes will not be reversed – there is a strong argument for revisiting our approach to trauma. We can no longer isolate traumatic events and believe that we have contained the psychological impact to the immediate vicinity. Nor can we follow the trail of emotions and connect readily with everyone who may be affected, offering appropriate psychological support, reassurance, and/or treatment. The simple truth is that modern technological advances mean that it has become harder, if not impossible, to meet the need by reacting to events. That is not to say we shouldn't act; the fact that we can't do everything doesn't mean we shouldn't do something. It introduces, though, the notion that we ought to do something else, something more fundamental.

It is this insight that directs us toward an engagement with adversity.

Stress

Although this chapter is not specifically about stress, it is important to consider it before we expand on the theory of adversity, and to touch on some key characteristics. Stress has been defined in many ways but fundamentally there is a common theme: it is a combined physical and mental response to a situation in which the demands exceed the resources, or, more accurately, when the perceived demands exceed the perceived resources. Taken at that level it is not hard to imagine that a sudden unexpected event may overwhelm people's belief in their ability to deal with the psychological pressure because it is something that people experience when they believe that they can't cope. Much has been written about this, not only about the actual definitions to be used but also about the underlying mechanisms that are associated with stress and that can create or diffuse it. In many ways, post-trauma stress is no different, except that it is linked to specific experiences and events; it is, in old psychiatric terms, an exogenous emotional reaction rather than an endogenous one. Psychiatrically it is more likely to be seen as a disorder only if, after some time has passed, the individual isn't returning to stability or if the stress leads to other psychiatric problems such as depression, anxiety, or psychosis. There is a strong body of opinion that sees the "stress" response to a traumatic incident as entirely normal, an essential part of survival. Stress in itself isn't the problem so much as the ability to bounce back. Purchase a packet of balloons and you have a number of items that have the potential to be functional so long as you inflate them; it is then, under pressure, that they perform their normal function. It is when you continually pump air into the balloon, without sometimes reducing the pressure, that the thing will burst. As an analogy it is not perfect, but it is an illustration of how people work; we are best when under pressure, we are dysfunctional when we have no pressure, but we can be broken if we have too much. This isn't a culturally specific issue; it is part of the human condition, part of our basic psychology. What does differ, between individuals and between cultures, is the way we deal with our stress; this is something that will be discussed in this chapter.

Although stress can materialize in any area of life, a great deal of work has been done to examine it in a work context and to understand exactly how it is incubated and how it can be prevented. The Health and Safety Executive (HSE) in the United Kingdom have published on this topic (2005), setting out guidelines to help employers and employees alike address the components of work that may lead to stress. In fact, HSE are increasingly referring to the standards not as "stress management" but as "good management," acknowledging that the behaviors and attributes that help to minimize problems are actually universals, rather than purely about stress. Whilst there is this excellent wealth of research, it is also true that many unusual, psychologically overwhelming events can happen at work, beyond the normal operational activity and outside of the scope of normal stress management interventions. Stress, trauma, and the workplace are interwoven but it is not always easy to predict how or when this will manifest itself. Part of the problem lies in the very fact that traumatic incidents are relatively unusual, and thus it can be hard to acquire the skill needed to cope purely by exposure and experience. A person for instance who works in a mortuary is probably quite accustomed to seeing dead bodies at work, whereas someone who works in a swimming pool is not. On the very rare occasions when a body is found in a swimming pool, we would anticipate that the associated trauma would be accompanied by a high degree of stress and have the potential to severely undermine an employee. A dead body at work may, or may not, be traumatic. On the other hand, being accustomed to dead bodies at work doesn't guarantee they will not be traumatic; in December 2010 at Karolinska Institute in Stockholm, medical students attending their first autopsy were distressed to discover that the corpse they were presented with was their former professor. Sometimes even the normal becomes abnormal because of particular circumstances. Likewise, there have been numerous examples of emergency responders becoming traumatized by an event even though they have witnessed many others; the suggestion is that, rather like the balloon analogy, there can be a cumulative effect of exposure to trauma that may ultimately lead to collapse.

Acknowledging that stress and trauma are potential hazards in the workplace puts the employer on notice; it puts a responsibility on the employer to take all reasonable steps to anticipate, to prevent, and to respond appropriately to them. This isn't merely an issue of good corporate responsibility; it is backed by legislation and by implied legal responsibilities.

One important message for both employers and employees comes from the UK Court of Appeal (*Hatton v. Sutherland*, 2002), who held that no job is inherently any more stressful than another. That is not to say that work might not be stressful, but it acknowledges that it is not the job so much as the individual reactions of employees that determine stress at work. Similarly, in another case (*Dickins v. 02*, 2008), the courts found that the employer was liable for the stress experienced by the employee even though there was nothing apparently inherently stressful in the task. What, then, causes one person to experience an event as stressful whilst another is relatively unmoved? When looking at the workplace and considering a strategy on stress and trauma, how can we identify those who are at risk? These are important questions for employers that lead to further discussion on suitable strategies for prevention and response.

Resilience

It has been said, perhaps simplistically, that the best way to deal with workplace traumatic stress is to not have any. At an organizational level this means building stress out of the system, creating a workplace that is psychologically healthy and stress free. As obvious as it sounds for the organization, there is also good evidence that there are steps that an individual can take to reduce the propensity for stress by increasing and improving resilience. The individual, like the organization, can build it out of the system. Most employers advocate a resilient and well workforce who can engage with the rigors of daily workplace life and thrive, although fewer organizations particularly know how to achieve this. Too many individuals, however, believe that resilience is a personality characteristic that one is born with or without; it can sometimes be difficult for people to accept that things can be different and they can acquire new ways of being, yet the evidence is strong to support this contention.

Although there is debate about a precise definition of "resilience," it is largely seen as the process that allows an individual to bounce back in the face of adversity. In fact, it is really something far more complex; resilience is a roadmap, an internalized version of the world that enables an individual to map out paths to the resources that help and sustain, that support cultural needs, and that promote wellness. Resilience takes into account the individual and cultural differences we mentioned earlier and transforms them into successful strategies for navigating a complex world and enhancing survival and subjective well-being. It is the fact that it is a roadmap, an acquired knowledge and ability rather than a trait, that demonstrates that resilience is not a personality characteristic so much as a behavioral process that can be taught and learned. And therein lies one of the keys to mastery.

It is not a chance occurrence that, globally, education takes place when we are young. Although learning is life-long, we invest significant time and energy with our children to provide them with the building blocks for future stability. Research into resilience shows that it is exactly at this age that the best outcomes are achieved. Resilience is a skill that is best acquired young, because it is more readily internalized, it helps to prevent some early-years traumas, and it gives a developing child a toolkit when faced with challenges.

Research can show some intriguing aspects to resilience. Wells (2010) looked at underlying resilience in older (over-65) people coping with life changes as part of the aging process. She identified four factors that were closely associated with enhanced resilience. Good mental health, good physical health, and strong social networks were all positively correlated. Perhaps counter-intuitively, so too was having a lower family income. Importantly, the link between physical health and resilience was seen to be positive but weak; other research suggests that reduced physical health does not automatically lead to reduced resilience. This research is, however, simply a backward glance at resilience that people have acquired by the time they are tested. Far more relevant for the workplace, and for the prevention or reduction of psychological trauma and stress, are the components and techniques that are part of the acquisition process for the individual.

The ability to develop narrative is one component of resilience. One of the features of traumatic events is that they are outside of the normal experience of the individual and often seem to defy meaning. A narrative approach looks at the situation and constructs a mental story to define and explain the event, to contextualize it, and to enable the individual to respond effectively. Cryulnik (2007) talks about narrative as a way of clearing the fog of trauma, of preventing mountains and molehills from being confused. The application of narrative is a post-hoc activity, implemented after a traumatic event, but the acquisition of the skills of narrative happens prior to exposure as a learning experience in its own right, something that is internalized and becomes part of the resilience roadmap.

Another important component is self-esteem. Seligman (1996) positions self-esteem as feeling good, as opposed to doing well. It emerges from mastery of tasks, from working successfully and winning. Taking this view, it is not self-esteem that is taught but success. Again, this is something that is internalized in readiness for its use when needed; it is certainly not something one can scramble to acquire in the face of adversity, but overcoming obstacles can enhance it for the next time it is needed.

Informed optimism – as opposed to naïve optimism – is another feature of resilience. This forms part of a positive attitude set that looks for good solutions and that believes that change is possible. Naïve optimism will jump into a fire because everything is sure to be fine, whereas informed optimism will look at a fire and be certain there is a way to work out how not to get burned. It is closely coupled with the absence of features like scapegoating, blame, and personal attacks, behaviors that are psychologically more about the survival of me rather than you. Some of these attitudes are clearly better developed from a young age, but they are also accessible to adults who might earlier have developed more aggressive, hostile ways of facing difficulties. Transformational learning can take place at any age, and there is a strong probability that this adds to self-esteem as the individual learns and masters new skills in resolution.

Some observers say that another element is having meaning beyond oneself. This is a view that is sometimes seized upon by people who believe religious faith is essential, but in fact it is simpler than that. It requires the individual to have a sense of self in relationship to an outer world; to many this would be called relatedness coupled with a moral compass. Charney and Nemeroff (2004) highlighted the ability to maintain meaningful confidential relationships with friends, family, and colleagues. This is also something that is acknowledged in suicide risk management – the better the individual's networks, the lower the suicide risk, whilst conversely individuals with no social connections are perceived to be at great risk. Connectedness and socialization are significant features in the workplace, where the nature of the relationships between people has a major impact on their collaborative effectiveness and their ability to work together to achieve the business mission and to overcome obstacles, including potentially traumatic events. Seeing oneself as part of a connected whole that also connects to other people comprises an attitude and a philosophy that, once learned and internalized, create the resilient survival culture that is central to minimizing psychological distress. It is also central to the management of adversity.

Organization as Organism

Connectedness to other people takes on a completely different aspect when the employer is brought into the equation. Although many people may see their organization as a collection of diverse people doing different things, some psychologists and social scientists take a different view; they see the organization as a living entity, an organism in its own right. It exists, it grows and develops, it interacts, it learns and changes, it has aspirations and goals, it likes and dislikes, and it shows many of the characteristics of a living being. Bakan (2004) provided a thought-provoking analysis of the way in which corporations develop an existence that is independent from that of any of the individuals within it, that survives even when the people within it come and go. Although his focus was on shareholder entities and accountability (or lack of it) in large organizations, the analogy is equally valid with all businesses, from sole traders with a handful of staff through to large employers with many thousands. Maintaining a perspective of the organization as organism allows us to consider completely different analyses of organizational development and also creates two distinct types of support for business, making a clear distinction between business analysts and organizational consultants. The argument would be that the latter could almost be defined as organizational psychotherapists.

In this context it is possible to see that an organization can suffer some similar experiences to people. It can be hurt; it can be happy; it can relate. It can be evaluated by people in the same way that we evaluate other people; we like it, we are drawn to it, we are antagonistic toward it. It nurtures us or it lets us down. It has a reputation. It can be traumatized.

Understanding this analysis allows us to take one more step. If an organization can be hurt, maybe it can also be protected from hurt or maybe it can be helped to recover. Perhaps some of the principles of adversity and resilience can be applied to the organization itself, as well as reactive post-traumatic stress interventions.

Personal Adversity

When does a molehill become a mountain, a drama become a crisis? A healthy approach to adversity requires us to understand the dividing line between shock and trauma, between the manageable and the unmanageable.

One of the problems that besets anybody working with human emotions is that the available vocabulary is large but the language commonly used is quite limited. We tend to polarize our emotions through language; people are "livid" rather than a bit upset, they are "depressed" rather than sad, they are "ecstatic" rather than quite happy, and they use extreme words for less extreme emotions. This is despite research by Schrauf and Sanchez (2004) that showed we have, across cultures, a wide range of functional emotion words at our disposal, with 50% relating to negative emotions, 30% to positive, and 20% to neutral. Psycholinguistically this supports the idea that we need more complex language to define problematic situations, because these are more likely to need a protective response from us. It can lead people to describe themselves and others as "traumatized" when in fact they are a bit shaken. This isn't to deny the importance of helpful interventions, or of relevant diagnostics, nor is it a denial that some people really are traumatized by experiences, but there is a risk that we use language inappropriately and thus deny the existence of a whole swathe of human emotions and reactions that fall between polar extremes. A dictionary definition (Dictionary.com, 2011) of adversity gives synonyms of "catastrophe, disaster, trouble, misery"; this highlights the way in which we use language to cover a wide range of meaning and it goes to the heart of the proposition that, in reality, adversity is not necessarily a catastrophe or a disaster but failing to deal with it effectively might lead to one of these states.

At an individual psychological level this is significant. If resilience is a roadmap, it is important that we don't forget we have a compass. A resilient person will swing into action, or might perhaps need some prompting, some reminding that he or she has the skills to do so. That person may use polarized emotional language in a crisis but needs encouragement to rein it in and identify the true feeling and the true capability. A person who is likely to become traumatized almost certainly doesn't have the building blocks of resilience in place, so no amount of prompting can produce positive action.

Seery, Holman, and Silver (2010) conducted a longitudinal study across a number of years, looking at people who had experienced adversity in their lives. They compared those with moderate amounts of adversity, those who had encountered frequent high adversity, and those who had experienced little or none. They found that a moderate amount of adversity fostered subsequent resilience, enhanced mental health and well-being, fewer post-traumatic stress symptoms, and higher life satisfaction. They were the people who were least likely to be affected by new adverse events and appeared to be functioning better than people who had experienced no adversity and those who had experienced repeated high levels. This finding advances the notion that, although exposure to adverse life events typically predicts subsequent negative effects on mental health and well-being, such that more adversity predicts worse outcomes, there is a point at which adversity has positive outcomes and leads to improved resilience and coping capacity.

Learning and applying the skills of resilience enable us to cope with moderate amounts of adversity whilst at the same time providing the framework for a better adjusted later life. It also allows us to recover more rapidly from severe disturbances.

Returning to the discussion of the speed and nature of mass communication, it becomes more obvious that building resilience is the optimal solution to the global sharing of emotion. Trauma and shock can be localized or it can have a global reach, but in either case the people who cope best will be those who have learned to overcome adversity by using their resilience skills. As a pre-emptive developmental activity, resilience creation is not only more effective and efficient than reactive post-trauma interventions but also creates at an individual level the ability to generalize learning and experience across a number of settings.

Looking to our employee in the mortuary, the way that person develops the resilience to deal with what might otherwise be severe adversity is through a combination of training and exposure. We should expect the employer to conduct a comprehensive induction and on-boarding process so that employees understand the nature of the work, the hazards that they might be exposed to, and the potential for seeing unpleasant things. This helps to build resilience whilst also evoking within the employee previous life experiences and training that help to add to the ability to cope. Gradual exposure to the hazards allows the individual to acclimatize, to rapidly acquire coping skills, and to identify areas that require further focus in order to promote confidence and competence.

Organizational Adversity

In the space of two years, one organization (Right Corecare, personal correspondence, 2011) reduced the number of workplace incidents that were classified as either "trauma" or "critical" from over 250 per annum to 6. This was achieved by overturning the prevailing traditional reactive approach and moving toward a resilience-focused system that had two key components. Managers were trained first in appropriate defusing skills coupled with clear guidance about roles and responsibilities, and the focus of support post-event was geared strongly toward supporting managers to be effective rather than sending an external provider on site, using brief telephone conversations to re-engage them with their training, their knowledge base, and their resilience. The outcome was that managers managing effectively greatly reduced the extent to which staff felt traumatized and, just as importantly, the organization returned to stability and full productivity more quickly. Part of this approach involved the development of improved induction for customer-facing staff; working in this arena exposes staff to abuse, criticism, and adversity in a number of ways. Although unjustified and unfair, it is the experience of many people who work with the public. Instead of responding to this every time it happened, declaring each incident of abuse a "critical incident," the organization worked more closely to help staff understand that, although the organization was trying hard to eliminate the problem, they may have to deal with these incidents sometimes. It was better to be forewarned and also to have clear information about where the reporting lines were, where the support would come from, and what the roles of colleagues and managers were; this enabled the organization to shift towards an adversity model rather than a trauma model.

Part of the responsibility of an organization under their duty of care to the workforce is to conduct regular health and safety assessments and to implement any indicated changes. For some businesses this is confined to a check for "trips and spills" hazards, whereas an appropriate and compliant assessment also features an analysis of psychosocial risks in the workplace. Understanding these allows organizations to work to support resilience in staff and to identify and head off some of the potential risks.

Organizational adversity isn't confined to the emotional. Organizations of all sizes have the potential to experience adverse circumstances; these may be in the shape of the events that typically lead to a trauma-focused intervention or may also be other types of situations that are difficult but not impossible for the organization to handle. Thinking again of the concept of the organization as organism, many of the sources of adversity to individuals also apply to organizations. The following case study, which is mostly true but has been altered to preserve confidentiality, illustrates this well:

> An organization discovered that two of its staff had committed a major fraud. Following an internal audit, the police were called and the pair were arrested and taken away. After a prolonged investigation, the organization was advised in February that a court case would commence in May. Recognizing the potential distress this would cause staff, particularly since there were angry investors who had lost large sums of money and there was a

> possibility that the company would not survive, they turned to an external employee assistance provider (EAP) to ask that they have trauma staff available in May, to support their people as the need arose.

Up to that point, this looks like a standard situation, with the company believing that they were being proactive in arranging the availability of trauma helpers so far in advance. In fact, further discussions helped to complete a different picture:

> The situation had placed the organization in jeopardy in a number of ways. The court case would bring a large amount of adverse public comment, having a potentially disastrous effect on the company's reputation. Although the two were cooperating with investigators, there was a chance that in court the company would be held to have failed in its regulatory responsibilities. If, in addition to the money lost through the fraud, the company was fined substantially, there would inevitably be job losses and site closures, and the entire enterprise could be at risk. The media had already started to show great interest in the case, and staff had received phone calls asking for information; some people felt they had been followed home, and others had been stopped in the street by customers and verbally abused.

The response of the EAP was to persuade the organization to move their focus from the forthcoming court case in May to the present moment. Rather than meet the trauma when it arose, there was a three-month period in which the problems could be anticipated and analyzed and helpful solutions implemented in advance. This involved working with other partners to look at public relations including reputation management and media handling, communicating clearly with shareholders and investors, and setting up a collaborative group to work together toward a positive outcome. It involved identifying individuals who may be at emotional risk and providing them with practical help in using resilience skills to deal with pressure situations. Groups of staff worked together to learn how to give clear, consistent, and approved messages to any inquirers. The HR team worked with a business consultant to understand how the organization was responding and behaving and what could be done to shift the focus toward a response to adversity rather than impending doom. There were other interventions that built on the partnership between the organization and the EAP, identifying the nature of the adversity and producing a response that mirrored the concept of informed optimism. The outcome was that the organization weathered the storm, hostile critics were brought into the fold, and the company was able to demonstrate in court that it had, in fact, taken all the precautions appropriate to prevent fraud and that the situation could not have been avoided by any known security system. Although, when the court case hit the press, there was some vociferous criticism and some harsh words, the time spent helping the organization to approach the situation in a resilient manner helped them absorb this without being traumatized.

In managing adversity, organizations often need to think of people, cost, governance, legislation, and reputation. A solution that could help them address all of this would be of more benefit than a more standard critical incident or post-trauma approach. This is true for organizations of all sizes. It takes the trauma out of "workplace trauma" and replaces it with adversity that can be managed.

Conclusion

There will always be a need for interventions that help people who have suffered traumatic shock, and also for those who develop more severe problems. The problem lies in the fact that the spread of the emotional impact of events is now much greater than it was maybe half a century ago. In law there is a concept of "remoteness" – a person cannot claim damages for an event that they were too many layers removed from. The problem facing us in the emotional world is that we are nearer and nearer the epicenter of events even though we are physically remote from them; "remoteness" doesn't apply to the emotional world.

It is unrealistic, ineffective, and inefficient to try to deliver post-trauma support to everyone who may somehow have been connected with an event. We need to see resilience as a key life skill, something that is built into the curriculum at school and home as children grow and develop the ability to deal with the slings and arrows of daily life. Recognizing the importance of this and helping to foster it means that, when adversity strikes, people and organizations will be better placed to cope, to survive, and even to thrive.

References

Bakan, J. (2004). *The corporation: The pathological pursuit of profit and power*. New York: Free Press.

BBC News. (2005, July 22). Shot man not connected to bombing. Retrieved from http://news.bbc.co.uk/2/hi/4711021.stm.

Buerk, M. (1984, October 23). BBC News. Retrieved from http://www.youtube.com.

Charney, D., & Nemeroff, C. B. (2004). *The peace of mind prescription*. Boston: Houghton Mifflin.

Cryulnik, B. (2007). *Talking of love on the edge of a precipice*. London: Penguin.

Dickins v. O2 plc. (2008). EWCA Civ 1144.

Dictionary.com. (2011). Adversity. Retrieved from http://www.dictionary.reference.com/browse/adversity.

Gleick, J. (1999). *Faster: The acceleration of just about everything*. New York: Pantheon.

Hatton v. Sutherland. (2002). ICR 613, CA.

Health and Safety Executive. (2005). Tackling stress: The management standards approach. Retrieved from http://www.hse.gov.uk.

Schrauf, R., & Sanchez, J. (2004). The preponderance of negative emotion words in the emotion lexicon. *Journal of Multilingual and Multicultural Development*, *25*(2/3), 266–284.

Seery, M.D., Holman, E. A., & Silver, R. C. (2010). Whatever does not kill us: Cumulative lifetime adversity, vulnerability, and resilience. *Journal of Personality and Social Psychology*, *99*(6), 1025–1041.

Seligman, M. E. P. (1996). *The optimistic child: Proven program to safeguard children from depression and build lifelong resilience*. New York: Harper.

Wells, M. (2010). Resilience in older adults living in rural, suburban and urban areas. *Journal of Rural Nursing and Health Care*, *10*(2). Retrieved from http://www.rno.org/journal/index.php/online-journal/article/viewFile/239/284.

Part C

New Understandings on Models of Trauma Support

11

The Role and Nature of Early Intervention: The Edinburgh Psychological First Aid and Early Intervention Programs

Chris Freeman and Patricia Graham

Introduction

One of the most controversial areas in the treatment of traumatic stress reactions has been the discussion on how individuals and groups should be helped in the acute aftermath of trauma exposure. There are two main areas of debate: firstly, what is the value in providing intervention such as psychological debriefing, psychological first aid, or very early cognitive behavioral psychotherapy? Secondly, what if any is the role of dissociation in acute trauma responses? This chapter will deal mainly with the first of these controversies, but we will touch briefly on the second. The chapter will first describe the nature and range of reactions to acute trauma; it will then review the scientific evidence for the role of intervention, if any, and finally describe in some detail what interventions might be appropriate.

Nearly all the work in this area has dealt with large-scale disasters, which have often attracted considerable media coverage, rather than individual accidents or events that have occurred in or around the workplace. Individuals who are involved in major disasters such as *Piper Alpha*, Lockerbie, the Paddington rail crash, the Zeebrugge ferry disaster, or the London bombings often have special services set up for them, whereas those suffering trauma in single or isolated settings often complain that no help was available, that their GP did not recognize their distress and did not know what to do, and that their work colleagues found it difficult to know what to do or say in the days and weeks after the event. This chapter is aimed mainly at such situations.

International Handbook of Workplace Trauma Support, First Edition. Edited by Rick Hughes, Andrew Kinder, and Cary L. Cooper.

What Symptoms and Syndromes Occur after Acute Traumatic Events?

Studies report very high levels of certain psychological symptoms in the first few hours to days after a very frightening or life-threatening event. Such acute stress reactions are normal and are characterized by reports of high levels of emotional numbing, a sense of reduced awareness of one's surroundings, derealization, and depersonalization. Most people report insomnia for at least the first few nights after such an event; they feel very on edge and over-aroused, and go over and over what happened to them. Such acute stress symptoms are extremely common; Rothbaum *et al.* (1992) reported that 94% of rape victims had such symptoms in the first two weeks after the rape. Riggs *et al.* (1995) reported that 70% of women and 50% of men had such symptoms in the first three weeks after an assault, but this had dropped to 21% for women and 0% for men at four months and in a study of people who met post-traumatic stress disorder (PTSD) criteria. Blanchard *et al.* (1996), in a follow-up of motor vehicle crash victims, found half had remitted at six months and two-thirds had done so at one year.

The Importance of Dissociation

A dissociative symptom can be alterations of awareness, out-of-body experiences, fugue states, and sometimes amnesia; these symptoms are part of the diagnosis of acute stress disorder but not part of the diagnosis of PTSD. The proposal is that this will be altered in the new DSM-5 (American Psychiatric Association, forthcoming) diagnostic system. The importance of mentioning it here is that it is one of the most puzzling and sometimes disturbing reactions both for the survivor and for the supporter or rescuer. Individuals may appear to be behaving quite oddly, in a dreamlike state, appearing numb or quite disengaged from the trauma they have just been through, observed, or witnessed. It is not clear whether such dissociative symptoms are a predictor of the future development of a more serious psychiatric disorder, such as PTSD, or whether in fact they are a protective mechanism allowing the individual to cope with overwhelming stress by being somewhat disengaged from it and perhaps by preventing the development of more deeply engrained trauma memories (see Chapter 16, this volume).

The Rise and Fall of Psychological Debriefing

The formal model of critical incident stress debriefing (CISD) was first described by Mitchell in 1983. It is clear that there has been a long history of very early intervention after traumatic events going back to attempts in World War One to treat shell-shocked soldiers very close to the front line; however, Mitchell's CISD was the most carefully described and structured intervention that had so far been developed. It was originally intended for use in emergency service personnel and was conducted in a group setting. The key elements of CISD are given in Table 11.1. By the late 1980s and early 1990s, CISD had become routinely used post trauma. This followed its use in several

Table 11.1 The seven stages of psychological debriefing

Stage 1: Introduction
The debriefer states that the purpose of the meeting is to review the participants' reactions to the trauma, to discuss them, and to identify methods of dealing with them to prevent future problems. It is made explicit that (a) participants are under no obligation to say anything except why they have come and what their role was vis-à-vis the traumatic event, (b) confidentiality is emphasized, and (c) the focus of the discussions is on the impressions and reactions of the participants.

Stage 2: Expectations and facts
Details of what actually happened are discussed in considerable detail without focusing on thoughts, impressions, and emotional reactions.

Stage 3: Thoughts and impressions
This stage aims to (a) construct a picture of what happened, (b) put individual reactions into perspective, and (c) help with the integration of traumatic memories. Sensory impressions in all five moralities are listed when the debriefer, for example, asks "what did you see, hear, touch, smell, and taste?"

Stage 4: Emotional reactions
Maybe the longest stage of the debriefing; the debriefer attempts to aid the release of emotions with questions about some of the common reactions during the trauma such as fear, helplessness, frustration, anger, and guilt.

Stage 5: Normalization
The debriefer aims to facilitate the view that such reactions are entirely normal, that different people may experience different reactions, and that not all emotions are experienced by each individual. Some emphasis is placed on what symptoms may appear or continue in future.

Stage 6: Future planning and coping
This stage focuses in managing future symptoms, discussing with family and friends, and addressing the need that some people may have for additional support.

Stage 7: Disengagement
Written information is often given at this stage and participants are encouraged to seek further psychological support if their symptoms do not decrease after 4–6 weeks or if their symptoms are getting worse before that.

high-profile major disasters. It came to be used not just for service and rescue personnel but also for serving soldiers and for civilians, and it was used in both group and individual forms. By the mid-1990s it was in regular and routine use in employee assistance programs (EAPs), the civilian National Health Service (NHS) and private practice, and police, fire, and ambulance services. It became the case that an employer who had failed to offer or organize CISD for an employee might be seen as in breach of duty of care.

All this has happened without any convincing evidence that CISD was effective. It is important to note that the intention of delivering CISD was to prevent future adverse consequences such as the development of post-traumatic stress disorder, depression, anxiety states, and so on. It was not simply an intervention to provide immediate support and solace after a frightening and traumatic event. The principle was that all those exposed to significant trauma should be offered CISD preferably within the first 72 hours after the trauma. By the mid-1990s there was increasing concern about the lack of scientific evidence to support the efficacy of CISD. When the International Society for Traumatic Stress Studies (ISTSS) produced the first edition of its practice guidelines in 2000, it concluded that there was, "No evidence to support the preventative value of debriefing delivered in a single session" (Bisson *et al.*, 2007). It did note that the evidence was limited and that there was a bias toward single-session psychological debriefing for individuals rather than group debriefing or more extended interventions over a period of days or weeks.

The second edition of the guidelines, Foa *et al.* (2009) had much more evidence to draw on and in particular had four new randomized control trials. They concluded, "Our findings support and strengthen the original conclusion that no evidence suggested that psychological debriefing is effective in the prevention of PTSD symptoms shortly after a traumatic event or in the prevention of longer term psychological sequelae." They also noted that other systematic reviews had come to similar conclusions: the National Institute for Health and Clinical Excellence (NICE) guidelines (2005) and a Cochrane Review (Rose, Bisson, & Wessely, 2005).

The issue of whether reviews of the efficacy of psychological debriefing have been biased toward single-session individual debriefing has been addressed, at least in part, by the Litz and Adler (2005) study. This was carried out on active-duty personnel, and at follow-up there were no differences between CISD, stress education, and no intervention on outcomes including PTSD, depression, general well-being, aggressive behavior, marital satisfaction, perceived organizational support, and morale. Like several other studies, those who had been debriefed did rate their satisfaction with CISD as high.

One interesting development is the suggestion that psychological debriefing would be helpful if it were given as early as possible. Two small studies have found positive results in this area: Campfield and Hills (2001) and an early study by Bunn and Clarke (1979) where individuals received psychological debriefing within 10 and 12 hours, respectively, post trauma.

Why might psychological debriefing do harm? In our psychological trauma service we started to receive a small, but steady, stream of referrals from the mid-1990s onward who had felt damaged by psychological debriefing. Interestingly this seemed to be much more about practical and operational issues than it was about the actual content of the debriefing session; for example, having to wait for the debriefer to arrive and not being allowed to go home, or being called back the day after the robbery or assault to a debriefing session instead of having the Saturday off at home. It was clear that what most people wanted to do was to go home, be with their loved ones, and not to have to stay at work or come back to work for some work-organized intervention. Others have raised the question as to whether the debriefing process itself somehow modifies the individual's adaption to an acute stress response in a way that increases the risk of PTSD and other psychological disorders. If most individuals have an acute stress response and nearly all recover without any intervention, it is clear that there is a powerful natural healing process occurring and that it is vital that this natural recovery is not interfered with. Is it possible that theoretical principles developed in quite separate areas of psychology have been applied inappropriately to trauma survivors? The idea that catharsis is a good thing and that expressing feelings associated with memories is an essential process of healing goes back to Freud and is a widely accepted process within most psychotherapies. However, it may not be appropriate in the hours and days after a traumatic event. Similarly, the idea of constructing a narrative is a key ingredient of many psychotherapies and may be helpful at a later stage when coming to terms with a trauma, but not in the immediate aftermath. Critical incident stress debriefing was initially a group process and has largely returned to those roots in the way that it is practiced today. The power of group processes in promoting a sense of belonging, a sense of community, by sharing emotions and stories is well established in group psychotherapy but again may not be appropriate in the immediate aftermath of trauma. Some individuals do not want to be part of a group, do not want to share, and may be quite effective in using avoidance as an initial coping strategy.

Given the powerful nature and direction of natural recovery after trauma, it is clearly important that, whatever we do, we do not disrupt this process. This has led to a somewhat nihilistic view that there is nothing we can or should do and that it is best to do nothing.

We do not know, in the early days after a trauma, who is going to do well and who is going to do badly. There has been a great deal of debate about the predictive power of dissociation, with a few studies suggesting that it may be protective but most indicating the opposite. There does seem clear evidence that those individuals who have an acute stress reaction, which then progresses to acute stress disorder, are at much higher risk of developing long-term problems including PTSD (an acute stress disorder merges into PTSD if the symptoms continue for four weeks or more). This has led to the development of screen-and-treat programs, the principle being that only those individuals who continue to have peri-traumatic and early dissociation symptoms after a few days or even weeks, or whose symptoms are escalating should be offered some sort of early intervention. The following principles are set out in the ISTSS guidelines:

1. Shortly after a traumatic event it is important that those affected should be provided, in an empathic manner, with practical pragmatic psychological support. Individuals should be provided with information about possible reactions, about what they can do to help themselves (coping strategies), and for accessing support from those around them (particularly families and community); and how, where, and when to access further help, if necessary (Bisson *et al.*, 2007).
2. It is important to make provision for the appropriate early support of individuals following a traumatic event. However, any early intervention approach should be based on an accurate and current assessment of needs prior to intervention. People cope with stress in different ways. No formal intervention should be mandated for all exposed to trauma. Use of trauma support should be voluntary except in cases where event-related impairment is a threat to an individual's own safety or the safety of others (Bisson *et al.*, 2007).
3. Strive to make interventions culturally sensitive, developmentally appropriate, and more related to the local formulation of problems and ways of coping.
4. Lack of distress and/or rapid recovery may not be a desired outcome. Ethnic, political, cultural, and economic factors may contribute to differing goals for functioning and identity, and providers should be sensitive to the particular motivations of each survivor.

5. Because of the dearth of evidence in early interventions, as much as possible, strive to evaluate whether early interventions are effective in enumerating specific outcomes, or whether new interventions should be designed to accomplish such objectives (Watson, 2007).

The Emergence of Psychological First Aid (PFA)

Although there is a general consensus from all recent guidelines that psychological debriefing should not be used, there is also a general consensus that in the immediate aftermath of trauma there should be some sort of psychological support available. There also seems to be general agreement that this should not be a psychological treatment, should not medicalize distress, and should offer practical and emotional support. There is no generally agreed definition of what PFA consists of, but there are some general themes which are set out in this section.

Psychological first aid is not designed to specifically prevent the development of post-traumatic stress disorder; it is designed to reduce the initial stress caused by the traumatic event and hopefully optimize successful adaption to both short- and long-term functioning. Singer (1982) proposed a nine-step model of PFA to help deal with the psychological consequences of exposure to major trauma.

1. Adopt a sensitive, sympathetic, and flexible attitude toward the wide variety of reactions that might be encountered.
2. Ensure that injured and frightened survivors are not left alone and children, in particular, are not separated from their parents.
3. Make gestures and provide tokens of a simple practical nature which communicate psychologically that the survivors are being cared for.
4. Encourage the verbal expression of feelings associated with the disaster experience.
5. Offer reassurance.
6. Convey accurate and responsible information to survivors, their loved ones, and the media, and dispel rumors as they emerge.
7. Refer individuals who are showing grossly abnormal, violent, or self-destructive behavior to psychiatric personnel, and transfer such persons to a specialist treatment center so as to minimize their disturbing influence on others.
8. Issue instructions in a confident, easy-to-follow manner.
9. Encourage survivors to participate in a simple useful task as soon as possible.

It is clear that Singer was aiming his intervention at survivors of a major disaster rather than individual trauma. There are clearly elements here that can readily be applied to work-based trauma including points 1, 2, 3, 5, 7, and 8. Beverly Raphael's model (1986) comprises 11 steps. She indicates that this should be implemented within the first few hours following a trauma; the steps are:

1. Comfort and console.
2. Protect from further threat.
3. Provide immediate care for physical necessities.

4. Help affected individuals become involved in goal-directed behavior.
5. Promote reunion with loved ones separated by the trauma.
6. Support individuals who have to identify the bodies of loved ones.
7. Accept the ventilation of feelings.
8. Structure the routine of the individuals to give a sense of order in the aftermath.
9. Promote group support networks.
10. Identify and refer individuals who need more traditional mental health care.
11. Ensure the individual is linked to an ongoing system of care and support.

It is interesting that in the ISTSS guidelines (Foa *et al.*, 2009), Raphael's model is regarded as debriefing, not as PFA, perhaps because of point 7.

In the aftermath of the September 2001 terrorist attacks on the United States, the US National Institute of Mental Health issued some guidance (2002). It is clear that they intended that PFA should adopt a much less interventionist approach than debriefing. They recognized that individuals clearly would benefit from immediate support and assistance but might not need a structured format that encouraged the disclosure of trauma memories. They recommended the immediate provision of shelter, safety, social support, and linkage to appropriate social networks and resources. They actively discouraged the facilitation of the disclosure of trauma experiences or related emotions.

What we have moved to now is an acceptance that the most common and "normal" response to trauma is recovery rather than the development of PTSD or other psychological injuries and that any early intervention should be toward supporting that recovery rather than pathologizing it.

The second principle is that there must be some robust mechanism for identifying those individuals who are not on the road to recovery and that such failure of recovery will become evident at approximately 4–6 weeks post trauma. Such individuals should be identified and offered early psychological treatment, with the best evidence to date being for brief, specially adapted cognitive-behavioral psychotherapy (Bryant *et al.*, 1998).

In the rest of this chapter, we set out in detail our particular model of PFA which has been specifically adapted for individual trauma and trauma occurring in the workplace. We also review the evidence for the early use of cognitive-behavioral therapy (CBT) but emphasize that this is not a first- but second-stage intervention and is only appropriate for those who have significant symptoms at 4–6 weeks.

The Edinburgh Early Intervention Model: E-EIM

The Edinburgh early intervention model (E-EIM) is a three-staged tiered model for early response following a traumatic event. The model consists of eight possible sessions over three stages with three potential exit points after sessions 2, 4, and 8. To complete all eight sessions can take between 11 and 19 weeks. Stage one is an abbreviated version of our model of PFA, Edinburgh Psychological First Aid. Stages two and three are our especially adapted version of interpersonal psychotherapy (IPT) for depression, IPT-Post Trauma (IPT-PT). We have adapted the original model of interpersonal psychotherapy for depression (Klerman *et al.*, 1984) which was developed to treat a diagnosable major depressive disorder. Our model, IPT-PT, has been developed specifically for individuals who are in distress and have symptoms due to a recent specific traumatic event.

The rationale for the development of our early intervention model post trauma is based on our acceptance, as described, that the most common and "normal" response to trauma is recovery; and our model is focused toward supporting that recovery rather than pathologizing it. A summary of some of the key themes that informed the development of E-EIM is as follows:

- There is now indisputable evidence for significant levels of distress following a range of different traumatic events within the first days and weeks of the event. As an example, dysphoric-depressed emotions, anxiety, and dissociative symptoms have all been reported to occur quite frequently during the first few days following an accident (Shalev *et al.*, 1996).
- Brewin and Lennard (1999) have demonstrated that risk factors operating during and/or after trauma, such as trauma severity, lack of social support, and additional life stress, have somewhat stronger effects than pre-trauma factors. It is the early days after a traumatic event which may offer a window of opportunity to ensure adequate use is made of both emotional and practical social support; these findings underpin our rationale for the adaptation of IPT to IPT-PT.
- Within the E-EIM, there is an acceptance that successful adaptation to a traumatic event will eventually involve an emotional engagement with the traumatic memories and that to intervene too early with formal psychological treatment may be counterproductive to that adaptation process.
- During the immediate aftermath of trauma, there will be a handling of the trauma, rather than treating a post-traumatic condition (Shalev & Ursano, 2003) and this is the very essence of E-EIM.
- Although *most* individuals who develop prolonged stress disorders show symptoms of distress in the early aftermath of traumatization, not all will do so (Rothbaum & Foa, 1993).
- Most instances of recovery from early responses to traumatic events will occur within the year following the trauma (Kessler *et al.*, 1995; Shalev *et al.*, 1997).

The early practical support advocated within E-EIM is not designed to be a replacement for the potential support offered by the individual's interpersonal world of friendships, family, or work colleagues, but it *is* designed to optimize that support. E-EIM is a series of individually tailored, practical, collaborative suggestions designed to supplement and enhance the potential support available from within the existing social support network and thereby optimize successful adaptation to the trauma. We do not claim that E-EIM will prevent the possible occurrence of acute stress disorder or post-traumatic stress disorder, as to date no empirical investigations have been conducted.

One of the most distressing elements of the immediate post-impact period of traumatic events can be facing losses, the recovery of which will involve grieving and re-adaptation. In the later stages of E-EIM, we use an adapted version of IPT, specifically designed to help in those processes. The original model of IPT (Klerman *et al.*, 1984) uses "grief" as a focus area when it is associated with major depressive disorder and therefore is described as being pathological or complicated grief. We make the distinction in IPT-PT that we are encouraging the normal process of grief, not treating a major depressive disorder associated with a complicated grief response.

The Edinburgh Early Intervention Model: An Overview of the Stages

Stage one

Two sessions (1 + 1).

1. Assessment of the nature of the trauma, the most distressing symptoms, and an impression of the individual's desire to work through the traumatic event.
2. Provision of basic psycho-education and normalization of the traumatic response, using our abbreviated Edinburgh psychological first aid model. Our abbreviated model consists of seven stages:
   a. Provide practical help, which may be simple and will vary according to the individual's needs.
   b. Comfort and console distress.
   c. Educate others on normal responses to trauma which involves two essential elements: *recognizing the range of reactions*, and *respecting and validating the normality of the post-trauma reaction*.
   d. Protect from further threat and distress.
   e. Furnish immediate care for physical needs.
   f. Provide support for real-world-based tasks.
   g. Facilitate reunion with loved ones.

E-EIM outlines specifically the detail required for each stage and provides photocopiable handouts for both worker and sufferer.

Stages two and three of E-EIM: interpersonal psychotherapy post trauma (IPT-PT). IPT-PT is based on the premise that psychological distress that occurs following a trauma, regardless of biological vulnerability or personality, does so within a psychosocial and interpersonal context. The structure for IPT-PT has been directly derived from interpersonal psychotherapy for depression (Kerman *et al.*, 1984). IPT-PT, just like full IPT, has three phases consisting of an assessment, middle, and termination. IPT-PT has not been subjected to any empirical research as yet, but has been presented as a poster at the ISIPT conference (Freeman, Graham, & Wheeler, 2006) (Table 11.2).

Stage Two has two sessions, both designed to map out and review the interpersonal network:

1. Pre-trauma network (based on months and years of interaction)
2. Post-trauma network (based on days or weeks of interaction)

Table 11.2 Brief outline of Stages 2 and 3 of Edinburgh Early Intervention Manual: IPT-PT

Session number	*Summary of session task*
1 (60 minutes)	Mapping out the interpersonal network prior to the trauma: There is guidance on exactly how to complete the interpersonal network in the E-EIM manual. Examples of questions asked include: 1. How many changes in the interpersonal world are inevitable changes as a result of the traumatic event? 2. What did the trauma actually do to the interpersonal network? 3. What has the individual's trauma reaction done to the way he or she now reacts to others in the interpersonal network?
2 (60–90 minutes)	Mapping out the interpersonal network post trauma: The systematic review of relationships following the trauma involves an exploration of the individual's important relationships with others – information should be gathered about each person who is important in the individual's life now. We are looking for any major disruptions to close personal relationships as a result of the trauma.
3 (60 minutes)	Initiating IPT:PT: 1. Dealing with the distress. 2. Relate distress to interpersonal context following the trauma. 3. Identify whether this is a transition or grief focus (related to traumatic event). 4. Monitor symptoms described through the Patient Health Questionnaire (PHQ-9) and to keep a record of such as an intertreatment task. 5. Making a narrative.
4 and 5 (60 minutes)	Focus area: 1. Grief 2. Role transitions Each problem area has a set of goals and strategies as outlined by Klerman *et al.* (1984); however, the IPT-PT model differs somewhat, and therefore is outlined in the detailed manual.
6 (60 minutes)	Termination of treatment: Explicit discussion of the ending of contract, and its implications for the individual and their interpersonal world and the Interpersonal Inventory, are drawn out. Help the individual move toward a recognition of his or her independent competence through redrawing the Interpersonal Inventory. Review the course of treatment and progress with an opportunity to evaluate future needs. Early warning signals are reviewed.

The important relationships in the individual's life are explored whilst gaining a thorough understanding of how the interpersonal world has changed as a direct result of the traumatic events.

Stage Three consists of four sessions of IPT:PT, the focus area (grief or role transition), and the ending session.

The overarching principles for all stages of E-EIM

- Don't do harm.
- Offer warmth, comfort, support, and advice to those who need it.
- Don't intrude unless asked by the individual.
- Make it clear that refusing help now doesn't jeopardize receiving help in the future.
- If you can't facilitate help in the future, don't offer it now (in other words, do not begin what you cannot finish, and do not dip in and then opt out).
- Validate the individual's experience and his or her reaction to it (nothing is too much, too little, right, or wrong at this stage).

The overarching objectives of the E-EIM

- To alleviate the effects of traumatic life events
- To decrease the stress attributable to those traumatic events
- To provide assistance with any practical concerns in the immediate aftermath of trauma and thereby prevent further discomfort and distress
- To optimize the potential emotional and practical support available from within the individual's interpersonal network

Professionals who have been specifically trained in the model of IPT-PT are able to deliver IPT-PT. Ideally, in the United Kingdom, practitioners would be accredited as IPT practitioners through either IPT-Scotland or IPT-UK. In order to train in the model of IPT in the United Kingdom, it is necessary to attend a recognized training course through IPT-Scotland or IPT-UK. The course usually takes up to five days. Training pathways and competency frameworks for all levels of training and supervision are easily accessible.

Throughout our recommendations, we urge caution in the consideration of any mandatory recommendations for any intervention following trauma, especially program-based manualized treatments, unless there is unequivocal evidence that can be quoted in its support. The Edinburgh early intervention model provides *guidance* as to how to instigate practical, flexible, individually tailored social support and minimize additional life stress.

The routine use of E-EIM: key points

- We clearly accept that the most common response to a traumatic event is one of recovery, and our model is focused toward supporting that recovery rather than pathologizing it.
- Not all individuals who develop prolonged reactions to trauma will demonstrate this early in the aftermath of trauma, although it is reasonable that survivors should be considered as being *at risk* for developing traumatic stress disorders.
- Not all individuals will react in the same way following a traumatic life event – there can be a myriad of responses, and therefore it is extremely important not to generalize an approach to those in the immediate aftermath of trauma.
- The survivor's progress toward recovery should be followed and clinical decisions made on the basis of longitudinal observations instead of cross-sectional examination.

- E-EIM should not be regarded as a psychological treatment program.
- Early and urgent needs of all individuals affected by trauma should be addressed.
- Intervention should be provided in the context of continuity of care and should be individually tailored to suit specific needs, which is the premise of E-EIM.
- E-EIM provides *guidance* as to how to instigate practical, flexible, individually tailored social support in a staged model with clear and detailed guidelines and clearly defined exit points guided by robust clinical outcome measures to aid decision making.

Conclusions

Most chapters of this type end with the conclusion that what is needed is more research before definite recommendations can be made. It is not clear that that is so in this area. There has already been sufficient research on psychological debriefing which has clearly shown it to be ineffective. In those trials there was a clearly defined set of outcome criteria, namely, the prevention of future psychiatric and psychological disorders such as PTSD. Psychological debriefing did not do this, and in some trials it increased the risk. Some of those trials also showed that asking individuals who had been traumatized about their satisfaction with the intervention was not a reliable guide to their risk of developing future adverse reactions.

As far as PFA is concerned, it is not universally agreed what, if any, the desired and measurable outcomes are. Psychological first aid is designed to at least provide a humane, supportive, and acceptable intervention for those who appear to need such support after a trauma. Because, by its very nature, PFA is designed to be individually tailored rather than applied uniformly across the board, and because PFA in different settings and with different workforces may have quite different desired outcomes, it is not self-evident that a whole tranche of randomized controlled trials of first aid interventions will be useful.

It may be that self-report is a more valid and perhaps the only or main way of evaluating PFA. When individuals are injured or traumatized in the workplace, they often complain that the early response they received was not appropriate or relevant to their needs. There was certainly a period in the mid-1990s where in our clinic we received a number of referrals where stress from a highly structured debriefing intervention was greater than that from the trauma it was designed to help. Outcomes such as satisfaction with the employer's response, relationship with the individual's line manager, and return to work might be appropriate measures which could be examined following the application of PFA. However, given that by its nature the intervention has to be individualized and can vary from doing almost nothing to a quite extensive package of support and care, it is unlikely that a randomized trial will be of help.

In this chapter we have discussed some of the issues in applying psychological first aid and have offered one particular model which combines an abbreviated first aid approach which can then progress to staged psychological help if required. It is a model that has been quite widely used in some employee assistance programs, but we do not claim it has any more or less evidence than other models of first aid. We hope it may be of use to you; you are welcome to use as much or as little of it as you wish.

The complete manuals are available free to download on http://www.ukpts.co.uk.

References

American Psychiatric Association. (Forthcoming). *Diagnostic and statistical manual of mental disorders* (5th ed.). Washington, DC: Author.

Bisson, J. I., Brayne, M., Ochberg, F., & Everley, G. (2007). Early psychosocial intervention following traumatic events. *American Journal of Psychiatry, 164*, 1016–1019.

Blanchard, E. B., Hickling, E. J., Barton, K. A., Taylor, A. E., Loos, W. R., & Jones-Alexander, J. (1996). One-year prospective follow-up of motor vehicle accident victims. *Behaviour Research and Therapy*, 34, 775–786.

Brewin, C., & Lennard, H. (1999). Effects of mode of writing on emotional narratives. *Journal of Traumatic Stress, 12*, 355–361.

Bryant, R. A., Harvey, A. G., Dang, S. T., Sackville, T., & Basten, C. (1998). Treatment of acute stress disorder: A comparison of cognitive-behavioural therapy and supportive counselling. *Journal of Consulting and Clinical Psychology, 66*(5), 862–866.

Bunn, T., & Clarke, A. (1979). Crisis intervention: An experimental study of the effects of a brief period of counselling on the anxiety of relatives of seriously injured or ill hospital patients. *British Journal of Medical Psychology, 52*, 191–195.

Campfield, K., & Hills, A. (2001). Effect of timing of critical incident stress debriefing (CISD) on posttraumatic symptoms. *Journal of Traumatic Stress, 14*(2), 327–340.

Foa, E. B., Keane, T. M., Friedman, M. J., & Cohen, J. A. (2009). *Effective treatments for PTSD* (2nd ed.). New York: Guilford Press.

Freeman, C., Graham, P., & Wheeler, E. (2006). Edinburgh early intervention model. Poster presented at the ISIPT Conference, Toronto, ON.

Kessler, R., Sonnega., A., Bromet, E., Hughes, M., & Nelson, C. (1995). Posttraumatic stress disorder in the National Comorbidity Survey. *Archives of General Psychiatry, 52*, 1048–1060.

Klerman, G., Weissman, M., Rounsaville, B., & Chevron, E. (1984). *Interpersonal psychotherapy of depression*. New York: Basic Books.

Litz, B., & Adler, A. (2005). [A controlled trial of group debriefing.] Unpublished raw data.

Mitchell, J. T. (1983). When disaster strikes. *Journal of Emergency Medical Services, 8*, 36–39.

National Institute for Health and Clinical Excellence (NICE). (2005). *Post-traumatic stress disorder (PTSD): The management of PTSD in adults and children in primary and secondary care*. London: Author. Retrieved from http://www.nice.org.uk/CG26

National Institute of Mental Health. (2002). *Mental health and mass violence: Evidence-based early psychological intervention for victims/survivors of mass violence. A workshop to reach consensus on best practices*. NIH Publication No. 02-5138. Washington, DC: Government Printing Office.

Raphael, B. (1986). *When disaster strikes*. New York: Basic Books.

Riggs, D. S., Rothbaum, B. O., & Foa, E. B. (1995). A prospective examination of symptoms of posttraumatic stress disorder in victims of non-sexual assault. *Journal of Interpersonal Violence, 10*, 201–214.

Rose, S., Bisson, J., & Wessely, S. (2001). *Psychological debriefing for preventing post traumatic stress disorder (PTSD)*. The Cochrane Library (3). Oxford, UK: Update Software.

Rothbaum, B., & Foa, E. (1993). Subtypes of posttraumatic stress disorder and duration of symptoms. In J. R. T. Davidson & E. B. Foa (Eds.), *Posttraumatic stress disorder: DSM-IV and beyond* (pp. 23–35). Washington, DC: American Psychiatric Press.

Rothbaum, B. O., Foa, E. B., Riggs, D. S., Murdock, T., & Walsh, W. (1992). A prospective examination of post-traumatic stress disorder in rape victims. *Journal of Traumatic Stress, 5*, 455–475.

Shalev, A., Freedman, S., Peri, T., Brandes, D., & Sahar, T. (1997). Predicting PTSD in trauma survivors: prospective evaluation of self-report and clinician administered instruments. *British Journal of Psychiatry, 170*, 558–564.

Shalev, A., Per, T., Canetti, L., & Schreiber, S. (1996). Predictors of PTSD in injured trauma survivors: A prospective study. *American Journal of Psychiatry, 153*, 219–225.

Shalev, A., & Ursano, R. (2003). Mapping the multidimensional picture of acute responses to traumatic stress. In R. Orner & U. Schnyder (Eds.), *Reconstructing early intervention after trauma: Innovations in the care of survivors.* Oxford: Oxford University Press.

Singer, T. (1982). An introduction to disaster: Some considerations of a psychological nature. *Aviation, Space and Environmental Medicine, 53*(3), 245–250.

Watson, P. J. (2007). Early intervention of trauma-related problems following mass trauma. In R. J. Ursano, C. S. Fullerton, L. Weisaeth, & U. B. Raphael (Eds.), *Text-book of disaster psychiatry* (pp. 121–139). Cambridge: Cambridge University Press.

12

An Organizational Approach to the Management of Potential Traumatic Events: Trauma Risk Management (TRiM) – the Development of a Peer Support Process from the Royal Navy to the Police and Emergency Services

Neil Greenberg and Marilyn Wignall

Background

The link between sending troops to war and their suffering psychological trauma has been known for centuries. However, it was not until after the end of World War Two that both academics and military leaders came to realize that mental disorders were not necessarily a result of inadequacy, a lack of backbone, or a lack of moral fiber. Over the past 60 years, there has been a substantial amount of research investigating military and civilian personnel who have been exposed to potentially traumatic events. We now know that everyone has a breaking point.

During the late 1990s, senior leaders within the Royal Marines, elite naval combat troops, decided that a structured post-incident management process would be helpful to maintain operational effectiveness, and it was from this idea that TRiM (Trauma Risk Management) was borne. Today TRIM is used by a large number of military and civilian organizations, and its current structure has developed over a number of years from an original idea devised by Major Norman Jones and Captain Peter Roberts (Jones, Roberts, & Greenberg, 2003), who were two mental health nurses with considerable experience of working with army personnel.

Before the year 2000, the UK military, like many other organizations, relied upon "trained debriefers" to provide support after traumatic incidents. However, as science

International Handbook of Workplace Trauma Support, First Edition. Edited by Rick Hughes, Andrew Kinder, and Cary L. Cooper.

critically examined the nature of psychological trauma, it became clear that most personnel who are exposed to traumatic events do not develop problems. Therefore it made little sense to "debrief" everyone in the aftermath of potentially traumatic events. Furthermore, there is now a good deal of evidence (such as that by van Emmerik *et al.*, 2002) demonstrating that formal psychological debriefing such as critical incident stress debriefing (CISD) is ineffective in preventing psychological injury. This negative view regarding the use of single-session psychological debriefing has been confirmed by the publication of the National Institute for Clinical Excellence (NICE) guidelines (2005) on the treatment of PTSD. The evidence also suggests that debriefing some individuals may increase, rather than decrease, their levels of distress. On the other hand, evidence suggests that non-CISD interventions, based upon practical support and not upon a single post-incident interview, appear to be beneficial after traumatic incidents.

There are numerous legal, moral, and economic reasons why organizations should consider the potential effects of "stress at work" upon their staff. After traumatic events, personnel who have been adversely affected are likely to be less effective at work and less likely to continue to serve unless dealt with effectively. However, the nature of occupational stressors can vary considerably. For instance, the difficulties faced by British soldiers in Afghanistan are substantially dissimilar from those faced by the emergency services or by journal editors. In 2009 the National Institute for Health and Clinical Excellence (NICE) published public health guidance on promoting mental well-being through productive and healthy working conditions which clarified the numerous financial, health, and legal reasons why organizations should do their utmost to enhance mental well-being. In 2008 a report for the Department for Work and Pensions and the Department of Health noted that impaired work efficiency, as a result of mental disorders, costs the United Kingdom £15.1 billion a year, with mental health–related absenteeism costing an additional £8.4 billion annually. On February 19 of the same year, an article by Matthew Hickley published in the *Daily Mail* reported that police forces lost over 30 000 working days in 2006 due to sickness absence attributed to stress. They claimed that the total stress-related sickness among the 140 000 officers in England and Wales in 2006 was equivalent to slightly more than two days per officer – at an average cost of £162 per day. Thus, whilst not all organizations predictably place their personnel in harm's way, in order to comply with the NICE guidance those that do so, such as the emergency services, journalistic organizations, and the military, need to consider how best to mitigate and manage the psychological sequalae of exposure to potentially traumatic events.

The Nature and Likely Effects of Potentially Traumatic Events (PTEs)

PTEs are characterized by their potential to cause damage to an individual's health, including mental health. However, whilst there is a well-demonstrated relationship between trauma exposure and the onset of mental health disorders, including, but not limited to, PTSD, most people do not become ill after exposure to traumatic events. For instance, a study of the London Ambulance Service personnel carried out by Misra *et al.* (2005) two months after the 2005 London Bombings found that 4% of respondents reported probable PTSD, a rate only marginally above the 3% rate found in the general UK population and comparable to the rate in UK military personnel.

Detailed analyses of factors associated with trauma-related psychological disorders suggest that post-event factors, including importantly social support and exposure to subsequent stressors, are more important predictors of psychological outcome than pre-trauma or peri-traumatic factors. Put another way, whether people who are exposed to a PTE will become psychological ill mostly depends on what happens to them after the event has occurred. This is important for a number of reasons. Firstly, it explains why efforts to "screen" out vulnerable people before a PTE happens are likely to be unhelpful, since it is impossible to "screen" for the level of post-incident support people will be provided with or what other subsequent stressors they might be exposed to. However, organizations which are cognizant of this should be able to proactively manage the post-incident period and thereby increase their organizational resilience. Secondly, from an organizational perspective, those close to the potentially distressed individual, including importantly managers and colleagues, are ideally placed to both provide high levels of social support after an incident has occurred and to manage how much pressure, at least from work, individuals will experience during the post-incident period.

There are, therefore, good scientific and economic arguments for organizations that predictably place their personnel in harm's way to provide effective and proactive post-PTE management for their personnel. However, doing so can be challenging. Firstly, in order to establish such mechanisms, organizations must be cognizant of which personnel have been exposed to traumatic incidents. Whilst major events often attract a manager's attention, less physically traumatic, and therefore less obvious, events may also be associated with the development of mental health problems; similarly, so may "near-miss" events. Secondly, the majority of people exposed to PTEs deal with them without suffering prolonged distress or developing formal psychiatric illnesses.

The evidence therefore suggests that organizations should direct the majority of their intense efforts toward supporting the small proportion of personnel who might benefit from them rather than the larger number who would not. This approach allows for those who do not need intensive support to avoid undue interference, optimizing the recovery environment by ensuring the provision of social support, and minimizing exposure to other stressors for those showing signs of distress and also directing those who do not recover to professional sources of help so that they can receive early and effective treatment according to nationally agreed protocols.

Post-incident Interventions

Older, single-session models of post-incident management, such as critical incident stress debriefing (CISD), failed to follow the above principles. When subjected to scientific scrutiny these interventions were found not only to lack effectiveness but also to have the potential to cause harm. For instance CISD, which was developed within the US emergency services, initially aimed to prevent the onset of post-traumatic stress disorder (PTSD) through the use of a single-session brief psychological intervention within the first few days after a PTE for everyone involved, as illustrated by Rose *et al.* in 2003. However, this "one-size-fits-all" approach does not take into account the high degree of variability in individuals' recovery trajectories; that is to say that those personnel who exhibit early post-incident distress frequently do not go on to develop post-incident illnesses and those who initially cope well can become ill over time.

For instance, in a recent case described by John Campbell-Beattie and David Mulhall (personal communication, 2011), an ex-military member of a police force developed PTSD many decades after leaving the military when a heavy object narrowly missed him as it fell from a building during a routine patrol. This is of considerable interest since this individual had experienced intense combat duties and witnessed enemy combatants being killed whilst in the military, but at that time he did not develop PTSD. Why might it be that this individual develops PTSD after a comparatively minor incident during his police work when he did not many years previously whilst experiencing intense traumatic events in combat?

Looking at resilience and coping strategies, there is some evidence to suggest that some people are more likely to develop PTSD than others: for instance, Brewin, Andrews, and Valentine (2000) found that people who have a prior history of mental health disorders are at increased risk of developing PTSD after exposure to traumatic incidents. An alternative hypothesis is that PTSD is a result of exceeding the limits of a person's normal coping mechanisms. In order to "survive" in such extreme environments, people who routinely work in high-threat environments learn to control or suppress their emotional responses to the traumatic events they experience. Many may become inured to trauma, often relieving stress through black humor, and often find the support of colleagues is critical. Within high-threat occupations, this culture of relative suppression and support is often found; such is the case with military personnel and police, fire, and ambulance services.

Since the majority of those exposed to traumatic incidents cope well with the exposure, NICE in 2005 suggest the use of "watchful waiting" for a month or so after an event followed by a formal psychological health check; one month was chosen as, in the main, this is sufficient time for most of those who are going to recover to have done so and for most of those who will become unwell to have developed sufficiently clear symptoms to allow their disorder to be diagnosed by a suitably trained practitioner. Following the principles of watchful waiting, an organization's initial response to their staff being exposed to a traumatic event should be one of heightened alertness for possible mental health problems in their staff, not an assumption that such disorders will inevitably occur.

What Are the Barriers to Employees Not Seeking Help for Traumatic Stress Disorders?

As highlighted in a recent legal case against the UK Ministry of Defense (MoD) by ex-service personnel who claimed to be suffering with psychological injuries as a result of their military service, McGeorge, Hacker Hughes, and Wessley (2006) found that stigma is a very real and significant issue for the UK Armed Forces. However,

Greenberg *et al.* (2009) found that stigma also acts as a significant barrier to care for nonmilitary organizations and should in fact be viewed as a societal issue and not one restricted to the military, emergency services, or other similar organizations. Stigma has been defined as something which sets an affected individual apart from others and can be split into internal stigma, or self-stigma, and external stigma. An example of the former is the belief that asking for help for mental health problems will lead to a premature end to one's career, and an example of the latter is the belief that people who suffer from mental health problems are universally weak and cannot be trusted. Stigma can act as a barrier to seeking help, although there are other practical barriers to care which exist, such as a lack of treatment services or being unable to take time off from other duties to attend for care.

A recent study by the University of Exeter by James (2010) found that police officers were acutely aware of being judged by the public, about how they "should" act as a police officer. The results suggested that being required to be police officers first and emotional and reactive individuals second is one of the "side effects" of wearing a uniform which may act as a psychological shield allowing police officers to act, whilst at work, in a manner which is at odds with their usual demeanor. The study also found that this "duality of persona" caused some police officers to be confused about when a personal line has been crossed. Those who continue to "suffer in silence" may become ineffective and inefficient officers, and may only seek help, if at all, when their occupational effectiveness and social life have been harmed to such a degree that treatment may be unable to return them to efficient function. As a result those that are eventually recognized in service may require early retirement whilst those who do not present at all may leave without any health-related benefits. One study by Clair (2006) reported that nearly half (43%) of police officers claimed exposure to trauma as a reason for their early retirement. It is notable that some studies like Miller (2006) have suggested that law enforcement personnel are more likely to struggle with divorce, substance abuse, spousal abuse, and suicide than the general population.

It is unclear which are the greatest barriers to employees seeking support for mental health problems, but there are several areas of commonality that have been described within emergency services and the armed forces: Green *et al.* (1990) found that these organizations share similar key experiences, namely, threat to life and exposure to grotesque death. An individual may be averse to seeking psychological help because they think it is unnecessary; for instance, O'Brien and Hughes (1991) found that personnel may consider post-traumatic symptoms to be an inevitable consequence of their job, or like Solomon (1989) found that they perceive professional help as irrelevant and useless. Whether this attitude is indicative of resilience or a lack of faith in the help available is debatable.

Koopman (1991) found that most organizations have "resistors," who are often managers or heads of organizations who are opposed to change at the risk of "sunk costs" (money already invested in the current system) despite, as Hoge, Ritschard, and Cooper (2002) found, staff proposing the need for change. Meyer (1979) found that a resistor's attitude is often contrary to the commonly stated view that people are the greatest resource. Resistors may engender a managerial versus staff divide which may in turn be responsible for the failures of intervention programs. The positive effects of senior officers, though, can be useful in organizations; as Jeannette and Scoboria (2008) found, evidence suggests their involvement in post-incident management such as TRiM increases its efficiency. Unfortunately, senior "resistor" managers may be unaware or uninterested in the value of post-incident support such as TRiM and may not recognize the need for mental health services. Rogers (2003) found that a senior manager who champions staff support may be able to influence key individuals, particularly other command staff, and can be crucial in determining an employee's decision to accept the legitimacy of a post-incident support program. Nowicki (2000) found that employees must believe that their cooperation with a service positively affects their performance evaluation, which only seniority can enforce. They must also develop an understanding of the repercussions for untreated PTSD such as long-term illness. For the rank and file this includes an awareness of the risks that they and their colleagues face regularly. For management this requires an understanding of the time, resources, and effort invested in their staff which can and will be wasted if mental health support is not initiated where necessary. However, it is noteworthy that Hoge *et al.* (2004) found that much of the stigma which prevents personnel from seeking help is internal, such as the fear of judgment by peers and leadership. It is likely that these beliefs are in part influenced by the attitudes of senior leaders and an organization's policies relating to the management of mental health issues in the workplace.

As a result of stigma, many employees who suffer mental health problems, as a result of their work or otherwise, are often hesitant to seek help. This is problematic as untreated mental health disorders are likely to lead to lessened productivity and decreased quality of life, and increase the chance that an individual will prematurely leave his or her employment.

However, whilst many employees report being concerned about the consequences of seeking help, doing so is actually highly unlikely to limit an individual's career options, except perhaps in the short term whilst they undergo treatment. In fact, in the United Kingdom, the Disability Discrimination Act (1995 and 2010) requires employers to make reasonable adjustments to take account of enduring mental health disorders; simply making those with mental health problems redundant or preventing them from being promoted is against the law. On the contrary, exhibiting persistently poor performance, secondary to mental health problems or otherwise, may well limit someone's career progression, and thus not seeking help may actually lead to the very outcomes that a hesitant distressed individual was trying to avoid. Seeking help for PTSD is extremely difficult, whoever you are, meaning that greater training and referral routes should be explored within high-risk professions.

Social support was found to be a strong theme for all the participants in the research carried out by the University of Exeter. Ozer *et al.* (2003) found that a perceived lack of social support is characteristic of PTSD. Similarly, McFarlane and Yehuda (1996) concluded that networks are vital to the healing process. Crisp *et al.* (2005) found that public stigmas of mental health are reducing and where these networks can be present staff may show early improvements and have a lesser chance of developing PTSD. Meta-analysis by Brewin *et al.* (2000) has revealed an absence of social support as the largest risk factor following post-trauma reactions, which supports the finding that social support can act as a protective barrier. The participants in the research were keen to identify the social support from colleagues who were receptive to each other and able to empathize with one another. In the armed forces it is bravado that impedes help, but camaraderie which can protect individuals from suffering alone.

The Origins of Trauma Risk Management

Responding to emerging concerns, in 1998 the Royal Marines Commandos began to establish a PTE management process attuned to their close-knit culture and the need for personnel to remain occupationally effective in highly challenging conditions. Although developed before the 2005 NICE PTSD management guidelines, the methodology of the TRiM system is very much in keeping with what NICE later suggested as best practice. TRiM is effectively a peer-delivered psychological first aid process delivered by TRiM practitioners who are nonmedical personnel trained to be able to monitor those exposed to PTEs in order to assess what support, if any, they might benefit from. Importantly, the TRiM process aims to promote organizational resilience by not assuming that individuals will become ill. Instead, personnel identified as suffering with an early, post-incident, psychological reaction are provided with supported management from within their department or sub-unit in a timely fashion. Also, within a TRiM-aware organization both managers and TRiM practitioners will reinforce the "normality" of early psychological symptoms and engender an expectation of recovery.

Distressed but not ill personnel should not be exposed to overly complex solutions but encouraged to seek immediate recourse to mental health care or evacuation. Within the military, these techniques are often referred to as PIES which stands for proximity, immediacy, expectancy, and simplicity. PIES refers to symptom management being delivered proximal to the workplace, including combat zones; delivered immediately to those who need it; delivered with the expectation of occupational recovery; and using simple rather than complicated solutions to emergent issues. The use of these principles has been found not only to have utility in the short term but also to prevent longer term difficulties. For instance, a 20-year follow-up of Israeli war veterans who suffered with acute stress reactions by Solomon and Mikulincer (2003) found that the more PIES principles were applied in the immediate aftermath of an individual suffering from an acute stress reaction, the better the person's outcome was 20 years later on.

Although TRiM was pioneered within the Royal Marines, it has since been adopted by a wide variety of nonmilitary organizations such as the Foreign and Commonwealth Office, the British Broadcasting Corporation (BBC), and a number of emergency services including Dorset Fire Service, several police forces, and the London Ambulance Service. TRiM aims to be NICE compliant in terms of meeting the aims of both the 2009 NICE public health guidance on promoting mental well-being through productive and healthy working conditions and the 2005 NICE guidance on the management of PTSD in adults and children in primary and secondary care.

TRiM: A Mechanism for Early Identification and Management of Potential Psychological Injuries

TRiM is not a mechanism which purports to prevent PTSD; instead, TRiM aims to provide an early indication of who may go on to develop acute psychological conditions and to empower managers to implement management plans which may help create the best possible conditions for psychological recovery to occur. TRiM practitioner training aims to equip nonmedical personnel to manage the psychological aftermath of a traumatic incident or series of incidents. Training covers a wide subject matter including psychological aspects of incident site management, how to plan for personnel's psychological needs after an event, how to conduct a semistructured risk assessment interview, and how to conduct basic psycho-educational briefings. The TRiM course is a combination of didactic teaching and role play. At the end of the initial 2.5-day course, participants will have learned why it is that some people develop mental health difficulties after a PTE and understand why traumatic events need to be addressed by organizations. Trainees should also understand the potential pitfalls in post-incident psychological management, identify the critical aspects of a traumatic event that have the greatest impact, decide when and if to intervene using the most appropriate strategy and time frame, and be able to identify the risk factors and behavior changes that may require a referral. The TRiM course teaches trainees how to carry out a 1:1 structured risk assessment and assist in a group risk assessment (up to 8 interviewees), both shortly after an event and again a month later, and how to use the information gained from the risk assessment to optimize the opportunity for personnel to recover well from being exposed to the PTE. The team leader's course focuses more on planning how to deal with the psychological aftermath of a PTE, carrying out risk assessments in small groups, and supervising TRiM practitioners.

How TRiM Works

Policies vary, but in most organizations, supervisors have the responsibility for looking after the welfare of their staff. Supervisors are expected to ensure that staff who have been exposed to a potentially traumatic incident are encouraged to speak in general terms about their role in the incident before they go off shift. Where an event is viewed as being significantly traumatic, a planning meeting takes place with the key managers involved with the incident as well as TRiM personnel. Sometimes a briefing takes place to inform all those deployed about the facts of the incident: this allows early rumor management as well as some brief psycho-education for those attending. Where appropriate TRiM risk assessment interviews will be arranged, although these will take place at least 72 hours after the incident to allow reactions to settle.

The TRiM risk assessment interview process aims to identify the presence of 10 evidence-based risk factors which are all known to be associated with the potential to develop longer term psychological problems (Table 12.1). TRiM practitioners are taught to avoid using the interview as a form of emotional catharsis, and indeed TRiM practitioners are taught to gently "shut down" the interview should an interviewee

Table 12.1 TRiM practitioners' list of risk factors for later psychological disorder

1. The person perceives that they were out of **control** during the event.
2. The person perceives that their **life was threatened** during the event.
3. The person **blames others** for what happened.
4. The person reports **shame and/or guilt** about their behavior during the event.
5. The person experienced **acute stress** following the event.
6. The person has been exposed to **substantial stress** since the event.
7. The person has had **problems with day-to-day activities** since the event.
8. The person has been involved in **previous traumatic events**.
9. The person has **poor social support** (family, friends, and/or unit support).
10. The person has been **drinking alcohol** excessively to cope with distress.

become increasingly distressed during the interview. As Sijbrandij *et al.* (2006) found, there is evidence that single-session debriefing had a poor outcome, at least in part because highly distressed people being debriefed do especially poorly when forced to retell the most traumatic aspects of their experience; this is often known as "retraumatization."

Whatever the outcome of the risk assessment interviews, the TRiM practitioners understand the benefits of ensuring that appropriate, effective post-incident support is put in place. During a TRiM interview, if any person is identified as being likely to develop longer term problems, TRiM practitioners aim to assist such highly distressed personnel to take up an early referral for a professional psychological health assessment; this may be through an occupational health department or another medical practitioner. Where the outcome of a TRiM interview is equivocal, TRiM practitioners are encouraged to discuss their concerns with the team manager or a more experienced TRiM practitioner, an occupational health practitioner, or indeed a mental health professional.

Within the Trauma Risk Management manual, TRiM practitioners are given guidance on how to conduct an assessment as laid out in the following subsections.

Overall

- Key aim is to answer all 10 risk assessment items.
- Use the planning meeting information as well as that gained from the interview.
- Important not to retraumatize or "harm" those being assessed.
- You MUST have enough information at the end.

Introduction

- Welcome and explain it's not a cure.
- Say what the TRiM process aims to achieve.
- Explain the timing.
- Say what to do "if you want to leave" (however, encourage participants not to do so).
- Confidentiality issues.

- Explain how the information will be used.
- Ask participants to "talk for yourself" when working with groups.

Before

- Establish rapport.
- Initial identification of the very stressed.
- Get the group talking amongst themselves (also known as "bouncing").
- Make it long enough but no more.
- Let them see it as normal conversation.
- Ask questions, but do not force answers.
- Reflect and summarize.

During

- Cover the initial risk assessment items (1–4).
- Use the "rewind the tape" analogy.
- Avoid blame and shame amongst the group (retraumatization).
- Ensure you get the overall picture or story of the trauma across.
- Stay away from enquiring about "feelings"; instead, try to focus on thoughts.

After

- Get them down from the "high" of the "during" stage.
- Ask about the time since (items 5–10).
- Set the scene for the future.

Conclusion

- Tell them there will be a follow-up and how to get hold of you.
- Feedback individually or as a group.
- Ask to feedback information to managers:
  - Education/normalization (leaflet?)
- Take any questions.
- Finish off the risk assessment scoring.

Whether an initial TRiM interview provides a low risk assessment, a second one is carried out after about a month in order to ascertain whether personnel have adjusted to and coped with the psychological aspects of the incident. Satisfactory adjustment would be taken as a substantial lowering of the TRiM risk assessment score at the second risk assessment interview and also agreement between the TRiM practitioner and the interviewee that any temporary problems were resolving. Whereas initial TRiM interviews may be conducted in groups of up to eight people, or individually, depending on the nature of the traumatic event, follow-up TRiM interviews are always carried out on a one-to-one basis. Those personnel who do not appear to have adjusted to the event are either referred on for help or monitored again if their adjustment had been slow. In such cases a third risk assessment interview to decide upon the need for referral would be carried out after another month or two. As indicated above, TRiM practitioners should have easy access to supervision at all times.

Real-life Experiences

All examples quoted in this section have the full approval of the persons and organizations concerned.

Police Sergeant JC requested the TRiM process for his team after they dealt with a fatal road traffic collision – he did so to support his team, never thinking he might benefit from the program's assistance. This is Sergeant JC's story:

> The TRiM process really helped me after I attended and supervised the scene of a fatal road traffic collision (FATAC). The circumstances – in terms of the age and sex of the deceased and the fact that the deceased's mother arrived at the scene – were very similar to a FATAC I had attended about eight years earlier whilst serving with another force. This similarity came to mind whilst at the scene, but I then focused on dealing with the incident.
>
> I contacted the control room to declare the FATAC, requested relevant resources, and requested that the incident was flagged for TRiM. I never dreamed that I would need their services, let alone those of my GP, the force medical advisor, or a counselor. After all, "the uniform takes over" at these types of incidents . . . doesn't it?
>
> After the incident I debriefed my staff and as usual we all said how bad it was, but we all declared we were just fine. I requested TRiM because it's policy for a supervisor to be responsible for their officers' welfare and it's good practice to flag such incidents. I wanted to support my staff, lead by example, and show them there was no stigma attached to the TRiM process. I was there for them, not for me.
>
> I then started thinking about the FATAC from eight years earlier. I didn't know why, but I found myself thinking about it more and more and unable to shake the memory.
>
> Two TRiM risk assessors, a fellow sergeant and a constable, visited my station 72 hours later. As a section we decided on a team TRiM debrief where we were open and honest with ourselves, each other, and the risk assessors. Unbeknown to me I set alarm bells ringing for the assessors (TRiM practitioners) who later called me back in, expressed their concerns, and referred me immediately to the employee assistance program.
>
> I couldn't believe it. What, me? I need help? No. They must have got it wrong. What would people think of me if they found out I couldn't "cope" after dealing with a FATAC? What would they think of me if they found out I was probably going to have counseling? What would they say if I went off sick?
>
> Knowing what I went through when I had all the necessary support in place, I dread to think what would have happened to me had TRiM not been available and had the assessors not flagged their concerns. I've been lucky. It wasn't lucky suffering as I did, but where I have been lucky is in discovering what fantastic

colleagues I have, particularly my section. I made the conscious decision not to hide what I was going through. I never believed this would happen to me and thought that if it can happen to me, then it can happen to anyone. I was surprised at how many colleagues contacted me and told me that they too had suffered from stress, anxiety, depression, and post-traumatic stress.

Now, nearly two years on I have learned a lot about myself; I know how to spot the "signs" and to deal with them. I now have a very positive and determined outlook in life, and I intend to enjoy it. I only wish that I had had access to TRiM after that FATAC 10 years ago.

Another example is the case Police Inspector HB, who spoke about his TRiM experience at a public meeting as follows:

On the night of Friday 11th April 2008 going into Saturday morning, I was working as the duty inspector. I was on patrol, single crewed, in an area just outside Torquay's night-time economy hot spot, when I was approached by a male who said another male nearby had been disturbing him and asking him for cigarettes whilst pointing to a knife in his trousers. He went on to claim that the second male was in company with a female.

Some minutes later I found a male fitting the description of the male with a knife in company with a female. I informed Comms [the police control room] I intended to stop and check him, but at that time there were no other units available locally. A cursory search revealed nothing after patting down legs and searching pockets. Still sensing something wasn't right, I asked the male to pull back the waistband of his trousers. As he went to comply he instead pulled out a 10-inch knife, held it at head high, pointed it toward me, and screamed in what can only be described as a feral manner.

I shouted something like "Drop the fucking knife." He didn't comply – I sprayed him [with CS gas], called code zero (which indicated I need urgent assistance), and he dropped; I cuffed him waited for the troops to arrive. Job done. . . . So I thought.

My section duly arrived and took him back to custody, trussed like a Christmas turkey in the back of a van, leaving me and my patrol car on the side of the road. I got in my car and started to drive back to the nick, but after about 200 yards just stopped. For some reason I had started to cry and just simply could not drive. To put this into context, I play rugby; I've been in scrapes before. I have even had to tackle someone armed with a loaded firearm. But on this occasion my coping mechanism had imploded. To my eternal shame I had to call up on the radio for an officer to come and collect me and drive me back to the nick.

When I got back, a little more composed, my team were exceptional and obviously made me a cuppa. But for some reason I was now functioning on a lower level than what I was used to. Being the one zero (district response inspector) meant I still had a whole host of custody reviews, missing people, and other matters to deal with as well as doing an arrest statement, etc., slowly. I finished my night shift at about 0900 hours and drove home; my wife had already gone to work. I barely slept, I felt very tense, I kept needing to go to the toilet, and I kept envisioning what had happened and what could have happened.

The following night, my section had actually appointed someone to double crew with me [result] and the night went well; I thought everything was great. But when I got home again and tried to sleep, I was in turmoil. Three days later, a sergeant and police constable spoke to me for some time about the incident. At the time I am not sure whether I knew I was undergoing the TRIM process, but just being able to talk openly about the incident, not in front of my immediate section, helped. It also put other things into perspective and helped me understand simple facts surrounding body changes at times of emotional trauma.

I don't want any of you to think TRIM is a cure, but I do want all of you to be aware of it. It helped me enormously, and there should be no shame whatsoever attached to using it. What I do want you to go away with, though, is "Look upwards as well as down. Your sergeant may offer you TRiM, or simply may just have a chat with you, but who is doing the same for them?" Be there for each other.

A final example of a grassroots view of TRiM is from a police constable who unwittingly participated in the TRiM process:

The 10th February 2001 was a turning point in my life; it was the last straw to almost 30 years of deceiving myself and everyone else that I was fine and nothing could bother me, that I could cope with anything that "the job" could throw at me, and that I would always manage. Most police officers in a whole lifetime in uniform might attend one or two monumental incidents that would have a profound effect upon them, but somehow the odds always seemed to be stacked up against me personally. It became a parade room joke that if I were on duty, some disaster would occur. I always tried to outwardly laugh off these light-hearted comments, but underneath I knew the reality only too well of the effect that those "once in a lifetime of service" incidents have on you. My service in the police began in the mid-1970s, and within two years I had probably seen more in my rural and isolated Cornish beats than many officers would experience before drawing their pensions. The loss of six fishermen on the trawler the *Ben Asdale* in hurricane-force storms near the Lizard was my first experience of looking down upon the faces of men fighting for their lives as our breaches-buoy rescue equipment was ripped from the cliff tops where we struggled in vain to help them. The desperate search for two colleagues in a missing panda car, swept into the sea by mountainous waves near Helston, became a national story as we recovered only one of our friend's bodies along with their vehicle the following morning, and not long afterward the attention of the nation was gripped again by the fate of the lifeboat the *Solomon Browne*; thank goodness the television cameras didn't focus upon a young officer assisting to recover the bodies of those brave young men who answered the call as the crew that tragic night.

The following years of my service continued to be marred; a road accident soon followed in a remote area of Dartmoor on a bad bend where all my might could not budge the door of the upturned car before it exploded in front of me, taking the lives of three teenagers with it. Other similar incidents occurred until one night, in a most isolated spot on the moors, I sat and comforted a woman as she died in my arms at the scene of another road accident, which had already claimed the life of the driver a few moments before I arrived. Clearly people were starting to notice that I wasn't coping at all well. My defenses might have seemed rather odd to most. I threw myself into my work, working way on past my hours and using rest days to take on more and more community work. Anything to avoid the reality of what rest, time to think, and the dark lonely nights would bring. I was booked off sick for a month to give me time to get better. That was the worst thing that could have happened. I was deprived of all my defense mechanisms I had created over those years of sleepless nights, night-time trauma, and terrible visions. I came back to work in an even worse state, having had 28 days to dwell on the guilt and remorse caused by my failure to save the lives of those I had let down. My mind had twisted the truth beyond description; the three youngsters looking at me with desperate faces was an image that regularly haunted my hours when I should have been sleeping. To say that I enjoyed perhaps one good night's sleep a month might have been an exaggeration – I had forgotten what it was like to wake up refreshed in the morning. A cot death in a remote village resulted in me having to lift the poor little child, only a few months old, from her bed. That night when my daughter, herself a similar age, cried out for her night-time feed I lifted her up and knew of little more until my wife found me collapsed on the floor, an uncontrollable wreck some time later.

As time went on, I found that 16th February each year was dreaded when it should have been a joyous occasion. I always managed to position myself so that I didn't have to see my daughter's face on the other side of the flames on her birthday cake while she looked at me with the flames flickering on her innocent face as she blew out the candles. She still doesn't know why daddy always went out of the room to "check on something" while mummy sliced the cake and covered up for his absence as he composed himself. I even applied to become a "Post-incident Colleague Supporter" in the hope of avoiding others from living the nightmare I found myself in, but the training course involved images of men on fire and, needless to say, as I was helped sobbing from the classroom by bemused colleagues, it was clear that something was seriously in need of being put right.

But the 10th February 2001 was when I decided that to enter a burning building in the vain and impossible hope of rescuing a trapped child might bring a solution to the living nightmare I was experiencing. As I crept further into the flaming abyss, I heard what I thought to be the voice of my daughter screaming out, and I found the child – a young lad – and pulled him out to safety before returning and bringing out his father just as the Fire Service arrived. As the ambulance crew took me, injured, to hospital, I caught sight of my burned face in a mirror, the living nightmare continued. I was hailed a hero, national accolade followed, and the pretense that all was "all right" continued. But now almost 30 years of lies and deceit, failure, and shame came to a head.

By chance I was talking to the force TRiM co-ordinator about something totally unrelated. It was clear that all was not right, and gently over a period of time I was encouraged to accept the support that privately I had wanted for so many years but denied would be of any use, because everything was fine. Over the next few weeks a friendly but firm hand supported me; never once did I receive platitudes and false sympathy, but instead tangible and very real support and nonjudgmental understand-

ing. I was at last being listened to without me feeling a freak. With a variety of seemingly extraordinary strategies it felt as though the burden of 30 years was lifting. I started to sleep at night, my pyjamas were no longer drenched in sweat in the morning, and I found myself able to look at those past incidents with a fresh eye. I remembered so much of those times as a person looking anew at each individual occasion. I had forgotten so much that my mind had shut out and at last I knew that if somebody had died it was okay to say, "I tried my best."

For the first time in over 25 years, I walked past the lifeboat station at Mousehole without fear of what my mind might do, the harbor at Porthleven near Helston was no longer to be avoided, the cliff tops above the Lizard peninsular had become sunny and beautiful once again, and recently I walked past that bad bend on an isolated road on Dartmoor. I stood and said goodbye at last to the three unfortunate teenagers who I tried my utmost to save, and at the same time I said goodbye to a lifetime of demons who had driven me not to be too bothered if the 10th February 2001 was the last time they would haunt me. I then returned home and watched my daughter, as she laughingly struggled to blow out 21 candles on her special birthday cake, so many candles spreading light and joy on her face, as it had never done before. She looked at me as I stayed in the room beside her for the first time in 21 years and smiled; I didn't need to explain . . . it was as if she knew why.

I sleep now, I look back on those various incidents and can talk about them now, I went out and bought a campervan and found that there's a whole life out there to live – because my life, and the will to live it, has been given back to me by a small word that appears on one of those multitude of posters on the parade room wall . . . TRiM.

The Evidence That TRiM Helps

The TRiM process has been subjected to a considerable amount of research, including a large randomized controlled trial carried out in the UK Armed Forces by Greenberg *et al.* (2010). Studies have also been done looking at whether TRiM can measure changes in psychological health after traumatic events and whether TRiM training can alter perceptions toward mental health problems. The summation of this and other research has shown TRiM to be highly acceptable to those whom it aims to help, that it does not cause harm, it improves organizational functioning, is able to measure changes in psychological health over time after a PTE, and is able to change trainee TRiM practitioners' attitudes toward dealing with stress in others.

In Devon and Cornwall Police, since the implementation of TRiM in the spring of 2006, there have been 2300 incidents where TRiM has been used in some way. Of these, 600 TRiM risk assessment interviews have been carried out. To date none of the staff seen at the initial TRiM assessment interview at 72 hours post incident have reported sick during the intervening period to the follow-up assessment interview one month later. They have been watched over whilst waiting, and this appears to have been sufficient.

The data are highly supportive of TRiM, although it is fair to say that TRiM is not a form of "penicillin" for PTSD. The NICE guidelines (2009) suggest that preventing

PTSD is not currently possible, but through the use of a peer support process, such as TRiM, organizations can maximize the opportunity for personnel to remain resilient when exposed to traumatic events and thereby keep them being productive at work whilst ensuring that the minority who need mental health support are encouraged to seek it.

Conclusion

TRiM is an innovative system of peer group traumatic stress management which has been robustly investigated and found to be acceptable to those who might benefit from it and, most importantly, to do no harm. From an organizational perspective there is also evidence that TRiM supports efficient organizational function and is perceived as being of substantial benefit for TRiM practitioners themselves. However, if TRiM is to be effective then it is essential that the system is implemented correctly and the right personnel are trained to become practitioners; and, once trained, they must remain up to date. As with other forms of intervention, it is important to ensure that the correct governance measures are put in place to ensure practitioners remain current and effective. However, such measures need to be simple enough to ensure compliance is possible in an otherwise busy organization. Furthermore, it is hoped that an organization's use of TRiM over time will help to reduce the stigma which surrounds the potentially detrimental effects of traumatic stress. Even when an organization is not routinely faced by threatening incidents, the regular exercising of the TRiM system should ensure that when such circumstances are encountered, all personnel are cognizant of potential effects of traumatic events and are best placed to remain resilient in the aftermath.

References

Brewin, C. R., Andrews, B., & Valentine, J. D. (2000). Meta-analysis of risk factors for posttraumatic stress disorder in trauma exposed adults. *Journal of Consulting and Clinical Psychology*, *68*, 748–766.

Clair, M. E. (2006). The relationship between critical incidents, hostility and PTSD symptoms in police officers. Unpublished data.

Crisp, A. H., Gelder, M., Goddard, E., & Meltzer, H. (2005). Stigmatisation of people with mental illnesses: A follow up study within the changing minds campaign of the royal college of psychiatrists. *World Psychiatry*, *4*, 106–113.

Department for Work and Pensions and the Department of Health. (2008). *Dame Carol Black's review of the health of Britain's working age population: working for a healthier tomorrow*. London: Author.

Green, B. L., Grace, M. C., Lindy, J. D., Gleser, G. C., & Leonard, A. (1990). Risk factors for PTSD and other diagnoses in a general sample of Vietnam veterans. *American Journal of Psychiatry*, *147*, 729–733.

Greenberg, N., Gould, M., Langston, V., & Brayne, M. (2009). Journalists' and media professionals' attitudes to PTSD and help-seeking: A descriptive study. *Journal of Mental Health*, *18*(6), 543–548.

Greenberg, N., Langston, V., Everitt, B., Iversen, A., Fear, N.T., Jones, N., *et al.* (2010). A cluster randomized controlled trial to determine the efficacy of Trauma Risk Management (TRiM) in a military population. *Journal of Traumatic Stress, 23*(4), 430–436.

Hickley, M. (2007, February 19). 1,000 police are off work with stress every day. *Daily Mail.* Retrieved from http://www.dailymail.co.uk/news/article-437217/1-000-police-work-stress-day.html.

Hoge, C. W., Castro, C. A., Messer, S. C., McGurk, D., Cotting, D. I., & Koffman, R. L. (2004). Combat duty in Iraq and Afghanistan, mental health problems, and barriers to care. *New England Journal of Medicine, 351*, 13–22.

Hoge, C. W., Ritschard, H. V., & Cooper, C. L. (2002). Obstacles to effective organizational change: The underlying reasons. *Leadership and Organization Development Journal, 23*, 6–15.

James, G. (2010). *Breaking down the barriers to support: Improving access to Trauma Risk Management (TRiM) in Devon and Cornwall Police.* Exeter, UK: Exeter University.

Jeannette, J. M., & Scoboria, A. (2008). Firefighter preferences regarding post-incident intervention. *Work and Stress, 22*, 314–326.

Jones, N., Roberts, P., & Greenberg, N (2003). Peer-group risk assessment: a post traumatic management strategy for hierarchical organisations. *Occupational Medicine, 53*, 469–475.

Koopman, A. (1991). *Transcultural management.* Malden, MA: Basil Blackwell.

McFarlane, A. C., & Yehuda, R. (1996). Resilience, vulnerability, and the course of posttraumatic reactions. In B. A. van der Kolk, A. C. McFarlane, & L. Weisaeth (Eds.), *Traumatic stress: The effect of overwhelming experience on mind, body, and society* (pp. 155–181). New York: Guilford.

McGeorge, T., Hacker Hughes, J., & Wessely, S. (2006) The MOD PTSD decision: A psychiatric perspective. *Occupational Health Review, 122*, 21–28.

Meyer, H. E. (1979). Personnel directors are the new corporate heroes. In T. H. Patten (Ed.), *Classics of personnel management.* Chicago, IL: Moore Publishing.

Miller, L. (2006). *Practical police psychology.* Chicago, IL: Charles C. Thomas.

Misra, M., Greenberg, N., Hutchinson, C., Brain, A., & Glozier, N. (2005). Psychological impact upon London Ambulance Service of the 2005 bombings. *Occupational Medicine (London), 59*(6), 428–433.

National Institute for Clinical Excellence (NICE). (2009). *Guidance for employers on promoting mental wellbeing through productive and healthy working conditions.* London: Author.

National Institute for Health and Clinical Excellence (NICE). (2005). *Post-traumatic stress disorder (PTSD): The management of PTSD in adults and children in primary and secondary care.* London: Author.

Nowicki, D. (2000). Mixed messages. In G. Alpert & A. Piquero (Eds.), *Community policing: Contemporary readings* (2nd ed.). Long Grove, IL: Waveland Press.

O'Brien, L. S., & Hughes, S. J. (1991). Symptoms of post traumatic stress disorder in Falklands veterans five years after the conflict. *British Journal of Psychiatry, 159*, 135–141.

Ozer, E. J., Best, S. R., Lipsey, T. L., & Weiss, D. S. (2003). Predictors of posttraumatic stress disorder and symptoms in adults: A meta-analysis. *Psychological Bulletin, 129*, 52–73.

Rogers, E. (2003). *Diffusion of innovations* (5th ed.). New York: Free Press.

Rose, S., Bisson, J., & Wessely, S. (2003). *Psychological debriefing for preventing post traumatic stress disorder (PTSD) (Cochrane Review).* The Cochrane Library (1). Oxford: Update Software.

Sijbrandij, M., Olff, M., Reitsma, J. B., Carlier, I. V., & Gersons, B. P. (2006). Emotional or educational debriefing after psychological trauma: Randomised controlled trial. *British Journal of Psychiatry, 189*, 150–155.

Solomon, Z. (1989). Untreated combat related PTSD: Why some Israeli veterans do not seek help. *Israeli Journal of Psychiatry and Related Sciences*, *26*, 111–123.

Solomon, Z., & Mikulincer, M. (2003). Trajectories of PTSD: A 20-year longitudinal study. *American Journal of Psychiatry*, *163*, 659–666.

Van Emmerik, A. A., Kamphuis, J. H., Hulsbosch, A. M., & Emmelkamp, P. M. (2002). Single session debriefing after psychological trauma: A meta-analysis. *Lancet*, *360*(9335), 766–771.

13

Evidence-Based Support for Work-related Trauma: The Royal Mail Group Experience

Jo Rick, Andrew Kinder, and Steven Boorman

Introduction

Starting in 2002 a unique piece of research was undertaken at Royal Mail Group (RMG). The study, which looked in detail at the impact of trauma support services and the management of workplace trauma from an organizational perspective, was the first of its kind in the United Kingdom. Over a two-year period more than 800 RMG workers exposed to a potentially traumatic incident in the course of their work were contacted and invited to take part in a study of their well-being. Measures of symptom levels were taken for the 13 months following the incident. Types of support received by workers and their absence levels were also recorded. The study findings provide some of the best available applied evidence about safe and effective trauma management from an organizational perspective (Rick, O'Regan & Kinder, 2006).

This chapter describes the evolution of the RMG approach to trauma management, details the structure and content of the current RMG trauma management intervention, and presents findings from the study. It concludes with a summary of good practice in management for workplace trauma.

RMG is one of the largest organizations in the United Kingdom with approximately 140 000 employees in a variety of business units. There is a wide range of job roles within RMG. Perhaps the most recognizable of these include delivering mail to householders, handling high-cash-value items in secure vans, and serving customers from post offices in the high street. However, behind this public face of RMG sits a considerable infrastructure that includes trains, airplanes, cars, vans, and trucks. RMG vehicles cover 500 million miles a year – the same as driving to Jupiter and back. In addition, RMG includes the administrative and logistic functions to manage such a large operation. The result is that virtually every type of occupation or job role can be found somewhere in the organization. Inevitably in an organization of this size and scope, there is the potential for employees to be harmed through accidents or incidents which they – or anyone – could experience in the course of a working day (e.g., a road

International Handbook of Workplace Trauma Support, First Edition. Edited by Rick Hughes, Andrew Kinder, and Cary L. Cooper.

traffic accident). However, RMG faces an additional challenge in that their workers can be subject to violent attack – verbal or physical assault, armed raids, and even hostage-taking situations – precisely because of the nature of their job.

Post Offices[1] operate in almost every town or village and are more numerous than all the banking and building society outlets in the United Kingdom combined. These offices deal with large sums of money as do the armored vans (run by the cash-handling business) that transport the money. Both post offices and armored vans are subject to bandit attacks and sometimes to hostage-taking incidents. Service delivery personnel – postmen and -women – can also experience potentially traumatizing incidents as a result of their work, from dog attacks to armed theft of delivery bags. Such traumatic incidents create a multitude of problems not only for individuals but also for the organization. These include:

- Personal distress of those directly or indirectly involved in the incident.
- Sickness absence of those directly or indirectly involved.
- Costs of providing trauma management and counseling.
- Costs of providing temporary cover for absent workers.
- Loss of money through theft.
- Costs incurred through damage to property or loss of business.
- Industrial relations problems if the organization is seen as failing to care for injured workers.
- Costs of security.

The need to protect staff and minimize the impact of traumatic incidents has long been well established within RMG. The expertise in trauma management within RMG and the incidence of a range of potentially traumatizing events combine to make RMG a potentially fruitful research environment for understanding how workers who are subject to traumatic stress can best be supported.

The Royal Mail Approach to Trauma Management: A Brief History

Early 1990s: the original concept

In the early 1990s a team working for the Occupational Health Services within the Post Office identified that an important area of clinical support was needed for those subject to violent attack. The team, including occupational health physicians and a consultant psychiatrist and managed by a consultant psychologist, was responsible for the development of the new trauma care program. This new model, influenced by the work of Dyregrov (1997) and Mitchell (1983, 1993), and based on the models of psychological debriefing that were emerging at the time, is described in detail elsewhere (Tehrani, 1995; Tehrani & Westlake, 1994).

The late 1990s: change and development

In the late 1990s two events coincided to influence the development of the original trauma care program.

First, the Post Office went through a series of major organizational changes. The entire business was reshaped and rebranded initially as Consignia, and subsequently as Royal Mail Group (RMG). One consequence of these changes was that the occupational health and employee support services that had previously been provided internally were outsourced to Atos Healthcare. The substantial reorganization of the businesses and changes to the provision of occupational health services necessitated a fundamental reassessment of service delivery models across RMG which included trauma care provision.

At around the same time research evidence was emerging that challenged widely held beliefs about the efficacy of psychological debriefing as an early intervention for psychological trauma. In 1998 the first of a series of systematic reviews of the evidence on psychological debriefing was published (Rose *et al.*, 2002). This review used standardized and well-established procedures to examine the evidence on psychological debriefing from the best available studies. The results of the review called into question the benefits of psychological debriefing approaches, concluding they at best were ineffective and at worst could increase trauma symptoms in some people. Subsequent updates of the original 1998 review have further confirmed these findings.

As a result of the RMG review of service provision and the emerging evidence on debriefing, trauma support services at RMG were reshaped and evolved into a new trauma management service which was implemented on a rolling basis across the organization. The main changes were to focus more on practical support and information and to re-establish trauma management procedures and monitoring in the newly reconfigured businesses and outsourced occupational health arrangements.

A key learning point from the review of trauma management services was that this is a new and evolving area of practice. The trauma management systems need to be continually reviewed to match the changing needs of the organization and to ensure that new evidence based on best clinical practice is assimilated into the service.

The trauma management service at RMG was further developed in line with the guidance for clinical practice in the National Health Service (NHS) published by the United Kingdom's National Institute for Health and Clinical Excellence (NICE) in 2005. This guidance

1. confirms the findings of earlier reviews that single-session debriefing should not be routine practice in the delivery of services;
2. highlights the need for social, practical, and emotional support in the aftermath of an incident;
3. recommends watchful waiting as an appropriate way of monitoring an individual's progress following exposure to a potentially traumatic incident; and
4. recommends eye movement desensitization and reprocessing (EMDR) and trauma-focused cognitive-behavioral therapy (TF-CBT) as treatments for PTSD

There are no specific guidelines for non-healthcare organizational approaches to managing trauma. Nonetheless the general guidance from NICE on responding to the early stages of psychological trauma applies equally in organizational settings. Numerous organizations employ peer support systems to provide practical, emotional, and social support post incident, often referred to as "buddy systems." Some of these systems aim to screen individuals for symptoms and assess the need for further specialist support or input.

One of the best evaluated of these systems is Trauma Risk Management (TRiM). Developed by the Royal Navy, TRiM is a peer-delivered support program for personnel who have experienced a traumatic event. It is based on the view that although post-traumatic illnesses cannot be prevented, they can be treated successfully and early identification of the problem is important in ensuring that those who require help are identified and offered appropriate support. The service is provided by trained serving service personnel.

A final aspect of the NICE guidance relevant to organizational procedures states that in providing immediate interventions for all survivors of traumatic incidents:

> All health and social care workers should be aware of the psychological impact of traumatic incidents in their immediate post-incident care of survivors and offer practical, social and emotional support to those involved.
>
> (NICE, 2005, p. 92)

Current Trauma Management Services at RMG: Support Post Trauma (SPoT)

The current RMG trauma management system has four stages:

1. Same-day crisis management and defusing.
2. Ongoing manager support (via the SPoT protocol).
3. Trauma counseling.
4. Specialist trauma treatment (including EMDR and TF-CBT).

Crisis management and defusing

Crisis management on the day of the incident aims to provide immediate practical, social, and emotional support in line with NICE recommendations. Same-day crisis management is an automatic process conducted by a supervisor or line manager within the same work period in which an incident occurs, or very soon after.

The primary aim is to ensure the immediate safety and well-being of staff, summoning the emergency services, and identifying those people who need further support. Taking accurate details of the incident is important, such as the names and contact details of those who were involved and whether further support is likely to be necessary.

Secondly, the manager will provide an opportunity for everyone involved in the incident to talk about what happened (i.e., defuse what has happened) and to identify how the operation can get back to normal as soon as possible. Giving specific roles to people in the location where the incident took place is one way of achieving this. Care needs to be taken if the location of the incident is a crime scene and forensic examinations need to take place before anything is changed.

Third, the manager will explain and hand out information on the further support available to workers. Leaflets, which include self-help material and a letter that

individuals can take to their GP, are provided which enable people to absorb information at their own speed. If appropriate, the manager will also cover referral direct to the outsourced occupational health provider.

Finally, the manager will make arrangements to return to the office to follow up the defusing and to undertake the SPoT meeting for workers who require extra help.

All people involved in the incident will be reminded about how to deal with media interest which is especially important with a large or severe traumatic incident.

SPoT meetings

Similar to the Royal Navy TRiM system, SPoT meetings are a peer support approach delivered by specially trained managers. SPoT meetings normally happen within three days and two weeks of the incident and are voluntary for any workers who feel they would like more support. Managers can offer or employees can request SPoT meetings at a later date if this is felt to be appropriate.

This meeting is a more formal way of providing workers with the opportunity to talk about what happened, and for the manager to provide further information and support. The SPoT meeting involves going through the facts of the incident. This meeting also forms part of the organization's "watchful waiting" approach as it gives these specially trained managers the opportunity to understand how individuals are reacting to their experiences and, if appropriate, arrange for further follow-up and support or offer an appointment for trauma counseling.

Trauma counseling

Where necessary, trauma counseling is normally provided within one month of the incident. This third stage of the trauma management process is available when a worker continues to experience trauma symptoms. It provides a more in-depth screening and specialist support. Workers can self-refer or be referred by their line manager, and the counseling is conducted by specialist counseling personnel via RMG's occupational health service providers, Atos Healthcare.

RMG's Employee Assistance Programme can also be used as a back-up resource where, for example, there are additional legal, social, or domestic issues that need support or where more extended trauma counseling support is required.

Both SPoT meetings and trauma counseling are normally used with individuals, although, on occasions, occupational health practitioners provide support to groups where this is appropriate.

Onward referral

In more severe cases where initial stages of the trauma management process have not resolved problems or where there is late referral of cases, the final stage of the trauma management process offers the opportunity to refer workers on to a number of different treatment options. In line with current NICE guidance, workers can be referred for treatments such as EMDR or TF-CBT dependent on symptoms and individual preferences. As part of this work, intensive trauma rehabilitation plans are put in place.

The RMG Study

The research was prompted by concerns regarding the lack of evidence about effective support for workplace trauma, and particularly the systematic reviews of evidence on the effects of psychological debriefing described in Chapter 1. Primarily the RMG research was concerned with answering the following question:

What can organizations do that is both safe and effective to support employees following a workplace trauma?

The study, funded by a consortia of organizations under the auspices of the British Occupational Health Research Foundation (BOHRF) and hosted by RMG and Atos Healthcare, set out to examine RMG workers' experiences of and reactions to workplace trauma over time and to look at what support they accessed and whether it helped reduce trauma symptoms and absence. The research was conducted by the Institute for Employment Studies and supported by the Communication Workers Union (CWU) and the Communication Managers Association (CMA).

A particular strength of this research is the "real-world" setting in which it takes place. This was an evaluation of an organizational innovation as it happened. The study was designed to take advantage of a naturally occurring change at RMG. The fundamental reorganization of the business and the reassessment of needs described earlier in the chapter led to the introduction of a new service delivery model for trauma management.

The trauma management program, developed by RMG both to meet the newly identified organizational needs and to conform to the emergent evidence on best practice, was being rolled out across the organization from 2002 onward. This intervention ultimately became known as Support Post Trauma (SPoT). During the course of the research, some workers received support post trauma via SPoT-trained managers, whereas others, in areas with no trained managers, did not. These circumstances provided the ideal opportunity for an applied study to compare outcomes for these two groups of workers. Ultimately this type of comparison enabled us to produce "real-world" evidence about different levels of support post trauma, and appropriate strategies and approaches to workplace trauma management.

Study Design

People and incidents

Workers eligible for inclusion in this research were either direct RMG employees engaged in a wide range of roles, or franchise holders. Direct RMG employees tended to work in service delivery or cash handling and distribution, although they included some directly employed Post Office counter workers. Franchise holders are the workers who provide the majority of Post Office services in the United Kingdom (i.e., post masters and post mistresses).

The aim of the research was to focus on those who had been exposed to a potentially traumatizing experience at work. There is no clear-cut definition of a traumatic event.

Rather such events are defined by the way individuals experience and respond to the situation and their symptoms following exposure to the event. The research therefore needed to identify workers who had been exposed to a potentially traumatic incident, and this was done through the Occupational Health (OH) reporting structures and accident management systems within RMG. These systems had been revised and re-implemented across the organization following the changes of the previous years. Incidents were notified and then tracked to ensure post-trauma support was appropriately implemented. These systems identified all workers who had experienced the following types of incidents, regardless of whether they self-referred to occupational health:

- Dog attacks.
- Accidents, falls, and road traffic accidents.
- Physical assaults.
- Robbery or attempted robbery.
- Armed robberies, hostage-taking situations, and ram raids.

Research process

Over a 24-month period, all RMG workers who had been exposed to a potentially traumatic incident during the course of their work were identified via RMG's OH incident-reporting systems. Once recorded, OH notified the independent research group with details of the incident. Provided that the incident had taken place in the last five days and the worker involved was not already in the sample, the workers were then contacted directly by the research group and invited to take part in the study.

Workers were tracked for a 13-month period and contacted at three time points: immediately post incident when they were invited to participate in the study; then at three and 13 months post incident. At each time point workers were asked a series of questions about their symptoms and the support they had received.

The survey included questions about the event itself and how the organization had responded, questions about symptoms associated with trauma, the impact of these symptoms on physical and emotional health, social support, and the experience of other incidents.

In addition to survey responses, linked data were collected on the type of incident experienced (via the incident-reporting procedures), details of any further OH involvement (via the OH trauma-tracking system), and outcome data from organizational records in the form of absence levels amongst employees and contract status amongst franchise holders.

Research Findings

Sample

Over the two years that recruitment to the study was open, a total of 837 workers were identified by the OH reporting systems as having been exposed to a potentially traumatizing experience. All participants were sent questionnaires immediately post incident and at 3 and 13 months. Overall, 55% responded to the survey at one or more time points.

There were no differences between workers responding to the survey and workers who did not respond in terms of sickness absence for the 12 months post incident (RMG employees only) or contract status (for franchise holders). The sample of employees responding to the survey was felt to be representative of RMG workers more broadly.

Experiences of trauma at work

Ninety-two percent of workers responding to the first questionnaire had directly experienced a potentially traumatizing experience. The most common experiences were a countersnatch or raid (affecting 38% of respondents) closely followed by an incident involving a weapon (32%) or physical assault (26%).

In 61% of all cases, workers had been required to give a witness statement on the day of the incident, and in 43% of cases had needed to leave work early or close the premises.

To gauge perceptions of the incident workers were asked both how usual or unusual this type of event was, and how upsetting they found it (Figure 13.1). A clear majority (78%) indicated that, overall, incidents were unusual, but one of the risks of the job. However, over half of those who had experienced a physical assault felt that such an experience was totally unexpected in the job.

When asked how upsetting they found the incident, around two in five workers overall described the incident they had experienced as very upsetting, with a further 18% describing it as totally devastating.

As can be seen from Figure 13.1, most workers felt that although their experience was rare, it was a known risk of the job and the largest group of workers described their experiences as "very upsetting."

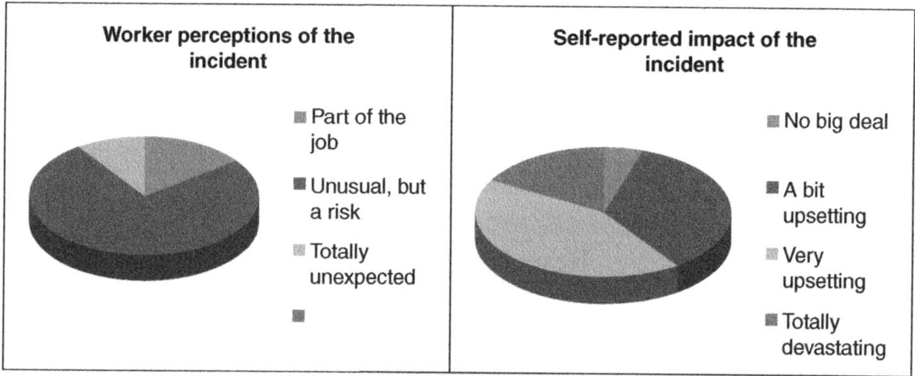

Figure 13.1 Perceived risk and impact of the incident.

Support

Workers were asked a series of questions relating to the type of support they had received following their exposure to a potentially traumatic incident.

The first level of support offered within RMG is that of defusing and crisis management. In 37% of cases workers report that the incident they experienced was attended by a network manager or a supervisor. In Post Office Networks and in Cash Handling and Delivery (CHD), evidence of this type of support was far higher (38–46%). However, for those working in service delivery there was limited evidence of supervisors or network managers attending incidents. This reflects the different risks in the separate business units, where the service delivery unit has fewer incidents and therefore had fewer staff trained in the RMG trauma management response.

In addition to attending incidents, part of the manager's role in defusing and crisis management is to provide support (e.g., assisting with statements) and think more widely about a worker's welfare (e.g., checking that they have support and won't be returning to an empty house).

Workers described considerable variation in the immediate levels of support they were offered. Overall, CHD workers reported highest levels of immediate support, with over half being offered help with witness statements to police or RMG security and two-fifths saying they were asked about support at home following the incident. In contrast only 16% of workers in Post Office Networks reported that they were either offered assistance with making a statement or asked about support after the incident. There appeared to be variation in the extent to which same-day support practices were embedded in the different business units at the time of the research.

SPoT meetings

The second level of support offered at RMG is SPoT, more formal support meetings organized by specially trained managers in the days or weeks following an incident. Workers responding to the survey were asked if they had been offered or taken part in a SPoT meeting following the incident.

Overall, just over half of the workers questioned had either taken part in a SPoT meeting or were planning to, or had been offered such a meeting but declined (participation in SPoT meetings is voluntary). Just under half of the sample said they had never heard of this type of meeting. The size of this latter group is not surprising given the aim of the research to contact some workers where RMG trauma management services had yet to be rolled out.

As with the first-level support services, there is considerable variation by business with those in CHD far more likely to have already taken part in a SPoT meeting (62%) than workers in Service Delivery (23%) and Post Office Networks (31%). In contrast, Service Delivery and Post Office Network workers were far more likely to say that they had never heard of this kind of support (61% and 41%, respectively).

The third level of support offered at RMG is more detailed trauma counseling provided by OH professionals. This type of support is generally only taken up by a far smaller proportion of those experiencing an incident – typically between 3 and 6%, as was the case here. Once again, CHD workers were far more likely to take up trauma counseling support from OH.

Satisfaction with Support

Workers were asked a series of questions about their satisfaction with the RMG response and the support they had received.

At each time point, workers were asked about their satisfaction with support they had received on the day of the incident (Figure 13.2). 55% rated support on the day as good or excellent, 29% rated it as neither good nor bad, but 16% rated support on the day as poor or very poor.

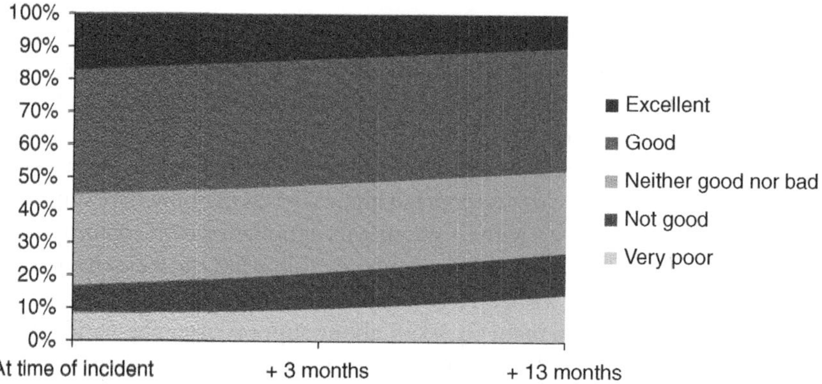

Figure 13.2 Satisfaction with first-day support.

Positive perceptions of help received were reasonably consistent over time. Forty-eight percent of employees continued to rate support on the day as good or excellent 13 months after the incident. Neutral views also declined only slightly over the time period. However, negative perceptions of support received on the day of the incident rose sharply from 16% at the time of the incident to 28% by 13 months post incident.

Workers were also asked about satisfaction with SPoT meetings. One hundred and sixteen participants attended a SPoT meeting during the course of the study. Overall participants in these meetings reported a very high degree of satisfaction with both the amount and quality of information that was provided during SPoT meetings.

Workers were also asked about how the information they received from the organization had helped. Those who had attended a SPoT meeting were more likely to report they:

- felt reassured about the symptoms they were experiencing,
- knew where to obtain further information, and
- knew where in the organization they could access further support.

A second set of questions asked about the impact of the support they had received from the organization since the incident. Again, those who had attended a SPoT meeting were more likely to report that

- they felt the company cared about their well-being, and
- the support enabled them to get "back to normal" quicker.

For some people, there were also improvements in confidence about going back to work.

The Impact of Support on Symptoms and Absence

So far, findings from research indicated that there were good levels of satisfaction amongst workers with both the information given and the type of support offered by RMG post incident. A key question for this research, as for many workplace health-related interventions, was whether or not the intervention reduced symptoms or sickness absence.

The Impact of Events (IES) scale (Weiss & Marmar, 1996) – a well validated and psychometrically sound scale – was used to measure symptom levels in workers exposed to an incident. Typically, people exposed to a trauma have high symptom levels in the days immediately after the event and these symptoms decline considerably over subsequent weeks, with only a minority of people experiencing problematic levels of symptoms by 4–8 weeks post trauma. The workers in this study followed this pattern, with significant drops in symptom levels by 3 months post trauma and again at 13 months post trauma.

Understanding whether and how the interventions in place at RMG to manage trauma impacted RMG workers involves looking at numerous different factors in combination. In this instance preliminary findings showed that in addition to the intervention itself (immediate support and SPoT meetings), a number of other factors were important, in particular the type of incident individuals had experienced and their perception of how well the organization had supported them. To understand these relationships, we wanted to look at interactions between

- The type of incident the individual had experienced,
- How they rated the support they received at the time or immediately afterward,
- Whether the individual attended a SPoT meeting,
- How they perceived the organization had supported them,
- Their symptom levels, and
- Absence over the 12 months post incident.

A statistical modeling technique (structural equation modeling) was used to look at the interactions between the different measures. This type of modeling approach enables you to look at complex interrelationships in data and to provide a "best-fit" model based on the data.

The relationships identified in this model (Figure 13.3) show two clear pathways:

- First, the type of incident directly affects symptom levels at three months and absence levels over the 12 months post incident. Workers who experience an incident perpetrated by another person (e.g., assault and armed raid) report higher

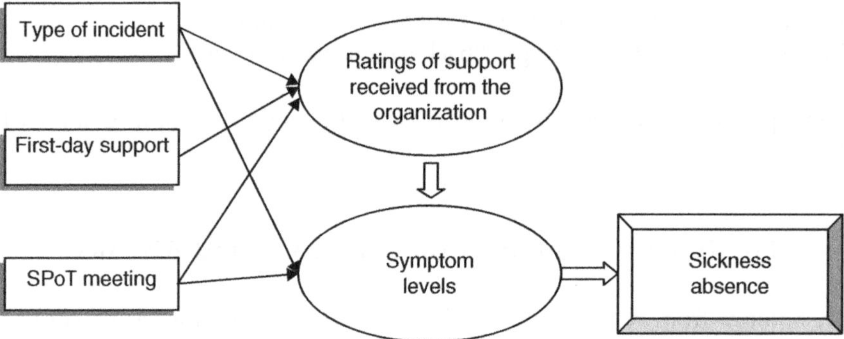

Figure 13.3 Relationships between type of incident, support, symptoms, and absence.

symptom levels at three months as opposed to those who experience an accident (e.g., a road traffic accident). Having higher symptom levels at three months is associated with higher absence at 12 months.

- In the second pathway it can be seen that both support on the first day and attending a SPoT meeting strongly influence the way an individual perceives the support they receive from the organization, their symptom levels, and subsequently their absence levels. Higher levels of support, particularly on the day of the incident, lead to higher perceived organizational support and lower symptom levels at the three-month follow-up. This in turn leads to lower absence levels for the 12 months post incident.

What these findings show is that the right type of support post incident (i.e., support that enhances a worker's perception of the way the organization is treating him or her) can help to reduce symptom levels in the medium term and absence in the longer term.

"Good" Support: What Is Effective and Safe?

The strongest relationship in the model discussed here is between support on the day of the incident and perceived organizational support. Both data from the questionnaires and interviews with workers indicated two aspects to this support: practical help and an empathic response.

Post incident, it is practical help that is most valued. This includes a range of activities such as

- assistance with bureaucracy or formal procedures such as helping with paperwork for accident reports or witness statements;
- assistance with security including helping to clear up and secure premises, and advising on security procedures; and
- guiding individuals through HR processes, explaining what they needed to do if they were taking time off.

The second element to post-incident help which was highly rated by workers was an empathic response from their supervisor or line manager, examples of which include:

- Displaying concern.
- Contacting family or friends.
- Accompanying the worker, for example to hospital or home.
- Ensuring the worker wasn't returning to an empty home.
- Following up in the days and weeks after the incident.

Two other aspects of good support highlighted by respondents were the immediacy of the response and a personal touch – form letters, whatever the source, were particularly poorly regarded.

SPoT meetings were the second aspect of good support. Respondents were asked a series of questions about the content of the SPoT meeting and their satisfaction with the process. Overall, the single most important element of the SPoT meeting was information about how to access further support in the organization – although very few workers went on to use this service.

Workers who rated the SPoT meeting as helpful were also likely to have had the SPoT process introduced and explained, identified their own symptoms following the incident, and had normal trauma reactions and coping mechanisms explained.

Implications for Practice

One of the contributions of the early thinking within the Post Office and continued within RMG is recognition of the need to have an "integrated" model of care so that there are arrangements in place for all aspects of the incident, including the hours immediately after the incident. How people feel they are treated at this early stage often sets up how they respond to later support interventions. For instance, if they feel supported and cared for during the immediate aftermath, this can have a beneficial effect in terms of how they feel the organization is generally supporting them. Often, the organization can provide reassurance that workers followed the right procedures and that it was not their "fault."

Organizations therefore need to have in place clear systems and procedures to deal effectively with the immediate crisis as well as during the days and weeks that follow (Chapter 21, this volume). The findings from this research show that workers' perceptions of the organizational response to incidents are central to how they feel about the incident and their subsequent symptom levels and absence. Getting support right can result in enduring positive perceptions of the organization's role, which in turn help recovery. The key to effective organizational intervention post trauma is early, practically focused, and empathic support from a line manager or supervisor.

Formal support from a trained manager (via SPoT meetings) was also found to play an important role in conveying information and concern from the organization. Those attending SPoT meetings gave the highest ratings of both perceived support and information provided after the incident. These research findings indicate that the SPoT approach to trauma management is a safe and effective mechanism for delivering continued support and information to workers post trauma.

Note

1. Although most Post Offices are now franchised, RMG still provides trauma care support.

References

Dyregrov, A. (1997). The process in psychological debriefings. *Journal of Traumatic Stress*, *10*(4), 589–605.

Mitchell, J. (1983). *Guidelines for psychological debriefing: Emergency management course manual*. Emmitsburg, MD: Federal Emergency Management Agency, Emergency Management Institute.

Mitchell, J. (1993). *Critical incident stress debriefing: An operations manual for CISD, defusing and other group crisis intervention services*. Ellicott City, MD: Chevron.

National Institute for Health and Clinical Excellence (NICE). (2005). *NICE Clinical Guideline 26: The management of PTSD in adults and children in primary and secondary care*. London: Author. Retrieved from http://www.nice.org.uk/nicemedia/live/10966/29771/29771.pdf

Rose, S. C., Bisson, J., Churchill, R., & Wessely, S. (2002). Psychological debriefing for preventing post traumatic stress disorder (PTSD). *Cochrane Database of Systematic Reviews*, (2), Art. No. CD000560. doi: 10.1002/14651858.CD000560

Rick, J., O'Regan, S., Kinder, A. (2006). Early Intervention Following Trauma a Controlled Longitudinal Study at Royal Mail Group, Report 435, Institute for Employment Studies.

Tehrani, N. (1995). An integrated response to trauma in three Post Office businesses. *Work and Stress*, *9*, 380–393.

Tehrani, N., & Westlake, R. (1994). Debriefing individuals affected by violence. *Counselling Psychology Quarterly*, *7*(3), 251–259.

Weiss, D. S., & Marmar, C. R. (1996). The Impact of Event Scale – Revised. In J. Wilson & T. M. Keane (Eds.), *Assessing psychological trauma and PTSD* (pp. 399–411). New York: Guilford.

14

The Development of a Practice Research Network and Its Use in the Evaluation of the "Rewind" Treatment of Psychological Trauma in Different Settings

William Andrews and Scott Miller

While clearly it would make no sense to either dismiss or ignore evidence about the effectiveness of psychological therapies, there is a continuing debate about what constitutes worthwhile and reliable evidence that can be translated into meaningful practice in the front line of treatment delivery. There is much debate around diagnosis. Post-traumatic stress disorder (PTSD) only entered the literature in the third edition of the *Diagnostic and Statistical Manual* (DSM-III; American Psychiatric Association, 1980). As a result, PTSD is commonly associated with life-threatening events, initially strongly related to war, but later including experiences of sexual abuse and many other sources of threat to life, such as road traffic accidents. However, a perception of "trauma" can arise from a wealth of circumstances affecting an individual or group of individuals, often not quite as dramatic in nature as major catastrophes or war zones. In the United Kingdom, the National Institute for Health and Clinical Excellence (NICE; 2005) recommends trauma-focused cognitive-behavioral therapy (TF-CBT) and eye movement desensitization and reprocessing (EMDR) as evidence-based treatments for psychological trauma. However, because of the hierarchical system of evaluation of evidence (discussed in this chapter), other innovative treatments that may prove of immense benefit to add to client choice of treatment are sometimes ignored and limited in availability. One such innovative treatment is commonly known as "rewind."

The Rewind Technique

The rewind technique is a non-intrusive psychological method for detraumatizing people, which can also be used for removing phobias. It is only performed once a person is in a state of deep relaxation. When fully relaxed, the client is encouraged to bring his

International Handbook of Workplace Trauma Support, First Edition. Edited by Rick Hughes, Andrew Kinder, and Cary L. Cooper.

or her anxiety to the surface, and then is calmed down again by being guided to recall or imagine a place where he or she feels totally safe and at ease. Their relaxed state is then deepened, and the client is asked to imagine that, in their special safe place, they have a TV set and a video or DVD player with a remote control facility. They are asked to imagine floating to one side, out of body, and to watch themselves watching the screen, without actually seeing the picture (creating a double dissociation). They watch themselves watching a "film" of the traumatic event that is still affecting them. The film begins at a point before the trauma occurred and ends at a point at which the trauma is over and they feel safe again. They are then asked, in their imagination, to float back into their body and experience themselves going swiftly backward through the trauma, from safe point to safe point, as if they were a character in a video that is being rewound. Then they watch the same images but as if on the TV screen while pressing the fast-forward button (dissociation). All this is repeated back and forth, at whatever speed feels comfortable, and as many times as needed, until the scenes evoke no emotion from the client. If the feared circumstance is one that will be confronted again in the future – for instance, driving a car or using a lift – they are asked, while still relaxed, to see themselves doing so confidently. The technique has the advantage of being nonvoyeuristic. Intimate details do not have to be made public. The underlying principles involved in processing the trauma memories using rewind are similar to techniques like EMDR but can offer certain advantages as the client is spared from needing to narrate the trauma memories. Practitioners trained in the use of rewind see psychological trauma as a continuum, from physiological arousal in response to a traumatic stimulus, through acute stress disorder, anxiety, depression, addictions, and PTSD. In other words, there are hosts of different responses to trauma, from a normal healthy response where a person makes a complete recovery, through to the manifestation of a variety of symptoms that may lead on to a variety of diagnoses.

Evolution of Rewind and Evaluation of the Existing Evidence

The rewind technique, originally known as the visual-kinesthetic dissociation technique and described as a treatment for phobias, evolved from neurolinguistic programming (NLP) (Grinder & Bandler, 1976). Muss (1991) described the adaption of the technique as an intervention for the treatment of psychological trauma with traumatized police officers. Hossack and Bentall (1996) described the use of the technique with five patients suffering from PTSD. The results conveyed that three out of the five patients had significantly improved, one patient showed partial improvement, and one patient showed no change at all. Carbonell and Figley (1999) investigated four PTSD treatments, including rewind, EMDR, thought field therapy, and traumatic incident reduction. A total of 39 participants were treated, and the results revealed that all four approaches demonstrated benefit. Griffin and Tyrrell (2001) described further modifications of the technique and have been training practitioners in this model for a decade. In a cohort study where 30 clients, assessed as suffering from psychological trauma, were followed over a 2-year period, Guy and Guy (2003) found that 40 percent of participants rated rewind as extremely successful, 53 percent rated it as successful, and none of the participants rated the method poor or as a failure. Poster presentations on the use of the rewind technique in trauma work with 34 war veterans (Bishop & O'Callaghan, 2010) and with 16 psychiatric patients who had a comorbid or main diagnosis of PTSD (Lohawala, 2009) have been presented at conferences. Both cohort studies demonstrated favorable results for the technique. In a mixed-methodology 4-year cohort study involving treatment of victims of the Troubles in Northern Ireland, Murphy (2007) found considerable benefit from the use of rewind. Extensive anecdotal evidence on the effectiveness of the technique has also been published (Griffin & Tyrell, 2008).

All of the above studies cited set out to study the use of rewind in deliberate experiments, albeit in cohort design with the absence of any sort of control. However, the observation in everyday psychological treatment delivery of a naturalistic occurrence of psychological trauma and the monitoring of treatment response through robust and systematic use of service user self-report instruments can yield a different sort of knowledge, both about the manifestations of psychological trauma and about effective treatments. A suitable vehicle for making such observation is a practice research network (PRN) (Parry *et al.*, in press).

Background to PRNs

Practice research networks are collaborations of practitioners committed to using their work-based settings as laboratories for the generation of practice-based knowledge. Mostly, the output from early examples of PRNs, such as the American Association of Marriage and Family Therapists, has been based on member surveys. Other PRNs have emerged where the focus is more about understanding process and outcome in treatment. For example, the Pennsylvania PRN emphasizes evaluation of the progress of clients in outpatient treatment and the UK Art Therapists' PRN develops practitioner-led research projects instead of relying on specialist researchers. The Clinical Outcomes in Routine Evaluation (CORE) PRN emerged from the early work of Michael Barkham and colleagues and placed emphasis on the evaluation of service delivery in primary care in the United Kingdom. Over time, as increasing amounts of data became available, benchmarks were developed for treatment in primary care settings, concerning acceptance rates into treatment, recovery and improvement rates, waiting times, and profiling of clients at assessment (Bewick *et al.*, 2006).

After an initial period of development throughout 2005 and 2006, The Human Givens Institute PRN (Andrews, 2007) was created to provide an opportunity to explore the results of therapy in everyday practice. The main aims of the HGIPRN were to actively promote and encourage outcome-informed research into the effectiveness of delivery of the human givens approach in a wide variety of settings, to provide a central point for collation of such research and to promote the concept of the "clinician researcher." It was anticipated that, as more and more robust data were accumulated, the gathering of evidence about the effectiveness of this approach would contribute to new knowledge and offer an alternative means via which innovative treatments could achieve recognition. Results from the initial pilot study at a general medical practice have been published (Andrews *et al.*, 2011).

Evaluation of "Rewind" Using the PRN

Within the PRN, it was possible to identify "trauma" cases because practitioners flag this up in the "identified problems and concerns" section of the Therapy Assessment Form filled in at assessment. Furthermore, some of the services participating in the PRN work exclusively with psychological trauma and receive referrals on that basis. Individual practitioners also focus exclusively on trauma work. This provided the opportunity to investigate how the subset of cases treated for psychological trauma progressed in treatment. With more than 400 trauma cases identified, the data revealed that over 75% of service users arrived at an agreed ending in treatment, over 90% had at least one additional measure beyond the first appointment, and over 70% of service users demonstrated significant improvement on at least one measure.

The Challenge of Evidence

> With respect to randomization I would paraphrase Churchill on democracy, that it is a terrible process that has little to recommend it except that it is better than the alternatives.
>
> (S. Hollon, personal communication, 2008)

It is a considerable challenge for evidence emerging from PRNs to gain acceptability and be treated seriously. Data gathered are viewed as "audit" as opposed to "research" because of the absence of any experimental manipulation. The randomized control trial (RCT) is considered the strongest design to test whether any given psychological treatment is better than placebo or other treatment under ideal conditions. It remains firmly at the top of the evidence hierarchy applied by bodies such as the National Institute of Health and Clinical Excellence (NICE), where the RCT or meta-analysis of RCTs is considered level-1 evidence. The RCT's strength lies in the fact that subjects are randomly assigned to the treatment or control group, increasing the likelihood that the groups will be comparable both in terms of variables that can be identified and measured, and in terms of variables that cannot be identified or measured but that are likely to be associated with improvement resulting from treatment. However, it also has shortcomings. An RCT may require that service users are unnaturally assigned to a treatment group they would not normally choose. Furthermore, participating practitioners may be proscribed from carrying out treatments that they would normally choose or they may be obliged to prescribe treatments that they would not normally use, simply to form a control arm for the study. Ethical issues can also arise in the conduction of RCT studies, particularly where waiting list controls are used and service users are denied access to treatment that might otherwise be made available. In addition, RCTs may ultimately be inadequate because they just do not satisfactorily generalize to the population at large and there may be selection bias generated by the very fact that those willing to participate in RCTs are not representative of service users in general. RCTs also tend to exclude participants suffering from comorbid conditions, thus creating data relating to unnaturally "pure" diagnostic categories.

As meta-analyses of RCTs are deemed to represent the highest form of evidence it is appropriate to consult meta-analytic findings about trauma treatments. A series of meta-analysis studies were conducted between 2005 and 2009 (Ehlers *et al.*, 2010). While their findings suggest that the treatment should focus on the traumatic memories and personal meaning of the trauma to the individual concerned, it was concluded that no one particular type of trauma-focused treatment demonstrated superiority over any other. To make matters even more confusing for providers of services looking to the literature for the best way of working with trauma, one controversial meta-analysis conducted by Benish and colleagues (Benish, Imel, & Wampold, 2008) found no difference in outcomes between any head-to-head comparison of bona fide treatments, whether trauma focused or not. This battle of the research giants leaves practitioners bemused. Clinical trials may have some significant limitations, such as strict entry criteria, in an attempt to limit participants to discrete diagnoses in order to control internal validity and rule out other explanations for variance. But the accompanying demand for strict adherence to manualized treatment protocols and intensive supervision also does not really seem to fit with reality for many practitioners, and it seems removed from the real world of everyday practice.

It would seem to make perfect sense to involve the consumers of the service in providing feedback about what they find helpful. The value of working with feedback on progress in treatment has long been advocated by well-known theorists such as Lambert *et al.* (2005) and Duncan, Miller, and Sparks (2004). In fact, several RCT studies have clearly demonstrated the value of working with feedback to cut down on dropout rates and help recognize clients at risk of a negative outcome. Furthermore, an emerging literature around excellence (see Box 14.1, "The Elephant in the Room") highlights the importance and significance of measuring one's baseline performance and continually soliciting feedback in order to make improvements. This is as relevant to psychological trauma treatment as to any other.

Box 1 The Elephant in the Room

Scott Miller *is the founder of the International Center for Clinical Excellence (ICCE), an international consortium of clinicians, researchers, and educators dedicated to promoting excellence in behavioral health services.*

During the last few decades, more than 10 000 "how-to" books on psychotherapy have been written. Enormous effort has gone into the development of evidence-based "treatments" in the last four decades, with, now, 145 manualized treatments officially approved for 51 of the 397 possible DSM diagnostic groups. This might lead us to believe that we are now clear on what treatment may work for a particular client with a particular psychopathology under what certain circumstances. Unfortunately, this does not seem to be the case. When bona fide treatments are compared in RCT studies, they largely produce similar outcomes. Furthermore, while any RCT study may establish differences over waiting list control, it fails to inform us of the service user experience of that particular treatment, why service users drop out, and why some service users in the control arms of these studies actually do recover. In short, as Robert Elliot (2001) suggests, the RCT is causally empty. But, perhaps more significantly, there is another confounding variable that puts a fly in the ointment, that is, the variability of results by therapist. It is an irritating fact for research that the "individual" delivering the therapy has an enormous influence on the outcome. Furthermore, the allegiance of treating practitioners to the particular model or orientation they are working with may actually be even more significant than the model itself. So, practitioner variability is a reality. Some practitioners get consistently better results than others, have fewer dropouts, and seem to be instrumental in bringing about more change for their clients. Another obvious and enormously important element with respect to therapeutic effectiveness concerns, naturally, the service users themselves. Establishing service user preferences for a particular method of working, presenting a cogent explanation that they can agree with, and offering strategies that fit well for them have all been shown to be important factors in determining outcome.

If it is true that practitioner variability is an important and significant factor in relation to outcomes for service users, it would make sense then to investigate what sort of characteristics, habits, or practices of the more successful practitioners might lead to better outcomes for service users. In addressing this issue, realizing that the existing psychotherapy literature failed to substantially investigate this, our research team in the ICCE took a different tack and looked to the expansive literature on excellence in general, across all fields of human endeavor. The investigation teased out three distinct features that provide the hallmarks of excellence regardless of the area of human activity investigated.

Know your baseline. If we do not currently know how we are doing, we cannot know where we are going. Although seemingly obvious, the fact is that the overwhelming majority of our profession does not know what sort of outcomes they get with their clients. They may "think" they know. In fact, researchers Hiatt and Hargrave (1995) demonstrated that, in a team of behavioral healthcare workers, the least effective believed they were as good as the most effective. What makes sense, then, is to know our baselines by measuring them. Thankfully, there is a wealth of well-validated brief self-report tools, such as the Clinical Outcomes in Routine Evaluation 10 question measure (CORE-10; Connell & Barkham, 2007) and the Outcome Rating Scale (ORS; Duncan, Miller & Sparks, 2004), available to assist us in this effort. Furthermore, the impact of soliciting for and obtaining feedback from service users on progress in treatment and on their perception of the therapeutic relationship has repeatedly demonstrated improved results from treatment, with lowering of dropout rates, better engagement, reduction of the number of sessions, and an overall larger impact of treatment measured by larger effect sizes.

Engage in deliberate practice. K. Anders Ericsson (2006), the expert on experts, defines deliberate practice as "effortful activity designed to improve individual target performance." The hallmarks of deliberate practice are captured by the acronym TAR (think, act, reflect). Applying this to our field, the superior performers might think more carefully about what approach is likely to be a better fit for their clients, having investigated what goals the service user is focused on achieving. Having acted to deliver those goals in partnership with the service user, the superior practitioner will reflect on what actually happened and adjust accordingly. She or he may solicit feedback from others, through supervision or peer support, to try to find additional ways of maintaining service user engagement.

Constantly solicit for feedback about progress. As with "knowing our baseline" (the first item on this list), using self-report measures session by session can provide us with feedback on ongoing progress in treatment and allow us to monitor the therapeutic alliance with our clients. It then provides us with the opportunity to make adjustments in treatment, or even "fail successfully" by realizing early on that the service user is not progressing in our care and may be more suited to a different provider or service.

It would seem sensible, then, to recommend that all treating clinicians should abide by these principles and be encouraged to solicit for feedback in treatment so our field can perhaps begin to learn more from those of us who seem to genuinely have more to teach us about improving outcomes for our service users.

What the PRN Tells Us That the RCT Does Not

One of the beauties of obtaining large quantities of data through the PRN is the opportunity this creates for exploring the relationships between measures that have been used across heterogeneous groups, mixed by age, gender, traumatic incident exposure, social support, medication, and complex ranges of confounding variables. In short, by virtue of the fact that the groups of cases are so mixed, PRN data offer a different type of control from that typically seen in RCT studies, where all these differences are controlled through random variation. In other words, by virtue of the variation across cases, randomization is naturally introduced. There are, however, no controls because the creation of controls through experimental manipulation would be against ethical practice, and practitioners feel it is incumbent upon them to begin treatment as soon as possible after service users present.

As the data gathered within the network are centrally collated and can be exported to computer software such as Excel™, it is possible to drill down into the data. Focusing on cases where psychological trauma was identified allowed examination of cases that met certain criteria. For instance, in cases where service user and provider had agreed when to end treatment, and where three different outcome measures, the Clinical Outcomes in Routine Evaluation Outcome Measure (CORE-OM), Outcome Rating Scale (ORS), and Impact of Events Scale – Extended (IES-E), had been used, regardless of which measure was used to look at change through treatment, there was strong evidence of large effect sizes suggesting considerable benefit from treatment. Drilling down further into the data, it was possible to filter out those cases recording extremely high levels of distress on these self-report measures. For example, in a cohort of 44 cases, the effect sizes suggesting considerable benefit from treatment were equally large and the final mean scores were in nonclinical ranges on all measures. Work is now in progress to benchmark the PRN trauma data with data available from RCT studies.

Empiricists who remain wedded to the sanctity of the RCT will inevitably make their criticisms. They can argue that there is no proper randomization of groups and there are no controls. There is no information about diagnosis. Treatments have not been manualized, and there is no way of knowing with certainty that measured change is due to the actual treatment delivered. However, it can be argued, in return, that, when one drills deeper into seminal studies about treatment of psychological trauma, not everything may be what it seems at first glance. For example, because, often, the control arms of RCT studies are wait list only, there is no way of knowing whether the change that came about in treatment was more to do with engagement with a treating practitioner offering a cogent rationale and a strong therapeutic alliance rather than the specific active ingredients of the therapy protocol itself. In any case, the broad agreement amongst academics that there are no differences between trauma-focused treatments and the controversy over the argument that there may well be no difference between *any* treatments for trauma, whether trauma focused or not, strongly suggests that the routine monitoring of progress or lack of progress in treatment simply makes sense.

Where leading academics seem to agree is that, in addition to the development and monitoring of a safe, respectful, and trusting therapeutic relationship where there is nurturing of hope and agreement about the goals and tasks of treatment, having a cogent rationale that is acceptable to service users and a set of treatment actions consistent with the rationale is very important. Pragmatically, as the meta-analytic research indicates, the predictors of PTSD (Brewin, Andrews, & Valentine, 2000) – namely, post-trauma life stress, trauma severity, and lack of social support, with lack of social support the greatest predictor –surely provide strong clues about what to focus on correcting in treatment.

In discussing a pragmatic approach to psychological enquiry, Dan Fishman (1999) concluded that pragmatism might offer a solution to the dilemmas about "evidence":

> Coming down from the lofty perch of ideological purity, pragmatism meets the world as we find it and asks: How can we improve it – not in some ideal way with a predetermined endpoint, but in a practical way in the here and now, within a context of the social, cultural, political, and economic realities we are given? (p. 292)

Pragmatic Solutions

A pragmatic approach to research inquiry into new and innovative treatments for psychological trauma should first and foremost put the feedback from the service user at the top of the agenda. The consumer of services should be the referee on effectiveness of interventions. To facilitate this, robust internationally recognized self-report measures should be used to track change in treatment on a sessional basis. Then, these results should be benchmarked against the results from published data in a similar manner to the way that treatment of depression in naturalistic settings has been benchmarked against clinical trials data (Minami *et al.*, 2008). It also makes sense to map innovative treatments onto existing approved treatments. For example, the rewind technique described in this chapter contains many elements already approved under the cognitive-behavioral therapy (CBT) umbrella of trauma-focused treatments. Designing yet more RCT studies with high internal validity will be unlikely to add new knowledge about such interventions, but gathering robust pre- and post-treatment data will give a strong indication if any approach seems to be working. However, a significant step further will be to encourage some PRN members to become involved in case study research to help elaborate on the specific elements and features of innovative treatments from multiple perspectives. The RCT data yield general information about treatments. But case study research has the potential to yield far more detail, unpacking the process of treatment at the micro level and helping the field to better understand more about both successful and, perhaps even more importantly, unsuccessful outcomes in particular practitioner–service user pairings.

Furthermore, having a large database of practitioners allows the identification of those practitioners who appear to consistently get superior results with their service users and provides the opportunity for qualitative research to explore the attitudes, beliefs, characteristics, and treatment approaches they take, in the hope that all learn more about how to help service users more effectively (see Box 14.2, "Practice-based Evidence: A 'Case Study Interview'").

Box 2 Practice-based Evidence: A "Case Study Interview"

Keith Guy is director of a national workplace counseling service that specializes in the treatment of psychological trauma. Keith carefully and robustly measures his outcomes with all his cases and has consistently demonstrated superior results since first contributing his data to the network in 2007. This is an excerpt from an interview with Keith conducted in January 2011:

> While I always felt that whilst it was useful for people to talk to one another, it almost felt there was something missing from my work. I was kind of like in the dark with people and I felt that it wasn't enough. I'd been on training courses previously, but to be honest with you, I'd been asleep.
>
> I was searching for something else to help in treating trauma. When I learned about the rewind technique, I thought, well, let's try this. And I think what we'd already determined was that a lot of the people that had been coming to us for occupational health counseling presented with stress, anxiety, depression, and all sorts of other labels that you might attach to it. But what we started to notice behind that was that trauma was a lot more common than we originally thought. So, we started working with rewind and then basically what happened was that people were actually coming back and saying that they were feeling better. I was suspicious at first. I thought to myself, what is happening here, what is it that's making this difference?

I remember the feeling when I was counseling prior to the human givens training. The clients used to give me stuff that used to overwhelm me. After the training, it was different. I just felt more positive about really actually being in a position to help, and this seemed to work for a significant proportion of people. People came back and said that they felt different and they felt better. That encouraged me, and it made me feel as if I can help, I can assist here. It spurred me on really. We know about all the stigma associated with mental health: long, enduring, and it's hopeless, you're never going to get better, you'll always be on medication. We wanted to move away from that.

Talking about psychological trauma, I see a lot of people who suffered in road traffic accidents, people who have been assaulted physically, either in their workplace or elsewhere. We've dealt with a lot of sexual abuse cases within the workplace, because of course they surface a lot in a workplace environment, strangely enough, not so much that sexual abuse takes places in the workplace, but it's happened in the past and there's something that's happened in the present that's kicked it off really. Reviewing my cases reminded me that much of what we have been dealing with has been really harrowing. I don't think now that I'm surprised by anything.

I can't work as a lot of people work. I make sure that there's always a big bank of time between each person that comes. That's really important. I think another thing that's important to me is that I always have to be prepared. I treat people as I want to be treated, as I would wish to be treated. I've just got those simple principles that are in my head. I've got no real formula. Sometimes I just do what I do on instinct.

Every client that I see, I always think it's a new beginning. It's blank. I think it's a human being here, let's start, let's talk to this person. I just work from what's in the room, from the seat of my pants if that makes sense, because I don't know what's coming. Sometimes I've said something before I know what I've said, I haven't policed it. It's just come out. That, then, somehow inspires me, if somebody says something and I react to it. So in a way I think that's the best kind of help sometimes. It never feels like real counseling to me.

I always remember one of the lecturers on a training course, and he said "to practice without theory is blind," and that's always stayed with me. I've got some basic principles in my head but they're only loose, and I just allow it to happen. I never want to be mechanical. I just want to be real with the person in the room.

It's not the dark that we're frightened of; it's actually coming out into the light.

I don't waffle. It's about cutting through to the heart of things. I like to get with people, get into the blood of it. I like to get my hands in the blood of their problem. I think there's a sense in other people that they know that there's a genuineness and I'm not afraid of being human with them and knowing what that's about. I've got something inside me. There's something in me that wants to do well in every case, even some of the cases that maybe a lot of other people have given up on. So I've got this driving force to do a really good job with the person, treating everybody new. Even after all these years, and it sounds crazy this, but just going to meet somebody new, I like it, I'm passionate about it. It's completely new when somebody comes through the door. I suppose the other thing is I tend not to work a long time with people.

To me that first session is crucial, and it's establishing that relationship, isn't it? It's the relationship that's important, the person that you're working with. I often talk about how the brain isn't a fixed organ, the wiring inside can change, can reconnect. But I let a lot of stuff come from the clients. Often I'll be saying to someone, "You're alive, you're here. Yes, you're in the bottom of the well, you're in the ditch, you've been there a long time and you keep climbing out and somebody kicks you back in, or you throw yourself back in for whatever reason. But you've got a lot of strength. You've got an enormous amount of strength because you're still here. You've been surviving and sometimes not very well, but you're here. And so you've got a lot about you, you've got a lot of strength, you've got some steel in you." I don't know whether that inspires people or not.

Again, it's probably wrong, but I always say at the end, the door's ajar. I always keep that door ajar, and they rarely come back. I'll often say to people, if it makes a bit of difference to you, then that's good enough, isn't it, just to make a little difference?

Practice-based studies can have high external validity in that they can be truly representative of "what is going on" out in the real-world settings of everyday practice and they can have rigor and robustness where practitioners commit to a very high standard of inclusiveness of all cases and all sessions.

Of course, there will always be limitations. There is no attempt to control participant selection, no randomization, and no control groups. It may not be clear precisely what is being studied. There will always be the need to mitigate the danger of cherry picking of results. But large sample sizes can add a new dimension to research and really give more confidence in the approach while also providing the opportunity for reflection when the results are less than what one might wish for.

While it may seem self-evident that a pluralistic approach to research has much to commend it, to date bodies such as NICE still place the greatest emphasis on RCT evidence. Paul Salkovskis, the architect of the hourglass model of psychotherapy research which led to the hierarchical model used in NICE Guidelines, observed as long ago as 2002 that "The risk inherent in the current practice of evidence-based mental health is that the field will degenerate into a parody, a kind of one-dimensional science, and there are signs that this has already occurred to some degree." Salkovskis (2002) suggested we needed to move away from this evidence hierarchy and develop a better model where all the different elements of research can contribute to new knowledge in a more balanced manner.

Furthermore, the ever-increasing body of research from neuroscience is explaining much about how unprocessed trauma memories control and affect function of the autonomic nervous system and how psychological trauma is truly a mind–body experience (Phan *et al.*, 2004). The recent shifts to experimentation with sensorimotor therapies, mindfulness practices, and body awareness therapies that work toward soothing the chaos in the right hemisphere of the brain lend themselves to investigation more easily through a PRN than through strictly controlled trials. Expert opinion certainly suggests that any treatment that works to sooth the autonomic nervous system and helps to manage that chaos is likely to be of benefit. This indeed is what has been shown in the PRN evaluation of the human givens approach, as the rewind treatment or variants thereof are used by most of the practitioners contributing data.

But there is still a long way to go, of course. Within a PRN, practitioners are volunteering to participate and so have volition with respect to *how* they will participate. In spite of the best practice recommendations, some practitioners choose to introduce a measure just at the beginning and end of treatment. More are better at using a measure session by session. Still others are excellent at using a menu of instruments, both trauma focused and more generic, as well as soliciting for feedback on the therapeutic alliance. Additionally, some practitioners who were willing to commit to comprehensive measurement during the initial six-month period feel it is too much effort to continue indefinitely with that level of engagement.

Thus obtaining practitioner engagement with outcome measures has proved challenging, and encouraging practitioners to measure the alliance even more so. To date, only a minority of practitioners choose to do so. We do recognize, though, that outcome measurement is a skill that needs to be taught, practiced, and developed over time. Training has been built into the Psychotherapy Master's program at Nottingham Trent University in the United Kingdom and is offered through the network. Considerable online training is made available free to members, and "peer group champions" across the United Kingdom are engaged in promoting and encouraging more and more practitioners to adopt these practices. The International Center for Clinical Excellence (http://www.internationalcenterforclinicalexcellence.com) is another very useful resource where practitioners across the globe are encouraged to support and learn from each other in a continuing quest to improve outcomes for service users.

The main goal of the PRN is to encourage practitioners to measure their outcomes in their work. By October 2010, data were available on over 2000 closed cases overall and on over 400 cases where psychological trauma was flagged up. The PRN supports this grassroots, front-line research effort and practitioners are encouraged to continue gathering data. The PRN has now moved beyond its original guild specificity and has a new title, the Pragmatic Research Network, of which the original HGIPRN is a part. This is to reflect a widening of horizons and to encourage international collaboration, with a central purpose of focus on research into the effectiveness of treatments in naturalistic settings across diverse populations. Practitioners of different orientations cooperating in data collection can all inform each other and hopefully improve the quality of the services offered to service users. It is very important to foster a spirit of transparency, openness, and respect for all modalities so that knowledge and learning can be shared. Finally, the wisdom and voice of service users need to be privileged to help guide practitioners about what really seems to help in their recovery path.

References

American Psychiatric Association. (1980). *Diagnostic and statistical manual of mental health disorders* (3rd ed.). Washington, DC: Author.

Andrews, W. (2007). Human Givens Institute Practice Research Network. Retrieved from http://www.hgiprn.org.

Andrews, W., Twigg, E., Minami, T., & Johnson, G. (2011). Piloting a practice research network: A 12-month evaluation of the human givens approach in primary care at a general medical practice. *Psychology and Psychotherapy: Theory, Research and Practice, 84*, 389–405.

Benish, S. G., Imel, Z. E., & Wampold, B. E. (2008). The relative efficacy of bona fide psychotherapies for treating post-traumatic stress disorder: A meta-analysis of direct comparisons. *Clinical Psychology Review, 28*, 746–758.

Bewick, B. M., Trusler, K., Mullen, T., Grant, S., & Mothersole, G. (2006). Routine outcome measure completion rates of the CORE OM in primary care psychological therapies and counselling. *Counselling and Psychotherapy Research, 6*, 50–59.

Bishop, P., & O'Callaghan, B. (2010). Human givens therapy for war veterans. Poster presentation at the BPS Annual Conference, Stratford.

Brewin, C. R., Andrews, B., & Valentine, J. D. (2000). Meta-analysis of risk factors for posttraumatic stress disorder in trauma-exposed adults. *Journal of Consulting and Clinical Psychology, 68*(5), 748–766.

Carbonell, J. L., & Figley, C. (1999). Promising PTSD treatment approaches: A systematic clinical demonstration of promising PTSD treatment approaches. *Traumatology, 5*(1), Art. 4.

Connell, J., & Barkham, M. (2007). *CORE-10 user manual* (Version 1.1). Rugby, UK: CORE System Trust & CORE Information Management Systems.

Duncan, B. L., Miller, S. D., & Sparks, J. A. (2004). *The heroic client: A revolutionary way to improve effectiveness through client directed outcome informed therapy.* San Francisco, CA: Jossey-Bass.

Ehlers, A., Bisson J., Clark, D. M., Creamer, M., Pilling, S., Richards, D., *et al.* (2010). Do all psychological treatments really work the same in posttraumatic stress disorder? *Clinical Psychology Review, 30*, 269–276.

Elliot, R. (2001). The effectiveness of humanistic therapies: A meta-analysis. In D. J. S. Cain (Ed.), *Humanistic psychotherapies: Handbook of research and practice.* Washington, DC: American Psychological Association.

Ericsson, K. A., Charness, N., Feltovich, P., & Hoffman, R. R. (2006). *Cambridge handbook on expertise and expert performance.* Cambridge, UK: Cambridge University Press.

Fishman, D. B. (1999). *The case for a pragmatic psychology.* New York, NY: New York University Press.

Griffin, J., & Tyrrell, I. (2001). *The shackled brain: How to release locked-in patterns of trauma.* East Sussex, UK: HG Publishing.

Griffin, J., & Tyrrell, I. (2008). *An idea in practice: Using the human givens approach.* Chalvington, UK: Human Givens Publishing.

Grinder, J., & Bandler, R. (1976). *Patterns of the hypnotic techniques of Milton H. Erickson, M.D.* (Vol. 1). Cupertino, CA: Meta.

Guy, K., & Guy, N. (2003). The fast cure for phobia and trauma: Evidence that it works. *Human Givens Journal, 9*(4), 31–35.

Hiatt, D., & Hargrave, G. E. (1995). The characteristics of highly effective therapists in managed behavioral provider networks. *Behavioral Healthcare Tomorrow, 4*, 19–22.

Hossack, A., & Bentall, R. P. (1996). Elimination of post traumatic symptomatology by relaxation and visual-kinesthetic dissociation. *Journal of Traumatic Stress, 9*(1), 99–110.

Lambert, M. J., Harmon, C., Slade, K., Whipple, J. L., & Hawkins, E. J. (2005). Providing feedback to psychotherapists on their patients' progress: clinical results and practice suggestions. *Journal of Clinical Psychology, 61*(2), 165–174.

Lohawala, A. (2009, April). Visual kinaesthetic dissociation therapy (rewind therapy) for post traumatic stress disorder. Poster presentation at the RCPsych Psychotherapy Faculty annual meeting, London.

Minami, T., Wampold, B., Serlin, R., Hamilton, E. G., Brown, G. S., & Kircher, J. C. (2008). Benchmarking the effictiveness of psychotherapy treatment for adult depression in a managed care environment: A preliminary study. *Journal of Counsulting and Clinical Psychology, 76*(1), 116–124.

Murphy, M. (2007) Testing treatment for trauma. *Human Givens Journal, 14*, 4.

Muss, D. C. (1991). A new technique for treating post-traumatic stress disorder. *British Journal of Clinical Psychology, 30* (Pt. 1), 91–92.

National Institute for Health and Clinical Excellence (NICE). (2005). *Management of post traumatic stress disorder in adults in primary, secondary and community care.* London: Author.

Parry, G., Castonguay, L., Borkovec, T., & Wolf, A. W. (In press). Practice research networks and psychological services research. In M. H. Barkham (Ed.), *A core approach to delivering practice-based evidence.* Oxford: John Wiley & Sons, Ltd.

Phan, K. L., Wager, T. D., Taylor, S. F., & Liberzon, I. (2004). Functional neuroimaging studies of human emotions. *CNS Spectrum, 9*(4), 258–266.

Salkovskis, P. M. (2002). Empirically grounded clinical interventions: Cognitive-behavioural therapy progresses through a multi-dimensional approach to clinical science. *Behavioural and Cognitive Psychotherapy, 30*, 3–9.

15

The Emergency Behaviour Officer (EBO): The Use of Accurate Behavioral Information in Emergency Preparedness and Response in Public and Private Sector Settings

Mooli Lahad, Ruvie Rogel, and Steven Crimando

Our global risk profile is rapidly changing. It seems clear; there will be more disasters and more people will be affected by them. Compounded by the speed and sophistication of our technological advances, as well as the ever evolving nature of terrorism, the new challenges in disaster management and humanitarian care become even more apparent. We are entering, or have already entered, the early stages of a new era in disasters, presenting new challenges and requiring a fresh look at ways to mitigate the impact of these events.

As always, future disasters will not recognize borders or boundaries. Such events are often multiregional or multinational events. The massive storm plodding across the Atlantic basin can devastate small island nations and then move along to level large swaths of a continent in its journey. A disease outbreak can quickly circle the globe, as can a cloud of volcanic ash or radioactive fallout. Countries, communities, and corporations are facing a growing number of mass emergencies (MEs) caused by natural disasters, as well as human-made disasters such as terrorism, wars, industrial accidents, and transportation catastrophes. The events in Japan during March 2011 are a striking example of a cascading, complex disaster, in which natural disasters (i.e., earthquake and tsunami) triggered a technological disaster (nuclear plant crisis). Major emergencies have become more frequent during the first decade of the twenty-first century, and they affect more and more people, disrupting essential services, slowing the process of sustainable human development, and disrupting the economies of regions and nations (World Health Organization, 2007a, 2007b).

International Handbook of Workplace Trauma Support, First Edition. Edited by Rick Hughes, Andrew Kinder, and Cary L. Cooper.

Lessons Learned

One of the first lessons we learned as early as 1980 was that decision makers, managers, and incident commanders often failed to "read" the developing situational map and lacked the capacity of monitoring and properly analyzing human behavior in emergencies. Many strategic and tactical decisions were based on false assumptions regarding public misbehavior (panic, looting, rioting, etc.) and leaders frequently relied too heavily upon the media as a credible or sole source of information. This "command post blindness" has been identified by us as a debilitating principle in many cases worldwide. It has been documented in many incidents around the globe, where the failure to understand and anticipate public behavior has led to flawed decisions resulting in the loss of many lives and increased suffering. Leaders and decision makers in both the public and private sectors have been slow to appreciate the importance to human behavior in developing overall situational awareness during crisis events.

A classic example of the misinterpretation of public behavior was the Hillsborough Football Stadium disaster in the United Kingdom in April 1989, in which 39 people were crushed to death by panicked football fans. This tragedy was deemed the result of officers in the control center failing to recognize that too many fans were being directed into a section of the stands that could not adequately accommodate their numbers. As the number of people in the tight space increased, the crowd reached a "flashpoint" and became a mob, "panicked and dangerous." Both command and tactical officers mistook the crowd's frenzied attempt to flee as the aggression of football hooligans trying to invade the pitch. Using dogs and aggressive police tactics to contain the mob, the situation escalated and led to the unfortunate consequences.

A more recent example of over-dependency on the media and failure to read the situation from a human behavior perspective can be found in the concluding comments of the US House of Representatives report on Hurricane Katrina. The committee called its report "a failure of initiative." The conclusion referred to Lt. Gen. H. Steven Blum, who told the Select Committee on October 27, 2006,

> We focused assets and resources based on situational awareness provided to us by the media ... frankly. And the media failed in their responsibility to get it right.... [W]e sent forces and capabilities to places that didn't need to go there in numbers that were far in excess of what was required, because they kept running the same roll over and over ... and the impression to us that were watching it was that the condition did not change. But the conditions were continually changing.
>
> (Davis *et al.*, 2006, p. 361)

This wasn't the worst information leaders relied on during Hurricane Katrina. More alarming is the fact that incident commanders accepted media reports of looting, shooting, and rape as facts, whereas the report says, "As discussed in our report, widely-distributed uncorroborated rumors caused resources to be deployed, and important time and energy wasted, chasing down the imaginary. Already traumatized people in the Superdome and elsewhere, listening to their transistor radios, were further panicked." "The sensational accounts delayed rescue and evacuation efforts already hampered by poor planning and a lack of coordination among local, state, and federal agencies," and "Government at all levels lost credibility due to inaccurate or unsubstantiated public statements made by officials regarding law and order. But it's clear, accurate reporting was among Katrina's many victims. If anyone rioted, it was the media. Many stories of rape, murder, and general lawlessness were at best unsubstantiated, at worst simply false" (Davis *et al.*, 2006, p. 360).

With increased disaster activity and an ever growing global population, the role of human behavior is gradually being acknowledged as critical to the management of such emergencies across all their stages (i.e., mitigation, preparedness, response, and recovery). Emergency managers in both community and corporate environments, including public sector emergency management agencies and private sector security, safety, and operational risk management programs, are increasingly required to acknowledge human factors and behavioral considerations within their operational procedures. In fact, acquaintance with the possible human behavior and the likely reactions of the affected population and of the involved emergency responders (organizations as well as individuals) is essential for decision makers and emergency managers in any setting.

Acknowledgment of the centrality of the behavioral aspects of major emergencies led 168 governments to adopt in January 2005 a 10-year plan to make the world safer from natural hazards at the World Conference on Disaster Reduction held in Kobe, Japan. The Hyogo Framework for Action (HFA, 2005) is a global blueprint for disaster risk reduction efforts aimed at improving the resilience of nations and communities to disasters.

The European Union (EU) Council of Home Affairs (2010) concluded in its report of June 2010 that all member states are encouraged to include, in their respective civil protection systems, psychosocial assistance for the victims of major disasters.

The council pointed out to empirical evidence that those affected by disaster may have psychosocial experiences affecting their personal health and the health of their community (Fraguas *et al.*, 2006; Gabriel *et al.*, 2007; Galea, Nandi, & Vlahov, 2005; Norris *et al.*, 2002; Whalley & Brewin, 2007). The council includes first responders – emergency service personnel and other staff categories present on site – as possibly being psychosocially affected by the impact of working in critical situations.

The concern over the role of the media, as an "amplifier" of the impact of major crisis events and disasters on the psychosocial well-being of the population, has been widely discussed (Lahad & Ben Nesher, 2008; Pfefferbaum *et al.*, 2001; Rosenfeld *et al.*, 2010; Ross, 2003) and additionally challenges emergency managers.

In the private sector settings, industrial standards for emergency and disaster management are beginning to run on a parallel track to those in the public sector. For example, in the United States, standards set by the National Fire Protection Agency (NFPA) for emergency preparedness and response require businesses to anticipate and appropriately respond to the behavioral aspects of all types of disasters and crisis event. "Employee assistance and support otherwise called human continuity, human impact, workforce continuity, human aspects of business continuity, and so forth, is the ability to provide assistance and support to the entity's employees and their families/significant others affected by the incident" (NFPA, 2010).

Following the terrorist attacks on the United States in September 2001, recommendations made by the 9/11 Commission involved the Voluntary Private Sector

Preparedness Accreditation and Certification Program law (PS-PREP). Under the management of the US Department of Homeland Security (DHS), this law established a voluntary preparedness certification and accreditation process for all types of businesses, as well as universities and healthcare organizations. Drawing upon established standards developed by the National Fire Protection Association (NFPA), the British Standards Institution (BSI), and ASIS International, the law seeks to create a more resilient and recoverable private sector. Such resilience incorporates an attention to the human impact of disasters and emergencies. In addition, DHS also created a Human Factors/Behavioral Sciences Division within the Science and Technology Directorate to address a range of human concerns in disasters and terrorism. One of the declared goals of the division is to "Enhance preparedness and mitigate impacts of catastrophic events by delivering capabilities that incorporate social, psychological and economic aspects of societal resilience" (DHS, 2011).

As the US homeland security field matures, the importance of human behavior is becoming an important area of attention.

The approach to addressing disaster-related human factors is also expanding from the clinical to tactical disciplines, with a growing understanding that human behavior in disasters must be understood and incorporated into planning, response, and recovery efforts. Leaders and decision makers need to accept the fact that these issues are embedded in every aspect of emergency management in both the public and private sectors. Such knowledge can no longer be the exclusive domain of mental health professionals and academics. Commanders in military and community disaster management, as well as those in risk management, security, and business continuity roles in private organizations, must possess a thorough understanding of how people are most likely to behave in different disaster scenarios. Failure to do so can lead to response actions that are ineffective, inappropriate, and, in some instances, dangerous.

The Importance of Understanding and Anticipating Disaster-Related Behaviors

For emergency plans, exercises, and real-time responses to be most effective, they must be based on accurate behavioral assumptions, that is, what people are most likely to do in an actual emergency. Whether leaders and decision makers are corporate executives, military commanders, emergency management officers, or front-line tactical responders, an understanding of human behavior is critical. In the popular Wiley Pathway Disaster Response and Recover textbooks, David McEntire (2006) has cautioned would-be emergency managers, "Having correct views of disaster behavior is one of the best ways to promote successful response and recovery operations" (p. 79). Understanding and anticipating human behavior across a variety of hazard scenarios and around the entire cycle of emergency planning are critical to developing sound policies, plans, and protocols. These same accurate behavioral assumptions must also be present in exercises to make them as realistic and useful as possible.

The learning curve necessary to form such accurate behavioral assumptions can be steep and beyond the reach of most in disaster response leadership positions. Such knowledge and experience are more common to behavioral and social science professionals not typically included in the planning or exercise phases of emergency management. To help educate, inform, and advise leaders and decision makers about disaster-related behaviors, the authors recommend the inclusion of an emergency behavior officer (EBO) as a formal member of both the public and private sector emergency management apparatus.

Public Behavior

The behavior of individuals and communities in past events may be indicative for future behavior in critical incidents and disasters. The radiation accident in Goiânia, Brazil, in September 1987 is a striking example of a mass population reaction to a critical event. In this incident a radioactive nuclear medicine source was scavenged from an abandoned medical clinic outside of the city and, once broken open, was subsequently handled by many people, resulting in four deaths and serious radioactive contamination of 249 other people. When word of the accidental release of cesium 137 (Ce-137) spread through the community, the Brazilian government was forced to open the Olympic soccer stadium as a medical screening site. An estimated 112 000 individual presented for radiation screening. Most remarkable was that out of the first 60 000 people who were medically examined, 5000 (8%) displayed the signs and symptoms of radiation sickness, although not one had been exposed or contaminated (Carvalho, 1989). The story of Goiânia is an example of a physical and medical crisis that quickly evolved into a large-scale behavioral emergency. In fact, the Goiânia incident resulted in a ratio of 500:1 of behavioral casualties to medical casualties (Roper & Sperb Leite, 1988).

Another useful example is the reaction of the Israeli population during the 30 days of armed conflict in July 2006 on the northern border between Israel with the Hezbollah forces in Lebanon. During that crisis, the ratio between physical injury and behavioral or stress-related injuries surging the hospitals' emergency rooms (ERs) in the affected areas was 1:25 to 1:40. In cases in which children were killed, the ratio increased to 1:51.

Regardless of their cause, disasters produce significant psychological causalities. In fact, the Israeli Ministry of Health proclaims that most casualties in disasters are stress-related casualties (Ben-Gershon, Grinshpoon, & Ponizovsky, 2005). Terrorism is a form of "disaster by design," in which the actor(s) create disasters specifically intended to produce the highest degree of psychological, economic, and social disruption. As such, acts of terrorism routinely produce much higher levels of behavioral causalities (Hall *et al.*, 2002).

Often referred to as the "worried well" effect, the prevalence of psychosomatic or psychogenic physical symptoms is also highly associated with public health emergencies. In exposures, real or perceived, to invisible hazards, such as chemical or radioactive materials, or biological scenarios involving microscopic bacteria or viral agents, those charged with planning for such crises must anticipate unique behavioral reactions, such as "multiple unexplained physical symptoms" (Pastel, 2001). Dramatic evidence of this type of behavioral response was apparent in the first wave of the H1N1 (swine flu) pandemic in 2009. On May 25 – the worst day of the swine flu outbreak – New York City hospital emergency departments saw more than 2500 patients with influenza-like illness (ILI) symptoms compared to only 150 on the same day in 2008. Very few of these patients, mainly driven by fear and misinterpretation of other symptoms (allergies, stress responses, etc.), required hospitalization. From the second to the third week of July 2009, the United Kingdom experienced a doubling of new H1N1 cases with more than 100 000 new cases triggering a run on pharmacies for gloves, masks, antibacterial gels, and thermometers. A key lesson learned in many countries was that the overwhelming surge for healthcare services or supplies crippled overall response efforts (Hine, 2010).

First Responders

Stress-related behaviors may also affect the performance of emergency responders either by secondary traumatization and the effects of compassion fatigue (Figley, 1995) or by experiencing the traumatic event simultaneously with the people they are responding to (Saakvitne, 2002).

Hurricane Katrina 2005 produced many examples of the behavioral impact on responders themselves. One such example is that, in October 2005, the New Orleans Police Department had fired 51 of its personnel for abandoning their posts. Some of those responders left their posts and were never heard from again. The chief of police was quoted saying, "It isn't representative of our department; we had a lot of heroes that stepped up after the storm" (Associated Press, 2005; *Fox News*, 2005). Another example is the story of Memorial Hospital in New Orleans during Katrina where flood water resulted in a power failure. Doctors and nurses were overstretched and overtired. Patients were dying, and evacuation seemed impossible. Some members of the medical staff started talking about injecting lethal drugs to patients, and eventually there were several cases of euthanasia that became the subject of a major two-year investigation by the Louisiana state attorney. The case was widely reported by the media (Fink, 2009).

In Katrina and other disasters, first responders often shared the same fate as the rest of the population. More than a million people left the New Orleans area before the storm. Evacuation orders and pleas were made by New Orleans Mayor Ray Nagin and President George W. Bush, urging the population to evacuate before Hurricane Katrina made landfall in New Orleans. More than 100 000 people refused to leave the city (out of a population of 455 000). CNN writer Palser (2005) described the behavior of the non-evacuating population as "gambling on their lives." Reasons for evacuation refusal ranged from a belief that there was enough protection from the storm, to a desire to protect property, to personal challenges, such as lack of funds for transportation, exacerbated by the fact that the date of the evacuation was one day before regular pay checks for many, or relying on past experiences such as the Hurricane Ivan evacuation that left memories of 6–10 hours of traffic lines and the illness and injury of older people stalled in those lines.

In these examples and many others, operational decisions may have been handled in substantial different ways if organizational leaders, emergency managers, and tactical commanders and responders were able to consult with psychosocial professionals to gain a better understanding of the human characteristics of the situations and the anticipated behavior of the population, first responders, and others involved.

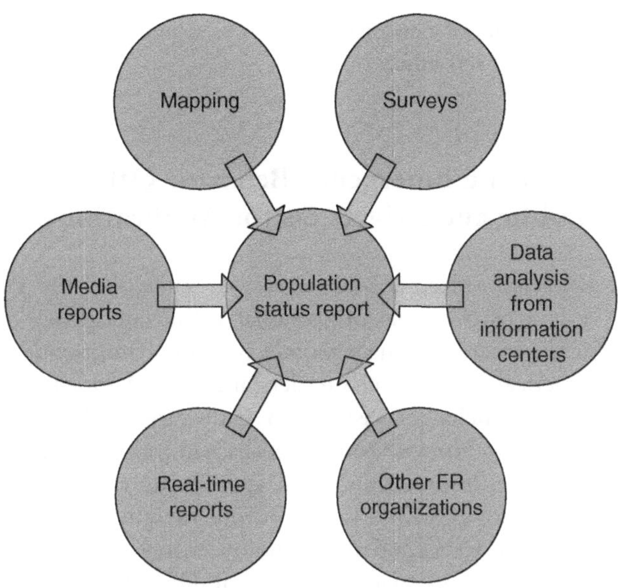

Figure 15.1 Sources of status report.

The Role of Emergency Behavior Management Systems (EBMS): Public and Private Sector Applications

In recognition of the critical role of behavior in all phases of emergency management (Figure 15.1) and the importance of mitigating the psychosocial consequences of disasters and mass violence, it is suggested that in concert with existing emergency management structures, a defined Emergency Behavioral Management System (EBMS), including defined roles for Emergency Behavioral Consultants (EBC), be further developed and integrated into the current models of emergency management.

To adequately incorporate mental health and behavioral factors into emergency management at the national as well as local levels, a system focusing on the psychosocial aspects of disasters must be developed. The task of the EBMS should include (1) advising policy makers, (2) building (resource) databases, (3) developing tools and methods to analyze information in "real time," and (4) coordinating agencies and services dealing with the public during normal times so that the response will be well organized and resources will be better utilized.

The EBMS should function at all levels: (1) headquarters (public and private sector decision makers), (2) intervention teams, and (3) ground level (in the community). Additional factors must be taken into consideration with terrorist attacks, including the public's reaction to the perpetrators (such as the wish for revenge, not always directed at only the perpetrators) and the altered perception of what constitutes a safe activity (such as traveling on a bus and eating in a café). Another aspect of psychosocial preparedness should be on the local-authority level, building upon existing services and adopting the concept of "helping the public to help themselves," that is, enhancing "motivational resiliency" and the concept of "assisted coping" (i.e., doing *with* rather than doing *for* people to foster a sense of self-efficacy).

The Role of the Emergency Behavior Officer (EBO): Public and Private Sector Applications

The emergency behavior officer (EBO), otherwise known as the public behavior consultant (PBC), is a psychosocial professional within an agency or unit, whose function is to advise the emergency management systems on improving the operational manager's and/or tactical commander's understanding and anticipation of the likely individual and collective behaviors associated with various disaster scenarios and potential post-incident psychosocial consequences and resulting needs, with recommendations for assistance, information about appropriate resources, and solutions. Another important task for the EBC is monitoring the affected population for the psychological effects on the forces and units, employees and organizations, and general population, as to improve coping with psychosocial difficulties.

As a member of a public or private sector emergency management team or at a unit command, the EBC should be well informed and inspect the preparatory process in both the development of behaviorally accurate plans and their implementation. Prior knowledge of the community (local population, workforce, and/or first responders) is important. Knowledge of the community or organization's resilience is a working tool.

The work of the EBO is based upon several fundamental assumptions:

1. *Continuity*: The role of the EBC is to help the individual, organization, or community maintain operational continuity, as well as cognitive, social, and historical continuity, so that the individuals, communities, and organizations are returned to pre-event levels of functioning as quickly as possible.
2. *Social systems*: EBCs work within multiple levels, multiple systems, and multiple disciplines and with short- and long-term implications. They are expected to synchronize and work together with others throughout all vulnerability circles.
3. *Empowerment*: EBCs are consultants, providing guidance for individuals and organizations to find means and coping mechanisms to assume responsibility and to be able to help themselves.
4. *Proactivity*: The work of EBCs is proactive and direct. Rather than waiting to observe the psychosocial consequences of a catastrophe, the EBC is able to use evidence-informed models of the behavioral response to disasters and apply the doctrines and concepts of the behavioral sciences to help leaders better predict and prepare for the human impact of events.
5. *Ripple effect*: The assistance processes will disseminate through the community, thereby increasing the numbers of individuals and units empowered and active.
6. *Communication*: Information and education are the main tools to work with the population and the responders. As such, the EBO may also work in collaboration with corporate and community risk and/or crisis communications professionals to

help craft behaviorally sensitive messages that may improve the overall response to the emergency or disaster.

7. *Intervention versus treatment*: In a crisis situation, the EBC focuses on the here and now. Disaster research indicates that only a small subset of individuals exposed to disasters go on to develop lasting mental health problems (Norris *et al.*, 2002). While it seems many people benefit from early and ongoing support in the wake of a disaster, most will not require traditional mental health treatment or therapy.
8. *Dignity*: Maintaining the individual or community dignity through empathy and support.
9. *Culture sensitivity*: Focusing on norms, rituals and cultural aspects of reaction to disasters as well as to bereavement and grief rituals. Creating networks with faith leaders and anthropologists whenever necessary.

Public Behavior Evaluation (PBE)

Public behavior evaluation (PBE) is an approach employed to formulate operational recommendations to support the essential needs of the population, within the different circles of vulnerability. The process is based on a status report followed by analysis of contributing factors, possibilities, and recommendation for action which is then translated to operational terms. The evaluation provides leaders and decision makers with a critical layer of situational awareness essential to guide ongoing operations.

A status report may be created through varying sources (see Figure 15.1). The data gathered from the various sources are then processed through verification processes, understanding of the implications, and drawing of conclusions and recommendations.

The EBC as part of the executive team, within military units, emergency management agencies, first responder units, or the corporate "situation room," can help guide the response to the needs of the population, those being:

- Physical rescue.
- Basic needs.
- Contact with family and colleagues.
- Orientation, information, and understanding.
- Control over the situation.
- Relating to property losses and damages (such as loss of homes).
- Restoration of basic assumptions about self – the world and faith.

As a consultant to the decision makers and management, the EBO's roles are to continuously monitor the behavioral status report, and to gather and synthesize information from all sources, such as information centers, telephone or internet help lines, and online resources. The EBC may consult with specialists both for behavioral situation analysis and also in order to forecast potential developments in the coming hours and days. The EBC may generate different scenarios, as well as alternative responses, for the management to discuss and decide. In some cases the EBC may advise on the use of media, internet, and other means of mass communication and the need to establish information centers and meeting points, sometimes referred to as family reunification centers.

The EBC in many instances may serve as a liaison between the community or workforce and local responders or army units and civil services such as those at the municipal level. The consultant may also help structure information to be used to help orient survivors and help them understand both the evolving situation and response. In the communications cycle, the EBC may also work alongside risk and crisis communications experts to reduce miscommunications and rumors. Sandman (Sandman, Weinstein, & Hallman, 1998) warns emergency risk communicators of likely mistakes that can occur in this process. In his comprehensive list of *do's* and *dont's*, Sandman stresses the importance of honesty and accuracy in delivering messages, acknowledging uncertainty and legitimate fears, and giving people things to do in a way that allows them choices. Covello (2001) suggests overt demonstration of empathy, competence, honesty, and commitment to create trust and credibility. These recommendations are of equal importance to the incident commander operating from the command post and the CEO orchestrating a corporate response to a critical incident from the boardroom.

The EBO serves as a consultant to the emergency managers or commanders on the development of messages to the public and means of their delivery, including the setup and choice of words. Sandman's formula; "risk = hazard + outrage" coined in the 1980s, is the cornerstone of the EBC's task in managing outrage through effective, strategic risk communication. Sandman *et al.* (1998) writes that "a growing body of research indicates that people assess risks according to metrics other than their technical seriousness: these factors include trust, control, voluntariness, dread, and familiarity."

The role of the EBC is to assess these "rage factors" in the community or organization and advise managers, commanders, and business leaders on proper ways of handling them.

As the media consultant to management, the EBO does not replace public information or public relations officers or other spokespersons. The role is to focus on crafting messages to the affected population that are accurate, are effective, and refrain from the use of trauma-provoking terminology (Ross, 2003). The EBC can guide leaders and decision makers in developing messages that foster resiliency and coping and help them to shape, reword, and reframe such information in a manner that bolsters public strength and endurance.

The EBO is also expected to advise those in leadership positions on how best to restore a sense of control over the situation and promote self-efficacy in the general population or the workforce. In this capacity the EBO is primarily a liaison between the affected population and the authorities or management as a staff member advising the decision-making panel. It is the EBC's responsibilities to ensure connectedness of the affected population with existing services, monitoring their changing needs and their adjustment so that these developments will be part of the status report.

EBCs as mental health professionals should monitor stress levels, advise, and, in some cases, provide direct stress management services to both leadership and tactical teams, as well as operation and situation room staff, including managers and commanders themselves, as needed. As it may not be apparent to commanders and other leaders, the EBC should provide feedback regarding levels of stress that may interfere with the leadership team's performance or lead to compassion fatigue or burnout. In the recovery phase, it is the EBC's role to initiate psychological debriefings or other interventions to promote support and closure interventions. The EBC may either facilitate such interventions themselves or, preferably, encourage leaders to provide such services with available stress control teams to support the organization's effective management of stress and its resumption of normal, pre-event functioning.

EBCs can play a critical role in military, government, industrial, or business organizations. In each organization they may take a unique characterization, and perhaps a different title, yet the role is always to consult and advise those tasked with emergency management, security, and risk management on the likely behavior of and mental health impact on the affected population during a disaster in order to help save life, property, and psychological well-being.

Conclusion

With trends toward an increasing number of disasters affecting an ever growing global population, the human impact of disasters is outpacing our collective ability to respond effectively with existing models of psychosocial assistance. Developments in the behavioral sciences, especially in the area of disaster-related behaviors, increase the potential usefulness of specialized behavioral advisors to emergency management leaders, decision makers, and responders to help them make accurate assumptions about how people will react to various threat scenarios around the entire emergency cycle. Policies, plans, and exercises that ignore the behavioral response of disaster victims and survivors, as well as responders and other helpers, invite failure. To be successful, leaders in both public and private emergency management roles must be aware of likely behavioral responses to disasters and terrorism, and practice decision making in the light of accurate behavioral expectations.

To facilitate the development of the Emergency Behavioral Management System (EBMS) and behavioral considerations into all levels of disaster response planning, we recommend:

- Initiatives to instill a public behavior approach in all models of intervention following disaster.
- Including a professional behavioral expert and/or team in the standardized framework of decision-making, management, headquarters, or command staff levels.
- Strengthening reciprocal relationships and the cooperation between the government administrations, social-community systems, and the defense and economic systems (i.e., public and private sectors).
- Developing a regional and district behavioral system for intervention, support, and rehabilitation that will be able to function even in the absence or delay in response from the local or federal government.
- Establishing behaviorally aware doctrine and protocol for local and national mass media during and following a disaster.
- Creating indices according to which community business and national resilience may be measured.

- Developing standardized measurements for assessing the efficacy of post-disaster interventions.
- Implementing a standard intervention kit based on an evaluation of which basic services or means are required to ensure the survival of the systems and individuals.
- Increasing focus on programs to further the concept of individual capability and community resilience for the private sector, family, and community.

All of the above must be predicated on accurate models of disaster-related human behavior, not myths, exaggeration, or speculation of how people are most likely to behave based on the representations of the mass media, movies, or the internet.

The reality, whereby more people are being affected by more disasters, requires an increasing demand on a finite number of disaster assistance resources, and this means we have to get it right the first time. Misestimating the human impact and response to catastrophic events can no longer be an option.

References

Associated Press. (2005, October 21). New Orleans police fire 51 for desertion. Retrieved from http://www.msnbc.msn.com/id/9855340/ns/us_news-katrina_the_long_road_back/.

Ben-Gershon, B., Grinshpoon, A., & Ponizovsky, A. (2005). Mental health services preparing for the psychological consequences of terrorism. *Journal of Aggression, Maltreatment & Trauma, 10*, 743–753.

Carvalho, A. B. (1989, November 6–10). The psychological impact of the radiological accident in Goiânia. In *Proceedings of the International Seminar Recovery Operations in the Event of a Nuclear Accident or Radiological Emergency* (IAEA-SM-316/18, pp. 463–477). Vienna: International Seminar Recovery Operations.

Covello, V. (2001). *Audiences and messages: Thinking them through.* Oak Ridge, TN: Oak Ridge Institute for Science and Education. Retrieved from http://www.orau.gov.

Davis, T., Rogers, H., Bonnila, H., Buyer, S., Myrick, S., Thornberry, S., *et al.* (2006). *A failure of initiative: Final report of the select bipartisan committee to investigate the preparation for and response to Hurricane Katrina.* Washington, DC: Government Printing Office.

European Union Council of Home Affairs. (2010, June 3). 3018th Justice and Home Affairs Council meeting. Luxembourg. Retrieved from http://www.consilium.europa.eu/uedocs/cms_Data/docs/pressdata/en/jha/114833.pdf.

Figley, C. R. (1995). Compassion fatigue as secondary traumatic stress disorder: An overview. *Compassion Fatigue, 1*, 20.

Fink, S. (2009, August 27). The deadly choices at Memorial. *ProPublica.* Retrieved from http://www.propublica.org/article/the-deadly-choices-at-memorial-826.

Fox News. (2005, October 30). N.O. police fire 51 for desertion. Retrieved from http://www.foxnews.com/story/0,2933,173879,00.html.

Fraguas, D., Teran, S., Conejo-Galindo, J., Medina, O., Cort n, E. S., & Ferrando, L. (2006). Posttraumatic stress disorder in victims of the March 11 attacks in Madrid admitted to a hospital emergency room: 6-month follow-up. *European Psychiatry, 21*, 143–151.

Gabriel, R., Ferrando, L., Cort n, E. S., Mingote, C., Camba, E., & Liria, A. F. (2007). Psychopathological consequences after a terrorist attack: An epidemiological study among victims, the general population, and police officers. *European Psychiatry, 22*, 339–346.

Galea, S., Nandi, A., & Vlahov, D. (2005). The epidemiology of post-traumatic stress disorder after disasters. *Epidemiologic Reviews, 27*, 78.

Hall, M. J., Norwood, A. E., Ursano, R. J., Fullerton, C. S., & Levinson, C. J. (2002). *Psychological and behavioral impacts of bioterrorism*. Emmitsburg, MD: National Emergency Training Center.

Hine, D. D. (2010). *The 2009 influenza pandemic: An independent review of the UK response to the 2009 pandemic influenza*. London: Cabinet Office.

Hyogo Framework for Action (HFA). (2005, January 18–22). Hyogo Framework for Action 2005–2015: Building the resilience of nations and communities to disasters. Paper presented at the World Conference on Disaster Reduction, Kobe, Hyogo, Japan. Retrieved from http://www.unisdr.org/eng/hfa/hfa.htm.

Lahad, M., & Ben Nesher, A. (2008) Community coping: Resilience models for preparation, intervention and rehabilitation in manmade and natural disasters. In K. Gow & D. Paton (Eds.), *Resilience: The phoenix of natural disasters* (pp. 195–208). Hauppauge, NY: Nova Science.

McEntire, D. (2006). *Wiley pathway disaster response and recovery*. Oxford: John Wiley & Sons, Inc.

National Fire Protection Association (NFPA). (2000). *Standard on Disaster/Emergency Management and Business Continuity Programs*. Quincy, MA: Author. Retrieved from http://www.nfpa.org.

Norris, F. H., Friedman, M. J., Watson, P. J., Byrne, C. M., Diaz, E., & Kaniasty, K. (2002). 60 000 disaster victims speak: Part I. An empirical review of the empirical literature, 1981–2001. *Psychiatry: Interpersonal and Biological Processes, 65*, 207–239.

Palser, B. (2005). *Hurricane Katrina: Aftermath of disaster*. Minneapolis, MN: Compass Point.

Pastel, R. (2001). Collective behaviors: Mass panic and outbreaks of multiple unexplained symptoms. *Military Medicine, 166*, 44–46.

Pfefferbaum, B., Nixon, S. J., Tivis, R. D., Doughty, D. E., Pynoos, R. S., & Gurwitch, R. H. (2001). Television exposure in children after a terrorist incident. *Psychiatry: Interpersonal and Biological Processes, 64*, 202–211.

Roper, L. D., & Sperb Leite, M. A. (1988). The Goiânia radiation incident: A failure of science and society. Retrieved from http://arts.bev.net/roperldavid/gri.htm

Rosenfeld, L. B., Caye, J. S., Lahad, M., & Gurwitch, R. H. (2010). *When their world falls apart*. Washington DC: NASW Press.

Ross, G. (2003). *Beyond the trauma vortex: The media's role in healing fear, terror, and violence: Workbook – beyond the trauma vortex*. Berkeley, CA: North Atlantic.

Saakvitne, K. W. (2002). Shared trauma: The therapist's increased vulnerability. *Psychoanalytic Dialogues, 12*, 443–449.

Sandman, P., Weinstein, D., & Hallman, W. (1998). Communications to reduce risk underestimation and overestimation. *Risk Decision and Policy, 3*(2), 93–108.

US Department of Homeland Security (DHS). (2011). Science and Technology Directorate Human Factors/Behavioral Sciences Division. Retrieved from http://www.dhs.gov/xabout/structure/gc_1224537081868.shtm.

Whalley, M. G., & Brewin, C. R. (2007). Mental health following terrorist attacks. *British Journal of Psychiatry, 190*, 94–96.

World Health Organization. (2007a). Mass casualty management system, strategies and guidelines for building health sector capacity. Retrieved from http://www.who.int/hac/techguidance/MCM_guidelines_inside_final.pdf.

World Health Organization. (2007b). Strategies and guidelines for building health sector capacity. Retrieved from http://www.who.int/hac/techguidance/MCM_guidelines_inside_final.pdf.

16

Trauma-related Dissociation in the Workplace

Onno van der Hart, Xiao Lu Wang, and Roger M. Solomon

All of us have our breaking-point. To some it comes sooner than to others.
– T. A. Ross (1941, p. 66)

Experiencing psychological trauma is possible in almost every profession, but some professions are more dangerous than others and thus may involve higher proportions of traumatized individuals and professionals. This is especially the case for professionals working as first responders, such as firefighters, medics and paramedics, police officers, and combat soldiers (Benedek, Fullerton, & Ursano, 2007). For example, studies of firefighters generally have found rates of work-related post-traumatic stress disorder (PTSD) ranging from 13% to 18% following 1–4 years after large-scale "response events" (Benedek *et al.*, 2007). Such figures probably mean that many more professionals have faced potentially traumatizing events in the workplace while not becoming traumatized or not to such a degree that they developed full-blown PTSD and/or other serious mental illnesses such as major depression. In this chapter we argue that dissociation is a major but too often overlooked feature of psychological trauma and thus of trauma-related disorders.

The understanding of the role of dissociation in workplace-related trauma may be helpful for those who offer psychological help, whether first aid or specialized psychotherapy, to those who have been exposed to work-related potentially traumatizing events. However, clear understanding of trauma-related dissociation is seriously hampered by a severe lack of conceptual clarity about its nature and its manifestations (Cardeña, 1994; Marshall, Spitzer, & Liebowitz, 1999; Van der Hart, Nijenhuis, & Steele, 2005). For example, the American Psychiatric Association's 1994 *Diagnostic and Statistical Manual of Mental Disorders* (DSM-IV) defines dissociation as "a disruption in the usually integrated functions of consciousness, memory, identity, or perception of the environment" (p. 477). Not only are the functions of movements and sensations lacking in this definition, but also it is very imprecise with regard to which psychological phenomena should be considered as dissociative in nature

International Handbook of Workplace Trauma Support, First Edition. Edited by Rick Hughes, Andrew Kinder, and Cary L. Cooper.

and which should not. In the literature, dissociation pertains to a wide variety of psychological phenomena and diverse theoretical viewpoints have described dissociation as a process, an intrapsychic structure, a psychological defense, and a deficit (e.g., Cardeña, 1994; Nijenhuis, 2004; Van der Hart, Nijenhuis, & Steele, 2006). For example, common alterations of consciousness such as intense absorption and imaginative involvement were originally distinguished from dissociative symptoms. However, the domain of dissociative symptoms has come to include almost any altered states of consciousness in most contemporary publications. The Dissociative Experiences Scale (DES; Bernstein & Putnam, 1986), the most widely used instrument to measure "dissociative" phenomena, is a prime example of this expansion of the domain of dissociative symptoms. The concept of peri-traumatic dissociation – pertaining to symptoms during or directly following traumatic experiences – also includes alterations in consciousness that differ from the original dissociative symptoms. This is exemplified in the Peritraumatic Dissociative Experiences Questionnaire (PDEQ; Marmar, Weiss, & Metzler, 1996). The view on trauma-related dissociation espoused in this chapter is based on the original nineteenth- and early twentieth-century understandings and further developed in the theory of structural dissociation of the personality (Nijenhuis, Van der Hart, & Steele, 2002; Steele, Van der Hart, & Nijenhuis, 2009b; Van der Hart *et al.*, 2006). Related to this theory, a definition will be presented that is conceptually clear and specific; neither under-inclusive nor over-inclusive; and theoretically, clinically, and scientifically useful. Using case examples, we discuss or present several forms of work-related traumatization that involve different levels of dissociative complexity with related treatment implications. A large number of references focus on combat-related trauma. However, these sources are also useful to shed light on dissociation in trauma experienced in other workplaces.

Historical Perspective

Originally, in French nineteenth-century psychiatry, dissociation referred to an undue division or compartmentalization of consciousness or personality – with the latter term indicating that dissociation pertains to more psychobiological phenomena than just consciousness. Some other terms in vogue were doubling of the personality, double consciousness, psychological disaggregation, and division of the personality (cf. Van der Hart & Dorahy, 2009).

This dissociation of the personality was observed in psychiatric patients, especially those with a history of childhood traumatization, but eventually also among individuals with work-related trauma. During the end of the nineteenth and beginning of the twentieth centuries, these patients were often described as suffering from hysteria, which can be considered as a broad class of trauma-related disorders. Janet (1907) gave perhaps the best definition of hysteria:

> A form of mental depression [i.e., lowering of the integrative capacity] characterized by a retraction of the field of consciousness and a tendency to the dissociation and emancipation of the systems of ideas and functions that constitute personality. (p. 332)

A major source of observation took place in World War One (WWI), when thousands of combat soldiers became acutely traumatized. Thus, the opening quote of this chapter was written in a book on the war neuroses by a military psychiatrist based on his experiences during this war which led him to believe that "no man probably can stand the stress of warfare for an unlimited time" (Ross, 1941, p. 66). The term "breaking-point" that Ross introduced to denote psychological trauma is an important one when one wants to understand dissociation as a major feature of trauma (as will be discussed in this chapter). It is for obvious reasons that his book was published only in 1941, at the beginning of World War Two, when the need for such knowledge acquired during WWI became acute.

For the same reason, Myers's book, *Shell-shock in France 1914–18*, was published in 1940 (Myers, 1940). Although he used the concepts of trauma and shock, it will become obvious that what he describes when acute traumatization occurs is that the individual experiences a breaking point. Myers first described that when this happens, the soldier has a certain loss of consciousness, "which may vary from a slight, momentary, almost imperceptible dizziness or 'clouding' to profound and lasting unconsciousness" (p. 66). In many cases, Myers could ascertain that the soldier's attention in such a condition was generally concentrated on the scene which produced his condition, during which he may experience "occasional outbursts of hallucinatory delirium." Myers proceeded to describe his observations in terms of different personalities – Janet's "systems of ideas and functions" – that are dissociated from each other. Following the theory of structural dissociation of the personality, we take the position that all of us have but one personality, however divided or fragmented it may be. Hence, our inclusion of "part of the personality" in the following quotes:

> At this stage, then, the normal personality is in abeyance. Even if it is capable of receiving impressions, it shows no signs of responding to them. The recent emotional experiences of the individual have the upper hand and determine his conduct: the normal has been replaced by what we may call the "emotional" [part of the] personality. (p. 67)

Thus, the individual's breaking point or acute trauma involves a division of his or her personality into at least two dissociative parts, one the so-called emotional part of the personality (EP) and the other the so-called apparently normal part of the personality (ANP):

> Gradually or suddenly an "apparently normal" [part of the] personality usually returns – normal save for the lack of all memory of events directly connected with the shock, normal save for the manifestation of other ("somatic") hysterical disorders indicative of mental dissociation. Now and again there occur alternations of the "emotional" and the "apparently normal" [parts of the personality], the return of the former being often heralded by severe headache, dizziness or by a hysterical convulsion [i.e., pseudo-epileptic seizure]. On its return, the "apparently normal" [part of the] personality may recall, as in a dream, the distressing experiences revived during the temporary intrusion of the "emotional" [part of the] personality. (p. 67)

Dissociation of the Personality in the Theory of Structural Dissociation

As the terms "breaking point" and "trauma" imply, we maintain that trauma-related dissociation is primarily an ongoing integrative *deficit* that results in a *structural dissociation* of the personality. Only secondarily is dissociation a psychological *defense* (Steele *et al.*, 2009b; Van der Hart *et al.*, 2006) and thus enhances an individual's chance to survival. Thus, the dissociation involved when an individual has reached his or her breaking point has been defined as (Nijenhuis & Van der Hart, 2011):

> a division of an individual's personality, that is, of the dynamic, biopsychosocial system as a whole that determines his or her characteristic mental and behavioural actions. This division of personality constitutes a core feature of trauma. It evolves when the individual lacks the capacity to integrate adverse experiences in part or in full, can support adaptation in this context, but commonly also implies adaptive limitations.
>
> The division involves two or more insufficiently integrated dynamic but excessively stable subsystems. These subsystems exert functions, and can encompass any number of different mental and behavioral actions and implied states. These subsystems and states can be latent, or activated in a sequence or in parallel. Each dissociative subsystem, that is, dissociative part of the personality, minimally includes its own, at least rudimentary first-person perspective. As each dissociative part, the individual can interact with other dissociative parts and other individuals, at least in principle. Dissociative parts maintain particular psychobiological boundaries that keep them divided, but that they can in principle dissolve. (p. 148)

Phenomenologically, dissociation of the personality manifests in dissociative symptoms that can be categorized as negative (functional losses such as aphonia, amnesia, and paralysis, and loss of certain skills such as reading) or positive (intrusions such as flashbacks or voices), and psychoform (symptoms such as amnesia, hearing voices, and thoughts being "put in" one's mind) or somatoform (symptoms such as anesthesia or tics, and bodily sensations related to trauma) (Nijenhuis & Van der Hart, 2011; Van der Hart *et al.*, 2005, 2006).

It should be noted that this definition of trauma-related dissociation certainly does not reflect a consensus in the field. As mentioned in this chapter, by using the concept in a much wider sense many authors, mainly from North America, consider alterations of consciousness, such as absorption, imaginary involvement, and lowering of consciousness, as also dissociative in nature. Our problem with this inclusion is that these phenomena may occur in relation to a division of the personality (see Myers's quote in the "Historical Perspective" section of this chapter) but not exclusively (Steele *et al.*, 2009a).

Following Myers (1940), the theory of structural dissociation of the personality (TSDP) distinguishes two prototypical personality subsystems or dissociative parts. One prototype, then, is called the *emotional part of the personality* (EP; Van der Hart *et al.*, 2006; Van der Hart, Nijenhuis, & Solomon, 2010). The individual patient as EP is fixated in sensorimotor and highly emotionally charged reenactments of traumatic experiences. In other words, the individual as EP has strongly associated traumatic memories and is living in *trauma time* (Van der Hart *et al.*, 2010). Primarily mediated by the innate mammalian action systems, also known as emotional operative systems (Panksepp, 1998) or motivational systems, of defense and attachment cry, EP's reenactments include action tendencies of defense against perceived or actual threat to the integrity of the body or to life itself, as well as action tendencies regarding the need for attachment and the fear of attachment loss (Liotti, 1999). That is, EP is basically fixated in traumatic memories that may pertain to breaking points at the workplace but in some cases may also involve (particular combinations of) childhood emotional, physical, and sexual abuse, childhood emotional neglect, and otherwise frightening and frightened parental caretaking and attachment.

The other prototype is called the *apparently normal part of the personality* (ANP; Myers, 1940; Van der Hart *et al.*, 2006, 2010). As ANP, the survivor experiences EP and at least some of EP's actions and contents as ego-dystonic and is fixated in avoidance of traumatic memories and often of inner experience in general. Mediated by action systems for functioning in daily life, such as exploration, social engagement, caregiving, play, energy regulation, sexuality, and reproduction, ANP focuses on the functions of these systems. As ANP, the individual may be aware of having a mental disorder such as PTSD but attempts to appear "normal." The fact that this normality is only apparent manifests in negative symptoms of detachment, numbing, and partial or, in rather exceptional cases, complete amnesia for the traumatic experience. Apparent normality also shows in recurrent re-experiencing of traumatic memories from EP and other intrusions such as ANP hearing EP's voice.

Levels of dissociation of the personality

TSDP distinguishes three levels of complexity. The division of the personality into a single ANP and a single EP involves *primary structural dissociation*, and characterizes simple post-traumatic dissociative disorders, including PTSD. This division of the personality implies the emergence of negative dissociative symptoms such as depersonalization and, sometimes, a degree of dissociative amnesia and anesthesia, as well as positive dissociative symptoms such as recurrent intrusions of traumatic memories (although these intrusions may start only after a latency period). The EP tends to have a rigid and extreme narrowed attentional focus, primarily concentrated on perceived threat that is over-interpreted and thus over-reacted to in light of the traumatic past. In primary dissociation EP usually develops a rudimentary (e.g., as seen in acute and simple PTSD) first-person perspective, which is more elaborated and autonomous in cases of complex PTSD, dissociative disorder not otherwise specified (DDNOS; subtype 1), and, particularly, dissociative identity disorder (DID).

Structural dissociation tends to be more complex when the breaking points are more severe and start earlier in life. In *secondary structural dissociation* there is also a single ANP, but more than one EP. This division of EPs may be based on the failed integration among relatively discrete subsystems of the action system of defense (e.g., fight, flight, freeze, or collapse, also described as tonic immobility or total submission) (Porges, 2007; Van der Hart *et al.*, 2006). The basic division at the level, during a severe breaking point, may include one ANP, one EP, and one observing part of the personality. Such experiences are probably rather common in traumatized combat soldiers and other first responders. Cloete (1972) related his own personal example of such a division experienced at a WWI battlefield when during a frontal attack he had to engage in hand-to-hand combat during which he killed his opponent, receiving a serious chest wound and a smashed left shoulder blade. When he returned to the lines, he had to go through mud and duckboard tracks between craters, many of them 10 feet deep and filled with mud and water:

> If I slipped I should drown. I still felt no pain but I was tired. At this point I became two men. My mind left my body, went ahead and stood on a hill. From there I watched quite objectively and with some amusement, the struggles of this body of mine staggering over the duckboards and wading through the mud where the duckboards were smashed. I watched it duck when a salvo of German shells came over. I saw it fall flat on its face when a concealed battery of our own whizzbangs opened up within a few yards of it. I saw it converse with the gunners ... The gunners were too busy to talk but a corporal gave my body some rum which seemed to strengthen it. I was most interested in the process. I then rejoined my body. The rum may have done it. (p. 242)

Implicit in this personal account is that it was an EP that made his way back to the line. Following the reunion of the observing part with "the body," Cloete still suffered from post-traumatic stress (i.e., dissociation between ANP and EP). Incidentally, this example seems to indicate that the emergence of an observing part may have survival value: while being traumatized – in Cloete's case, also physically – there is still some part of the personality that is able to orient itself, detach overwhelming emotions and bodily states, and direct the individual toward safety.

We consider secondary structural dissociation, usually involving many more EPs than in this example, to be mainly relegated to complex PTSD (cf. Van der Hart *et al.*, 2005), trauma-related borderline personality disorder, and DDNOS subtype 1. Different (groups of) EPs may be related to different types of traumatization, while some of them may have been involved in more than one type, such as sexual abuse by the father and emotional and physical abuse by the mother.

Finally, tertiary dissociation involves not only more than one EP, but also more than one ANP. Division of ANP may occur as certain inescapable aspects of daily life become saliently associated with traumatizing events such that they tend to reactivate traumatic memories. The patient's personality becomes increasingly divided in attempts to maintain functioning while avoiding traumatic memories, or has never included an integration of action systems for functioning in daily life as well as for defense. This division of ANP thus tends to occur along different action systems of daily life and characterizes patients with DID, who, as a rule, have a history of chronic childhood traumatization. In a few DID patients who have an extremely low integrative capacity and in whom dissociation of the personality has become strongly habituated, new ANPs may also evolve to cope with minor frustrations of life.

Workplace-related traumatizing events and levels of structural dissociation

There are roughly three types of traumatization at the workplace that interact with levels of structural dissociation: (1) single traumatizing events or critical incidents, (2) cumulative (potentially) traumatizing events, and (3) critical incidents reactivating existing traumatic memories and related EPs.

Single traumatizing events or critical incidents. A single traumatizing event has the effect of pushing the individual to his or her breaking point. In work settings, post-traumatic stress reactions have been documented among workers having experienced a criminal assault or sustained an injury (Benedek *et al.*, 2007;Mitchell & Everly, 2000). Direct physical harm is not the only condition for experiencing one's breaking point: witnessing a bank robbery or a violent accident at the site of work may also mark the onset of post-traumatic stress or PTSD (MacDonald *et al.*, 2003). In most of these cases, the breaking point involved primary structural dissociation (i.e., the development of a rudimentary EP and an ANP as the "major shareholder").

A soldier was in an armored car when it went over an improvised explosive device. The car exploded, and the soldier blacked out. When he came to, the car was upended and he was hanging upside down outside the car which, along with smoke, dust, and scattered debris, gave him an "eerie perspective." His immediate belief was that he was dead. However, his physical wounds were minor. In the aftermath he experienced nightmares and flashbacks concerning the incident, and when he was triggered (i.e., reminded of the incident by internal or external stimuli), he would feel as if he were "dead." This critical incident was a breaking point, resulting in the formation of an EP with its traumatic memory of the explosion *and* the pathogenic kernel statement "I am dead." Treatment focused on the incident, resulting in the resolution of the kernel statement and the integration of the EP and its traumatic memories with the ANP. With his "completely normal personality thus at last obtained" (Myers, 1940, p. 69; and discussed further in this chapter), the soldier had again become fully functional.

Cumulative (potentially) traumatizing events. In certain professions, workers may be confronted with an accumulation of critical incidents or potentially traumatizing events, which gradually burden them and eventually cause them to experience their breaking point. Common examples are firefighters (Wagner, Heinrich, & Ehlert, 1998) and police officers whose daily work involves frequent contact with physical violence and dangerous situations (Robinson, Sigman, & Wilson, 1997). Although there is still a lack of research, our hypothesis is that *some* of these situations may involve more than one EP, thus pertain to secondary structural dissociation.

A policeman worked the scene of an auto accident involving multiple casualties who had been burned. The accident was the breaking point for the officer and it continued to bother him after his role was completed. The smell particularly lingered in his mind as well as the images of the contorted bodies. In the days and weeks following, he felt emotionally exhausted and depressed, and also experienced agitation and difficulty sleeping, and avoided going to accidents. His history revealed a number of "cracks" in his personality, that is, significant critical incidents or potentially traumatizing events in his 12-year career, including an incident that happened his first year where he responded to a sudden infant death (SID) situation, a previous auto fatality where the victim burned to death and the officer tried to rescue him but could not, and a particularly gruesome accident where the officer noticed he was walking on a deceased victim's brain which had spilled out of the head. A divorce several years before had also made him feel more vulnerable to the stressors of the job. Treatment addressed not only the precipitating "breaking point" incident, but also the other critical incidents that contributed to his vulnerability. Treatment also addressed issues related to his divorce, which also could be considered a factor in this vulnerability. He was able to resume his career.

Critical incidents reactivating existing traumatic memories and related EPs. For some workers the impact of potentially traumatizing events may become traumatic experiences because existing traumatic memories and related EPs are reactivated. An example pertains to one of two police officers who were attacked by a very large man with a knife. He pushed one police officer to the ground and jumped on top of the other. This second officer was able to hold on to the knife with both hands while the man continually hit him, and then started to strangle him. The officer who was initially knocked down joined the fight, but was unable to stop the man. Fearing his partner was about to be killed, the officer shot the man, which ended the struggle. The officer who had been beaten by the man was traumatized by the situation. After a week off, he returned to work. However, he froze once in a situation where he thought there would be violence, and he found himself avoiding calls where there was the potential for violence. After several weeks of continued difficulty, the officer received sick leave and psychological treatment. During the history-taking phase, memories of being frightened in childhood and related EPs were identified. These included memories of his mother clapping her hands very loud in his face when she was angry, scaring him badly. Thus, not only was there an EP resulting from the recent life-threatening situation, but also there were EPs originating from traumatizing experiences in childhood. Initial interventions focused on dealing with the recent event were helpful, but did not alleviate the officer's sense of vulnerability and sense of powerlessness. Only after EPs emanating from childhood and their traumatic memories were integrated, did he experience complete symptom relief and was returned to full duty.

Certain critical incidents that might be merely aversive to most people may in vulnerable workers reactivate traumatic memories and related EPs pertaining to earlier traumatization. For example, sexist language and looks on the work floor may trigger traumatic memories of childhood sexual abuse and EPs such as extreme flight parts or fight parts, and/or highly submissive child parts. And an aggressive, condescending look of one's boss or colleague may reactivate memories – and related EPs – of severely threatening hostility and violence from one's father or a past perpetrator.

A 32-year-old woman with DID had an extremely competent ANP working as an assistant manager in a large company. Her new male chauvinist boss often treated her in such a condescending way that it triggered extremely frightened child EPs and their traumatic memories of emotional and physical abuse by her sadistic father. She couldn't function at her job anymore. After a long sick leave she found a job in a different company, where she functioned well but was still vulnerable for such triggers and others, such as an apparently cold stare by a female colleague that could trigger EPs still suffering from the mother's distant and rejecting attitude toward her as a child and adolescent. Long-term individual psychotherapy – such as described in Van der Hart *et al.* (2006) – was needed for her to develop more resilience in such situations.

It should be noted that this classification is far from clear-cut. In assessing the type of traumatization and involved dissociation of the individual, clinicians should take into account the worker's "background stress" in relation to the latest traumatizing events. For example, a single, acute traumatizing event may better denote the breaking point of a veteran or police officer witnessing a particularly violent crime when his or her previous encounters with aggression were relatively mild and less stress provoking. Indeed, one ANP and one single EP may have developed. Likewise, emergency workers involved in an acute catastrophe may show different stress reactions depending on their past trauma histories, related to various levels of structural dissociation of the personality.

In addition to the risk of direct traumatization in emergency services following natural or man-made disasters or accidents, public health and public safety workers, such as medics and paramedics, social workers, and the like, might also be under the risk of vicarious traumatizations which often take place in "caring" work with traumatized people (Figley, 1995;McCann & Pearlman, 1990). It is conceptualized as an interactive process between the caregiver and the client which unfolds through empathetic engagement. For example, in crisis situations, medics and paramedics could be confronted with potentially traumatizing images, sounds, and smells, or even violent or aggressive behaviors from distressed survivors, patients, or their family members and friends, while they could also be vicariously traumatized through empathizing with survivors and clients and being exposed to their traumatic experiences. The latter situation could turn into a breaking point, be it a single-episode, cumulative, or precipitative experience. The clinical consequence of vicarious traumatization is often used interchangeably with secondary traumatic stress disorder and "compassion fatigue" (Dunkley & Whelan, 2006; Figley, 1995), also called the "cost of caring".

Maintenance of dissociation of the personality

Different parts of the personality maintain more or less permeable boundaries that keep them divided. These boundaries depend on mental actions of these parts, and these actions are open to change, at least in principle, so that the boundaries can become more permeable or disappear altogether. We have hypothesized that the boundaries among dissociative parts essentially relate to phobias of traumatic memories – with ANP being frightened and avoidant of the intrusions of these memories – and phobias that these parts have towards each other (Van der Hart *et al.*, 2006).

These phobias may develop when social support in the aftermath of trauma is lacking, for instance when others deny the experience or prohibit talking about it. This may be the case in many workplaces, where talking about fears and vulnerability is a cultural "taboo." Thus, in the military PTSD has also been associated with lack of support in the aftermath of traumatizing events (King *et al.*, 1998). The phobias are also maintained by pathogenic kernel statements, that is, uncritically accepted verbal formulas based on feelings, prejudice, suggestion, and restricted view of self and others such as "There is something wrong with me if I feel vulnerable and scared," "I will go crazy if I start to feel," "It was no big deal" (which reflects a minimalization or denial of the emotional impact), and "It is all my fault," reflecting the tendency to over-take responsibility and blame oneself for things beyond one's control. The phobias are also maintained by avoidant behavioral actions. For example, some patients use alcohol, drugs, or medication, lowering their level of consciousness to avoid painful bodily and emotional trauma-related feelings or memories; and some hurt themselves to cover them up. These phobic mental and behavioral actions can be seen as *substitute actions*, that is, less adaptive actions that substitute for the more efficient but also (much) more difficult actions of integrating traumatic memories and various parts of the personality. Treatment, then, should assist the individual to overcome the various phobias that, so far, have been pivotal in maintaining dissociation of the personality (as will be discussed in this chapter).

Dissociation Measurements

Clinicians who fail to use instruments measuring dissociative phenomena and disorders and to take a detailed trauma history may be unaware that such complexities can exist. When, in these cases, interventions that are appropriate for simple post-traumatic stress (with symptoms conceptualized in TSDP as belonging to one EP and its traumatic memory) may open a veritable Pandora's box, and therapeutic and adaptive disaster may ensue. For instance, the target traumatic memory reactivates chains of the EPs' other traumatic memories, involving a major crisis, including parts' attempts at self-harm or suicide or rapid, uncontrollable switching among parts. Instruments have been developed to measure dissociative symptoms occurring during or right after the individual's breaking point, that is, peri-traumatic dissociation, as well as for chronic dissociative symptoms and dissociative disorders.

Peri-traumatic dissociation

Some individuals report experiences during or immediately after a potentially traumatizing event that have been referred to as peri-traumatic dissociation (e.g., Marmar *et al.*, 1996). The most widely used instrument used to measure peri-traumatic dissociation is the PDEQ (Marmar *et al.*, 1996; Marmar, Weiss, & Metzler, 1998). The PDEQ is a 10-item instrument, of which rater and self-rating versions exist. Examples of items, to be endorsed on 5-point scales, are "I had moments of losing track of what was going on – I 'blanked out'" or "I spaced out or in some way felt that I was not part of what going on," and "My sense of time changed – things seemed to be happening in slow motion." Although the results are far from unequivocal, many studies, such as those among individuals with critical incidents in the workplace, including emergency medical service workers, paramedics, fire fighters, police officers, war veterans, and transportation workers (Marmar *et al.*, 1998), indicated a positive relationship between peri-traumatic dissociation and post-traumatic stress later in life. Thus, a meta-analysis of 59 independent studies, with 83 outcomes, found a significant positive correlation of 0.401 between peritraumatic dissociation and posttraumatic stress (Lensvelt-Mulders *et al.*, 2008). Interestingly, it was also found that the longitudinal studies reported a stronger positive relationship than the retrospective studies. It should be noted, however, that there are also studies that did not find such a relationship.

Earlier in this chapter, we pointed out that there exists in the field much confusion about the nature of dissociation, which is also observable in the items of the PDEQ. For instance, what is the reason to regard the PDEQ item mentioned here about changed sense of time as a dissociative phenomenon? The same would go for an item such as "reduced awareness." Indeed, many other instruments measuring peri-traumatic dissociation do not include both items (Van der Hart *et al.*, 2008). Furthermore, experiences referring to somatoform dissociation, such as paralysis and loss of motor control, seem to be missing altogether. The Somatoform Dissociation Questionnaire – Peritraumatic (SDQ-P; Nijenhuis & Van der Hart, 1998) is an attempt to remedy this lacuna.

Chronic dissociative symptoms

The Dissociative Experiences Scale (DES; Bernstein & Putnam, 1986) is the most often used instrument measuring psychoform dissociation (amnesia, depersonalization, or derealization) as well as absorption and imaginary involvement, and has good reliability and clinical validity (Frischholz *et al.*, 1990). The DES is a 28-item self-report questionnaire on which participants indicate what percentage of time (0–100%) each statement applies to them. Examples of statements are "Some people occasionally find themselves at a place while they do not have a clue how they got there" and "Some people are occasionally told that they don't recognize their friends or family members." Total scores are calculated by averaging the 28 item scores. The DES is a screening instrument, not a diagnostic one. In the literature, various cutoff scores have been suggested (Foote *et al.*, 2006). For instance, in the Netherlands, the cutoff score demanding a diagnostic assessment for DSM-IV dissociative disorders is 25. Average scores for people with DDNOS are around 40, and for people with DIS around 50 (Draijer & Boon, 1993). See Van IJzendoorn and Schuengel (1996) for an overview of average DES scores among a wide range of mental disorders.

The Somatoform Dissociation Questionnaire (SDQ-20; Nijenhuis, 2004; Nijenhuis *et al.*, 1996) measures, as its name indicates, somatoform dissociative symptoms. The SDQ-20 is a 20-item self-report questionnaire using 5-point scales to indicate to which degree presented statements apply. Examples of statements are "It sometimes happens that I have pain while urinating" and "It sometimes happens that I cannot hear anything for a while (as if I were deaf)." Total scores are the sum of the 20 item-scores ranging from 20 to 100. The scale has high reliability and good construct validity (Nijenhuis *et al.*, 1996, 1998). Including five items of the SDQ-20, the SDQ-5 has been developed as a screening instrument and a cutoff score of 8 is used for DSM-IV dissociative disorders (Nijenhuis, 2004; Nijenhuis *et al.*, 1997).

Dissociative disorders

Before discussing the two most often used diagnostic instruments for the DSM-IV dissociative disorders, it should be mentioned that its classification of dissociative disorders is rather arbitrary and has been criticized (Dell & O'Neil, 2009; Van der Hart *et al.*, 2006). One problematic issue is that the DSM-IV does not include somatoform dissociative phenomena, such as stupor, pseudo-epileptic seizures, dissociative paralysis, and dissociative anesthesia, under the dissociative disorders, while the ICD-10 classifies them as such under the heading of "Dissociative Disorders of Movement and Sensation" (World Health Organization, 1992). A further concern is that the DSM-IV does not recognize the dissociative nature of other trauma-related disorders, such as

PTSD, complex PTSD, and trauma-related borderline personality disorder (e.g., Van der Hart *et al.*, 2006).

A major diagnostic instrument is the Interview for DSM-III-R Dissociative Disorders (SCID-D-R; Steinberg, 1994), a semistructured diagnostic interview to be used by trained clinicians who are highly knowledgeable about mental disorders and psychopathology. Questions are asked, among other things, about five symptoms groups: amnesia, depersonalization, derealization, identity confusion, and identity alteration. The SCID-D is the most often used and a reliable and valid instrument for the diagnosis and exclusion of DSM-IV dissociative disorders (Brand, Armstrong, & Loewenstein, 2006). However, this instrument also has limitations, such as the absence of accurate descriptions of the qualitative differences among the various dissociative symptoms and the lack of questions pertaining to the often occurring comorbidity with DID and DDNOS (Boon & Draijer, 1995).

A second, also often used diagnostic instrument, is the Dissociative Disorders Interview Schedule (DDIS; Ross, 1997), a 131-item structured interview that includes questions about childhood abuse. The DDIS has good concurrent validity with the DES and the SCID-D (Ross, Duffy, & Ellason, 2002). Finally, a very promising instrument measuring so-called pathological dissociation is the Multidimensional Inventory of Dissociation (MID; Dell, 2006), a 218-item, self-administered, multiscale instrument that correlates strongly with other measures of dissociation.

Structural Dissociation and Treatment Principles

If we assume that the essence of a breaking point or trauma is the development and maintenance of dissociation of the personality, then interventions essentially should consist of strategies that involve its resolution, whatever the treatment approach used. Myers (1940) had already formulated this with regard to acutely traumatized WWI combat soldiers:

> [T]he treatment to be recommended . . . consists in restoring the "emotional" [part of the] personality deprived of its pathological, distracted, uncontrolled character, and in effecting its union with the "apparently normal" [part of the] personality hitherto ignorant of the emotional experiences in question. When this re-integration has taken place, it becomes immediately obvious that the "apparently normal" [part of the] personality differed widely in psychical appearance and behaviour, as well as mentally, from the completely normal personality thus at last obtained. Headaches and dreams disappear; the circulatory and digestive symptoms become normal; even the reflexes may change; and all hysterical [i.e., dissociative] symptoms are banished. (pp. 68–69)

In relatively simple cases, this "re-integration" of the personality involving some degree of primary dissociation and relatively mild forms of secondary dissociation can be reached either by some form of trauma support, such as those described throughout this book, or by treatment approaches originally developed for simple PTSD, such as eye movement desensitization and reprocessing's (EMDR) basic protocol (see Chapter 17), prolonged exposure, and cognitive restructuring therapy, that directly focus on the integration or processing of the traumatic memory and implied rudimentary EP.

However, in more complex cases, understanding of the dissociation of the personality among various dissociative parts is recommended. Also, the therapist should be knowledgeable about the mental actions involved in the integration of traumatic memories and dissociative parts, as well as about phase-oriented treatment that is the "standard of care" in the treatment of secondary and tertiary structural dissociation (e.g., Brown, Scheflin, & Hammond, 1998; Van der Hart *et al.*, 2006).

Phase-oriented treatment

The treatment of traumatized individuals with more complex forms of traumatization and structural dissociation needs to be based on careful screening and diagnostic procedures, including those mentioned in this chapter with regard to dissociative symptoms and dissociative disorders, as well as on a thorough history taking (cf. Van der Hart *et al.*, 2006). The recommendation of phase-oriented treatment is based on consistent clinical observations that the more severe the traumatization and the more complex the dissociation of the personality, the more survivors need to develop skills with regard to functioning in daily life before facing the most difficult challenge of integrating their traumatic memories, for example using EMDR to promote the integration or processing of traumatic memories, and the further integration of their personality. These skills can pertain to a wide variety of domains, such as emotion regulation, self-care, energy management, planning and adaptive execution of daily life activities, social interaction, assertiveness, reflection, and constructive communication among different dissociative parts. The development of these skills is part and parcel of the first phase of a usually three-phase treatment model consisting of (1) stabilization, symptom reduction, and skills training; (2) treatment of traumatic memories; and (3) personality (re)integration and rehabilitation. The model takes the form of a spiral, in which these different treatment phases can be alternated according to the needs of the patient. The reader is referred to the clinical literature for further orientation on the treatment principles involved (Brown *et al.*, 1998; Chu, 1998; Gelinas, 2003; Van der Hart *et al.*, 2006). It should be noted that some traumatized individuals are endowed with a lower integrative capacity than others. These patients need (far) more preparation before attempts to integrate traumatic memories are undertaken (if ever), and tend to have a less favorable prognosis.

Although not discussed in this chapter, the importance of group-level and/or organization-level support must be acknowledged. The organization needs to foster a caring work environment, acknowledge the normality of post-traumatic reactions, and provide appropriate time and resources for impacted workers to recover.

Integration: synthesis and realization

The ultimate goal of treatment for trauma survivors, as well as those traumatized in the workplace, is thus fostering mental health, which includes increased adaptive functioning including in the workplace. (However, sometimes the individual traumatized in the workplace concludes that finding other jobs is needed for maintaining his or her mental health.) Mental health has been described as "a high capacity for integration, which unites a broad range of psychological phenomena within one personality"

(Janet, 1889, p. 460). In simple cases, the integration of the traumatic memory and related EP(s) fosters mental health. In complex cases, the skills training mentioned in this chapter and additional work with dissociative parts are needed for this purpose.

We distinguish two main levels of integration (Van der Hart *et al.*, 2006, 2010). *Synthesis* pertains to those basic integrative mental and behavioral actions through which experiences, such as sensory perceptions, movements, thoughts, affects, memories, and a sense of self, are bound together (linked) and differentiated (distinguished from each other).

Realization consists of higher order integrative actions, which are based on these lower order integrative actions. It is defined as developing a high degree of personal awareness of reality as it is, accepting it, and reflectively and creatively adapting to it. Ownership (i.e., personal awareness and acceptance of experience as one's own) is defined as *personification*: for example, "That happened to *me* and I am aware of how it helped shape who I am," and "These are *my* feelings and *my* actions." Dissociative individuals do not sufficiently own or *personify* their inner and outer experiences. Full realization also requires *presentification*, defined as being in the present with a synthesis of all one's personified experiences – past, present, and anticipated future – at the ready to support reflective decision making and adaptive action. Well-integrated individuals remain grounded in the present when they remember originally traumatizing events, and experience the recall as an autobiographical narrative memory rather than a reliving of the past. Traumatized individuals thus can be described as suffering from a syndrome of nonrealization.

Conclusion

The point of departure of this chapter was the contention that psychological trauma, experiencing one's breaking point, by definition involves a dissociation of the personality into two or more subsystems (i.e., dissociative parts), at least one part being the "main shareholder" that functions in daily life and at least one part, living in trauma time, which can be triggered and then start re-experiencing the trauma. Contrary to what is often assumed, the dissociative symptoms stemming from this division are not only negative but also positive, including intrusions of reactivated traumatic memories. Thus, even simple PTSD related to trauma in the workplace, characterized by primary structural dissociation, should be regarded as a dissociative disorder. However, having one EP or several EPs does not mean per se that the person meets the criteria for any psychological disorder, let alone a DSM-IV dissociative disorder. In all cases, the treatment goal should be, at least, a resolution of the dissociation, that is, a (re-)integration of the personality. This is what fostering traumatized workers' "mental health" is all about. Not only is the trauma resolved but also the worker's integrative capacity and flexibility are increased, resulting in greater adaptive functioning on the job. Clinicians treating individuals traumatized in the workplace need to be alert to the fact that they may suffer from more than primary dissociation of the personality. Thus, taking a history, which may reveal previous traumatization, and assessing for secondary and above dissociation are important considerations in developing an appropriate treatment plan. Being knowledgeable about dissociation of the personality may help therapists to understand the complex inner world of survivors of successive or chronic traumatization, as manifested in complex trauma-related disorders, including complex dissociative disorders. Such understanding can guide the application of phase-oriented treatment of these individuals, specifying how much stabilization is needed, and identifying the successive problems that need to be resolved, including the major challenge of the integration of traumatic memories, in order to promote mental health.

References

American Psychiatric Association. (1994). *Diagnostic and statistical manual of mental disorders* (4th ed.). Washington, DC: Author.

Benedek, D. A., Fullerton, C., & Ursano, R. J., (2007). First responders: Mental health consequences of natural and man-made disasters for public health and public safety workers. *Annual Review of Public Health, 28*, 55–68.

Bernstein, E. M., & Putnam, F. W., (1986). Development, reliability, and validity of a dissociation scale. *Journal of Nervous and Mental Disease, 174*, 727–735.

Boon, S., & Draijer, N., (1995). *Screening en diagnostiek van dissociatieve stoornissen.* Lisse, Netherlands: Swets & Zeitlinger.

Brand, B. L., Armstrong, J. G., & Loewenstein, R. J. (2006). Psychological assessment of patients with dissociative identity disorder. *Psychiatric Clinics North America, 29*, 145–168.

Brown, D., Scheflin, A. W., & Hammond, C. D. (1998). *Memory, trauma treatment, and the law.* New York: Norton.

Cardeña, E. (1994). The domain of dissocation. In S. J. Lynn & J. W. Rhue (Eds.), *Dissociation: Clinical and theoretical perspectives* (pp. 15–31). New York: Guilford Press.

Chu, J. A. (1998). *Rebuilding shattered lives: The responsible treatment of complex posttraumatic stress and dissociative disorders* (Second rev. edition published in 2011). New York: Guilford Press.

Cloete, S. (1972). *A Victorian son: An autobiography.* London: Collins.

Dell, P. F. (2006). The Multidimensional Inventory of Dissociation (MID): A comprehensive measure of pathological dissociation. *Journal of Trauma and Dissociation, 7*(2), 77–106.

Dell, P. F., & O'Neil, J. A. (Eds.) (2009). *Dissociation and the dissociative disorders: DSM-V and beyond.* New York: Routledge.

Draijer, N., & Boon, S. (1993), Trauma, dissociation and dissociative disorders. In S. Boon & N. Draijer (Eds.), *Multiple personality disorder in the Netherlands* (pp. 177–194). Amsterdam: Swets & Zeitlinger.

Dunkley, J., & Whelan, T. A. (2006). Vicarious traumatisation: Current status and future directions. *British Journal of Guidance and Counselling, 34*, 107–116.

Figley, C. R. (Ed.) (1995). *Compassion fatigue: Coping with secondary traumatic stress disorder in those who treat the traumatized.* New York: Brunner/Mazel.

Foote, B., Smolin, Y., Kaplan, M., Legatt, M. E., & Lipschitz, D. (2006). Prevalence of dissociative disorders in psychiatric outpatients. *American Journal of Psychiatry, 163*, 623–629.

Frischholz, E. J., Braun, B. G., Sachs, G. R., Hopkins, L., Schaeffer, D. M., Lewis, J., *et al.* (1990). The Dissociative Experiences Scale: Further replication and validation. *Dissociation, 3*, 151–153.

Gelinas, D. J. (2003). Integrating EMDR into phase-oriented treatment for trauma. *Journal of Trauma and Dissociation, 4*, 91–135.

Janet, P. (1889). *L'automatisme psychologique.* Paris: Félix Alcan.

Janet, P. (1907). *The major symptoms of hysteria.* New York: Macmillan.

King, I. A., King, D. W., Fairbank, J. A., Keane, T. M., & Adams, G. A. (1998). Resilience–recovery factors in post-traumatic stress disorder among female and male Vietnam veterans: Hardiness, postwar social support, and additional stressful life events. *Journal of Personality and Social Psychology, 74*, 420–434.

Lensvelt-Mulders, G., Van der Hart, O., Van Ochten, J. M., Van Son, M. J. M., Steele, K., & Breeman, L. (2008). Relations among peritraumatic dissociation and posttraumatic stress: A meta-analysis. *Clinical Psychology Review, 28*, 1138–1151.

Liotti, G. (1999). Disorganization of attachment as a model for understanding dissociative psychopathology. In J. Solomon & C. George (Eds.), *Attachment disorganization* (pp. 297–317). New York: Guilford.

MacDonald, H. A., Colotla, V., Flamer, S., & Karlinsky, H. (2003). Posttraumatic stress disorder (PTSD) in the workplace: A descriptive study of workers experiencing PTSD resulting from work injury. *Journal of Occupational Rehabilitation, 13*(2), 63–77.

Marmar, C., Weiss, D. S., & Metzler, T. J. (1996). Characteristics of emergency services personnel related to peritraumatic dissociation during critical incident exposure. *American Journal of Psychiatry, 153*, 94–102.

Marmar, C. R., Weiss, D. S., & Metzler, T. J. (1998). Peritraumatic dissociation and posttraumatic stress disorder. In J. D. Bremner & C. R. Marmar (Eds.), *Trauma, memory, and dissociation* (pp. 229–252) Washington, DC: American Psychiatric Press.

Marshall, R. D., Spitzer, R., & Liebowitz, M. R. (1999). Review and critique of the new DSM-IV diagnosis of acute stress disorder. *American Journal of Psychiatry, 156*, 1677–1685.

McCann, I. L., & Pearlman, L. A. (1990). Vicarious traumatization: A framework the psychological effects of working with victims. *Journal of Traumatic Stress, 3*, 131–149.

Mitchell, J. T., & Everly, G. S. (2000). Critical incident stress management and critical incident stress debriefings: Evolutions, effects and outcomes. In B. Raphael & J. P. Wilson (Eds.), *Psychological debriefing: Theory, practice and evidence* (pp. 77–90). Cambridge: Cambridge University Press.

Myers, C. S. (1940). *Shell shock in France 1914-1918.* Cambridge: Cambridge University Press.

Nijenhuis, E. R. S. (2004). *Somatoform dissociation: phenomena, measurement & theoretical issues.* New York: Norton.

Nijenhuis, E. R. S., Spinhoven, P., Van Dyck, R., Van der Hart, O., & Vanderlinden, J. (1996). The development and psychometric characteristics of the Somatoform Dissociation Questionnaire (SDQ-20). *Journal of Nervous and Mental Disease, 184*, 688–694.

Nijenhuis, E. R. S., Spinhoven, P., Van Dyck, R., Van der Hart, O., & Vanderlinden, J. (1997). The development of the somatoform dissociation questionnaire (SDQ-5) as a screening instrument for dissociative disorders. *Acta Psychiatrica Scandinavica, 96*, 311–318.

Nijenhuis, E. R. S., Spinhoven, P., Van Dyck, R., Van der Hart, O., & Vanderlinden, J. (1998). Psychometric characteristics of the Somatoform Dissociation Questionnaire: A replication study. *Psychotherapy and Psychosomatics, 67*, 17–23.

Nijenhuis, E. R. S., & Van der Hart, O. (1998). *Somatoform Dissociation Questionnaire-Peritraumatic.* Unpublished manuscript, Utrecht University. See also http://www.enijenhuis.nl.

Nijenhuis, E. R. S., & Van der Hart, O. (2011). Dissociation in trauma: A new definition and comparison with previous formulations. *Journal of Trauma and Dissociation, 12*, 416–445.

Nijenhuis, E. R. S., Van der Hart, O., & Steele, K. (2002). The emerging psychobiology of trauma-related dissociation and dissociative disorders. In H. D'Haenen, J. A. den Boer, & P. Willner (Eds.), *Biological psychiatry* (pp. 1079–1098). London: John Wiley & Sons, Ltd.

Panksepp, J. (1998). *Affective neuroscience: The foundations of human and animal emotions.* New York: Oxford University Press.

Porges, S. W. (2007). The polyvagal perspective. *Biological Psychology*, *74*, 116–143.

Robinson, H. M., Sigman, M. R., & Wilson, J. P. (1997). Duty-related stressors and PTSD symptoms in suburban law enforcement officers. *Psychological Reports*, *81*, 835–845.

Ross, C. A. (1997). *Dissociative identity disorder: Diagnosis, clinical features and treatment of multiple personality* (2nd ed.). New York: John Wiley & Sons, Inc.

Ross, C. A., Duffy, C. M. M., & Ellason, J. W. (2002). Prevalence, reliability, and validity of dissociative disorders in an inpatient setting. *Journal of Trauma and Dissociation*, *3*(1), 7–17.

Ross, T. A. (1941). *War neuroses*. Baltimore: Williams & Wilkins.

Steele, K., Dorahy, M., Van der Hart, O., & Nijenhuis, E. R. S. (2009a). Dissociation versus alterations in consciousness: Related but different concepts. In P. F. Dell & J. A. O'Neil (Eds.), *Dissociation and the dissociative disorders: DSM-V and beyond* (pp. 155–170). New York: Routledge.

Steele, K., Van der Hart, O., & Nijenhuis, E. R. S. (2009b). The theory of trauma-related structural dissociation of the personality. In P. F. Dell & J. A. O'Neil (Eds.), *Dissociation and the dissociative disorders: DSM-V and beyond* (pp. 239–258). New York: Routledge.

Steinberg, M. (1994). *Structured clinical interview for DSM-IV Dissociative Disorders-Revised (SCID-D-R)*. Washington, DC: American Psychiatric Press.

Van der Hart, O., & Dorahy, M. (2009). Dissociation: History of a concept. In P. F. Dell & J. A. O'Neil (Eds.), *Dissociation and the dissociative disorders: DSM-V and beyond* (pp. 2–26) New York: Routledge.

Van der Hart, O., Nijenhuis, E. R. S., & Solomon, R. (2010). Dissociation of the personality in complex trauma-related disorders and EMDR: Theoretical considerations. *Journal of EMDR Practice and Research*, *4*(2), 76–92.

Van der Hart, O., Nijenhuis, E. R. S., & Steele, K. (2005). Dissociation: An insufficiently recognized major feature of complex PTSD. *Journal of Traumatic Stress*, *18*, 413–424.

Van der Hart, O., Nijenhuis, E. R. S., & Steele, K. (2006). *The haunted self: Structural dissociation and the treatment of chronic traumatization*. New York: Norton.

Van der Hart, O., Van Ochten, J., Van Son, M.J.M., Steele, K., & Lensvelt-Mulders, G. (2008). Relations among peritraumatic dissociation and posttraumatic stress: A critical review. *Journal of Trauma and Dissociation*, *9*(4), 481–505.

Van IJzendoorn, M. H., & Schuengel, C. (1996). The measurement of dissociation and dissociative types: A taxometric analysis of dissociative experiences. *Clinical Psychology Review*, *16*, 365–382.

Wagner, D., Heinrichs, M., & Ehlert, U. (1998). Prevalence of symptoms of posttraumatic stress disorder in German professional firefighters. *American Journal of Psychiatry*, *155*, 1727–1732.

World Health Organization. (1992). *The ICD-10 classification of mental and behavioural disorders. Clinical descriptions and diagnostic guidelines*. Geneva: Author.

Part D

The Theory and Practice of Post-trauma Support

17

Utilization of EMDR in the Treatment of Workplace Trauma

Roger Solomon and Isabel Fernandez

Eye movement desensitization and reprocessing (EMDR) is an empirically supported psychotherapeutic approach for treating trauma, which is also applicable to a wide range of other experientially based clinical complaints. It is guided by the adaptive information processing model (AIP), which conceptualizes the effects of traumatic experiences in terms of dysfunctional memory networks in a physiologically based information-processing system. It is composed of eight phases and a three-prong methodology that (a) processes past memories that underlie current problems, (b) identifies and processes present triggers that elicit disturbance, and (c) incorporates positive templates into the client's repertoire for adaptive future behaviors. Research indicates that EMDR can be used to treat trauma in the workplace, both effectively and efficiency, resulting not only in symptom reduction but also in an increase in resiliency. The methodology will be illustrated by case examples.

Utilization of EMDR to Treat Workplace Trauma

Post-traumatic stress disorder (PTSD) can develop after an event where a person is confronted with, or witnesses, actual or threatened death or serious injury to oneself or another. Major symptoms include intrusiveness of images of the event, avoidance of situations and stimuli that can evoke these feelings, and hyperarousal which can provoke startle response, concentration difficulties, sleep disorders, and autonomic reactions. When trauma occurs at the workplace, personnel can continually be exposed to triggers and lose competency, skills, and the ability to function professionally. Psychological trauma not only is personally devastating to the people involved but also can result in financial losses to the companies that have invested in their training and development. In recent years, there has been much interest in defining which psychological treatments can be considered effective for the treatment of trauma-related disorders. EMDR is one psychotherapeutic approach that has been found to be effective

International Handbook of Workplace Trauma Support, First Edition. Edited by Rick Hughes, Andrew Kinder, and Cary L. Cooper.

(Bisson & Andrew, 2007). It is a therapeutic approach that emphasizes the brain's intrinsic information-processing system and how memories are stored. EMDR was introduced in 1989 with a randomized controlled study showing substantial treatment effects on post-traumatic stress (Shapiro, 1989a, 1989b, 2001). Over the past 20 years, more than 20 randomized studies have established the efficacy of EMDR with a wide range of populations (see Bisson & Andrew, 2007). Currently, EMDR is recognized and recommended as an effective treatment for trauma in numerous international guidelines (e.g., American Psychiatric Association, 2004; Bisson & Andrew, 2007; Department of Veterans Affairs & Department of Defense, 2004; National Institute for Clinical Excellence (NICE), 2005). Further, EMDR has also been shown to be effective with a wide range of disorders that are caused or exacerbated by experiential factors (see Shapiro, 2001, 2007; Solomon & Shapiro, 2008). Therapeutic effects are derived from processing targeted distressing memories, which results in an adaptive resolution that promotes psychological health (Shapiro, 1995, 2001). As predicted by the adaptive information processing (AIP) model that guides EMDR practice, clinical studies have demonstrated that beneficial effects result from processing the experiences that underlie current problems.

EMDR is an integrative psychotherapeutic approach composed of eight phases and a three-prong methodology that (a) processes past memories that underlie current problems, (b) identifies and processes present triggers that elicit disturbance, and (c) incorporates positive memory templates into the client's repertoire for adaptive future behaviors. Bilateral stimulation in the form of eye movements, taps, or tones is used as part of the procedures. Specific physiological effects of eye movements during EMDR treatment sessions have been found (Elofsson *et al.*, 2008; Sack *et al.*, 2008; Wilson *et al.*, 2001). The research suggests that eye movements result in an increase in parasympathetic activity and a decrease in psychophysiological arousal. Similar physiological results were found following one session of EMDR, evidenced by lowered heart rate and skin conductance. Working with traumatized refugees, Sondergaard and Elofsson (2008) found eye movements had physiological effects, namely a de-arousal with increased finger temperature and changes in the balance between the parasympathetic and sympathetic autonomic nervous system. Studies have also found that eye movements enhanced retrieval of episodic memories in laboratory studies (Christman *et al.*, 2003) and increased attentional flexibility (Kuiken *et al.*, 2001–2002). In other studies, the eye movements have been found to decrease vividness and emotionality of negative and positive memories (Barrowcliff *et al.*, 2004; Gunter & Bodner, 2008; Hornsvelt *et al.*, 2010; Kavanagh *et al.*, 2001; Maxfield, 2008; Sharpley, Montgomery, & Scalzo, 1996; van den Hout *et al.*, 2001).

Further research needs to determine if the change in vividness and emotionality precedes or follows the physiological de-arousal and whether these occur together or are separate elements. However, several hypotheses have been advanced regarding the mechanism of action related to the bilateral stimulation. These include the orienting response (MacCulloch & Feldman, 1996), increased interhemispheric coherence (Propper *et al.*, 2007), the same neurological mechanisms involved in rapid eye movement sleep (Stickgold, 2002, 2008) and working memory (Andrade, Kavanagh, & Baddeley, 1997; Maxfield, 2008). The apparent desensitization effects reported in various studies are predicted by all these hypotheses. However, additional research is needed to identify the actual mechanism of actions and to determine the interaction of various mechanisms.

EMDR and Workplace Trauma

EMDR has been successful in treating workplace trauma. Railroad employees who had experienced a "person-under-train" accident or had been assaulted at work were given five sessions of EMDR (Hogberg *et al.*, 2007a). Results showed remission of PTSD in 67% compared to 11% in the wait list control. Significant effects were documented in Global Assessment of Function (GAF) and Hamilton Depression (HAM-D) score. A follow-up study (Hogberg *et al.*, 2007b) showed that 70% of the EMDR treatment group no longer met PTSD criteria at an eight-month follow-up, and 65% no longer met the criteria at 35 months. The authors explained that the drop in percentage at follow-up was as a result of relapsing due to severe life crises. Twenty-five percent of participants had full working capacity before treatment, whereas 83% had full capacity at 35 months follow-up. The authors concluded that EMDR treatment had an enduring, positive effect on PTSD in civilian adult trauma victims.

Another study with railroad engineers involved in crossing accidents looked at the effectiveness of a two-day support program (Solomon & Kaufman, 2002). The program resulted in significantly lower Impact of Event scores (IES). However, engineers who received EMDR, as part of the program, had significantly greater reductions in IES scores. EMDR has also been utilized in treating other occupational trauma, such as law enforcement critical incidents (McNally & Solomon, 1999; Solomon, 1995, 2002; Wilson *et al.*, 2001), military combat (Russell, 2008; Solomon *et al.*, 2009), bank employees involved in hold-ups (Rost, Hoffman, & Wheeler, 2009), and mining accidents (Blore, 1997).

Adaptive Information Processing Model

EMDR treatment is guided by the AIP model, which emphasizes both memory networks and the physiological information-processing system (Shapiro, 2001), which transfers experiences into physically encoded memories that are stored in associative memory networks. These memory networks provide an important basis for the person's interpretation of new experiences and significantly influence his or her current perception, behavior, and feelings. When working appropriately, the innate information-processing system "metabolizes" or "digests" new experiences. Incoming sensory perceptions are integrated and connected to related information that is stored in existing memory networks, allowing us to make sense of our experience. What is useful is learned, stored in memory networks with appropriate emotions, and made available to guide the person in the future (Shapiro, 2001).

However, high levels of disturbance can disrupt processing mechanisms and cause the experiences to be stored in excitatory, distressing, state-specific forms (Shapiro, 1995, 2001). These unprocessed memories are stored in isolation, and contain the affects, thoughts, sensations, and behavioral responses that were encoded at the time of the event (Shapiro, 2001, 2007). Being stored in this dysfunctional way prevents them from connecting to more adaptive information and being assimilated within more comprehensive memory networks. When triggered by internal or external reminders or triggers, nightmares, flashbacks, and other symptoms of trauma may result. The lack of adequate assimilation means that the client is still reacting emotionally and behaviorally in ways consistent with the earlier disturbing event. Trauma, therefore, can be treated by targeting the event(s) that have been stored dysfunctionally in the nervous system (Shapiro, 2001).

The goals of the procedures and protocols of EMDR are to access the dysfunctionally stored experiences and stimulate the innate information-processing system in such a way that these isolated memories link up to and are assimilated into new or currently existing functional neurological networks. Clinical observations of resolution and recovery (including integration of the targeted memory) observed in EMDR treatment sessions show a rapid progression of intrapsychic connections, as emotions, insights, sensations, and memories emerge and change with each new set of bilateral stimulation (directed lateral eye movements, or alternating taps or tones). Successful EMDR treatment results in the targeted memory no longer being stored in an isolated state, becoming appropriately integrated with the larger comprehensive memory networks comprising the totality of the individual's life experience. Hence, processing is thought to involve the forging of new associations and connections enabling learning to take place. The memory is now stored in a new adaptive form, able to be recalled and verbalized without the negative affect and physical sensations that characterized their previous psychological condition. Once processed and adaptively stored, experiences that previously caused disturbance and maladaptive responses become the foundation of resilience and self-empowerment. In addition to a decline in symptoms, clients give evidence of a comprehensive perceptual and psychological reorganization. When negative experiences are successfully processed, adaptive, self-enhancing perspectives emerge, which can then start to generalize across the memory network that contains the maladaptive information. Such changes are readily observable in the remediation of trauma (see Bisson & Andrew, 2007, for review). Physiological changes have also been evidenced by neuroimaging studies (e.g., Bossini, Fagiolini, & Castrogiovanni, 2007; Lansing *et al.*, 2005; Levin, Lazrove, & van de Kolk, 1999; Oh & Choi, 2007; Ohta ni *et al.*, 2009; Pagani *et al.*, 2007).

The basis of EMDR's effects appears to be different from extinction (Lee, Taylor, & Drummond, 2006; Rogers & Silver, 2002), and may result from a reconsolidation of the memory (Solomon & Shapiro, 2008; Suzuki *et al.*, 2004). The accessing and reprocessing of the original memory result in the memory being re-stored in an altered form by a process that may be similar to what occurs during rapid eye movement sleep (Elofsson *et al.*, 2008; Shapiro, 1995, 2001; Stickgold, 2002, 2008). Reconsolidation might explain how EMDR can produce a lasting elimination of chronic pain by processing the salient underlying memories, in contrast to the simple pain management that results from other forms of therapy (Ray & Zbik, 2001). With the form altered, reconsolidation of the memory also decreases the likelihood of relapse. Further, unlike extinction-based therapies (Craske, Hermans, & Vansteenwegen, 2006) the type of associative process used in EMDR fosters a generalization effect, allowing multiple memories to be addressed simultaneously during a single processing session

(Shapiro, 1995, 2001). Rost *et al.* (2009) provided EMDR to bank employees and a transportation worker who had suffered repeated acute traumatization. Their results showed that EMDR effectively reduced their symptoms. In addition, the treatment appears to have provided them with a protective buffer in situations of ongoing workplace violence. They found that after EMDR treatment, the victims had less symptom reactivation when involved in a subsequent traumatic event as compared to victims who received only a debriefing procedure. This outcome is consistent with the predictions of the AIP model. If the dysfunctional stored memories have been processed and assimilated into adaptive memory networks, then the learning that has taken place becomes the functional basis for interpretation and response to any newly encountered situation. Basically, a new experience of a similar disturbing event will automatically connect with the same adaptive networks and the individual will respond with a sense of resourcefulness and resilience derived from the processed information.

EMDR and Resilience

An important aspect of EMDR is the incorporation of adaptive cognitions as part of the therapeutic process. Negative images, cognitions, and affect transmute into an adaptive perspective, with positive, self-enhancing cognitions evolving. The Rost *et al.* (2009) study, mentioned in this chapter, illustrates how EMDR is a paradigm of resilience, with those treated being able to respond more adaptively to a future trauma. Resilience is the positive capacity of people to cope with stress. It has been described as a dynamic process where people exhibit positive behavioral adaptation when they encounter significant adversity or trauma (Luthar, Cicchetti, & Becker, 2000). Similarly, Antonovsky (1987) proposed the concept of "sense of coherence" (SOC), which is a global orientation pertaining to feelings of confidence, and the ability to comprehend a stressful situation and utilize available resources and adaptive behavior. The SOC is achieved through an individual's belief that his or her life is comprehensible (rational, predictable, structured, and understandable), manageable (the perception that adequate resources exist to deal with adversity), and meaningful (adversity is viewed as a challenge worthy of engagement). Hence, SOC, in AIP terms, is a manifestation of the adaptive information networks that include the fully processed memories of previously traumatizing events. Future stressful situations would stimulate the adaptively stored experiences, thereby providing the base of stability, comprehension, and manageability. Processing of traumatic memories, resulting in adaptive resolution and assimilation into the memory network, enhances a person's SOC. Further, these adaptively stored experiences also underlie the availability of coping resources or strategies inherent in Rosenbaum's (1983) formulation of "Learned Resourcefulness." Learned resourcefulness is the cognitive and behavioral strategies acquired through conditioning, modeling, life experience, and instruction. The strategies, which AIP posits to be the result of adaptively processed memories of past experiences, are used to deal with potentially disturbing inner events and minimize distressing thoughts, feelings, and impulses in order to maintain adaptive functioning. Another study illustrating resilience was conducted by Zaghrout-Hodali, Alissa, and Dodgson (2008). They worked with a group of Palestinian children who were fired upon from a military watchtower close by while playing in their yard. In the first two sessions, using group EMDR, the children responded well and showed recovery from their somatic and behavioral symptoms. However, between sessions 2 and 3 of the four-session EMDR treatment the children experienced another traumatic experience where they were held in one room in their house by people in dark clothes and balaclavas, whom they believed to be the Israeli military. Although previous experience with serial traumatization after treatment with other therapies had led the clinicians to expect the children to have relapsed, they discovered that the level of disturbance experienced by these children after this second event was actually less than their disturbance had been to the previous event. Although the children recognized the trauma, they felt able to deal with it and respond effectively to events in their internal and external worlds. This resilience was reflected in their narrative of the second event, which was largely told as an account of an unpleasant memory, not as a reliving of the experience.

Three-Pronged Protocol

EMDR is a three-pronged protocol. Processing the past traumatic memories often has generalization effects to present situations that evoke distress. However, present triggers can still evoke distress and may need to be processed separately. Processing past memories and present triggers may result in the cessation of symptoms and an increase in positive affects, images, and thoughts associated with efficacy and self-confidence. To further enhance these results, and deal with anticipatory anxiety, positive future templates (Shapiro, 1995, 2001) can be laid down, as "Case Example: Railroad Engineer" illustrates.

Case Example: Railroad Engineer

A collision occurred between a train and a gasoline truck, resulting in an explosion that killed the truck driver and injured the train engineer. Naturally, this was very traumatic for the engineer. Because of his fear that a similar incident might occur someday, he became anxious and hesitant on the job. He was provided with EMDR four months after the incident, when he was physically able to work. First, the memory of the accident was processed with EMDR, which took one session. His negative cognition was "I am vulnerable," and at the end of the session the positive cognition that had evolved was "I am safe." At the next session, present triggers (e.g., seeing trucks or cars that look as if they may not stop at a crossing in time) were processed. The negative cognition of "I am powerless" transformed to a positive cognition of "I have some control and am capable of responding." The following session the engineer reported feeling better but he was worried about another significant accident occurring. Future templates were installed in which the engineer imagined the occurrence of a similar incident, coupled with his chosen adaptive behaviors (e.g., how he would respond, where he would place himself in the engine for maximum safety, and how he would brace himself for impact). Upon returning to work he behaviorally rehearsed what he would do in the event of another accident. Within a month, he was involved in another crossing accident, this one involving a truck. For the engineer, the event was a tragedy, but was not traumatic (i.e., an overwhelming experience). He reacted as he had mentally envisioned during the future template, and as he had practiced, and felt in control during the incident.

This example illustrates how EMDR seems to be effective in promoting resilience in the face of additional crises. Processing the negative memory and present triggers resulted in symptom reduction, and a future template was installed that set the stage for adaptive behavior in the consequent critical incident. Especially in occupations where traumatizing situations can happen again, it is important to plan for future adversity. This can be accomplished by processing the critical incident and present triggers, assisted by the laying down of future templates for adaptive behavior.

Eight-Phase Protocol

Below, the eight phases of EMDR will be discussed. The goal of these phases is to access and process experiences that are contributing to current difficulties (Table 17.1).

The first phase is *history taking*. The therapist obtains the information needed to design a treatment plan. The clinician evaluates the entire clinical picture, including the dysfunctional behaviors, symptoms, and characteristics that need to be addressed. The clinician then determines the specific memories that need to be reprocessed, including the events that initially set the pathology in motion, the present triggers that stimulate the dysfunctional material, and the kinds of positive behaviors and attitudes important for adaptive future functioning. Further, the client's suitability for EMDR needs to be evaluated because the reprocessing of traumatic material may precipitate intense emotions. The client's ability to deal with high levels of disturbance, personal stability, and life constraints is evaluated.

Case Example: Police Officer

The tactical squad of a national police department responded to a school shooting where children were killed. They entered a large room where there were six dead children. It was particularly difficult for the officers to contend with the constant ringing of the children's cell phones. . . . The officers knew it was the parents calling the children, but the officers could not answer the children's phones. There was one boy who was still alive. The tactical commander attended to the boy, and spoke to him, trying to comfort him. The boy died minutes before ambulance personnel arrived. The officer, in this case study example, who had small children, was distressed by the cell phones ringing, knowing how worried he would be if his children did not answer their phones. Even worse was the sense of guilt he felt regarding the boy who was initially alive. The commander felt that the one child they could have helped died because the emergency services did not get there in time. He later reported he often felt distant from his children, withdrawing from them when he was reminded about the school tragedy.

Eighteen months after the incident, this officer, other team members, and officers involved in other traumatic incidents came to a three-day program focusing on coping with traumatic events, based on the Federal Bureau of Investigation's (FBI) Post Critical Incident Seminar (McNally & Solomon, 1999; Solomon *et al.*, 2008; Solomon & Kaufman, 2002). EMDR, on a voluntary basis, was part of this program. The officer's personal history showed no major traumas in his past. He had been involved in many dangerous situations, as a member of the tactical team, and had experienced trauma reactions, but they had subsided with time, and did not bother him in the present. His Impact of Event Scale Revised (IES-R) (Weiss & Marmar, 1997), handed out at the beginning of the seminar, was clinically significant at 36.

The second phase is *client preparation*. This involves establishing a therapeutic alliance, explaining the EMDR process and its effects, dealing with the client's concerns, and teaching the client relaxation techniques for coping with high levels of emotions. Informed consent about the possibility of intense emotions being evoked is obtained. The preparation phase also includes psychosocial education regarding EMDR and the procedures involved, and explaining what can realistically be expected.

Case Example: Police Officer – Continued

"Preparation" consisted of participants telling their stories to the large group [of those officers who had attended], small-group discussion, psychosocial education on trauma, and information on EMDR. Participants, on a voluntary basis, were invited to sign up for an EMDR session. This officer did sign up for a session. Further explanation and assessment for appropriateness took place before the implementation of the next phases of EMDR.

The third phase is *assessment*. In this phase the clinician identifies the components of the target to be treated and takes baseline measures before reprocessing begins. The client is asked to select the image that represents the worst part of the memory. Then the therapist assists the client in identifying the negative cognition that expresses the dysfunctional, negative self-attribution related to participation in the event. Then, a positive cognition or a more rational, realistic, and empowering self-assessment is identified. While utilized later to replace the negative cognition in the installation phase (phase 5), the initial purpose of the positive cognition is to provide a therapeutic direction. To provide a baseline measurement, the client is asked to report how valid the positive cognition feels on a 7-point Validity of Cognition (VOC) Scale, with 1 being *it feels totally false* and 7 being *it feels totally true*. The client and therapist also explore the emotions and physical sensations associated with the traumatic experience. The client is asked to rate the intensity of the emotion on a 10-point Subjective Units of Disturbance (SUD) Scale, with 0 being *neutral* and 10 being *the worst it could be*, to provide a baseline from which to assess changes during the procedure.

Table 17.1 Overview of EMDR treatment

Phase	*Purpose*	*Procedures*
Client history	• Obtain background information	• Standard history taking questionnaires and diagnostic psychometrics
	• Identify suitability for EMDR treatment	• Review of criteria and resources
	• Identify processing targets from positive and negative events in client's life	• Questions regarding (1) past events that have laid the groundwork for the pathology, (2) current triggers, and (3) future needs
Preparation	• Prepare appropriate clients for EMDR processing of targets Stabilize and increase access to positive affects	• Education regarding the symptom picture • Metaphors and techniques that foster stabilization and a sense of personal self-mastery and control
Assessment	• Access the target for EMDR processing by stimulating primary aspects of the memory	• Elicit the image, negative belief currently held, desired positive belief, current emotion, physical sensation, and baseline measures
Desensitization	• Process experiences and triggers toward an adaptive resolution (0 SUD level)	• Process past, present, and future
	• Fully process all channels to allow a complete assimilation of memories	• Standardized EMDR protocols that allow the spontaneous emergence of insights, emotions, physical sensations, and other memories
	• Incorporate templates for positive experiences	• "Cognitive interweave" to open blocked processing by elicitation of more adaptive information
Installation	• Increase connections to positive cognitive networks	• Identify the best positive cognition (initial or emergent)
	• Increase generalization effects within associated memories	• Enhance the validity of the desired positive belief to a 7 VOC
Body Scan	• Complete processing of any residual disturbance associated with the target	• Concentration on and processing of any residual physical sensations
Closure	• Ensure client stability at the completion of an EMDR session and between sessions	• Use of guided imagery or self-control techniques if needed • Briefing regarding expectations and behavioral reports between sessions
Reevaluation	• Evaluation of treatment effects	• Explore what has emerged since last session
	• Ensure comprehensive processing over time	• Re-access memory from last session • Evaluation of integration within a larger social system

Source: Shapiro (2005).

Case Example: Police Officer – Continued

The officer's image was seeing the boy on the ground as he was dying. The negative cognition was "It's my fault" and the positive cognition was "I did the best I could," with a VOC of 3. He had emotions of sadness and guilt, with a SUD of 7, and tension in his neck and shoulders (which he reported he experienced frequently).

The next three phases have to do with the processing of the dysfunctionally stored information. During these phases, there is simultaneous remediation of negative affect, cognitive restructuring, and the generation of insights that can guide the client in the future. The individual phases are designated according to the elements that are used to determine treatment effects. For instance, phase 4, "Desensitization," uses the Subjective Units of Disturbance (SUD) Scale, "Installation" uses the Validity of Cognition (VOC) Scale, and the "Body Scan" (where the client is asked to scan his or her body for sensations) uses the evaluation of the body sensations. Bilateral stimulation such as sets of eye movement (where the client tracks the clinician's fingers back and forth across their visual field), alternate taps on the client's palms, or use of auditory tones are utilized to stimulate information processing according to appropriate protocols.

The fourth phase is *desensitization*. This phase focuses on the client's negative affect with clinical effects measured by the SUD Scale. While the client holds in mind the visual image, the negative cognition, and the sensations associated with the image, processing is activated during a focused clinician–client interaction involving sets of eye movement (or other stimulation) until the SUD level is reduced to 0, or higher if that is appropriate to client circumstances. The clinician is trained to stay out of the way of the process as much as possible to allow an unimpeded and undistorted associative process to occur. This allows relevant connections to be made. The manifestations of the dysfunctionally stored memory (e.g., image, thoughts and sound, emotions, sensations, and beliefs) change and transmute during the processing to an adaptive resolution.

Case Example: Police Officer – Continued

During the desensitization phase the officer's associations included describing the boy, other aspects of the crime scene, feelings of guilt for not getting there sooner, and how he was reminded of his own children – and then the realization that it was not his children who were killed. It is not uncommon for first responders to identify the victims with their own family members. He also had associations pertaining to the team's tactics, which were appropriate and competent. Finally he went back to the scene with the boy, with the thought "He waited for us to come before he died . . . he did not die alone." The officer felt sad about the boy who died in his presence, but felt it was a meaningful moment where he could perhaps provide some comfort to the boy before he died. After that he had a sense of peace, with a SUD of 0. He still had feelings of sadness about the incident and the boy, but he was calm when thinking about the situation and his feelings. A SUD of 0 does not mean that a person feels good about the incident or what happened – but that there is no undue disturbance when thinking about the incident.

Phase 5 is *installation*, where the positive cognition is paired with the memory. According to the AIP model, the transmutation of the dysfunctionally stored information involves not only a lack of distress (e.g., desensitization) but also the memory assimilating into an adaptive framework so that, when the memory is accessed, the client has an adaptive perspective with an enhanced sense of self. This phase is a continuation of the processing that took place during the desensitization phase where the client achieved a more adaptive perspective. In this phase, transmutation of the memory continues as the most appropriate positive cognition is identified and enhanced to increase connections to positive cognitive networks and increase generalization effects within associated memories. The phase is completed when the client can think of the memory and the positive cognition with a VoC of 7 (perhaps lower if it is ecological).

Case Example: Police Officer – Continued

For the police officer, the positive cognition that was installed was "I did the best I could, and we performed well," with a VOC of 7.

Phase 6 is the *body scan*. In this phase the client identifies and processes residual sensations to complete the metabolization of the memory. Processing may involve the memory moving from implicit to episodic and then semantic memory (Siegel, 2002; Stickgold, 2002). Sensations that are identified during this phase may be linked to further dysfunctionally stored information (e.g., implicit or episodic memory) that needs to be processed. Hence, the body scan is a final check to evaluate if processing is complete. Processing is considered complete when the client is able to bring up the memory and positive cognition without bodily tension, or is able to have ecological emotions (e.g., appropriate to the situation and circumstance).

Case Example: Police Officer – Continued

The police officer had some slight feelings of discomfort during the Body Scan, which dissipated with several sets of bilateral stimulation.

Phase 7 is *closure*. The client must be returned to a state of emotional equilibrium at the end of the session, whether or not the reprocessing is complete. Relaxation and other coping skills learned during the Preparation phase can be utilized when the client is experiencing discomfort. The client is briefed as to the possibility of other memories, feelings, or images emerging as the material continues to process between sessions. The client is asked to keep a journal or a log so that what comes up may be discussed in the next session.

Case Example: Police Officer – Continued

The officer was told that other emotions and thoughts regarding this situation and others may emerge, that this is normal, and that he could utilize the stress reduction and coping strategies taught during the workshop. Further, he was advised he could call the program coordinators and/or mental health professionals at any time.

Phase 8 is *re-evaluation*. In the next session, treatment results are reviewed to ensure complete treatment effects. The log is examined and the client is asked to re-access the material previously worked on to see if there are any reverberations of the already reprocessed experience that need to be addressed.

Case Example: Police Officer – Continued

A follow-up took place a week after the seminar and then a month later. The officer reported that when he arrived home after the seminar, he told his children, "Daddy is healed." Furthermore, he said the pain he often experienced in his neck and shoulders was gone, and had not returned. He no longer experienced trauma symptoms. His IES-R scores were subclinical at one month (score of 12).

Conclusion

The adaptive information processing (AIP) model that guides EMDR practice conceptualizes present problems as the result of dysfunctionally stored memories that are unable to integrate within the individual's comprehensive memory networks. When these experiences have been fully processed, what is useful is learned, stored with appropriate affect, and available to guide the person adaptively in the future. Hence, EMDR can be very useful in treating trauma in the workplace. However, EMDR extends beyond traumatic events to any distressing event that continues to haunt an individual. When memories of perceived failures and flaws have not been fully processed, they incorporate the negative perspectives, affects, and physical sensations that were part of the event, which in turn hamper current functioning and result in feelings of lack of worth and resourcefulness. EMDR processing allows individuals to be liberated from the dysfunctional ties of the past, and function instead in an adaptive, positive, and coherent manner.

The goal of psychological treatment for work-related trauma is not only to resolve current problems emanating from the trauma, but also to enhance functioning and the ability to deal appropriately with future adversity. EMDR is a therapeutic framework that can increase a person's capacity to deal with current life problems and future challenges by integrating distressing events, with an emphasis on the installation of positive cognitions, and adaptive future templates. For populations who will continue to be exposed to trauma on a frequent basis, such as emergency service personnel, trauma not only is personally devastating but also has implications for the people with whom it comes into contact. The inability to deal adaptively with stressful situations can often result in misguided actions that can harm others. When their traumas are processed and integrated, resilience can be restored, and an ability to deal with present circumstances in an adaptive way can be enhanced, rather than one's response resulting from past traumas that have been triggered.

References

American Psychiatric, Association. (2004). *Practice guideline for the treatment of patients with acute stress disorder and posttraumatic stress disorder*. Arlington, VA: Author.

Andrade, J., Kavanagh, D., & Baddeley, A. (1997). Eye-movements and visual imagery: A working memory approach to the treatment of post-traumatic stress disorder. *British Journal of Clinical Psychology, 36*, 209–223.

Antonovsky, A. (1987). *Unraveling the mystery of health: How people manage stress and stay well*. San Francisco: Jossey-Bass.

Barrowcliff, A. L., Gray, N. S., Freeman , T. C. A., & MacCulloch, M. J. (2004). Eye-movements reduce the vividness, emotional valence and electrodermal arousal associated with negative autobiographical memories. *Journal of Forensic Psychiatry and Psychology, 15*, 325–345.

Bisson, J., & Andrew, M. (2007). Psychological treatment of post-traumatic stress disorder (PTSD). *Cochrane Database of Systematic Reviews*, Issue 3, Art. No. CD003388. doi: 10.1002/14651858. CD003388. pub3.

Blore, D. C. (1997). Reflections on "a day when the whole world seemed to be darkened." *Changes: An International Journal of Psychology and Psychiatry, 15*, 89–95.

Bossini, L., Fagiolini, A., & Castrogiovanni, P. (2007). Neuroanatomical changes after EMDR in posttraumatic stress disorder. *Journal of Neuropsychiatry and Clinical Neuroscience, 19*, 457–458.

Christman, S. D., Garvey, K. J., Propper, R. E., & Phaneuf, K. A. (2003). Bilateral eye movements enhance the retrieval of episodic memories. *Neuropsychology, 17*, 221–229.

Craske, M. G., Hermans, D., & Vansteenwegen, D. (Eds.). (2006). *Fear and learning: From basic processes to clinical implications*. Washington, DC: APA Press.

Department of Veterans Affairs & Department of Defense. (2004). *VA/DoD clinical practice guideline for the management of post-traumatic stress*. Office of Quality and Performance publication 10Q-CPG/PTSD-04. Washington, DC: Veterans Health Administration, Department of Veterans Affairs and Health Affairs, Department of Defense.

Elofsson, U. O. E., von Scheele, B., Theorell, T., & Sondergaard, H. P. (2008). Physiological correlates of eye movement desensitization and reprocessing. *Journal of Anxiety Disorders, 22*, 622–634.

Hogberg, G., Pagani, M., Sundin, O., Soares, J., Aberg-Wistedt, A., Tarnell, B., *et al.* (2007a). On treatment with eye movement desensitization and reprocessing of chronic post-traumatic stress disorder in public transportation workers: A randomized controlled study. *Nordic Journal of Psychiatry, 61*, 54–61.

Hogberg, G., Pagani, M., Sundin, O., Soares, J., Aberg-Wistedt, A., Tarnell, B., *et al.* (2008b). Treatment of post-traumatic stress disorder with eye movement desensitization and reprocessing: Outcome is stable in 35-month follow-up. *Journal of Psychiatry Research, 159*, 101–108.

Kuiken, D., Bears, M., Miall, D., & Smith, L. (2001–2002). Eye movement desensitization reprocessing facilitates attentional orienting. *Imagination, Cognition and Personality, 21*, (1), 3–20.

Lansing, K., Amen, D. G., Hanks, C., & Rudy, L. (2005). High-resolution brain SPECT imaging and eye movement desensitization and reprocessing in police officers with PTSD. *The Journal of Neuropsychiatry and Clinical Neurosciences, 17*, 526–532.

Lee, C., Taylor, G., & Drummond, P. D. (2006). The active ingredient in EMDR: Is it traditional exposure or dual focus of attention? *Clinical Psychology and Psychotherapy 13*, 97–107.

Levin, P., Lazrove, S., & van der Kolk, B. A. (1999). What psychological testing and neuro-imaging tell us about the treatment of posttraumatic stress disorder (PTSD) by eye movement desensitization and reprocessing (EMDR). *Journal of Anxiety Disorders, 13*, 159–172.

Luthar, S. S., Cicchetti, D., & Becker, B. (2000). The construct of resilience: A critical evaluation and guidelines for future work. *Child Development, 71*, 543–562.

MacCulloch, M. J., & Feldman, P. (1996). Eye movement desensitization treatment utilizes the positive visceral element of the investigatory reflex to inhibit the memories of post-traumatic stress disorder: A theoretical analysis. *British Journal of Psychiatry, 169*, 571–579.

Maxfield, L. (2008). Considering mechanisms of action in EMDR. *Journal of EMDR Practice and Research, 2*, 234–238.

National Institute for Clinical Excellence (NICE)., (2005). *Post traumatic stress disorder (PTSD): The management of adults and children in primary and secondary care.* London: Author.

Oh, D-H., & Choi, J. (2007). Changes in the regional cerebral perfusion after eye movement desensitization and reprocessing: A SPECT study of two cases. *Journal of EMDR Practice and research, 1*(1), 24–30.

Ohta ni, T., Matsuo, K., Kasai, K., Kato, T., & Kato, N. (2009). Hemodynamic responses of eye movement desensitization and reprocessing in posttraumatic stress disorder. *Neuroscience Research, 65*, 375–383.

Pagani, M., Hogberg, G., Salmaso, D., Nardo, D., Sundin, O., Jonsson, C., *et al.* (2007). Effects of EMDR psychotherapy on 99mTc-HMPAO distribution in occupation-related post-traumatic stress disorder. *Nuclear Medicine Communications, 28*, 757–765.

Ray, A. L., & Zbik, A. (2001). Cognitive behavioral therapies and beyond. In C. D. Tollison, J. R. Satterhwaite, & J. W. Tollison (Eds.), *Practical pain management* (3rd ed., pp. 189–208). Philadelphia: Lippincott.

Rogers, S., & Silver, S. M. (2002). Is EMDR an exposure therapy? A review of trauma protocols. Journal of Clinical Psychology, *58*, 43–59.

Rosenbaum, M. (1983). Learned resourcefulness as a behavioral repertoire for the self-regulation of internal events: Issues and speculations. *Perspectives on Behavior Therapy in the Eighties, 9*, 54–73.

Rost, C., Hofmann, A., & Wheeler, K. (2009). EMDR treatment of workplace trauma. *Journal of EMDR Practice and Research, 3*, 80–90.

Russell, M. C. (2008). Treating traumatic amputation-related phantom limb pain. *Clinical Case Studies, 7*, 136–153.

Sack, M., Lempa, W., Steinmetz, A., Lamprecht, F., & Hofmann, A. (2008). Alterations in autonomic tone during trauma exposure using eye movement desensitization and reprocessing (EMDR) – results of a preliminary investigation. *Journal of Anxiety Disorders, 22*, 1264–1271.

Shapiro, F. (1989a). Efficacy of the eye movement desensitization procedure in the treatment of traumatic memories. *Journal of Traumatic Stress, 2*(2), 199–223.

Shapiro, F. (1989b). Eye movement desensitization: A new treatment for post-traumatic stress disorder. *Journal of Behavior Therapy and Experimental Psychiatry, 20*, 211–217.

Shapiro, F. (1995). *Eye movement desensitization and reprocessing: Basic principles, protocols, and procedures.* New York: Guilford Press.

Shapiro, F. (2001). *Eye movement desensitization and reprocessing: Basic principles, protocols and procedures* (2nd ed.). New York: Guilford Press.

Shapiro, F. (2007). EMDR, adaptive information processing, and case conceptualization. *Journal of EMDR Practice and Research, 1*, 68–87.

Solomon, R. M. (1995). Critical incident stress management in law enforcement. In G. S. Everly (Ed.), *Innovations in disaster and trauma psychology* (pp. 123–157). Ellicott City, MD: Chevron.

Solomon, R. (2002). Treatment of violated assumptive worlds with EMDR. In J. Kaufman (Ed.), *Loss of the assumptive world* (pp. 117–126). New York: Brunner-Routledge.

Solomon, R., Gruler, A., Shurgart, S., & Skidmore, E. (2008 , Spring/Summer). Post deployment seminar. *Life Net, a Publication of the International Critical Incident Stress Foundation.*

Solomon, R., & Kaufman, T. (2002). A peer support workshop for the treatment of traumatic stress of railroad personnel: Contributions of eye movement desensitization and reprocessing (EMDR). *Journal of Brief Therapy, 2*(1), 27–34.

Solomon, R. W., & Shapiro, F. (2008). EMDR and the adaptive information processing model: Potential mechanisms of change. *Journal of EMDR Practice and Research, 2*, 315–325.

Stickgold, R. (2002). EMDR: A putative neurobiological mechanism of action. *Journal of Clinical Psychology, 58*, 61–75.

Stickgold, R. (2008). Sleep-dependent memory processing and EMDR action. *Journal of EMDR Practice and Research, 2*, 289–299.

Suzuki, A., Josselyn, S. A., Frankland, P. W., Masushige, S., Silva, A. J., & Kida, S. (2004). Memory reconsolidation and extinction have distinct temporal and biochemical signatures. *Journal of Neuroscience, 24*, 4787–4795.

Van den Hout, M., Muris, P., Salemink, E., & Kindt, M. (2001). Autobiographical memories become less vivid and emotional after eye movements. *British Journal of Clinical Psychology, 40*, 121–130.

Weiss, D., & Marmar, C. (1997). The Impact of Event Scale – Revised. In J. Wilson & T. Keane (Eds.), *Assessing psychological trauma and PTSD.* New York: Guilford Press.

Wilson, S. A., Becker, L. A., Tinker, R. H., & Logan, C. R. (2001). Stress management with law enforcement personnel: A controlled outcome study of EMDR versus a traditional stress management program. *International Journal of Stress Management, 8*, 179–200.

Zaghrout-Hodali, M., Alissa, F., & Dodgson, P. W. (2008). Building resilience and dismantling fear: EMDR group protocol with children in an area of ongoing trauma. *Journal of EMDR Practice and Research, 2*, 106–113.

18

Trauma Inoculation: Mindful Preparation for the Unexpected

Gordon Turnbull, Rebekah Lwin, and Stuart McNab

Like beauty, stress is, in large part, in the eye of the beholder.

– Meichelbaum (2007)

Inoculate. *Impregnate (person, animal, with virus or germs of disease) to induce milder form of it & so safeguard person against its attacks; implant (disease etc.) thus (on, into, person etc); insert (bud, scion) in plant, treat (plant). Hence or cogn. (L. ўcular f. oculus eye, bud) engraft.*

– The Concise Oxford Dictionary (4th ed., 1951)

Introduction

This chapter comprises three sections, the first a discussion of the concept of stress inoculation, Stress Inoculation Training (SIT) as originally developed by Meichenbaum, and emerging neurobiological evidence supporting the notion of "inoculation" and the protective effects of prior "safe" experience. The latter two sections briefly present ideas exploring the potential for compassionate mind training and mindfulness meditation, respectively, as alternative or adjunctive approaches to preparing for potential traumatic experience.

Stress and Stress Inoculation Training

The concept of stress described by Lazarus and Folkman (1984) is a model that proposes that stress will occur when the perceived challenges of a situation exceed the perceived resources of the system (individual, family, group, or community) to meet those demands. According to this perspective, the stressed state represents a state of imbalance, and that stress does not belong to the environment alone or to the individual.

The concept of inoculation derives from the use of vaccines in medicine whereby exposure to a small or weak amount of "disease" can protect against the effects of later strong exposure through the body's capacity to produce antibodies in readiness and thus protect against the potentially overwhelming effects of later attack. This medical

International Handbook of Workplace Trauma Support, First Edition. Edited by Rick Hughes, Andrew Kinder, and Cary L. Cooper.

application has been extended in more recent years to psychosocial applications such as the treatment of stress (Meichenbaum, 1985, 1997; Meichenbaum and Jaremko, 1983), anxiety (see Saunders *et al.*, 1996, for a review), fears and phobias (Marks, 1987), and psychological trauma (Basoglu *et al.*, 1997; Meichenbaum, 1996a, 1996b). Descriptions vary and make reference to "inoculating" (Meichenbaum and Novaco, 1977), "steeling" (Rutter, 2006), "immunizing" (Hannum, Rossellini, and Seligman, 1976), "thriving" (O'Leary and Ickovics, 1995) "toughening" (Dienstbier, 1989; Smith, 1980), and "preparedness" (Basoglu *et al.*, 1997).

The principal of "inoculation" in clinical application draws on evidence, from animal and human studies, of the reported beneficial effect of repeated brief exposure to minor stressors for resilience building (see Gunnar *et al.*, 2009; Lyons *et al.*, 2009; Parker *et al.*, 2004; Rutter, 1981, 2006).

Such studies indicate that experiencing and overcoming minor challenges can enhance coping and increase resistance to the negative effects of later, more significant challenges. It is important to emphasize, however, that the level of intensity of the stressor must not overwhelm resources if it is to have an inoculating effect.

Stress inoculation training

Stress inoculation training (SIT) developed during an attempt to integrate the results of research into the comparative influences on the role of cognitive and emotional factors in psychological coping processes as cognitive-behavioral strategies evolved (Meichenbaum, 1977). Meichenbaum has been involved in its development as a mainstream way of preventing stress reactions and reducing their intensity for over 30 years and has become its most powerful advocate.

SIT is a psychotherapy method designed to help individuals to prepare themselves in advance to handle stressful events successfully and with a minimum of upset. It is an alternative to exposure-based therapies which some patients find too overwhelming. It is a cognitive-behavioral approach representing a more graduated stress inoculation training. The use of the term *inoculation* in SIT is based on the idea that the therapist or trainer is preparing individuals to become resistant to the effects of stressors in much the same way that a vaccination works by making individuals resistant to the effects of particular infectious diseases. Stress inoculation training has been used as a therapy to help individuals cope after exposure to stressful events and trauma and also as a preventative strategy against future and ongoing stressors, to enhance and improve already established coping skills (*grafting*), and also to develop new ones (*implanting*).

In SIT, individuals are educated about stressful situations and the general nature of stress, the negative outcomes they may be vulnerable to experiencing when confronted with stress, and steps they can take to avoid those negative outcomes. At the conclusion of SIT, individuals should feel that they are able to anticipate pitfalls that may occur during an event, and have a workable and practical plan in place to help themselves avoid those pitfalls. Stress inoculation training is therefore designed to provide a sense of mastery over stress by teaching a range of coping skills and then providing an opportunity to practice those skills in a graduated way, a process of *inoculation,* in both clinical settings and nonclinical settings.

SIT consists of three interlocking and overlapping phases:

1. An initial conceptual educational phase.
2. A skills acquisition and skills consolidation phase.
3. An application and follow-through (simulation) phase.

Implementation of these three phases varies depending on the nature of the stressors and the existing resources and coping capacity of the individual. The nature of the stressors likely to be faced needs to be taken into account. Are the stressors to be faced acute and time-limited such as a medical procedure, or are they prolonged ongoing repetitive stressors such as working in a highly stressful occupation or living in a high-risk violent environment?

Initial conceptual education phase. In the initial conceptual education phase, the therapist educates the individual about the general nature of stress and explains important concepts such as appraisal and cognitive distortion that play a key role in shaping stress reactions. The concept that individuals can inadvertently perceive their stressor(s) to be more threatening than they objectively are through the unconscious deployment of unhelpful coping habits is made clear. Finally, the therapist or trainer works to develop a clear understanding of the nature of stressors that are facing the individual.

A key component of psycho-education in the SIT conceptualization phase is the concept that stressors actually represent creative opportunities and puzzles to be solved rather than irritating or threatening obstacles. Individuals should be helped to differentiate between aspects of their stressors which are changeable and aspects that cannot change, so that coping efforts can be adjusted accordingly and energy is judiciously applied. If chronic stressors are to be faced, such as in soldiers facing combat or peacekeeping duties or individuals challenged by the real possibility of being held hostage, it might be useful to inoculate strategies to deal with short-term, intermediate, and long-term coping.

Individuals in high-risk roles going into high-risk environments who face the real possibility of psychological traumatization need the special inoculation of understanding that intrusive recollections of their traumatic experience(s) represent the mind's struggle to process new, unfamiliar, and extremely challenging information about survival and death and that the flashbacks and nightmares are the results of that struggle. Such individuals need to be able to *normalize* the recollections and see them as memories of what actually happened rather than evidence that they are *going mad.* They will be less likely to try to conceal the flashbacks from their comrades if they know that they are natural phenomena. Such individuals also need to know about *dissociation* in all its forms and to see it as the mind's way to "switch off" from reality for a while and not to perceive the disconnection as permanent and a loss of oneself. *Freezing* and *tonic immobilization* is not something that is familiar to action-oriented individuals but they need to be inoculated with the new skill of recognizing that this frighteningly helpless development can happen in certain circumstances and is also part of our range of natural psychological defense mechanisms, although at the extreme limits (van der Kolk *et al.*, 1996).

As neurobiology becomes increasingly sophisticated and reveals the connections between behavioral and emotional reactions and underlying neurobiological function, the sense of predictability of reactions when challenged by stressors grows and the older, more metaphysical notions of *neurasthenia* and *lack of moral fiber* (LMF) weaken (Shephard, 2000).

Skills acquisition and skills consolidation phase. The skills acquisition and skills consolidation phase is the second phase of SIT. The particular choice of skills taught is very important and must be individually tailored to the needs of individuals or groups of individuals. It has to take into account existing strengths and vulnerabilities. The range includes emotional regulatory skills, autonomic regulation, vagosympathetic regulation, relaxation training, cognitive appraisal and restructuring, problem-solving skills, communication, and socialization skills. This is not intended to be a complete list of the skills that may be taught, and this emphasizes the principle of uniqueness and meeting individual needs.

This chapter provides an opportunity to introduce some new and exciting ideas for treatment of emotional dysregulation by using body-focused methods such as biofeedback. There are a number of compelling strands of academic research to indicate that information generated by the body, and the heart in particular, is important in determining what happens in the brain in terms of its:

1. Cognitive function – including perception, clarity of thought, memory, and creativity.
2. Regulation of hormonal function.
3. Regulation of autonomic function (including blood pressure).
4. Involvement in mood, emotion, and general quality of life experience.

Much of the information generated by the body, and the heart in particular, is sent to the brain via the nerves of the autonomic nervous system. One of the most rapidly developing fields of psychological research is the treatment of psychological trauma. This has been helped by the acceptance of the diagnosis of post-traumatic stress disorder (PTSD). Treatment of PTSD has started to recognize that the traditional "talking therapies" may have limited effectiveness and this has resulted in approaches that treat trauma by focusing on the body in approaches such as eye movement desensitization and reprocessing (EMDR) and somatic experiencing.

For more than a decade there has been a growing realization that psychology has been focused almost exclusively on solving problems and negative emotional states. This has encouraged the new field of research focused on what can be done to positively enhance an individual's performance through the generation of positive emotional states. This work obviously has significance for the further development of SIT in a world where traumatic stress has become a growing problem.

In the last decade, the Dalai Lama and the Tibetan monk community have engaged significantly with the neuroscience community. This interface between science and spirituality has produced some fascinating insights. This has revealed that the persistent practice of positive emotional states in meditative practice can produce very significant improvements in cognitive function and physiological performance which can ultimately result in an alteration of brain structure (neuroplasticity).

The heart-rate variability (HRV) research has shown that electrical signals generated by the heart predict risk of death and disease in infants and adults, in normal individuals and those with disease. Heart-rate variability has also been shown to affect cognitive performance and mood. In particular, biofeedback has been shown to improve a wide variety of symptoms and has been widely applied to a number of different medical conditions.

Brain function can be improved by improving cardiac coherence (Watkins, 1997, 2011). Brain function is adversely affected by stress reactions, and in the modern world a primary task is to learn how to manage the challenges that we face. This is the same challenge that SIT has been charged with responsibility for, for the past 30 years. We need to learn how to prevent a "frontal lobe shutdown." HRV can be adapted to enhance brain function. Rather than the HRV signal fluctuating widely in a chaotic way, the electrical input to the brain can be adjusted in a much more stable and orderly fashion. This harmonious signal is known as *cardiac coherence*, and individuals can learn to generate this type of signal even when they are under pressure. Cardiac coherence means that the heart rate still varies but in a much more stable and orderly fashion. For example, it may vary between 60 and 80 beats per minute in a consistent repeating pattern with the distance between the heart rate peaks or troughs remaining constant. This stable pattern gives rise to the classic sinusoidal curve. There is variability but it is predictable and rhythmic compared with the unpredictable, erratic variability seen in a chaotic heart rate pattern. This has been shown to coincide with enhanced brain function (Cohen and Benjamin, 2006).

In stress and trauma inoculation training, autonomic nervous system regulation is essential (Rothschild, 1996). The brain responds to extreme stress, trauma, and threat by releasing hormones to prepare for defense, activating the sympathetic branch (SNS) of the autonomic nervous system (ANS), preparing the body and the brain for fight or flight, increasing respiration and heart rate, sending blood away from the skin and into the muscles, and so on. When threat is imminent or prolonged, the brain can also release hormones to activate the parasympathetic branch (PNS) of the ANS, and freezing and tonic immobility can result (Gallup and Maser, 1977; van der Kolk, 1994).

Post-traumatic stress disorder and other chronic stress reactions may involve chronic activation in the SNS and PNS as the brain continues to respond as if under stress, trauma, or threat and continues to prepare the body for fight or flight or to freeze. This chronic ANS activity is probably responsible for many of the symptoms of chronic stress reactions. Breaking this cycle is an important step in the treatment of PTSD but *sensorimotor* training is also a useful way to build up physical resilience through autonomic regulation. The application of an awareness of *braking techniques* or *applying the brakes* is a powerful tool in SIT especially as a preventative measure in individuals who are likely to be exposed to extreme danger in the course of their work.

Application and follow-through phase. The application and follow-through phase is the third and final phase of SIT. In this phase the therapist or trainer provides the individual to practice coping skills including encouragement to use a range of simulation methods to help to increase the authenticity of practicing coping strategies. This will include visualization exercises, rehearsal of coping procedures in the form of imagery rehearsal (coping modeling), behavioral rehearsal (role playing), as well as in vivo graduated exposure and simple repetitious behavioral practice of coping routines until they become over-learned, familiar, comfortable, and automatic.

SIT has been conducted with individuals, couples, and groups (both small and large). The length of intervention can be as short as 20 minutes or as long as 40 one-hour-per-week or twice-weekly training sessions. In most cases, SIT consists of 8–15 sessions, plus booster and follow-up sessions, conducted over a 3–12-month period.

SIT can be delivered by training others to conduct the intervention (e.g., nurses, probation officers, and police officers training other police officers). Stress inoculation training also recognizes that individually experienced stress can be a reflection of endemic or institutional stress and, in such circumstances, can be unavoidable. Stress inoculation training has helped individuals to reset environmental settings by working with others to address environmental stressors. Examples have been:

- Hospital staff for medical patients (Kendall, 1983).
- Coaches for athletes (Smith, 1980).
- Drill instructors for recruits (Novaco *et al.*, 1983).

SIT has been used successfully in the treatment of:

Acute time-limited stressors such as preparation for specific medical procedures (e.g., surgery, dental examinations, biopsies, and cardiac catheterization) and performance evaluation (e.g., academic examinations).

Chronic intermittent stressors such as military combat and episodic physical conditions like recurrent headaches or repetitive evaluations and ongoing competitive performances such as musical or athletic competitions.

Chronic continual stressors such as debilitating medical or psychiatric illnesses, physical disabilities resulting from exposure to traumatic events (e.g., burns, spinal cord injuries, traumatic brain injuries), or exposure to prolonged distress, including marital or family discord, urban violence, poverty, and racism as well as exposure to persistent occupational dangers and stressors in professions such as police work, nursing, and teaching.

Stressor sequence results from exposure to stressful or traumatic events such as terrorist attack, divorce, a natural disaster that results in a major loss of resources, or exposure to stressors that require transitional adjustments (e.g., death of a loved one, or becoming unemployed). (Elliot and Eisdorfer, 1982; Meichenbaum, 1996b, 2007).

An example of SIT being used as a preventative method. Weisaeth (1994) studied the unique demands placed on United Nations (UN) troops and identified the UN soldiers' stress syndrome. This was characterized by:

- Fear of death and injury.
- Exposure to grotesque images and to sights of carnage and deprivation involving innocents.
- Uncertainty about rules of engagement for using deadly force.
- Enforced passivity of peacekeeping troops when facing danger can result in a "helplessness so severe that the UN soldier's self-respect was damaged."

- Persistent fear of losing control of their anger.
- Continuing suppression of anger, exposure to danger and mass deaths, low control and high responsibility, time and group pressures, lack of information, high risk of failure or responsibility stress, continuing taunts, all which cumulatively contributed to UN soldiers' somatic and psychosomatic symptoms and to psychiatric signs of distress.

Research has demonstrated increasing rates of PTSD in peacekeeping troops when they return home (Danieli, 2002). Weisaeth (1994) observed that the enforced passivity which is counter to the way soldiers have been trained to react, continuing uncertainty about what they can and cannot do, and frequently not being able to see the consequences of their efforts can lead to high levels of feelings of helplessness and group demoralization.

Weisaeth (1994) suggested strategies to reduce the likelihood of such debilitating consequences:

- The importance of good leadership.
- Group cohesion.
- Proactive activities such as exercise, debriefings, and SIT aimed at training UN soldiers to address complex, uncertain, provocative situations they may find themselves in and to maintain a sense of being in control of such situations.

Weisaeth (1984) described the way that SIT was used in training UN soldiers for peacekeeping duties: to enhance resilience and prevent stress reactions, including PTSD. This practical example demonstrated the use of the three phases.

Initial education phase

- Encouraging small groups of soldiers to discuss the nature of stress and coping and detection of warning signs in selves and others.
- Identifying potential stressors.
- Recognition of personal and interpersonal coping skills.
- Information about their mission (to reduce uncertainty or speculation).
- Learning the nature of the culture of the mission region.
- Learning the history of the specific conflict.
- Viewing videotape footage before going to the mission site with the focus on coping rather than exposure to grotesque scenes ... positive outcomes such as a food convoy getting through, skirmishes stopped or avoided, and so on.

Skills-training phase

- Focused on behavioral, cognitive, and affective regulation skills needed to enhance individual and group coping skills.
- Physical and mental relaxation skills.
- Problem solving.
- Coping with provocations taught and practiced.

- UN soldiers were taught to engage in self-dialogue in provocation situations of verbal and physical taunts such as "You are a coward," "UN . . . United Nothing," and "You are not a man if you can watch women and children being killed." Under such conditions the UN soldiers were taught to ask themselves the question "What are they trying to make me do?"
- Weisaeth (1994) observed that it was vital that each UN soldier had sufficient self-respect to maintain an acceptable self-image regardless of the situation and that, when facing provocations, the UN soldiers may find it helpful to say to themselves, "This is nothing about me, but perhaps something about them."

Meichenbaum and Novaco (1977) and Novaco *et al.* (1983) described how SIT could be applied on a preventative basis with police officers and military recruits, and in the Meichenbaum and Novaco–directed application phase of SIT with police officers, actors were hired to taunt and tease the officers who had been pre-trained to deal with such events. The officers practiced in groups how to deal with such provocations. Later, Meichenbaum (1994) described how the SIT training of Israeli military recruits included those already trained as trainers.

Stress Inoculation Training has also been used to prepare workers for occupational traumas and rescuers to deal with the demands of their tasks (Ersland, Weisath, and Sund, 1989).

In considering such training programs, it is important to keep in mind that when well-trained rescuers are unable to implement and perform their skills, they are particularly at risk of developing PTSD. For example, Weisaeth (1994) reported that among well-trained sailors who were unable to rescue passengers during a ship's fire that resulted in 159 deaths, 50% of the crew suffered PTSD and other stress-related disorders two years after the fire.

In evaluating the role of SIT training and experience, Weisaeth (1994) concluded that if a person is given adequate education and training he or she is very likely to be reliable and act rationally, even when facing extreme stress. Stress inoculation of employees in high-risk occupations has been shown to increase resilience.

Neurobiology

Since the earlier work of Meichenbaum, Weisaeth, Novaco, and others, much has been learned about the neurobiological dimension of stress reactions and about modern stress and trauma inoculation training needs to take these new discoveries into account.

For example, Rutter (2006) took a penetrating look at the interplay between nature and nurture, setting out to explain how genes might influence behavior and how this might be important in understanding the causal pathways leading to the development of certain behavioral traits and psychiatric disorders. In the field of trauma inoculation, the key to understanding individual differences in susceptibility to stress reactions will come from insights into how the effect of genetic variation interacts with environmental exposure over the life span.

White (2005) has gathered 12 presentations on biopsychosocial medicine delivered at a conference in London in 2002 which are influential in acknowledging the limits of biomedicine.

Goldberg and Goodyer (2005) explored the origins and course of common mental disorders and provided a thorough developmental look at the interplay between biological and psychological factors in determining vulnerability to stressful life events. Goldberg and Goodyer's conclusions need to be taken into account in modern stress inoculation aimed at strengthening resilience. These are:

- Genes are important in determining the initial setting of resilience versus vulnerability; good attachment causes a *decrease* in levels of hypothalamic-pituitary-adrenal axis (HPA) responsiveness. Many other environmental changes cause an *increase.*
- Those with abnormal 5-hydoxytryptamine transporter (5HTT) genes *and* low maternal responsiveness have poor attachment.
- Neglect and abuse in childhood cause long-lasting inflammatory changes and vulnerability to later depression.
- Those with abnormal 5HTT genes are most vulnerable to adult stressors.

New brain chemicals are now being discovered that have the effect of modulating stress reactions. This actually fits in very well with the core biological principle of *homeostatic balance.* When faced with danger the brain initiates a chemical reaction to set up the fight-or-flight reaction. When the hypothalamus releases corticotrophin-releasing hormone (CRH) to cause the pituitary gland to release adrenocorticotrophic hormone (ACTH) to stimulate the release of cortisol from the adrenal glands, the stress reaction is being triggered. It is now known that a series of chemicals in the brain dampen this process, thereby modulating the stress response to perceived danger. These chemicals include neuropeptide Y and galanin which modulate the effects of CRH. Another hormone, dehydroepsiandrosterone (DHEA), is released from the adrenal glands to modulate the effects of the stress hormone cortisol. In the future, drugs or psychotherapy might stimulate production of these stress-modulating hormones to enhance resilience (Wheway *et al.*, 2005).

Acute stress reactions

Learning about acute stress reactions (ASRs) in a theoretical or descriptive way before exposure to a stressor is a useful thing, but it is also sometimes useful for those who are at risk of experiencing ASRs in the field to actually experience it, especially as SIT is designed to boost resilience and the reactions, if they occur at all, will be reduced in intensity. Acute stress reactions (as defined by the World Health Organization, 1992) following exposure to trauma are normal, common, and usually self-limiting and affect more than 50 percent of those exposed to trauma. They represent adaptive psychobiological processes involved in the assimilation of new information with an intense survival emphasis. Acute stress reactions contain elements of post-traumatic stress, depression, grief, anger, high arousal, hypervigilance, and dissociation and follow exposure to threat to life and limb. Acute stress reactions create a state of "high alert" designed to optimize survival and work well because fewer than half the number of ASRs progress to post-traumatic stress disorder (PTSD). Prior personal experience of effectively managing ASRs promotes self-reassurance and understanding of the process in others and can be of advantage to certain groups such as news gatherers, aid workers, soldiers, and rescue workers enhancing their appreciation of the mental conditioning of traumatized individuals whom they will inevitably meet in the course of their work (Mills, Rees, and Turnbull, 2002).

Reflections

The huge stimulus to the development of SIT to boost resilience before exposure to stressful or traumatizing environments is a testimony to the energy and insightfulness of Meichenbaum. As new discoveries have been made in the world of the behavioral sciences and neuroscience, SIT has evolved and is still in the process of evolving.

Slightly different applications have also developed, such as Psychological First Aid (National Center for PTSD, 2005), Critical Incident Stress Management (Mitchell, 1983), Trauma Incident Management (TRiM – UK Royal Marines), and the new military initiatives to deal with stress issues pre-deployment, during deployment, and post deployment (e.g., Battlemind and US Army), perhaps with different populations in mind but they all have, at their core, the principle of ***inoculation*** to enhance the skills for coping with stress by grafting and implanting.

Also encouraging are ideas that are emerging from the contemplative sciences and Eastern philosophies, specifically mindfulness meditation (Follette, Palm, and Pearson, 2006; Kabat-Zinn, 1990) and compassion-focused approaches (Neff, 2003; Neff, Kirkpatrick, & Rude, 2006; Neff & McGhee, 2009; Gilbert, 2009; Gilbert and Procter, 2006t; Lee, 2009). These are briefly discussed in the remainder of this chapter.

Self-compassion and its Potential to Protect

Compassion is defined in various ways, broadly; it is as an orientation of mind that is able to recognize pain, acknowledge its presence, and meet it with kindness, equanimity, nonjudgment, and patience (Feldman and Kuyken, 2011). For a theoretical analysis and review of compassion, see Goetz, Keltner, and Simon-Thomas (2010).

Self-compassion involves directing this kindness and nonjudgment toward oneself. According to Neff (2003a), self-compassion comprises three fundamental elements:

1. Being open and accepting of one's pain, suffering, and inadequacies and extending kindness and understanding to oneself rather than harsh judgment or criticism.
2. Perceiving one's experiences as part of the broader human condition to which we all belong rather than separating and isolating oneself from this common experience.
3. Holding one's painful thoughts and feelings in balanced awareness rather than over-identifying with them.

The cultivation of self-compassion as a buffer for the negative experience of traumatic events is not, as yet, reported as an adopted approach for trauma–stress work preparation. However, it is proposed here that Neff's (2003a) conceptualization of self-compassion could be effectively utilized in this way. This could be both with respect to the direct experience of key emotions associated with traumatic experience such as fear, horror, and helplessness (in Buddhist philosophy, compassion may be regarded as an "antidote" to fear and offers an alternative to aversion and avoidance; Feldman and Kuyken, 2011); and with reported other primary and secondary emotions which play an influential, and sometimes independent, role in the development and maintenance of post-traumatic symptoms following traumatic events (Brewin, Andrews, and Rose, 2000; Grey and Holmes, 2008; Grey, Holmes, and Brewin, 2001; Hathaway, Boals, and Banks, 2010; Ozer *et al.*, 2003). These additional peri-traumatic and secondary emotions can include self-attacking and self-denigrating sentiments such as self-criticism (Harman and Lee, 2010), shame (DePrince, Chu, & Pineda, 2011; Harman and Lee, 2010; Lee, Scragg, & Turner, 2001), and guilt (Lee *et al.*, 2001; Holmes, Grey, & Young, 2005) and can be powerfully present even in the absence of fear and attributable blame, and in the presence of courageous and effective execution of duty and human endeavor.

Holmes *et al.* (2005), for example, in their exploration of cognitive themes emerging through elicitation of hotspots of trauma memories, identified various emotions linked to moments of peak distress. Hotspots are memories of moments of the traumatic event that are associated with the highest levels of emotional distress. In only 50 percent of the cases were the key emotions identified as fear, horror, or helplessness, and the authors suggest that the key reason for the high emotional intensity may be the "negative meaning of these moments relative to the person's view of themselves, as well as any threat to their physical integrity" (2005, p. 15). They infer that their data support the need to consider the variety of peri-traumatically encoded emotions and cognitions when considering treatment for PTSD. This raises implications for models of therapeutic intervention, as these are most commonly fear focused. It also raises implications for inoculation approaches such as SIT, as this too focuses primarily on the stress and fear response.

We propose that effective trauma–stress inoculation might also benefit from pre-trauma training designed to specifically address the potential development of self-critical thinking following trauma. Such adjunctive approaches may both mitigate the severity of symptoms and perhaps even avert its development. Training directed at enabling and enhancing self-compassion may be one approach.

Evidence to support this hypothesis can be found in clinical reports and studies on self-compassion in various of areas of mental health. Neff (2003b), for example, in her development of the Self-compassion Scale, found that higher self-compassion was associated with greater psychological well-being; less depression, anxiety, rumination, and thought suppression; and greater life satisfaction and social relatedness. Elsewhere self-compassion has been associated with softening the impact of negative events (Germer, 2009), less self-evaluative anxiety in relation to perceived personal weakness (Neff, Kirkpatrick, and Rude, 2007), less avoidance behaviors (Thomson & Waltz, 2008), resistance to emotional and cognitive self-attack (Kelly, Zuroff, and Shapira, 2009), ability to foster capacity to engage painful thoughts and emotions (Leary *et al.*, 2007), and physiological benefits on stress-related systems (Uchino, Cacioppo, and Kiecolt-Glaser, 1996). Recent neurobiological evidence also proposes a possible neural mechanism for the effects of self-compassion and self-reassurance (Longe *et al.*, 2010).

Interestingly, with respect to extrapolation of self-compassion research findings to trauma-risk professions, Leary *et al.* (2007) in their series of five studies exploring self-compassion in undergraduate students found greater willingness of those rated as self-compassionate people to accept responsibility for their role in negative events but, also, to not judge themselves too harshly and to retain perspective and equanimity in the aftermath of the worst things that happened. They were also reported as being less likely than those who were low in self-compassion to ruminate about their mistakes or their reactions to unpleasant interpersonal events. Self-compassion was associated with lower negative emotions in the face of real, remembered, and imagined events. The authors conclude that self-compassion is an important construct that can moderate reaction to distressing situations involving failure, rejection, embarrassment, and other negative events.

It is important to emphasize that self-compassion is not equivalent to dispassion. Self-compassion is an active acknowledgment and acceptance of difficulty and distress, not indifference to it. In the face of such difficulty, however, self-compassion directs principles of common humanity to oneself as well as to others. Both self-compassion and mindfulness foster an attitude of nonresistance to "what is," and an acknowledgment of how things are in the present, and promote "being with" and tolerance of the moment-to-moment experience of difficult thoughts, memories, emotions, and related internal physical sensations. Self-compassion may thus enhance one's capacity to help stay the course of difficult emotions and sensations and allow experience of their natural processes through to more positive psychological well-being rather than resorting to avoidance behaviors (Leary *et al.*, 2007; Neff, Kirkpatrick & Rude, 2007; Thomson & Waltz, 2008).

With respect to the notion of inoculation, this exploration of the psychological and neurobiological literature highlights a convincing association between self-compassion and cognitive and emotional experiences commonly encountered following traumatic events. While none of the studies identify a causal link, Lee *et al.* (2001) present a clinical model with a possible mechanism for how self-compassion may help build adaptive schemas in the face of traumatic experience, and there is sufficient research evidence to postulate that self-compassion training may mitigate some of the effects of psychological trauma, particularly those thoughts, feelings, and actions arising from self-criticism and perceived shame and guilt.

A training approach that shows promise in this respect is compassionate mind training (CMT; Gilbert, 2009). CMT was originally conceptualized as a therapeutic approach for individuals caught in destructive cycles of self-criticism and self-attacking (Gilbert, 2009; Gilbert and Irons, 2005; Gilbert and Procter, 2006). The concept has been developed in the field of trauma therapy by Lee (2009), although empirical research on CMT as treatment for PTSD is limited.

The theory of CMT draws on evolutionary biology and the neurobiological systems in humans that are linked to emotions and emotional reactions. Three interdependent affect systems are described, one associated with threat and two associated with positive affect: drive and affiliation/care. Each system is linked to emotions such as fear, anxiety, and anger (threat); achieving, striving, and excitement (drive); and feeling connected, cared for, and compassionate (affiliative). Each system is also linked to different neurobiological pathways. Gilbert proposes that many psychological difficulties arise when these systems get out of balance and, in particular, when the "threat" system becomes too dominant such that fear overrides the ability to feel safe and content. He suggests two ways that individuals respond to a sense of threat and not feeling safe, either through the drive system and actions that counter the threat or via self-compassion which serves to activate and enhance the "affiliative" system thereby promoting self-soothing and emotional regulation and diminishing the "threat" system.

The objectives of CMT are primarily threefold. The first is psycho-education, knowledge of the psycho-physiology of emotion and the powerful influence of the threat system and its consequent effects. The second is raising individuals' awareness of their internal emotional landscape, the influence of the different emotional systems, and which state of mind is being experienced. The third is the development of skills and personal resources that foster self-compassion and an ability to self-reassure and self-regulate emotion. The training places great emphasis on the use of compassionate imagery and role-play to develop and practice these skills. Such practice also enables rescripting of shame-based or self-critical narratives and ruminations, and places the individual as the powerful and responsible person for changing the course of experience.

Applying Gilbert's ideas to PTSD, specifically shame and self-critical thinking in PTSD, Lee (2009) and Harman and Lee (2010) have suggested that individuals who develop PTSD may have associated high levels of shame or self-critical thoughts that are self-reinforcing and contribute to a sense of powerlessness and defenselessness thus fueling the ongoing sense of threat which is central in PTSD. As well as a positive association between shame and self-criticism in individuals with PTSD, Harman and Lee (2010) also found a negative association between shame and self-reassurance, thus those individuals who rated high in self-criticism were also unable to reassure themselves of their worth or safety.

Although CMT has not been researched specifically as a preparation for traumatic experiences, explorations of such associations may inform trauma inoculation training, perhaps as an adjunct to SIT, by way of directing individuals to self-adaptive ways to address their fear and other reactions to the external threats posed by future traumatic events and to the internal threats posed by the negative reaction of the self toward the self through self-criticism, self-blame, and shame. Such training might also serve to protect against not only acute reactions to trauma but also chronic and delayed-onset presentations which may be associated with allostatic overload (McEwen and Stellar, 1993) and may involve different vulnerability and adaptive mechanisms (Andrews *et al.*, 2009).

In summary, the complexity of emotional and physiological reaction following a traumatic event incorporates more than fear and perceived risk to personal integrity. Self-critical responses and perceived shame and guilt may also predominate and influence both the onset and maintenance of post-trauma symptoms. It is proposed that trauma inoculation approaches that consider the development of skills in self-reassurance and self-compassion may be an important adjunct to the preparatory training of those who work in trauma-risk environments. Such training may actively promote greater resilience. An adaptation of CMT (Gilbert, 2009) may provide the basis for such inoculation training. Another related approach is mindfulness; this too incorporates compassion as a fundamental principle but additionally attends to cultivating skills of moment-by-moment attention and awareness of emotions, thoughts, and physical sensations, enabling their recognition early in their emergence and developing skills, through practice, to respond to these effectively and appropriately without harm to self or others.

Cultivating Mindfulness

This section of the chapter is concerned with the role of mindfulness as an alternative to traditional stress inoculation and will explore the nature and history of mindfulness, its crossing from an eastern faith-based context to a western scientific context, its current application as a form of stress inoculation, and the findings of the research undertaken.

The word mindfulness spatters the academic and popular press with many interpretations of its meaning, its process, and its context. The family tree of mindfulness is, however, clearly delineated within Buddhism (although many faiths utilize meditation). The term is a translation of a Pali word "sati," which is itself difficult to define but involves both attention and awareness.

Mindfulness is seen within Buddhism as experiential and hence any term used to refer to it is only a metaphor or concept which tries to capture the process. Buddhism identifies in the Four Noble Truths that life involves suffering (dukkha) or disappointment as humans' expectations of life are rarely fully met. The reason for this suffering is desire or grasping, wanting what you do not or cannot have. Buddha identified a way of ending this suffering which involves living an ethical life and undertaking meditation.

Thus meditation (a process of mindfulness) was originally seen by the Buddha as a way of living one's life within the context of ethical and skillful behavior to aid in the reduction of suffering. Differing traditions of Buddhism further contextualized this process, thus within Tibetan Buddhism the concept of reincarnation is particularly significant and living this life well may enable progressing from sentient existence (everyday life as we know it) to nirvana.

In 1979 mindfulness began to be used in a Western, medical context to help particular groups of people who were suffering, but suffering chronic pain. Jon Kabat-Zinn and his colleagues (Kabat-Zinn, Lipworth, & Burney, 1985) created what came to be called mindfulness-based stress reduction (MBSR) at the University of Massachusetts Medical Center. MBSR is an eight-week program which takes place within a group and involves sitting meditation, mindfulness of the breath (focusing on the breathing and as one's mind wanders repeatedly, bringing the attention back to the breath), the use of a body scan (gradual sensitization to differing parts of the body), hatha yoga exercises, group discussion, and presentations about the nature of stress.

Underpinning the in-group activities are seven attitudinal positions that participants are encouraged to develop: *nonjudging*, or not evaluating their experience but simply witnessing it; *patience*, or accepting that things will take the time they take; adopting a beginner's *mind*, a position of approaching all experience as if for the first time; *trusting in oneself* and in your intuition; adopting a position of *nonstriving*, focusing on process not outcome (not trying to get somewhere but, rather, to know the "where" you are in); *acceptance*, or recognizing things as they are rather than denying them; and *letting go*, not fighting against your experience or striving to change it, but simply seeing it for what it is (Kabat-Zinn, 1990). Kabat-Zinn saw participants making transformations in how they saw their experiences and achieving a greater balance in their lives. Research by Kabat-Zinn *et al.* (1985) which involved 90 people who attended MBSR programs with chronic pain showed that 72% reported at least a 33% reduction in pain and 61% reported a 50% reduction. MBSR has subsequently been used with a wide variety of areas of illnesses including cancer (Speca *et al.*, 2000), fibromyalgia (Kaplan,

Goldenberg, and Galvin-Nadeau, 1993), binge eating (Kristeller, Baer, and Quillian-Wolever, 2006), and hypertension (Roth, 1997).

In 1995 Teasdale, Segal, and Williams developed mindfulness-based cognitive therapy (MBCT), a modification of MBSR which incorporated aspects of cognitive therapy and was specifically designed for those with recurrent bouts of depression. It had been identified that although cognitive therapy had proved successful in working with those who had experienced their first depressive episode, it was unsuccessful where the episodes had been repeated on three or more occasions.

> The focus of MBCT is to teach individuals to become aware of thoughts and feelings and to relate to them in a wider, decentred perspective as "mental events" rather than as aspects of the self or as necessarily accurate reflections of reality. (Teasdale *et al.*, 2000, p. 616)

Mindfulness-based cognitive therapy, like MBSR, is an eight-week program based around training participants in the use of the body scan and sitting meditation, which incorporates presentations on areas such as automatic negative thoughts. A randomized controlled trial in 2000 with 145 patients who had three or more recurrent depressive episodes and undertook an MBCT program found a 50% reduction in the relapse rate (Teasdale *et al.*, 2000) These impressive results led to MBCT becoming the UK National Institute for Clinical Excellence (NICE)–recommended treatment for recurrent depression.

Mindfulness has also been incorporated into acceptance and commitment therapy (ACT) and dialectical behavior therapy (DBT). Within ACT, clients learn to observe their experiences rather than live them and the attitudinal positions mentioned in this section, of nonjudging, acceptance, and letting go, are emphasized. In DBT mindfulness skills are taught in a year-long group with similar attitudinal positions being highlighted. Clients are encouraged to accept who they are, accept their past and present, whilst at the same time moving toward change.

Some therapists are now incorporating mindfulness into individual therapy (Bien, 2006), and mindfulness is being used in therapist training particularly to aid attention, congruence, empathy, and self-compassion (Shapiro, Brown, and Biegel, 2007). Work has also been undertaken in schools training young children in mindfulness practice (Richhart and Perkins, 2000).

In order to consider the mechanism of change that may be at work when mindfulness skills are learned, it is useful to consider a functional definition. Bishop *et al.* (2004) have defined mindfulness, and it is seen as having two components:

> The first component involves the self-regulation of attention so that it is maintained on immediate experience, thereby allowing for increased recognition of mental events in the present moment. The second component involves adopting a particular orientation toward one's experience in the present moment, an orientation that is characterized by curiosity, openness and acceptance. (p. 232)

Baer (2003) reviews the literature and identifies five mechanisms of change in mindfulness practice: exposure, cognitive change, self-management, relaxation, and acceptance. Exposure is concerned with mindfulness training enhancing an individual's ability to stay with an experience (attention maintained on immediate experience) which may be painful and not to try to avoid it. "Staying with" can facilitate a new experience, one in which an individual can learn that they will not be overwhelmed by pain and they can tolerate it. Over a period of time this exposure can enable new cognitive processes and new responses to painful or unpleasant stimuli as well as to pleasant experiences (a new orientation toward one's experience), thus cognitive change can take place and thoughts can be seen as not necessarily the truth but rather a representation of a truth. This change may bring about adaptive responses to previously emotionally disabling stimuli. Self-management is seen as the process by which mindfulness practice can enable new forms of coping strategies which, because of the increased awareness of bodily and cognitive responses, can be used earlier (i.e., the early warning mechanism is heightened). Relaxation is often seen by those less familiar with mindfulness practice and meditation as the primary objective. However, although focusing on the present and reducing rumination about the past or future can be relaxing and thus beneficial, mindfulness training is much more than relaxation; it is about achieving a fundamental shift in awareness and in responding to the present moment. Finally, the attitudinal position of acceptance is seen as a, and maybe *the*, crucial mechanism of change. The word "acceptance" may, however, be problematic as its familiar interpretation may suggest changing one's view of the stimuli, that is, making something painful feel neutral or comfortable rather than fully experiencing the feeling of pain with no attempt to change that experience. The response involves recognizing what is rather than trying to make it what would be preferred.

A further key component of mindfulness training is its potential impact on working memory capacity. Working memory is concerned with the day-to-day management of cognitive tasks and with the regulation of emotions. It has been shown that experiencing high levels of stress in a military context in those in the pre-deployment phase can reduce the capacity of the working memory and hence reduce the ability to undertake cognitive tasks and to successfully manage emotions (Bolton *et al.*, 2001). Jha *et al.* (2010) investigated whether mindfulness training could provide some protection against this process. This research thus focuses not on assisting those who have developed symptoms related to their traumatic experience but rather on protecting those about to experience or currently experiencing high levels of stress. The outcome of the research was promising, showing that mindfulness training may protect against these functional impairments. The mindfulness program utilized was an adaptation of MBSR and was modified to included military-related aspects of mindfulness-based mind fitness training (MMFT). The researchers also conclude that:

> Although this study was concerned with exploring whether MT [mindfulness training] might provide prophylaxis from functional impairments experienced in specific military contexts, the protective effects of MT observed herein suggest that MT might be effective as a resilience training protocol more generally. (p. 62)

Further research by Stanley *et al.* (2011) highlighted that mindfulness training in the military context may enable participants to modulate the effects of stress and to enhance their cognitive functioning whilst experiencing high levels of stress.

Mindfulness training may thus be seen as not only a treatment modality across an increasing range of conditions but also an intervention strategy to build resilience in those who know they are about to experience high levels of stress. As such, mindfulness training may be viewed as having the potential to inoculate those whose employment will inevitably have periods of high stress and possibly sudden traumatic incidents which they need to managed. The military experience suggests an accompanying educational process (also present in MBSR and MBCT) which further enhances the mindfulness training. Further research is needed to explore a range of occupational settings where adaptations of mindfulness training can be tested with specific focus on the impact on perceived stress and working memory.

Conclusion

This chapter has aimed to provide a description of trauma inoculation as originally conceived in the development of stress inoculation training (SIT) and further to contextualize this within a broader discussion of the neurobiology of the stress response and trauma reactions.

The SIT approach prepares, through education, skills acquisition, and graded or simulated practice, for the experience of expected anxiety and fear arising from a traumatic event. It is further proposed, however, that there may be benefit in preparing for the less expected reactions, among these the peri-traumatic and secondary reactions of self-criticism, shame, and guilt. These have been reported to be significant predictors of PTSD risk and may carry an independent and allostatic influence that maintains post-trauma symptoms and their consequent impact on health, social relationships, and psychological well-being.

Training in self-compassion and mindfulness meditation are therefore presented as other, potentially beneficial approaches to trauma protection. These share the aim of achieving a state of present-centered awareness, acknowledgment and acceptance of how things are (rather than how they might have been or fears of what they might become), nonjudgment, and self-compassion. The aim is not striving to fix or resist the consequences of a stressful, painful, or traumatic event but rather to be open to the presence of these and meet them with self-care and self-reassurance. In this way, the effects of self-compassion may moderate reactions to negative events through the development of a kind and constructive self-to-self relationship rather than one of self-recrimination or self-denigration. The additional benefits of mindfulness may enable early recognition and acknowledgment, without ruminative involvement, of the thoughts, feelings, and sensations that give rise the maintenance of unpleasant post-trauma experiences.

Central to all these approaches is the fundamental belief that there is inherent potential within individuals to be resilient, heal, and grow.

References

Andrews, B., Brewin, C. R., Stewart, L., Philpott, R., & Hejdenberg, J. (2009). Comparison of immediate-onset and delayed-onset posttraumatic stress disorder in military veterans. *Journal of Abnormal Psychology, 118*(4), 767–777.

Baer, R. A. (2003). Mindfulness training as a clinical intervention: A conceptual and empirical review. *Clinical Psychology: Science and Practice*, *10*(2), 125–142.

Basoglu, M., Mineka, S., Paker, M., Aker, T., Livanou, M., & Gok, S. (1997). Psychological preparedness for trauma as a protective factor in survivors of torture. *Psychological Medicine*, *27*, 1421–1433.

Bien, T. (2006). *Mindful therapy*. Boston, MA: Wisdom.

Bishop, S. R., Lau, M., Shapiro, S., Carlson, L., Anderson, N. D., Carmody, J., *et al.* (2004). Mindfulness: A proposed operational definition. *Clinical Psychology: Science and Practice*, *11*(3), 230–241.

Bolton, E., Litz, B., Britt, T., Adler, A., & Roemer, L. (2001). Reports of prior exposure to potentially traumatic events and PTSD in troops poised for deployment. *Journal of Traumatic Stress*, *14*, 249–256.

Brewin, C. R., Andrews, B., & Rose, S. (2000). Fear, helplessness, and horror in posttraumatic stress disorder: Investigating *DSM–IV* Criterion A2 in victims of violent crime. *Journal of Traumatic Stress*, *13*, 499–509.

Cohen, H., & Benjamin, J. (2006). Power spectrum analysis and cardiovascular morbidity in anxiety disorders. *Autonomic Neuroscience: Basic and Clinical*, *128*, 1–8.

Dienstbier, R. A. (1989). Arousal and physiological toughness: Implications for mental and physical health. *Psychological Review*, *96*, 84–100.

Elliot, G. R., & Eisdorfer, C. (1982). *Stress and human health*. New York, NY: Springer.

Ersland, S., Weisath, L., & Sund, A. (1989). The stress upon rescuers involved in an oil rig disaster, "Alexander L. Kielland, 1980." *Acta Psychiatrica Scandinavica*, *335*, 38–49.

Feldman, C., & Kuyken, W. (2011). Compassion in the lanscape of suffering. *Contemporary Buddhism*, *12*(1), 143–155.

Follette, V., Palm, K. M., & Pearson, A. M. (2006). Mindfulness and trauma: Implications for treatment. *Journal of Rational-Emotive and Cognitive-Behavior Therapy*, *24*(1), 45–61.

Gallup, G. G., Jr., & Maser, J. D. (1977). Tonic immobility: Evolutionary underpinnings of human catalepsy and catatonia. In M. E. P. Seligman, E. P. Martin, & J. D. Maser (Eds.), *Psychopathology experimental models*. San Francisco, CA: Freeman.

Germer, C. K. (2009). *The mindful path to self-compassion: Freeing yourself from destructive thoughts and emotions*. New York, NY: Guilford Press.

Gilbert, P. (2009). *The compassionate mind*. London: Constable.

Gilbert, P., & Irons, C. (2005). Focused therapies for shame and self-attacking, using cognitive, behavioural, emotional, imagery and compassionate mind training. In P. Gilbert (Ed.), *Compassion: Conceptualisations, research and use in psychotherapy* (pp. 263–325). London: Brunner-Routledge.

Gilbert, P., & Procter, S. (2006). Compassionate mind training for people with high shame and self-criticism: Overview and pilot study of a group therapy approach. *Clinical Psychology and Psychotherapy*, *13*, 353–379.

Goetz, J. L., Keltner, D., & Simon-Thomas, E. (2010). Compassion: An evolutionary analysis and empirical review. *Psychological Bulletin*, *136*(3), 351–374.

Goldberg, D., & Goodyer, I. (2005). *The origins and course of common mental disorders*. London: Taylor & Francis.

Grey, N., & Holmes, E. A. (2008). "Hotspots" in trauma memories in the treatment of post-traumatic stress disorder: A replication. *Memory*, *16*(7), 788–796.

Grey, N., Holmes, E. A., & Brewin, C. R. (2001). Peritraumatic emotional "hot spots" in memory. *Behavioural and Cognitive Psychotherapy*, *29*, 357–362.

Gunnar, M. R., Frenn, K., Wewerka, S. S., & van Ryzin, M. J. (2009). Moderate versus severe early life stress: Associations with stress reactivity and regulation in 10–12 year old children. *Psychoneuroendocrinology*, *34*, 62–75.

Hannum, R. D., Rosellini, R. A., & Seligman, M. E. P. (1976). Learned helplessness in the rat: Retention and immunization. *Developmental Psychology, 12*, 449–454.

Harman, R., & Lee, D. (2010). The role of shame and self-critical thinking in the development and maintenance of current threat in post-traumatic stress disorder. *Clinical Psychology and Psychotherapy, 17*, 13–24.

Hathaway, L. M., Boals, A., & Banks, J. B. (2010). PTSD symptoms and dominant emotional response to a traumatic event: An examination of DSM-IV Criterion A2. *Anxiety, Stress, & Coping, 23*(1), 119–126.

Holmes, E. A., Grey, N., & Young, K. A. D. (2005). Intrusive images and "hotspots" of trauma memories in posttraumatic stress disorder: An exploratory investigation of emotions and cognitive themes. *Journal of Behavior Therapy and Experimental Psychiatry, 36*(1), 3–17.

Jha, A. P., Stanley, E. A., Kiyonaga, A., Wong, L., & Gelfand, L. (2010). Examining the protective effects of mindfulness training on working memory capacity and affective experience. *Emotion, 10*(1), 54–64. Retrieved from http://www.amishi.com/lab/wp-content/uploads/jha_stanley_etal_emotion_2010.pdf.

Kabat-Zinn, J. (1990). *Full catastrophe living: Using the wisdom of your body and mind to face stress, pain and illness.* New York, NY: Delacourt.

Kabat-Zinn, J., Lipworth, L., & Burney, R. (1985). The clinical use of mindfulness meditation for the self-regulation of chronic pain. *Journal of Behavioural Medicine, 8*, 163–190.

Kaplan, K. H., Goldenberg, D. L., & Galvin-Nadeau, M. (1993). The impact of a meditation-based stress reduction program on fibromyalgia. *General Hospital Psychiatry, 15*, 284–289.

Kelly, A. C., Zuroff, D. C., & Shapira, L. B. (2009). Soothing oneself and resisting self-attacks: The treatment of two intrapersonal deficits in depression vulnerability. *Cognitive Therapy and Research, 33*, 301–313.

Kendall, P. (1983). Stressful medical procedures: Cognitive-behavioral strategies for stress management and prevention. In D. Meichenbaum & M. Jaremko (Eds.), *Stress reduction and prevention.* New York, NY: Plenum Press.

Kristeller, J. L., Baer, R. A., & Quillian-Wolever, R. (2006). Mindfulness-based approaches to eating disorders. In R. A. Baer (Ed.), *Mindfulness-based treatment approaches: Clinician's guide to evidence based applications.* San Diego, CA: Elsevier Academic Press.

Lazarus, R. S., & Folkman, S. (1984). *Stress, appraisal and coping.* New York, NY: Springer.

Leary, M. R., Tate, E. B., Adams, C. E., Allen, A. B., & Hancock, J. (2007). Self-compassion and reactions to unpleasant self-relevant events: The implications of treating oneself kindly. *Journal of Personality and Social Psychology, 92*(5), 887–904.

Lee, D. (2009). Compassion-focussed cognitive therapy for shame-based trauma memories and flashbacks in post-traumatic stress disorder. In N. Grey (Ed.), *A casebook of cognitive therapy for traumatic reactions.* Hove, UK: Routledge.

Lee, D. A., Scragg, P., & Turner, S. W. (2001). The role of shame and guilt in reactions to traumatic events: A clinical formulation of shame-based and guilt-based PTSD. *British Journal of Medical Psychology, 74*, 451–466.

Longe, O., Maratos, F. A., Gilbert, P., Evans, G., Volker, F., Rockliff, H., *et al.* (2009). Having a word with yourself: Neural correlates of self-criticism and self-reassurance. *NeuroImage, 49*, 1849–1856.

Lyons, D. M., Parker, K. J., Katz, M., & Schatzberg, A. F. (2009). Developmental cascades linking stress inoculation, arousal regulation, and resilience. *Frontiers in Behavioural Neuroscience, 3*(Art. 32), 1–6.

Marks, I. M. (1987). *Fears, phobias, and rituals: Panic, anxiety and their disorders.* Oxford: Oxford University Press.

McEwen, B. S., & Stellar, E. (1993). *Stress and the individual: Mechanisms leading to disease. Archives of Internal Medicine, 153*, 2093–2101.

Meichenbaum, D. (1977). *Cognitive behavioural modification: An integrative approach.* New York, NY: Plenum Press.

Meichenbaum, D. (1985). *Stress inoculation training.* Boston: Allyn & Bacon.

Meichenbaum, D. (1994). Disasters, stress and cognition. Paper presented at the NATO Conference on Stress, Coping and Disaster, Bonas, France.

Meichenbaum, D. (1996a). Stress inoculation training for coping with stressors. *The Clinical Psychologist, 49,* 4–7.

Meichenbaum, D. (1996b). *Treating adults with post-traumatic stress disorder.* Waterloo, ON: Institute Press.

Meichenbaum, D. (1997). Stress inoculation training: A preventative and treatment approach. In P. M. Lehrer, R. L. Woolfolk, & W. S. Sime (Eds.), *Principles and practice of stress management* (3rd ed.). New York, NY: Guilford Press.

Meichenbaum, D., & Jaremko, M. E. (Eds.). (1983). *Stress reduction and prevention.* New York, NY: Plenum Press.

Meichenbaum, D., & Novaco, R. (1977). Stress inoculation: A preventative approach. In C. Speilberger & I. Sarason (Eds.), *Stress and anxiety* (Vol. 5). New York, NY: Halstead Press.

Mills, B., Rees, P., & Turnbull, G. J. (2002). Centurion: Shielding journalists and aid workers. In Y. Danieli (Ed.), *Sharing the front line and the back hills international protectors and providers: Peacekeepers, humanitarian aid workers and the media in the midst of crisis* (pp. 323–330). Amityville, NY: Baywood.

Mitchell, J. T. (1983). When disaster strikes. *Journal of Emergency Medical Services, 8,* 36–39.

National Child Traumatic Stress Network and National Center for PTSD. (2005). *Psychological first aid: Field operations guide.* Los Angeles, CA: Author.

Neff, K. D. (2003a). Self-compassion: An alternative conceptualization of a healthy attitude toward oneself. *Self and Identity, 2,* 85–101.

Neff, K. D. (2003b). The development and validation of a scale to measure self-compassion. *Self and Identity, 2,* 223–250.

Neff, K. D., Kirkpatrick, K., & Rude, S. S. (2007). Self-compassion and its link to adaptive psychological functioning. *Journal of Research in Personality, 41,* 139–154.

Neff, K. D., & McGehee, P. (2010). Self-compassion and psychological resilience among adolescents and young adults. *Self and Identity, 9,* 225–240.

Novaco, R., Cook, T. M., & Sarason, I. (1983). Military recruit training: An arena for stress-coping skills. In D. Meichenbaum & M. Jerenko (Eds.), *Stress reduction and prevention.* New York, NY: Plenum.

O'Leary, V. E., & Ickovics, J. R. (1995). Resilience and thriving in response to challenge: an opportunity for a paradigm shift in women's health. *Women's Health, 1,* 121–142.

Ozer, E. J., Best, S. R., Lipsey, T. L., & Weiss, D. S. (2003). Predictors of posttraumatic stress disorder and symptoms in adults: A meta-analysis. *Psychological Bulletin, 129,* 52–73.

Parker, K. J., Buckmaster, C. L., Schatzberg, A. F., & Lyons, D. M. (2004). Prospective investigation of stress inoculation in young monkeys. *Archives of General Psychiatry, 61,* 933–941.

Richhart, R., & Perkins, D. N. (2000). Life in the mindful classroom: Nurturing the disposition of mindfulness. *Journal of Social Issues, 56*(1), 27–47.

Roth, B. (1997). Mindfulness-based stress reduction in the inner city. *Advances, 8,* 50–58.

Rothschild, B. (2000). *The body remembers: The psychophysiology of trauma and trauma treatment.* New York, NY: Norton.

Rutter, M. (1981). Stress, coping and development: Some issues and some questions. *Journal of Child Psychology and Psychiatry, 22,* 323–356.

Rutter, M. (2006a). *Genes and behaviour: Nature–nurture interplay explained.* Oxford: Blackwell.

Rutter, M. (2006b). Implications of resilience concepts for scientific understanding. *Annals of New Academy of Sciences, 1094*(1), 1–12.

Saunders, T., Driskell, J. E., Johnston, J. H., & Salas, E. (1996). The effects of stress inoculation training on anxiety and performance. *Journal of Occupational Health Psychology, 1*(2), 170–186.

Shapiro, S. L., Brown, K. L., & Biegel, G. M. (2007). Teaching self-care to caregivers: Effects of mindfulness-based stress reduction on mental health of therapists in training. *Training and Education in Professional Psychology, 1*(2), 105–115.

Shephard, B. (2000). *A war of nerves: Soldiers and psychiatrists 1914-1994.* London: Jonathan Cape.

Smith, R. E. (1980). A cognitive-affective approach to stress management training for athletes. In C. H. Nadeau, W. R. Halliwell, K. M. Newell, & G. C. Roberts (Eds.), *Psychology of motor behavior and sport.* Champaign, IL: Human Kinetics.

Speca, M., Carlson, L., Goodey, E., & Angen, M. (2000). A randomized wait-list controlled trial: The effects of a mindfulness meditation-based stress reduction program on mood and symptoms of stress in cancer outpatients. *Psychosomatic Medicine, 62*, 613–622.

Stanley, E. A., Schaldach, J. M., Kiyonaga, A., & Jha, A. P. (2011). Mindfulness-based mind fitness training: a case study of a high-stress predeployment military cohort. *Cognitive and Behavioural Practice.* doi: 10 1016/i.cbpra.2010.08.002.

Teasdale., J. D., Segal, Z. V., & Williams, J. M. G. (1995). How does cognitive therapy prevent depressive relapse and why should attentional control (mindfulness training) help? *Behaviour Research and Therapy, 33*, 25–39.

Teasdale, J. D., Segal, Z. V., Williams, J. M. G., Ridgeway, V. A., Soulsby, J. M., & Lau, M. A. (2000). Prevention of relapse/recurrence in major depression by mindfulness-based cognitive therapy. *Journal of Consulting and Clinical Psychology, 68*(4), 615–623.

Thompson, B. L., & Waltz, J. (2008). Self-compassion and PTSD symptom severity. *Journal of Traumatic Stress, 21*(6), 556–558.

Uchino, B. N., Cacioppo, J. T., & Kiecolt-Glaser, J. K. (1996). The relationship between social support and physiological processes: A review with emphasis on underlying mechanisms and implications for health. *Psychological Bulletin, 119*(3), 488–531.

van der Kolk, B. A. (1994). The body knows the score. *Harvard Psychiatric Review, 1.*

van der Kolk, B. A., McFarlane, A. C., & Weisaeth, L. (Eds.). (1996). *Traumatic stress: The effects of overwhelming experience on mind, body and society.* New York, NY: Guildford Press.

Watkins, A. D. (1997). *Mind–body medicine: A clinician's guide to psychoimmunology.* London: Churchill Livingstone.

Watkins, A. D. (2011). The electrical heart: Energy in cardiac health and disease. In D. F. Mayor & M. S. Micozzi (Eds.), *Energy medicine east and west: A natural history of qi.* New York, NY: Elsevier.

Weisaeth, L. (1994). Preventive intervention. Paper presented at the NATO Conference on Stress, Coping and Disaster, Bonas, France.

Wheway, J., Mackay, C. R., Newton, R. A., Sainsbury, A., Boey, D., Herzog, H., *et al.* (2005). A fundamental bimodal role for neuropeptide Y1 receptor in the immune system. *Journal of Experimental Medicine, 202*(11), 1527–1538.

White, P. (2005). *Biopsychosocial medicine: An integrated approach to understanding illness.* Oxford: Oxford University Press.

World Health Organization. (1992). *The international classification of mental and behavioural disorders, international classification of diseases* (ICD-10; 10th ed.) Geneva: Author.

19

How Employee Assistance Programs (EAPs) Respond to Trauma Support and Critical Incident Management: An International Focus

Mandy Rutter

Overview

I'd like to thank you so much for the excellent response you provided to us during this entire event. From the moment I contacted you, right to the winding down and debriefing, your participation was professional, compassionate and full of common sense.

– HR manager, oil and gas industry

This compliment was given to a group of employee assistance program (EAP) therapists assisting a group of employees and their families who were severely traumatized following street rioting and political disturbance in their home country of Libya. The therapists traveled out from the UK, a safe neighboring country, and spent a week with the employees and their families offering a wide range of supportive post-trauma interventions.

However, this particular compliment came, not from the traumatized employees or members of their families, but from their human resources (HR) manager across the other side of the world, in Canada, who had requested therapeutic intervention on behalf of his staff.

In essence, this is the crux of how EAPs respond to trauma support and critical incident management. The managers and directors within an organization have a legal and moral duty of care for their employees, who may have been involved in a bank raid, industrial accident, earthquake, flood, rioting, sudden death, or any unexpected and

International Handbook of Workplace Trauma Support, First Edition. Edited by Rick Hughes, Andrew Kinder, and Cary L. Cooper.

traumatizing incident. These managers then employ the services of the EAP to assist them with that duty of care and to offer effective and appropriate post-trauma intervention.

The EAP response to organizations experiencing trauma always has two clients: firstly, the managers requesting the intervention, who have a set of expectations for the EAP intervention; and, secondly, the employees receiving the intervention who are at various stages of distress and recovery.

As with any therapeutic intervention that is requested on behalf of a third person (or group of people), it is the role of the provider of the therapy (in this case, the EAP) to manage and contain the expectations of the person making the request. Whether it is a mother requesting therapy for her child, a husband requesting therapy for his wife, or a manager requesting EAP trauma support for his team, it is an essential part of the EAP process to understand as far as possible the spoken and unspoken agenda of the manager prior to organizing the post-trauma intervention.

The manager's agenda may be hugely varied, and not necessarily synergetic with the agenda of the employees. For example, the manager may request that the EAP trauma support:

- Occurs immediately after the incident has occurred.
- Lasts for only half a day.
- Focuses on specific employees.
- Ensures that employees don't sue the company or talk to the media.
- Ensures that all employees return to work as soon as possible.
- Reports back all information about the employee's health and well-being.

The employees however, may want to:

- Talk with their friends and family prior to talking with the EAP.
- Receive support from agencies outside the EAP.
- Have a confidential discussion with the EAP.
- Not talk about the incident at all.
- Have the option of receiving help when they choose to.

Hence the current EAP response to trauma and critical incident management aims to meet both the requirements of the organization, and the needs of the employees affected by the incident. If this wasn't enough of a tall order, the third component of the EAP response is compliance to clinical guidelines and best clinical practice derived from current research.

Each incident poses different logistical, individual, and managerial needs, and hence the content for the top two circles in Figure 19.1 will hold different content depending on the incident; however, the bottom circle remains consistent and it is therefore the skill and adaptability of the EAP to fulfill the clinical guidelines on trauma intervention, despite the varying needs of employees and managers.

Fortunately, most EAPs now have a vast array of resources available to them, in terms of both technology and qualified trauma personnel, and so can provide online, onsite,

Figure 19.1 The three interlocking priorities of EAP trauma response.

telephone, and face-to-face support for referring managers and employees. The Employee Assistance Service from Health Canada (2011) demonstrates this in their description of services.

> Your EAP can provide you as an employer with a trauma response service that guides you through the preparation of emergency response plans, trains your HR staff on how to support employees after an incident, ensures your business continuity plans are employee centric, offers immediate on-site trauma support and useful literature in a number of different languages, and assists you to put into practice the significant lessons learnt following the disaster. Your EAP provides you with complete reassurance, before, during and after a disaster.

This chapter will set the historic scene for EAP post-trauma response, by explaining the history and function of an EAP, and how EAPs began to deliver post-trauma response both in the United Kingdom and across the globe. We will hear from a number of EAP providers who describe the nature of their post-trauma response and some of the benefits and challenges of shifting from group debriefing to a psychological first aid (PFA) approach. The chapter will then describe some of the future opportunities for EAP trauma response, and outline some of the challenges faced internationally when attempting to deliver an EAP post-trauma response. Case studies and interviews with EAP professionals will be used to highlight issues.

Employee Assistance Programs: A Brief History of Developments

EAPs have been in existence for nearly 100 years and provide a mechanism for offering counseling and other forms of assistance, advice, and information to employees on a systematic and uniform basis, and to recognized standards. An EAP is also a strategic intervention designed to produce organizational benefit. EAPs can assist organizations to address performance and well-being for both teams and individuals.

The UK Employee Assistance Professionals Association (UK EAPA) explains that EAPs are intentionally defined more by what they achieve, rather than by what they are, in order to leave maximum room for tailoring services to meet the needs of each organization.

> EAPs have the resources to reach those employees people who would not otherwise have access to the psychological support and information from which they may benefit.
>
> (UK EAPA, 2010)

EAPs originated in the United States in 1917. Two companies, R. M. Macy and Co. and the Northern State Power Company, recognized the need to provide some support and assistance to employees who had work performance problems often relating to a personal situations. This recognition to offer a service to employees was developed in the 1930s and 1940s, when Prohibition ended and employers were concerned about the effect of alcohol on work performance. At this time, EAPs were called Occupational Alcohol Programs.

> The EAP movement began . . . with one recovering alcoholic worker sharing his recovery with another.
>
> (Dickman, 1988)

Gradually these programs evolved and began treating mental, emotional, and financial problems as well as those problems caused by alcohol and drug use. The tremendous growth in EAPs, however, began in the early 1970s. In 1972, the Occupational Programs Office of the Federal Institute on Alcohol Abuse and Alcoholism offered federal grants to help increase the number of programs (Masi, 1987).

During the next five years, the EAP concept spread rapidly to many organizations throughout the United States; for example, by 1982, more than 12 000 New York State employees and their families had received support and advice from the EAP, and by 1996 there were 35 000 EAP referrals per year for counseling and an additional 60 000 requests for information and follow-up, just from the US State Department.

Today, over 96% of Fortune 500 companies provide EAP services for their workplace. The International Employee Assistance Professionals Association (International EAPA) has over 7000 member organizations, including 800 from 30 different countries.

EAPs started in the United Kingdom during the mid-1980s and were mainly small independent organizations usually led by a psychologist. Even in those early days, customers were very varied and included commercial, local authority, health services, and uniform organizations who were looking for EAPs to assist with their duty of care and health and safety legislation. The EAP was essentially remedial in focus and provided a 24-hour, seven-days-a-week telephone helpline for support for a range of personal and work-related problems. Being available 24/7 meant that employees had equal access to the service even if they were shift workers. A central principle of EAP services throughout its short history is that all employees of an organization, whatever their position, have equal access to the service.

Colin Grange (2005) described the early EAPs as follows:

> In addition to telephone counselling, EAPs also provide limited-session face-to-face counselling near to where the employee lives or works. To provide this service EAPs need to have a network of affiliate/associate counsellors who can work for the EAP provider on a fee-for-service basis. To cover the needs of their clients EAP providers had to recruit and enter into a contractual relationship with a significant number of counsellors. (p. 4)

Hence, the clinical resources of EAPs became enormous and incredibly diverse. One UK-based EAP has over 950 affiliate counselors, an additional 250 psychologists, and other mental health professionals geographically dispersed all over the United Kingdom. Those EAPs who provide international support have developed networks of counselors and psychologists all over the world.

Although EAPs were largely clinically focused, being owned and managed by clinicians, other information was provided as part of the EAP service such as debt, legal, and employee rights and responsibilities. Indeed, throughout the 1980s and 1990s (before the ubiquitous access to the Internet), the majority of calls to an EAP were for information rather than counseling. However, the managerial focus for EAPs remained on the clinical services, and (as discussed later on in the chapter) it is interesting to note that EAPs began to explore trauma work during the mid-1980s.

As well as services for the individual employees, EAPs needed to understand and communicate with the customer organization. The managers and directors were paying for the service, and the performance of the organization could be maximized by appropriate and relevant information being fed back to the manager about their employees who were calling in. This "account management" consisted of statistic information regarding the number of callers and themes of their calls. Hence if 80% of callers from one department were describing stress symptoms caused by their workplace situation, the EAP was able to feed this back to the manager, and discuss options to solve the problem such as by providing stress management or resilience training.

It was these discussions that promoted the EAP as the experts in the mental health of the employees and encouraged managers and directors to call on the EAP if they identified workplace distress such as redundancy, restructuring, or bullying. The organization's relationship with the EAP became a two-way consultative process, which could start one of two ways, explained through the following examples and diagrams.

Example A: Company X

An employee from company X is being bullied at work and calls the EAP helpline for support and information. The EAP counselor takes the call and records the details of the enquiry and the nature of the support provided to the employee, on a secure database. The EAP clinical manager then reviews all the calls from company X and informs the EAP account manager for company X that there is a theme of bullying in one particular location. The account manager's information does not identify any particular caller, and hence the strict boundaries and protocols of confidentiality are maintained. The account manager and clinical manager will then offer consultancy and advice to company X on how to develop strategies to manage bullying in the workplace. The aim is that the organization manager reviews and strengthens the policies on bullying in the workplace and hence the employees at company X experience a more productive workplace environment. There are many examples of this process having a positive effect in the workplace: high numbers of callers describing large amounts of debt, leading to the workplace introducing lunchtime education seminars on money management, and high numbers of callers describing stress when returning to work after maternity leave, prompting the organization to introduce flexible working and a buddy system for all staff returning after maternity leave.

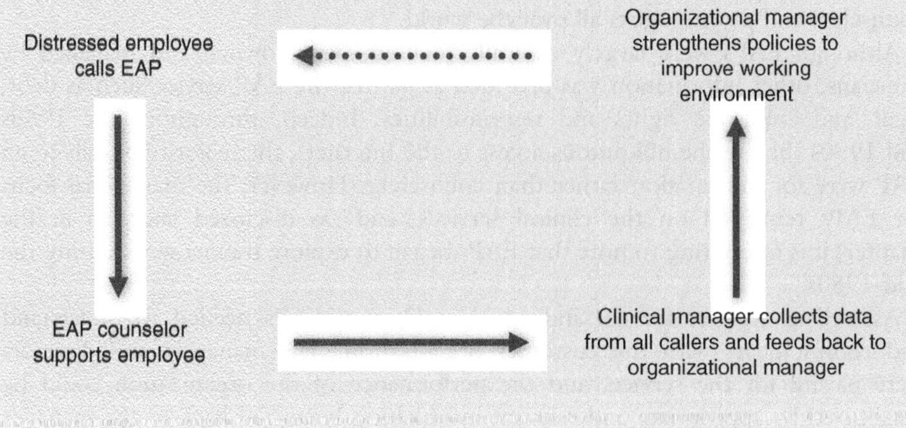

Example B: Company Y

An HR manager from company Y contacts the EAP clinical manager as he is worried that impending redundancies will cause huge distress for a group of long-serving employees. The clinical manager discusses a range of options including a counselor to visit company Y after the redundancy notices are given. The EAP counselor attends company Y and employees are invited to have a confidential discussion to gain support at a potentially very distressing time. Hence, company Y is fulfilling its moral and legal duty of care to employees who may be at risk of psychological distress following their redundancy notices.

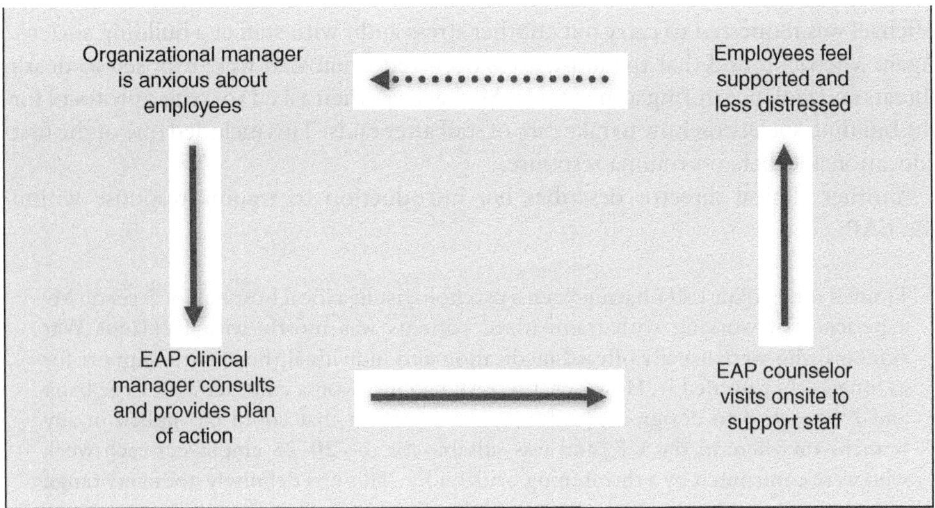

Whilst the EAP was primarily set up as a service that supports employees and managers when there is a problem, the more recent developments for EAPs are concerned with daily aspects of employee's lives before they become problems, for example organizing care for elderly relatives and providing advice on nurseries and childcare. In addition, on the mental health front there is a movement toward coaching and resilience training, hence providing a service for employees who are feeling well and healthy but want to continue to develop their strengths and talents. The latest developments for EAP are mainly focused on linking with other health professionals such as occupational health, health visitors, and life coaches, and embracing technology in order to offer e-counseling, therapeutic chat rooms, health information podcasts, and online health risk assessments.

As EAP services evolved and developed, it can be seen how the EAP held the position as the experts in employee mental health and subsequently were in a strong position to offer trauma support provision.

EAPs and Trauma Support

EAPs and trauma response has had a history speckled by incidents and disasters.

Michael Reddy (chairman of ICAS, 1985–2007) recalls how when he started ICAS in 1985, critical incident response was not part of the service "not because it wasn't important, but because it just wasn't part of the conversations I was having with employers and customers of the service."

However, part of the ICAS service was organizational stress audits, and Michael was asked by a rail company to assess the levels of stress in a group of train drivers. Michael found that the main stress issue for train drivers was coping with suicides on the line. When feeding back these findings to the HR managers, it became apparent to Michael that a different system was required by the EAP for supporting employees with unexpected traumatic situations. The stress of unexpected events was echoed when

Michael was requested to carry out another stress audit with staff at a building society. Again Michael found that the most stress occurred when staff were exposed to death threats and violence during armed raids. Michael was then asked to write a protocol for the building society on how to take care of staff after raids. This included one of the first educational leaflets on trauma response.

Another clinical director describes her introduction to trauma response within the EAP:

> I joined the EAP in 1991 having been a psychologist in a local hospital for 5 years. My experience of working with traumatized patients was mostly with Falklands War veterans who were usually offered medication and individual therapeutic support for as long as they needed it. However, the EAP had just won a contract with large bank and I was asked to design a post-trauma intervention that could be applied in any branch, anywhere in the UK, and was suitable for the 20–25 employees each week who were confronted by a threatening bank raider. This was definitely not in my range of experience.

Stephen Galliano (EAP clinical director for ICAS 1991–2001) was also confronted by a trauma out of his range of experience. He recalls how on his first day at work, he was asked to go to large railway station in London where there had just been a bomb explosion.

> ICAS was a newly established EAP, successful in providing telephone counseling, life management services, and face-to-face counseling; we needed to expand our reputation and hence when we got a call from the railway station asking us to come immediately after a large bomb, we knew we had to respond, so I headed down to London, not imagining what I was heading to. I knew I had to help as there were so many distressed staff and managers, but there was no structure, procedures, or coordination at that time. It was about being calm, available, and able to juggle questions from managers and tears from employees all within the same very short space of time. Whist the procedure was haphazard, the personal support in the live "here-and-now" environment was invaluable. We were cast in the role of experts, and it became our job to live up to that and "sort the staff out." The rail managers expected that if staff were upset, then a session with the psychologist or counselor would cure them, and often it did temporarily but it wasn't necessarily for our training at that time, but purely because we listened in a clam, trusting, and confidential manner, oh and we didn't panic.

All UK EAPs began to see an increase in requests for support for traumatized employees. It wasn't just bank raids and explosions but a very wide range of incidents as described by an EAP clinical manager working in 1990.

> I had started as an EAP clinical manager, expecting to be supervising counselors, but once we started working with the high street retail outlets we were getting calls on a daily basis from late-night stores who had been raided, off-licenses who had been robbed, and pubs where violent outbursts had led to employees being abused and traumatized. In order to contain the traumatic incidents we set up a specific department which was initially two of us in 1993, but very quickly grew to require five full-time clinical managers responding to death, suicide, murder, assaults, hostage situations, exposure to asbestos, plane crashes,

prison riots, and more. We actually had difficulty believing that this level of trauma occurred in the workplace.

This individual story echoed a national and international picture in the late 1980s and 1990s, which saw a huge rise in the number of crisis and disasters for example in the United Kingdom:

1985: Bradford football stadium fire
1987: Sinking of *Herald of Free Enterprise* passenger ferry
1988: Lockerbie bombing (Pan Am)
1988: *Piper Alpha* oil rig fire
1989: Kegworth plane crash
1989: Hillsborough football stadium disaster

And, internationally:

1982: Air Florida airline crash in Washington, DC
1984: Poison gas leak in Bhopal, India
1989: Students massacred in Tiananmen Square, China
1995: Oklahoma City bombing
1999: Mass shooting at Columbine High School

These disasters highlighted the inefficiencies of local authorities, workplaces, and governments in terms of their ability to cope with mass fatalities and mass injuries. The lack of an authoritative, compassionate, and organized response to victims' families is still talked about today, some 15 years after many of these incidents.

With this rise of public crises and disasters, and in 1980 the acceptance of post-traumatic stress disorder (PTSD) as a recognized psychiatric condition, it was of paramount importance that EAPs internationally developed an effective post-trauma response that assisted organizations with their duty of care to employees (i.e., to minimize the stress caused by traumatic incidents and to maximize the potential for employees to recover).

One of the key issues for an EAP back in the 1990s, and remains today, is to provide a service that has the ability to be replicated across any number of geographic locations, to any range of customers, and at any time of the day or night; that is cost effective; and where consistency of quality can be measured. In order to fulfill all these requirements, EAPs needed to be able to train their counseling affiliates across the country in a consistent and appropriate trauma response that could be delivered in any workplace environment.

EAP and Group Debriefing

The United Kingdom looked to the US EAP market which had begun to embrace Jeffery Mitchell's Critical Incident Stress Debriefing (CISD). Mitchell's structured six-step individual or group debriefing process that took place 36–72 hours after the incident fitted the requirements of the EAP perfectly. It appeared to be relatively clinically robust and transparent, and was easily replicated, and EAP affiliate counselors could be trained in the model.

Mitchell noted that:

> A single debriefing session will generally alleviate the acute stress responses which appear at the scene and immediately afterwards and will eliminate or at least inhibit, delayed stress reactions.
>
> (Mitchell, 1983)

> This can be achieved because the group process allows participants to learn and understand the facts and perspectives of other responders, creates a safe environment where responders can share their story and feelings, and creates a sense of psychological closure regarding the event.
>
> (Everly & Mitchell, 1999)

EAP customers liked the structured process of group debriefing. Managers could understand how it worked and liked the convenience and straightforwardness of phoning the EAP when an incident had happened, and a few days later a trauma counselor came and took all the employees who had been involved in the incident through a structured group process. Those employees who were still distressed after the debriefing were referred into the EAP, and received individual support through the usual route. Hence the employer's duty of care was fulfilled as they had used the EAP to provide an intervention that apparently minimized the risk of future post-trauma distress, and the EAP had demonstrated its value as the experts in psychological health.

> Individuals, employers and clinicians found group debriefing useful for a variety of reasons, not least of which was that it appeared to be a universal and "expert" approach. The illusion of the clinician having all the knowledge and all the answers was upheld in the debriefing approach. In addition, at some level, debriefing encouraged workplace managers and colleagues to leave trauma recovery to the professionals. The tears, the distress and the anger were all to be left for the onsite counselor to manage. Recovery of the traumatic incident was focused on those staff directly affected. Recovery was not considered as involving the group, team or organization.
>
> (Rutter, 2007)

Some EAPs made their own adjustments to the CISD model, and incorporated individual sessions with the debriefer after the group process, and provided a follow-up intervention six weeks after the initial group, to ensure that the duty of care was complete.

For approximately 10 years, group debriefing (or versions of it) was the backbone of EAP post-trauma support, and was used across an incredibly wide range of workplaces, locations, and countries. In 1999, two EAPs were called upon to support people involved in the London Paddington rail crash; one EAP supported all the staff from the rail companies, and the other EAP supported the passengers. Over 500 passengers attended debriefing groups following this incident. In 2001 there was a large explosion at an oil refinery, and the EAP for the oil company sent 10 trained counseling affiliates to the refinery to run 25 debriefing groups.

After the oil refinery explosion, the site manager was interviewed about the EAP intervention and commented:

> It is when a crisis happens that the EAP really comes into its own, the visibility of the EAP rises dramatically as here you are, on our premises, doing the things that you do well, which is supporting distressed staff. You tell us you do this all the time over the phone, but it is only when a trauma happens that we actually see you doing it in front of our eyes. We like the consistent group process, but, to be honest, we don't mind what you do as long as it works. You are the experts, and we expect you to do what is within national clinical guidelines; that's why we employ you.
>
> (Oil refinery site manager, interview, March 2011)

The debriefing process translated across the globe, and as EAPs began to explore the international market, they continued to use group debriefing as their main post-trauma intervention. The US and UK EAPs developed links with counselors and psychologists in Spain, Italy, France, Belgium, Switzerland, Russia, Portugal, Czech Republic, South Africa, Mexico, and Argentina, who were all expected to use the group debriefing process.

However, the universality of group debriefing as the EAP response to trauma was not to remain. A number of factors contributed to some major challenges for group debriefing. The most significant challenge was the publication of the NICE guidelines on PTSD in 2005, which, although it was guidance for health professionals within the health service, did provide evidence that debriefing was not as clinically robust as it was first thought to be. In addition, it could be said that in their desire to find a one-size-fits-all approach to post-trauma support the EAP world was slow to recognize that group debriefing had been designed for an intact, homogeneous working group, and that the corporate, global workplace was far from homogeneous or particularly intact.

In addition, the practitioners delivering the debriefing at the customer sites were also questioning its value. Here Sylvia Dewis, an EAP trauma counselor, describes her experience:

> I worked for an EAP who asked me to run a debriefing group for a group of staff at a dairy who were attacked by armed and masked men on a Saturday afternoon, after all the milk floats had delivered their weeks takings to the dairy. There were only four staff involved, but one member of staff had her eight-year-old son in the office during the raid, so it was very frightening for all concerned. The debriefing group lasted over four hours, going over the details of what happened. As the facilitator I felt traumatized at the end of it, and rather than being able to forget and move on, now even seven years later I can remember the minute details of what happened at that incident. After that I told the EAP I wasn't doing any more group debriefings, as I didn't believe it was a helpful process.
>
> (Sylvia Dewis, interview, 2004)

Other practitioners explained how they changed the focus of the debriefing group as other employees who weren't at the incident wanted to come along and support their colleagues, or the occupational health manager wanted to come into the group, or the group just wanted to talk about their impending redundancies rather than the trauma they had just gone through.

The crux for one EAP came when the medical director of their customer sent an email to the EAP clinical manager saying, "We've heard debriefing is inert at best and can be harmful – what are you as our EAP going to offer us now?"

Education, Resilience, and Psychological First Aid

Fortunately, many EAPs had begun to explore alternatives to group debriefing prior to the publication of the NICE guidelines. One US EAP described how they were contracted to support a number of employees caught up in the September 11, 2001, collapse of the Twin Towers in New York. They described how they physically could not get employees together for a debriefing, and many of the counselors in New York City were also victims of this tragedy. The US EAP therefore contacted their partners in the United Kingdom to request assistance. The UK EAP designed a more individual approach to trauma which was accepted by the New York company, and hence began contacting all employees by telephone and offering a supportive, educational discussion that allowed the employee to offload their experiences to the telephone counselor if they chose to, or to keep the call short if they didn't require external support. In addition, all employees were asked what support they would like from the workplace. The answers to this question were fed back to the organization, and this information provided a plan of action for the managers. This piece of work also extended to the organizational managers, who received a supportive phone call from the UK EAP as well. The information collected from the 30 managers was incredibly important in steering interventions for managers. Many of the managers were very anxious to provide emotional support for their staff, but felt exhausted and stressed. They felt they had to work for long hours in order to demonstrate to their staff that it was OK to come back to work. They were aware that as managers they provided consistency for their staff, and it was their attitude and behavior that would ultimately determine if their staff successfully returned to work.

Here was an example of the EAP utilizing its wealth of resources (i.e., international relationships and large numbers of telephone counselors) to offer post-trauma support to employees which was then fed back into the organization so that the overall recovery environment for employees and their managers became more positive and supportive.

In addition to the original telephone call just a few days after September 11, the UK EAP telephone counselors made a follow-up telephone call six weeks later, to both employees and managers, in order to monitor how the employees were recovering and again to invite them to suggest ways that the organization could improve its support. Again this information was fed back. In addition to the telephone calls, the US and UK EAP worked together to develop a specifically tailored information leaflet that informed employees of likely post-trauma reactions and provided information about how to take care of themselves and others during this difficult time.

This intervention set the scene for a range of educational and information leaflets to be produced for employees and their managers on specific traumatic situations, such as:

Recovering from trauma
Understanding suicide in the workplace
Managing death in the workplace
Tips and scripts for communicating distressing messages at work
How to stay resilient through difficult times

All of these leaflets are regularly used by EAPs today as part of their post-trauma response. A recent trauma leaflet was adapted for Japanese employees who had just experienced the tsunami and nuclear explosion in 2011.

Since 2006, the philosophy of the EAP approach to trauma response has shifted. Clinical guidelines and trauma research have directed EAPs to focus on delivering "practical support in an empathic manner" (NICE, 2005); to "enhance the capacity of existing networks, both formal and informal" (WHO, 2005); and to think of employees as resilient people, rather than victims who will all require post-trauma treatment.

Bonanno (2004) explains:

> The majority of people exposed to trauma and grief showed a resilient response and that the essential factors associated with resilience were emotional ties to family, social relationships and external support systems.

How this gets translated into EAP response is interesting; the response may look the same as the group debriefing response, with the customer making a telephone call to the EAP clinical managers and then a counselor coming onsite, but the nature of the telephone call and what the counselor actually does onsite are very different from group debriefing.

Geoff Holmes (2005) from a UK EAP, Care First, explains:

> One of the primary tasks of the EAP is to provide managers with support that empowers them to maintain control of a situation. Sometimes that support will be focused on helping them understand that individuals will all have their own reactions and that it is normal to be upset, tearful and possibly angry when difficult things happen. Often there is a myth to be exploded, for example that counselors coming on-site can make everything better quickly, thus avoiding the painful period of individuals working through natural reactions to trauma. Sometimes a great deal of this tutoring of managers can be done on the phone when an incident has taken place. Often an on-site visit from a counselor aimed at meeting with managers and providing some basic education about trauma and checking out the range of responses evident in teams, will suffice.

A number of EAPs now call their post-trauma response psychological first aid (Chapter 11, this volume), which aptly describes a process that recognizes that most employees are likely to be temporarily destabilized by a trauma, but with sufficient information, practical and emotional support, and access to appropriate helping agencies, it is likely that the employee will recover. For the few employees who need further help, PFA incorporates a triage or assessment process that utilizes the individual support mechanisms of the EAP.

Many EAPs recognized that when a workplace trauma occurs, there were a variety of audiences for PFA, compared with group debriefing that only had one audience (i.e., the staff involved in the incident). The PFA approach offered education, support, guidance, and assessment to the following groups:

Managers: On how to respond to their staff in order to provide the most supportive recovery environment, and how to look after themselves during a stressful time period.

Staff involved in the incident: This could be either group or individual sessions inviting staff to speak about the trauma in their own free and flexible way and offering appropriate coping strategies that linked staff with their own natural resilience.

Staff not directly involved in the incident but part of the workplace team who have a significant role to play in creating peer support and maintaining a positive work environment.

There may also be *family and friends of employees* affected by a workplace trauma for whom the organization feels they have a duty of care, and who may require psychological first aid to identify the necessity of further involvement.

This more flexible, educational approach to trauma response has opened the door for many other conversations to take place between the EAP and the customer. The PFA response to trauma views employees as healthy, optimistic people who are generally resilient in the face of adversity. If this is the case, there are a few interesting questions for EAPs to address in relation to proactive rather than reactive responses to trauma, for example:

Does the EAP have a role in educating managers on how to support their staff in the event of an incident?

Does the EAP have a role in educating staff in high-risk roles to develop resilience so that the impact of a traumatic event is lessened?

Does the EAP have a strategic role in assisting its customers to develop business continuity plans which include support to staff during a crisis?

The answer to all of these questions is yes; however, it depends on the sophistication and maturity of the relationship that the EAP has with its customers. Much of this work is already taking place in the UK EAP market. One EAP has been part of a national working group to produce a British Standard Document on the Human Aspects of Business Continuity (PD 25111) which draws attention to the importance of preparing staff physically and psychologically for a traumatic event, should it occur.

Challenges to EAP and International Trauma Response

Whilst there are many new opportunities arising for the EAP and its customers in relation to education, training, and resiliency models of post-trauma interventions, there are also some challenges.

The first challenge comes from within the EAP resource itself. Some affiliate counselors who were very committed to structured group debriefing have struggled to take on a more flexible attitude and approach. Psychological first aid asks clinicians to be responsive at an intuitive, spontaneous, human, and pragmatic level in a whole range of workplace situations. This could mean supporting a very tearful and frightened employee in the women's restroom as she is too embarrassed to be crying on the shop floor, or it could be walking a group of staff back in to their shop for the first time after their colleague was shot behind the counter. It could be educating a chief executive on how to talk to the parents of an employee who died at work, or writing a script for a communications manager who has to inform the staff that their colleague committed suicide. The training for such a role incorporates a whole range of skills which links therapy to disciplines such as community work and social work.

The second major challenge comes from our international perspective on trauma response. The large multinational EAP customers are often looking for an EAP service that has identical features in every part of the world. Hence the trauma response in India is required to be the same as the trauma response in Switzerland or Canada.

Jenny McFarlane, an EAP international development manager, explains:

> We have an ongoing dialogue with our multinational companies about the importance of providing therapeutic support that matches the language and culture of the people receiving it. For example, many countries link trauma and traumatic recovery with their religious faith, and it is important that our trauma therapists are knowledgeable and understand this. We want to develop credibility with a very wide audience of managers and employees, and this means a lot of listening and not making assumptions about our way always being the only way
>
> (Jenny McFarlane, interview, April 2011)

In order to highlight some of these difficulties, a recent roundtable discussion was held with EAP trauma therapists from five different countries who described some of the challenges in delivering onsite trauma response to employees in their countries. The countries represented at the table were the Czech Republic, Portugal, Poland, India, Mexico, and Spain. These commentaries are not stereotyping all employees in the specific country; they are describing the experiences of the therapists.

In the Czech Republic, Lior Behar, an EAP therapist, felt that employees tended to underestimate the size and impact of the problem, but also underestimated the chances of resolving it. The face-to-face contact in the workplace was often difficult, as employees on the whole preferred to email or text. Also, a lot of weight was placed on the knowledge of the therapist, and hence for trauma work, where the aim is to empower employees and link them with their own support networks, the persona of the "expert" is difficult to dismiss.

> Often when we go onsite we will be the first therapist that employees have ever come into contact with, and employees want us to judge their symptoms, and expect us to give an opinion about what is right. Often, however, employees believe that nothing can help them. We therefore have to be cautious and clear, and set small realistic goals, so that employees can see signs of their own recovery.
>
> (Lior Behar, interview, June 2011)

In Portugal, the EAP therapist described how there was no expectation that the working environment will be supportive, and the duty of care is not well developed. There appears to be a strong separation between "work life" and "private life" so a personal problem tends to be taken care of in a private setting. There is no expectation from employees that the organization will assist with financial, legal, or personal problems and hence the usage of EAPs is very small at the moment. However, one advantage of having large multinationals in Portugal is that they are bringing with them the concept of EAP and trauma support, and hence when employees do use it they are very satisfied.

The therapist from Poland summarized the clients he had worked with in Poland as shy and suspicious about psychologists and counselors and as difficult to help because they believed they should find their own solutions. He found it difficult to offer coaching, guidance, or education as clients felt they should come up with their own solutions and wanted the therapist to affirm their ideas. The suggestion of onsite support after trauma would probably be well received the employers in Poland, but the therapist felt that the employees, whilst thinking it was a good idea, would be unlikely to use the support.

In India, the therapist felt that the main place for employees to take any personal problems is the family. There are a lot of pressures from within the family to keep problems at home and not take them into the workplace. In addition, counseling as a profession is not well recognized or understood, and as there are nearly 30 languages in India, it may be difficult to find a therapist who always speaks in the same language of origin as the employees. However, there have some successes when the counselor has been placed inside the workplace and the employees visit the in-house counselor rather than traveling to other premises. This development bodes well for onsite trauma support, although many employees in India have family and religious rituals to follow after a trauma and would not naturally look to the workplace as a place of support.

Deborah Loffler, an EAP manager in Mexico, described some of the tensions when providing on-site support in Mexico:

> When we send a mental health professional to provide an on-site support, the employees will often want to talk about all the personal issues that are troubling them, and the therapist has to be keep firm boundaries about what can be achieved within the time allocated for the therapist visit. Employees often think we have come to "treat" them and view a therapist as someone to offload personal problems, rather than support them with workplace issues. This can make it difficult to encourage resilience and encourage staff to take control. The employees want the therapist to take control. In addition, many employees live in neighborhoods that are very violent, and their workplace is the only secure place in their lives, as the organization protects their security. However, when a trauma happens at work they often feel they have no safe place in their lives and this can increase the need for additional therapeutic work with employees on top of psychological first aid.
>
> (Deborah Loffler, interview, June 2011)

The therapist from Spain described EAPs as relatively unknown in Spain, covering approximately 500 000 people out of a working population of 20 million employees.

In addition, employees often had a strong family network, and were encouraged to deal with difficulties within the home, only going to a psychologist "when completely crazy." Trauma response in Spain was usually organized by public health departments, and therefore the therapist in Spain did not attend many traumatic incidents in the workplace.

Even with this very small number of countries profiled here, the cultural issues as described are enormously diverse and require a sensitive culturally specific post-trauma program in order to be accepted and appropriate.

Summary

Employee assistance programs have established themselves as the experts in employee psychological health, and are able to utilize this knowledge to assist the organization to develop strategies and policies to strengthen overall performance and well-being in the workforce. They are therefore in an ideal position to offer the organization support when a traumatic incident occurs that impacts their staff. The initial post-trauma support was derived from emergency services settings and was a formal, structured group process that was appreciated by both employers and employees across the globe. Whilst many countries still hold debriefing as the main backbone of EAP trauma response, the United Kingdom and parts of the United States identified that this process proved to be unreliable in assisting employees to recover, and a more practical, educational, and individually tailored approach was utilized. The PFA approach opened up a range of possibilities for trauma support that enabled the EAP to converse with its customers about proactive resilience-based support as well as reactive trauma-led support.

An interesting measure of how far EAP and trauma response have come in 22 years is to look at the International EAPA conference program. Back in 1989 the EAPA conference attempted to run a workshop entitled "Critical Incident Stress and Trauma Response in the Workplace – The Role of Employee Assistance Response," but out of the 1500 attendees at the conference, there were barely 100 at the workshop. This year, nine workshops are being run at the four-day EAPA conference, all focusing on different aspects of trauma and resilience in the workplace.

Acknowledgments

I would like to thank the following people who have contributed to this chapter with generosity and enthusiasm: Eugene Farrell, Mark Winwood, Jenny MacFarlane, Stephen Galliano, Michael Reddy, Andrew Kinder, Geoff Holmes, Deborah Loffler, Anjana Jaisingh, Jacek Lelon, Manuel Sommer, Mike Schofield, John Riley, Lior Behar, Elena Mulero, and Sylvia Dewis.

References

Bonanno, G. (2004). Loss, trauma and human resilience. *American Psychologist, 59*, 20–28.

Dickman, F. (1988). *Employee assistance programs: A basic text*. Springfield, IL: Charles C. Thomas.

Everly, G. S., Jr., & Mitchell, J. T. (1999). *Critical incident stress management (CISM): A new era and standard of care in crisis intervention* (2nd ed.). Ellicott City. MD: Chevron.

Grange, C. (2005, Summer). The development of employee assistance programmes in the UK: A personal view. *Counselling at Work Journal*, 2–5.

Health Canada. (2011). Employee assistance service. Retrieved from http://www.hc-Sc.gc.ca.

Holmes, G. (2005, Summer). In the business of trauma. *Counselling at Work Journal*, 20–21.

Masi, D. (1987). *Drug free workplace: A guide for supervisors.* Washington, DC: Bureau of National Affairs.

Mitchell, J. T. (1983). When disaster strikes: The critical incident stress debriefing process. *Journal of Emergency Medical Services*, *8*, 36–39.

National Institute for Health and Clinical Excellence (NICE). (2005). *Post-traumatic stress disorder (PTSD): The management of PTSD in adults and children in primary and secondary care.* London: Author.

Rutter, M. (2007, February). From sympathy to empathy: Organisations learn how to respond to trauma. *Counselling at Work*, 9–11.

UK Employee Assistance Professionals Association (UK EAPA). (2010). Welcome to the UK EAP Association. Retrieved from http://www.eapa.org.uk.

20

Training Resilience for High-risk Environments: Towards a Strength-based Approach within the Military

Sylvie Boermans, Roos Delahaij, J.E. (Hans) Korteling, and Martin Euwema

Preface

Stress and resilience are inevitable parts of a soldier's life, particularly in combat or war. Training military personnel to be resilient and capable of coping under high (combat) stress has been core business in most armies. Traditionally, this training was directed at maintaining physical performance under (combat) stress: the ability to fight. Nowadays, growing attention is given to physical, mental, and moral resilience, with a focus on short-term as well as on long-term adjustment in order to prevent PTSD and other stress-related symptoms after deployment.

The twenty-first century has brought new challenges for military organizations, particularly in Military Operations Other Than War (MOOTW). These missions bring new stressors and strains, and require new forms of training. This chapter deals with the question of how peacekeepers can best be prepared to deal with the psychological demands of the operational environment. To this end, the concept of resilience is of special relevance to the military, as well as to other high-risk occupations. We therefore provide an overview of what is currently known about resilience under stressful work environments. We specifically address the combination of *internal* as well as *external* resources for enhancing resilience. We review current evidence-based training and intervention methods to enhance resilience and provide examples of how resilience can be enhanced.

Resilience: Essential for Military Peacekeepers

Peacekeeping is not a soldiers' job, but only a soldier can do it.
UN Secretary-General Dag Hammarskjöld (1954–1961)

International Handbook of Workplace Trauma Support, First Edition. Edited by Rick Hughes, Andrew Kinder, and Cary L. Cooper.

MOOTW have undergone a tremendous metamorphosis and have become increasingly complex and diffuse. Where peacekeepers were initially deployed to post-conflict areas and had a strictly neutral role, today, they are deployed at various stages of a conflict, ranging from low-hostility areas to full-scale combat zones, with or without consent by warring parties and local military groups and warlords. As a result, their classic peacekeeper role has become the exception rather than the rule. Peacekeepers now have to be able to integrate two seemingly competing roles: the role of peacekeeper with the classic role of warrior (Broesder *et al.*, submitted). These developments expose soldiers to a new array of stressful demands that are as much psychological as military or diplomatic.

It is widely accepted that operational demands may negatively affect the well-being of these professionals. Researchers and practitioners have therefore mainly focused on *avoiding* risk factors that have been associated with deployment-related pathology. However, it has recently come to researchers' attention that even though most soldiers face major challenges and stressors, they do not develop mental health problems after deployment (Dickstein *et al.*, 2010). Moreover, the majority look back on their deployment as a positive experience in which they learned a lot about themselves, made friends for life, gained new understanding of personal values and priorities, and gained the opportunity to meaningfully contribute to peace and violence prevention. And for most soldiers, these positive effects outweigh the negative ones (Mouthaan, Euwema, & Weerts, 2005; Newby *et al.*, 2005; Parmak, Euwema, & Mylle, submitted; Schok *et al.*, 2008). These positive responses are attributed to the resilience of these professionals. Insights into these resilient responses are thus important as they offer an alternative pathway to successful adaptation by *strengthening* resilience factors that enable soldiers to successfully deal with operational demands.

In this chapter, we concentrate on the question of whether resilience can be cultivated. We first briefly consider how resilience is conceptualized in the context of the military. We then explore *internal* and *external* resources for resilience and will specifically address the combination of these resources for enhancing resilience. Finally, we review studies of resilience interventions in stressful work environments and discuss currently used intervention paradigms. Although the specific stressors differ, the combination of repeated Critical Incidents (CI) and chronic stressors are also typical for other first responders such as the police, emergency workers, or fire fighters (see, for an overview, Benedek, Fullerton, & Ursano, 2007). We will therefore also focus on other high-risk occupations.

Combining Internal Capacities and External Resources

We begin with defining resilience. In an effort to integrate the rapidly accumulating resilience research, Reich and colleagues (2010) recently published a comprehensive work on it. They concluded that resilience is best defined as the outcome of successful adaptation to hardships. Two equally important components are central to the meaning of resilience: *recovery* and *sustainability*. Recovery focuses on the healing of emotional wounds. It is indicated by the thoroughness and velocity of time needed to return to a former, more balanced level of functioning. This does not mean that a resilient recovery is without its emotional scars, but psychological and behavioral functioning is beyond what may be expected given the circumstances. Sustainability, on the other hand, refers to the capacity to maintain positive engagements with the environment and to maintain positive well-being while meeting the demands of the environment. It moves beyond the mere capacity to maintain competence when exposed to stressful events to also include the sustaining of personal interests in goals that give life meaning and bring feelings of pleasure.

These researchers also concluded that resilience should not be seen as a static or trait-like capacity. Instead, successful adaptation to hardship involves the dynamic interplay between *internal capacities* and *external resources*. Internal capacities are natural character strengths someone possesses that enhance positive psychological functioning. External resources describe those aspects of the social environment that empower individuals' capacity to respond in a positive way to adversity. This two-dimensional approach is important for developing resilience, as we will discuss in this chapter; interventions can and should be directed at strengthening internal capacities as well as environmental resources.

An important insight of resilience research is that the presence of positive affective states is not the same as the absence of negative affective states; both can coexist. Moreover, research shows that positive emotional engagements buffer against the negative effects of stress on well-being and health. Experiencing positive emotions after a stressful event for instance, accelerates physiological recovery and buffers against the development of depression (Frederickson *et al.*, 2003; Zautra, Johnson, & Davis, 2005). And individuals who derive a high sense of meaning from their work are less burdened by high job demands as compared to those who are cynical about their work (e.g., Bakker *et al.*, 2007; Britt & Bliese, 2003).

Recognizing the relevance of resilience for the well-being of military personnel and mission success, the concept of resilience has grabbed the attention of the military organization. Based on the work by Reich and colleagues (2010), we use the concepts of recovery and sustainability to define military resilience. We include the ability to maintain optimal performance during an acute stress situation, as this is a crucial aspect of military work. The capacity to *sustain* combat motivation and a sense of being able to meaningfully contribute to the mission are especially relevant when confronted with violence by the local people, continually changing rules of engagement, or boredom. Recovery is of vital importance during deployment as there is evidently the potential of being exposed to repeated CI's. Being able to swiftly return to an optimal level of functioning is pivotal when confronted with the next CI. Growth is especially important when confronted with significant setbacks and/or when coming to terms with deployment experiences. Hence, we define "military resilience" as ***the ability to maintain optimal performance during acute situations, positively recover afterward, and sustain combat motivation while meeting the demands of operational demands.***

In the next paragraphs, we consider internal attributes and external resources that enable soldiers to respond with resilience.

Internal capacities

We begin by describing personal capacities that have been linked with resilience. Indeed, some people just seem to be better able to sustain psychological and behavioral functioning when faced by challenging and demanding situations, and to recover afterward. Researchers have started to investigate positive individual capacities, and a clear picture is emerging as to how resilient soldiers are characterized.

Self-confidence, optimism, and perceived control. Personal attributes that have repeatedly proven their value for military resilience are self-confidence, optimism, and feelings of control (e.g. Bartone, 1999; Gilbar, Ben-Zur, & Lubin, 2010; King *et al.*, 1998; Pietrzak *et al.*, 2010). First of all, these cognitive attributes enable soldiers to maintain optimal performance during acute situations because they have confidence in their skills to control the situational demands (Schok, Kleber, & Lensvelt-Mulders, 2010). Feelings of control and confidence empower one to take action. As such, resilient soldiers proactively face difficulties with courage and perseverance and do not give up when faced with failure. They efficiently down-regulate negative affect, enabling them to stay focused on their task and swiftly and effectively take action to get control over the situation. Successful mastery in turn strengthens resilience as it enhances confidence in one's capabilities, creating a positive feedback loop (Benight & Bandura, 2004).

Secondly, these attributes enable soldiers to sustain positive affective engagements during stressful times. The most powerful way through which self-confidence, optimism, and feelings of control seem to cultivate positive affective engagements is through construing positive meaning from adversity. By positively reframing difficulties and using humor, soldiers are able to cope with operational stressors and maintain combat motivation (Riolli & Savicki, 2010). Indeed, Britt, Adler, & Bartone (2001) showed that soldiers who are characterized by these attributes perceive their deployment work as more meaningful than soldiers who do not possess these attributes.

Finally, these attributes are also related to recovering from adversity (Britt *et al.*, 2001; Schok *et al.*, 2010). The capacity to construe positive meaning from difficulties is crucial for restoring or even improving psychological capacities after exposure to a critical incident as well as long-term adjustment in coming to terms with deployment experiences. By finding personal relevance, soldiers are able to restore their self-esteem, regain a sense of mastery, and maintain a positive worldview and optimistic outlook on life. As a result, they are able to continue or resume active coping to overcome difficulties and facilitate connectedness with others and the world.

Interpersonal skills. Military work involves close coordination and team efforts to achieve mission objectives. As such, interpersonal conflicts are seen as an especially debilitating stressor of soldiers' resilience. When soldiers are confronted with high pressure to perform, even a minor argument among soldiers can have a critical impact on team performance. Possessing strong interpersonal and communication skills that are necessary for effective teamwork is therefore an important resilience capacity. In addition, these skills are also important for promoting access to social support in times of stress (Skodol, 2010).

Physical fitness. Physical fitness has always been crucial for operational effectiveness to sustain performance in physically demanding environments. Besides the importance of physical fitness for sustaining optimal performance, it has also been positively related to mood and self-confidence and has been linked to neurobiological effects that promote resilience (Cotman & Berchtold, 2002). Taylor *et al.* (2008), for instance, showed that physical fitness buffered against the psychological impact of a stressful mock captivity exercise during US military survival training.

External resources

As mentioned, the qualities of the person alone are not sufficient to predict resilience but also depend on empowering external resources. Especially the social environment can provide psychosocial resources that enhance soldiers' resilience. In the next paragraphs, we discuss the role of the team, leadership, family, and organization as psychosocial resources for individual resilience.

Team aspects. In MOOTW, military teams operate while dispersed over relatively large areas and may rapidly switch locations. Several teams work together to achieve a shared goal in which every team is assigned a specific function. As such, the ability to work and live together as a team is crucial for operational effectiveness, individual survival, and the maintaining of personal well-being.

Morale and unit cohesion. A crucial component of military resilience that is specifically related to the team is morale. Morale refers to a solider's enthusiasm for and persistence in working towards the goals and tasks of his or her team (Manning, 1991). As such, military organizations agree that it is the driving force for obtaining mission success. Indeed, research has related morale with higher levels of operational performance (Britt & Dickinson, 2006), putting in extra job efforts, organizational commitment, and combat readiness (Boxmeer *et al.*, 2010), and with finding more benefits after deployment (Britt *et al.*, 2007). In addition, morale has been found to buffer against the negative effects of work-related stressors on work–family conflicts (Britt *et al.*, 2005) and the development of duty-related PTSD after deployment (Iverson *et al.*, 2008).

A prerequisite for morale is team cohesion (Boxmeer *et al.*, 2010). First of all, cohesion provides soldiers with a shared reality enabling them to make sense of their experiences and sustain meaningful engagements. Shared experiences of threats in combat, fraternal comradeships, team optimism, encouragement, and good humor boost morale (Mouthaan, Euwema, & Weerts, 2005). Another way in which cohesion contributes to morale is through team performance. Cohesive teams are characterized by trust and teamwork which provide soldiers with confidence in their personal capabilities and joint team efforts to successfully deal with situational demands, in turn enhancing morale and team performance (Chen *et al.*, 2009; Jex & Bliese, 1999; Stetz, Stetz, & Bliese, 2006).

Conversely, a lack of cohesion may enhance psychological strain and decrease morale (Boxmeer *et al.*, 2010; Britt & Bliese, 2001; Britt *et al.*, 2007). When faced with an acute-stress situation, situational demands may become insurmountable and compromise problem-focused coping abilities. Reflecting on the situation, members may become disillusioned about the team's abilities, making them vulnerable to the development of psychopathology. Strengthening the team is therefore an important way for preventing combat breakdown and enhancing positive psychological adaptation.

Leadership. The importance of leaders for soldiers' motivation and performance is widely accepted within the military organization. Leaders directly influence the resilience of their soldiers by addressing their physical needs by providing them with good equipment and living conditions. In addition, deployments are often characterized by highly unstructured tasks or uncertain relationships with corroborating parties, forming a risk for maintaining morale and operational effectiveness. Leaders can buffer these negative effects by being attentive to interpersonal and morale-related team issues. Indeed, team members who feel supported by their leader also report less interpersonal conflicts (e.g., Bliese & Britt, 2002; Bliese & Halverson, 1998; Cole & Bedeian, 2007; Griffith, 2002).

Leaders also play a key role in meaning-making processes. They are the primary source for interpreting information and, as such, have a strong influence on making sense of stressful experiences (Bartone, 2006; Britt *et al.*, 2004). During a CI, leaders have to interpret the situation and make decisions in the pursuit of the desired goals. Leaders can help their team to develop optimistic outlooks by emphasizing the collective responsibilities of team members for the safety and well-being of others and by conveying the proper utilization of their resources and desired outcomes.

Finally, Leaders can sort their positive effects by facilitating group processes. They can provide their team with specific learning opportunities and feedback, and by encouraging and coaching them about the use of knowledge and skills. By emphasizing the shared values in the team and showing that they have trust in their capabilities, leaders empower members' confidence in both their personal capabilities and their group capabilities (Shamir *et al.*, 1998, 2000).

Organization. The military organization directly influences soldiers' resilience through its policies and training programs. Soldiers who feel well prepared for their deployment task are better able to deal with operational demands (Bartone, 2006; Gilbar, Ben-Zur, & Lubin, 2010; Renshaw, 2011; Shamir *et al.*, 1998). The organization can provide soldiers with a sense of purpose and meaningfulness by expressing mission objectives, and by providing task directions and priorities of assignments (Siebold, 2007).

Organizational culture influences military resilience by determining the accepted ways of coping. Dolan and Ender (2008), for instance, noted that among US soldiers, drinking and seeking social support are widely accepted strategies to cope with stress. Ben-Ari (1998) observed that controlling emotions is central to officers' identities in the Israeli Defense Forces. Likewise, Le Scanff and Taugis (2002) identified an organizational norm within the police force that made employees refrain from showing or admitting fear or anxiety, because this was perceived as weak. Thus, some emotion-focused coping strategies, such as venting of emotions, seem to be less accepted in organizations like the military and police force.

Organizational resources also refer to the availability and quality of instrumental resources that are crucial for their physical safety. Consider, for example, the effect of a lack of air support for military teams in hostile areas, or the lack of proper vehicle protection against improvised explosive devices (IEDs) on soldiers' confidence in their ability to manage stressful and threatening situations.

Family support. Soldiers are often deployed to remote locations and separated from their families for long periods of time. Besides the difficulty of missing loved ones, family members themselves can be very distressed, and soldiers are often concerned that their families worry about their well-being. Strong and cohesive families promote soldiers' resilience because they possess shared and empowering beliefs that facilitate options for problem resolution, healing, and growth. This enables soldiers to sustain combat motivation and facilitates their recovery in the aftermath of deployment (King *et al.*, 1998; Pietrzak *et al.*, 2010; Walsh *et al.*, 2003).

The importance of family cohesion was demonstrated in an elegant longitudinal study by Benotsch *et al.* (2000). They directly investigated the effects of internal attributes and family cohesion on the development of PTSD symptoms after deployment to the Gulf War. Poor coping skills and low family cohesion before deployment predicted PTSD symptoms after deployment. Moreover, they also showed that soldiers with already high levels of PTSD symptoms before deployment even amplified the use of poor coping skills and decreased family cohesion.

State-of-the-Art Interventions

Military organizations are currently developing preventive interventions that are explicitly based on a strength-based approach to positive adaptation. The US Army, for instance, has recently initiated their Comprehensive Soldier Fitness program; Australia recently launched their BattleSMART (Self Management and Resilience Training); the UK Army uses a peer support system Trauma Risk Management (i.e. TRiM) to empower easy access to social support; and the Dutch defense provides in-theater leadership advice on how to enhance or sustain resilience. This part of the chapter considers the interventions that have been designed to enhance resilience.

To determine the effectiveness of intervention approaches and gain insights into which resilience resources offer promising targets for intervention, we performed a systematic literature search on evidence-based resilience interventions. This yielded a total of 19 effect studies within the police and military domain. As can be seen from Table 20.1 and Table 20.2, we distinguished interventions that were designed to strengthen personal attributes from interventions that were designed to strenghten external resources. In the following, we first consider the methods that are used to train resilience. Next, we discuss the resources of resilience that have been targeted by these interventions. We also describe four existing military training programs that illustrate ways to promote resilience within the military.

Strength-based intervention

The concept of a strength-based approach has become a popular term in everyday discourse. As such, it becomes more important to clarify what it actually means. A strength-based approach aims to capitalize on strengths and resources that someone already possesses. A strength-based approach does not avoid risk or problems, but shifts the attention toward identifying what works for an individual to effectively deal with difficulties.

We identified three commonly used methods to train resilience: (1) a cognitive or knowledge-based approach to training, (2) a purely practice-based approach, and

Table 20.1 Effect studies on enhancing personal capacities

Method/Phase	*Intervention*	*Targeted resilience resources*	*Effects*	*Authors*
Cognitive-based approach *Recruit training*	Large-group counseling	- Recognizing and verbalizing psychological difficulties - Enhancing psychological safety	Feelings anger decreased; feelings of pleasantness increased	Rocco *et al.* (1975)
	Stress management discussion group	- Expectation management - Realistic appraisal and adaptive coping strategies	Positive state of mind increased; distress decreased	Cohn *et al.* (2008)
Pre-deployment training	Pre-deployment stress debriefing	- Stress awareness and stress reduction - Enhancing morale - Accessing external resources	No clear effects	Sharpley *et al.* (2007)
During deployment	Critical Incident Stress Debriefing	- Stress awareness - Adaptive coping strategies - Fostering emotional sharing	Positive effects only for those with high stress exposure	Adler *et al.* (2008b)
	Stress management education group	- Stress awareness - Adaptive coping strategies	No clear effects	Adler *et al.* (2008a)
Post-deployment	Critical Incident Stress Debriefing	- Stress awareness - Adaptive coping strategies	No clear effects	Adler *et al.* (2009)
	Battlemind (debriefing and training)	- Stress awareness - Recognizing psychological difficulties - Enhancing interpersonal trust - Accessing external resources - Positive reframing and adaptive coping strategies	Positive effects only for those with high stress exposure	Adler *et al.* (2009)

During regular operational work	Internet-based self-help training	- Stress awareness and stress reduction - Healthy lifestyle - Positive reframing and adaptive coping strategies - Interpersonal skills	Stress decreased	Williams *et al.* (2010)
	Stress management education group	- Stress awareness and stress reduction - Healthy lifestyle	Anxiety decreased	Le Scanff and Taugis (2002)
Cognitive- and practice-based approach	Stress Inoculation Training (SIT)	- Stress tolerance - Adaptive coping strategies	State and situational anxiety decrease performance increased under stress	Saunders *et al.* (1996)
Recruit training				
	Mental imagery training*	- Stress awareness and stress reduction - Adaptive coping strategies	Performance increased; negative mood and stress decreased; no effect on positive mood	Backman *et al.* (1997) Arnetz *et al.* (2009)
Pre-deployment training	Mindfulness training	- Cognitive control - Negative emotion regulation	Working memory capacity increased, but only for those who practiced a lot in mindfulness exercises	Jha *et al.* (2010)
During regular operational work	Integrative Training of Emotional Competencies (iTEC)	- Emotional awareness - Emotion regulation	Acceptance and tolerance of negative emotions increase; no effect on negative affect; positive affect increased	Berking, Meier, and Wupperman (2010)
Practice-based intervention	Graduated training: performance under gradual increase of realistic stressors	- Self-efficacy	Performance and psychological functioning increased	Zach *et al.* (2007)

(*Continued*)

Table 20.1 (*Continued*)

Method/Phase	*Intervention*	*Targeted resilience resources*	*Effects*	*Authors*
Recruit training		- Commitment - Challenge appraisal - Sense of control - Adaptive coping strategies		
Eye movement desensitizing reprocessing (EMDR)	EMDR training	- Cognitive processing of emotions	Psychological difficulties decreased; no effect on general psychological functioning	Wilson *et al.* (2001)
During regular operational work				

*Two studies.

(3) a combination of cognitive- and practice-based approaches to training. Cognitive or knowledge-based interventions aim at enhancing awareness and attitudes by providing information on a certain topic. Examples are briefs, discussions, and computer-based training. Skill-based interventions only use actual experience as a learning method. It is important to note that "skills" refer to not only behavioral skills but also mental skills (e. g., meditation and positive reframing). It is based on the idea that practicing in a real or simulated setting allows trainees to develop and integrate skills into their existing set of capacities. Skill-based methods are for example games and simulations, behavior modeling, case studies, role-playing, or cognitive exercises. Finally, some interventions used a combination of cognitive- and skill-based methods.

Strengthening internal capacities

The majority of the effect studies concerned interventions that targeted personal resilience resources (Table 20.1). Most of these interventions used a cognitive-based approach and were implemented at all stages during the career cycle. The most widely applied intervention method in high-risk occupations is "group psychological debriefing." An alternative approach that is gaining interest in military organizations is interventions that are cognitive and skills-based. One study was purely based on the practice of skills under increasing level of stressors. Finally, one study was found that investigated the effectiveness of behavior-based training and eye movement desensitization and reprocessing (EMDR)–based training. This intervention aimed to decrease daily work stress that results from inadequately processed emotional experiences.

Strength-based approach. Most interventions focused on enhancing awareness of stress and providing strategies that have been shown to ameliorate stress. Down-regulation of stress or enhancing stress resistance is of course of vital importance for optimal performance during critical incidents, but, as mentioned, resilience is not only characterized by the absence of stress or stress tolerance: it also involves positive meaningful experiences. Positive reframing and interpersonal skills were the most targeted resilience capacities. Focusing on goal-directed skills is important as it provides an important way to enhance resilience, especially when people are "stuck" in patterns of maladaptive functioning. However, the promotion of positive goals and outcomes does not seem to be a reflected or explicit target in a stress management approach to resilience. Only one psychological debriefing was found that was specifically based on insights from resilience research (Adler *et al.*, 2009). This briefing is therefore considered in more detail in Box 20.1. One cognitive and practice-based intervention was found that specifically used a strength-based approach to resilience (Jha *et al.*, 2010). To provide an example of this type of intervention, "Resilience XL" is considered in Box 20.2.

Effectiveness. Except for cognitive-based stress debriefing, all interventions positively affected different aspects of resilience capacities ranging from decreased stress to enhanced working memory capacity to enhanced psychological hardiness and performance. This is promising as it indicates that individual resilience capacities can indeed be cultivated. Given the paucity of effect studies, it remains difficult to draw conclusions

Box 20.1 Battlemind Debriefing

When returning from deployment, soldiers have to come to terms with deployment experience and transition back into their home environment. Soldiers have to adapt to new work duties, family life, and coming to terms with difficult deployment experiences. Post-deployment Battlemind Debriefings aim at enhancing transition from deployment to home. Battlemind has recently been developed by the US Army and refers to "the soldier's inner strength to face fear and adversity with courage" (US Army, 2008), which is composed of self-confidence and mental toughness. Battlemind Debriefings do not focus on traumatic events but specifically focus on positive adaptation. Psychological transition difficulties are positively reframed as a natural consequence of having developed effective coping skills related to deployment, and debriefings focus on how these skills can be adapted (Adler *et al.*, 2009).

about which method is best for enhancing soldiers' resilience capacities. The effectiveness of a cognitive-based approach seems to be most controversial. Some scholars have even suggested that debriefing may have a detrimental effect on the recovery processes (Bonanno, 2004; Emmerik *et al.*, 2002). These findings also suggest that actual practice might be a crucial aspect for training resilience capacities. This is supported by a study by Jha *et al.* (2010), who found that their intervention was only effective when soldiers practiced more often. As such, they concluded that practice time was the critical determinant for skill development. A meta-analysis on the SIT also showed that skills development improves as the number of training sessions increased (Saunders, Driskell, & Salas, 1996). Moreover, this study showed that knowledge and practice have effects on different aspects of resilience and are therefore equally important for the development of training protocols. More research needs to be done when developing training protocols.

Individual versus group intervention. All interventions used a collaborative training protocol, except for the EMDR intervention. This is based on the idea that group training facilitates individual learning processes, highlighting the importance of feedback as a key aspect for learning. However, in reviewing the intervention methods, facilitating interaction between participants was not a focus of the training. Although most interventions provide information on the importance of social support and interpersonal skills, it seems that interventions may enhance resilience even more by capitalizing on group interactions.

Strengthening external resources

Only three effect studies were found that targeted external resources (Table 20.2). All were cognitive- and practice-based interventions. Two of these studies aimed at developing effective leadership, and one intervention was aimed at enhancing peer support.

Supporting research findings on the importance of leadership for military resilience, leadership training indeed proved to be an effective way to increase personal resources. However, most military organizations do not provide their military leaders with

Table 20.2 Effect studies on strengthening environmental resources

Resources	*Intervention*	*Effects on resources*	*Effects on individual resilience*	*Authors*
Leaders	Transformational Leadership Training*	Leaders had more acceptance of group goals, appreciation of teamwork, reward contingency, and individual consideration	Self-efficacy increased among direct followers; performance increased among indirect followers; lower attrition rates	Dvir and Shamir (2003) Hardy *et al.* (2010)
Peers	Trauma Risk Management (TRiM)	Peers had more recognition of psychological difficulties, and offered more social support	No effect on psychological health to those exposed to a critical incident or on mental health stigma	Frappel-Cooke *et al.* (2010)

*Two studies.

Box 20.2 The Resilience XL Program

TNO, the Netherlands' organization for applied scientific research, recently developed the Resilience XL program for Navy recruits to enhance positive adaptation during recruit training. Resilience XL training comprises a cognitive-behavioral approach that covers cognitive, behavioral, and interpersonal aspects of functioning. Each aspect covers specific skills that have been shown to enhance resilience.

The program is integrated with Navy basic military training. At the beginning of basic training, established groups of recruits participate in an interactive 1-day workshop to increase understanding and awareness on resilience. As one of the main reasons for quitting basic Navy training is that basic training is not what recruits expected, there is a focus on expectation management as a way of dealing with the realities of basic training. Recruits discuss current expectations and are shown a short video of students from previous recruit training who share their own experience of the training. Recruits are also encouraged to actively manage their expectations by asking questions and supporting other members in doing so.

In addition, special attention is given to group processes. During a group-discussion, recruits consider the topics "coping with difficult situations," "supporting group members," "instructor responsibilities," and "recruit responsibilities." These discussions enable groups to develop a shared language to talk about difficult topics, and facilitate the access to social support.

Stressful exercises during Navy basic training allow recruits to directly practice newly learned skills. Instructors explicitly encourage recruits to reflect upon their experiences and actively try to use new coping strategies. Finally, recruits participate in two "reinforce" sessions of 2 hours. In these sessions, recruits are encouraged to share and reflect upon their past experiences. These "reinforce" moments are planned in a period that is known to be stressful, to enable recruits to reflect on immediate experiences.

training on how to manage the stressors of their team members, or enhance resilience in a structural way (Adler *et al.*, 2008a). Training leaders how to facilitate group processes and cohesion therefore seems highly fruitful. An example of an intervention that aims to raise leadership awareness about levels of moral and cohesion is described in Box 20.3.

The fact that we found only three effect studies highlights the need for more research on interventions that target the external resources of the individual. Moreover, new interventions need to be developed that use the full range of external resources.

Box 20.3 Morale Monitor of Dutch Military Teams

In the Netherlands' army, military leaders work closely together with psychological support professionals to enhance and maintain morale and to capitalize on strengths within teams during the deployment cycle. To this end, the Defense Services Center Behavioral Sciences developed a practical measure that provides specific in-time information to military leaders concerning the level of morale of their team.

The measure assesses resources that have been shown to be important for morale at the level of the individual, the unit, the organization, and leadership. Resources at the level of the individual are self-confidence, job-satisfaction, and home front support. Resources at the level of the unit are identification with the unit, cohesion, and respect for each other. And, finally, resources at the level of the organization are weapons and equipment, operational support, familiarity with the mission and terrain, living conditions, and communication with the home front. The measure also assesses a set of stressors that are specific for the environment. This way, morale and distress are assessed separately, making it possible to capitalize on strengths and to address signs of psychological distress.

Morale and distress are structurally assessed and analyzed at the level of the unit during pre-deployment training and during deployment. The results of the assessments are immediately fed back to the commanding officer. The scores on the different resources determine the advice that is given on how to maintain or boost morale. By measuring psychological distress and potential stressors, it is possible to detect psychological distress and its causes in specific units across the deployment cycle. Officers recognize their best performing and most cohesive teams, and the outcomes of the assessment. Most relevant in this respect is that it seems that capitalizing on or addressing unit-level resources by investing in the quality of leadership may be highly effective and efficient to promote individual and group well-being.

For instance, although most interventions stress the importance of social support and team cohesion, except for TRiM, we found no mention of research on interventions that explicitly aim at increasing collective team resilience, enhance organizational processes, or aims to increase family resilience.

Conclusion

In this chapter, we introduced the concept of resilience as especially relevant for soldiers as they nowadays operate in cumulative stressful environments. Resilience is different from traditional approaches to building, maintaining, and restoring soldiers' adaptation capabilities, because it focuses on positive adjustment besides the absence of pathology after a potentially traumatic event. We introduced the definition of military resilience, the ability to maintain optimal performance during acute situations, positively recover afterward, and sustain combat motivation under chronic stressful circumstances. Whether a soldier is resilient depends on his or her available internal and external resources. Several internal and external resources have been identified. However, not many studies have investigated the interplay between internal and external resources. More knowledge on the combined effects of these resources could provide valuable insights in how to best enhance military resilience.

We discussed existing resilience interventions for personnel in high-risk occupations. Most interventions focused on individual resources and were based on cognitive principles that aim to enhance awareness and knowledge that will enable a soldier to better cope with stressful situations. Although these interventions addressed the positive adaptation perspective of resilience, the full range of resources has yet to be capitalized upon. In addition, only a few interventions explicitly aim to enhance resilience through external resources. Future interventions should include the positive adaptation perspective and address external resources to enhance effectiveness of resilience interventions.

References

Adler, A. B., Bliese, P. D., McGurk, D., Hoge, C. W., & Castro, C. A. (2009). Battlemind debriefing and battlemind training as early interventions with soldiers returning from Iraq: Randomization by platoon. *Journal of Consulting and Clinical Psychology*, *77*, 928–940.

Adler, A. B., Cawkill, P., Berg, C. van den., Avers, P., Puente, J., & Cuvelier, Y. (2008a). International military leaders' survey on operational stress. *Military Medicine*, *173*, 10–16.

Adler, A. B., Litz, B. T., Castro, C. A., Suvak, M., Thoma, J. L., & Burrell, L. (2008b). A group randomized trial of critical incident stress debriefing provided to U. S. peacekeepers. *Journal of Traumatic Stress*, *21*, 253–263.

Arnetz, B. B., Nevedal, D. C., Lumley, M. A., Backman, L., & Lublin, A. (2009). Trauma resilience training for police: Psychophysiological and performance effects. *Journal of Police and Criminal Psychology*, *24*, 1–9.

Bakker, A. B., Hakanen, J. J., Demerouti, E., & Xanthaloupou, D. (2007). Job resources boost work engagement, particularly when job demands are high. *Journal of Educational Psychology*. *99*, 274–284.

Bartone, P. T. (1999). Hardiness protects against war-related stress in army reserve forces. *Consulting Psychology Journal: Practice and Research, 51,* 72–82.

Bartone, P. T. (2006). Resilience under military operational stress: Can leaders influence hardiness? *Military Psychology, 16,* 131–148.

Ben-Ari, E. (1998). *Mastering soldiers: Conflicts, emotions, and the enemy in an Israeli military unit.* New York: Berghahn.

Benedek, D. M., Fullerton, C., & Ursano, R. J. (2007). First responders: Mental health consequences of natural and human-made disasters for public health and public safety workers. *Annual Review of Public Health, 28,* 55–68.

Benight, C. C., & Bandura, A. (2004). Social cognitive theory of posttraumatic recovery: The role of perceived self-efficacy. *Behaviour Research and Therapy, 42,* 1129–1148.

Benotsch, E. G., Brailey, K., Vasterling, J. J., Uddo, M., Constans, J. I., & Sutker, P. B. (2000). War zone stress, personal and environmental resources, and PTSD symptoms in Gulf War veterans: A longitudinal perspective. *Journal of Abnormal Psychology, 109,* 205–213.

Berking, M., Meier, C., & Wupperman, P. (2010). Enhancing emotion-regulation skills in police officers: Results of a pilot controlled study. *Behavior Therapy, 41,* 329–339.

Boxmeer, F., van, Verwijs, C., Euwema, M., & Dalenberg, S. (2010). Assessing morale and psychological distress during modern military operations. *Unpublished manuscript.*

Britt, T. W., Adler, A. B., & Bartone, P. T. (2001). Deriving benefits from stressful events: The role of engagement in meaningful work and hardiness. *Journal of Occupational Health Psychology, 6,* 53–63.

Britt, T. W., & Bliese, P. D. (2003). Testing the stress-buffering effects of self engagement among soldiers on a military operation. *Journal of Personality, 71*(2), 245–265.

Britt, T. W., Davison, J., Bliese, P. D., & Castro, C. A. (2004). How leaders can influence the impact that stressors have on soldiers. *Military Medicine, 169,* 541–545.

Britt, T. W., Dickinson, J. M., Moore, D., Castro, C. A., & Adler, A. B. (2007). Correlates and consequences of morale versus depression under stressful conditions. *Journal of Occupational Health Psychology, 12,* 34–47.

Broesder, W., Vogelaar, A., Euwema, M., & op den Buijs, T.(Submitted). The peacekeeping warrior.

Cole, M. S., & Bedeian, A. G. (2007). Leadership consensus as a cross-level contextual moderator of emotional exhaustion–work commitment relationship. *The Leadership Quarterly, 18,* 447–462.

Cotman, C. W., & Berchtold, N. C. (2002). Exercise: A behavioral intervention to enhance brain health and plasticity. *Trends in Neuroscience, 25,* 295–301.

Dickstein, B. D., Suvak, M., Litz, B. T., & Adler, A. B. (2010). Heterogeneity in the course of post-traumatic stress disorder: Trajectories of symptomatology. *Journal of Traumatic Stress, 23,* 331–339.

Dolan, C. A., & Ender, M. G. (2008). The coping paradox: Work, stress and coping in the U.S. *Army. Military Psychology, 20,* 151–169.

Dvir, T., & Shamir. B. (2003). Follower developmental characteristics as predicting transformational leadership: A longitudinal field study. *The Leadership Quarterly, 14,* 327–344.

Emmerik, A. A. P., Kamphuis, J. H., Hulsbosch, A. M., & Emmelkamp, P. M. G. (2002). Single session debriefing after psychological trauma: A meta-analysis. *The Lancet, 360,* 766–771.

Frappel-Cooke, W., Gulina, M., & Green, K. (2010). Does Trauma Risk Management reduce psychological distress in deployed troops? *Occupational Medicine (London), 60,* 645–650.

Frederickson, B. L., Tugade, M., Waugh, C. E., & Larkin, G. R. (2003). What good are positive emotions in crises? A prospective study of resilience and emotions following the terrorist attacks in the United States on September 11th, 2001. *Personality Processes and Individual Differences, 84,* 365–376.

Gilbar, O., Ben Zur, H., & Lubin, G. (2010). Coping, mastery, stress appraisals, mental preparation, and unit cohesion predicting distress and performance: A longitudinal study of soldiers undertaking evacuation tasks. *Anxiety Stress Coping, 23,* 547–562.

Griffith, J. (2002). Multilevel analysis of cohesion's relation to stress, well-being, identification, disintegration, and perceived combat readiness. *Military Psychology, 14,* 217–239.

Hardy, L., Arthur, C. A., Jones, G., Shariff, A., Munnoch, K., Isaacs, I. *et al.* (2010). The relationship between transformational leadership behaviors, psychological, and training outcomes in elite military recruits. *The Leadership Quarterly, 21,* 20–32.

Jex, S. M., & Bliese, P. D. (1999). Efficacy beliefs as a moderator of the impact of work-related stressors: A multilevel study. *Journal of Applied Psychology, 84,* 349–361.

Jha, A. P., Stanley, E. A., Kiyonaga, A., Wong, L., & Gelfand, L. (2010). Examining the protective effects of mindfulness training on working memory capacity and affective experience. *Emotion, 10,* 54–64.

King, L. A., King, D. W., Fairbank, J. A., Keane, T. M., & Adams, G. A. (1998). Resilience–recovery factors in post-traumatic stress disorder among female and male Vietnam veterans: Hardiness, postwar social support and additional life events. *Journal of Personality and Social Psychology, 74,* 420–424.

Le Scanff, C., & Taugis, J. (2002). Stress management for police special forces. *Journal of Applied Sport Psychology, 14,* 330–343.

Mouthaan, J., Euwema, M. C., & Weerts, J. (2005). Band of brothers in United Nations peacekeeping: Social bonding among Dutch peacekeeping veterans. *Military Psychology, 17,* 101–114.

Newby, J. H., McCarroll, J. E., Ursano, R. J., Fan, Z., Shigemura, J., & Tucker-Harris, Y. (2005). Positive and negative consequences of deployment. *Military Medicine, 170,* 815–819.

Parmak, M., Euwema, M. C., & Mylle, J. J. C. (Submitted). Situational adaptation: Soldiers' behavioural tendencies modify during a combat deployment.

Pietrzak, R. H., Johnson, D. C., Goldstein, M. B., Malley, J. C., Rivers, A. J., Morgan, C. A., *et al.* (2010). Psychosocial buffers of traumatic stress, depressive symptoms, and psychosocial difficulties in veterans of Operations Enduring Freedom and Iraqi Freedom: The role of resilience, unit support, and post-deployment social support. *Journal of Affective Disorders, 120,* 188–192.

Reich, J. W., Zautra, J. W., & Hall, J. S. (2010). *Handbook of adult resilience.* New York, NY: Guilford Press.

Renshaw, K. D. (2011). An integrated model of risk and protective factors for post-deployment PTSD symptoms in OEF/OIF era combat veterans. *Journal of Affective Disorders, 128,* 321–326.

Riolli, L., & Savicki, V. (2010). Coping effectiveness and coping diversity under traumatic stress. *International Journal of Stress Management, 17,* 97–113.

Saunders, T., Driskell, J. E., Johnston, J. H., & Salas, E. (1996). The effect of stress inoculation training on anxiety and performance. *Journal of Occupational Health Psychology, 1,* 170–186.

Schok, M. L., Kleber, R. J., & Boeije, H. R. (2010). Men with a mission: Veterans' meanings of peacekeeping in Cambodia. *Journal of Loss and Trauma, 15,* 279–303.

Schok, M. L., Kleber, R. J., Elands, M., & Weerts, J. M. P. (2008). Meaning as a mission: A review of empirical studies on appraisals of war and peacekeeping experiences. *Clinical Psychology Review, 28,* 357–365.

Shamir, B., Brainin, E., Zakay, E., & Popper, M. (2000). Perceived combat readiness as collective efficacy: Individual- and group-level analysis. *Military Psychology, 12*, 105–119.

Shamir, B., Zakay, E., Brainin, E., & Popper, M. (1998). Correlates of charismatic leader behavior in military units: Subordinates' attitudes, unit characteristics, and superior appraisals of leader performance. *Academy of Management Journal, 41*, 387–409.

Sharpley, J. G., Fear, N. T., Greenberg, N., Jones, M., & Wessely, S. (2008). Pre-deployment stress briefing: does it have an effect? *Occupational Medicine, 58*, 30–34.

Skodol, A. E. (2010). A resilient personality. In J. W. Reich, A. J. Zautra, & J. S. Hall (Eds.), *Handbook of adult resilience* (pp. 112–125). New York, NY: Guilford Press.

Stetz, T. A., Stetz, M. C., & Bliese, P. D. (2006). The importance of self-efficacy in the moderating effects of social support on stressor-strain relationships. *Work and Stress, 20*, 49–59.

Taylor, M. K., Markham, A. E., Reis, J. P., Padilla, G. A., Potterat, E. G., Drummond, S. P. A., *et al.* (2008). Physical fitness influences stress reactions to extreme military training. *Military Medicine, 173*, 738–742.

Walsch, F. (2002). A family resilience framework: Innovative practice applications. *Family Relations, 51*, 130–137.

Williams, R. A., Hagerty, B. M., Brasington, S. J., Clem, J. B., & Williams, D. A. (2010). Stress gym: Feasibility of deploying a web-enhanced behavioral self-management program for stress in a military setting. *Military Medicine, 175*, 487–493.

Wilson, S. A., Tinker, R. H., Becker, L. A., & Logan, C. R. (2001). Stress management with law enforcement personnel: A controlled outcome study of EMDR versus a traditional stress management program. *International Journal of Stress Management, 8*, 179–200.

Zach, S., Raviv, S., & Inbar, R. (2007). The benefits of a graduated training program for security officers on physical performance in stressful situations. *International Journal of Stress Management, 14*, 350–369.

Zautra, A., Johnson, L., & Davis, M. (2005). Positive affect as a source of resilience for women in chronic pain. *Journal of Consulting and Clinical Psychology, 73*, 212–220.

Part E

The Organizational Response to Trauma Support

21

Preparing for and Managing Trauma within Organizations: How to Rehabilitate Employees Back to Work

Andrew Kinder and Jo Rick

Introduction

The focus of this chapter is to explore what an organization needs to put in place to respond to traumatic incidents and outlines the role of the line manager as well as human resources (HR) and occupational health (OH). This chapter is aimed at organizations where frequent traumatic incidents occur, although it is equally applicable to organizations where there is an unexpected trauma (e.g., a road traffic accident, witnessing an incident such as a fatality or the sudden death of a colleague, or witnessing a terrorist atrocity). Within this context, the definition of a traumatic event is where an individual has experienced a threat to his or her life or physical well-being or has suffered extreme emotional disturbance due to witnessing an incident that involves death or severe threat to another person's life.

The chapter identifies and draws on existing frameworks for risk management and rehabilitation that can be applied to the management and rehabilitation of employees following a traumatic experience.

Well-Being and Work

The workplace is important to us not just economically but also for our psychological and emotional well-being. It is often said that we spend most of our time at work, more time than for leisure or for domestic activities. A relatively recent review of the relationship between work and well-being (Waddell and Burton, 2006) highlighted that, in general, "Work is good for you." On the other side of the coin, worklessness or

International Handbook of Workplace Trauma Support, First Edition. Edited by Rick Hughes, Andrew Kinder, and Cary L. Cooper.

unemployment is certainly not good for you and is associated with various mental health conditions and lower quality of life.

Waddell and Burton (2006) summarize the evidence in their review as follows:

> There is a strong evidence base showing that work is generally good for physical and mental health and well-being . . . the beneficial effects of work outweigh the risks of work, and are greater than the harmful effects of long-term unemployment or prolonged sickness absence. Work is generally good for health and well-being.

Work, therefore, plays an essential part in promoting health and well-being, and the overwhelming majority of the working-age population are fit to work. There are those in work who may through illness or injury become unfit for their normal job for a while. Nevertheless, they will almost always become fit for work, and many in a relatively short time. The problem is that they may not be fit initially for their normal work, and it is the change in attitude by employers, employees, and healthcare professionals that will allow them to resume useful economic activity.

The Need for Rehabilitation

In 2008 a major review of the health of the working population was carried out in the United Kingdom. The review, led by Dame Carol Black, found that the economic cost of sickness absence and worklessness associated with ill health amongst those of working age is over £100 billion a year in the United Kingdom. Her report estimated that around 175 million working days were lost due to illness in 2006 which equates to approximately £11 billion per year in sickness absence costs, with long-term sickness absence contributing up to 75% of absence costs. The review identified that long-term absences accounted for only 5% of all episodes of absence but, when taken together, these long-term absence cases accounted for over 40% of the total working time lost. Much of this absence relates to mental health conditions. The review concluded that sickness absence not only is a drain on an organization's resources but also presents a public health and economic challenge.

The Role of Work in Recovery

Studies have found that the benefits for employees who return to work include a return to a sense of normality, increased self-esteem and self-respect, higher mental health and social inclusion, as well as financial independence. An organization which has effective return-to-work policies and procedures which are implemented successfully can reap benefits not just in terms of the employee's well-being but also for the organization's efficiency and productivity. A successful rehabilitation of an employee cuts out sick absence or ill health retirement costs and removes the need to rehire and retrain new people. Even when weighed against the management costs of supporting the employee back to work on a rehabilitation program and, for a temporary period, giving a reduced workload or hours worked, evidence from health settings makes the financial case for effective rehabilitation very strong. For example, recent work by the UK National Institute for Health and Clinical Excellence (NICE) found that, although based on a limited number of studies, international evidence for the effectiveness of interventions to return people to work following back problems were highly likely to be cost-effective.

There has also been a shift in the United Kingdom to focus more on rehabilitation and assisting people back into work. An example of this is the "fit note." The UK government has recently introduced a new system to encourage those employees on long-term sickness absence to attempt an earlier return to work. They have replaced the old "sick note" with "fit notes" which give the GP the opportunity to say that the employee *may be fit* for work *if* the employer can accommodate adjustments to the employee's duties or hours of attendance. There is therefore an expectation that the line manager will be able to respond to such requests and they, in turn, will likely need additional support from OH and HR to ensure the organization effectively follows through these adjustments.

However, the rehabilitation of those who have been on long-term sickness absence can still pose significant challenges to both the employer and employee alike. Line managers as well as health professionals and human resources have a crucial role to play here, but it is line managers who need particular support and direction.

Additionally, there is a lack of specific and up-to-date guidance to aid workplace responses to trauma, and PTSD specifically. However, frameworks do exist which can provide a basis for thinking about what organizations need in place, for instance, in relation to legal guidance, risk assessment methodology, medical and psychological treatment, and best practice in relation to workplace rehabilitation. This chapter takes these existing frameworks and then goes through how they can be formulated in practice and within policy to aid those employees who have been subject to traumatic stress at work.

Legal Context

There are legal duties that any organization needs to comply with. Although these duties vary depending on the country's judicial and legal systems, most countries would expect the employer to provide a safe working environment. This would extend to physical as well as psychological considerations. Liability is likely to occur where the employer can reasonably foresee harm occurring due to an issue at work and does not take reasonable action to address it. Although the precise legal definitions and context will vary depending on the country, the following questions can be helpful to work through to decide whether or not there is some liability using the example of traumatic stress:

- *Has the employee developed a real psychiatric injury?* For example, the employee may have experienced a traumatic incident at work and been diagnosed with PTSD as a result.
- *Were the working conditions such as to give rise to a real risk of the employee incurring psychiatric harm?* For example, as the employee experienced the traumatic incident at work, the organization would have to argue that the incident did not cause the harm and that the employee had a pre-existing psychiatric condition – although the employee could argue that this pre-existing condition was made worse by the trauma.

- *If so, did the employer know, or ought they (with reasonable care) to have known, that their employee was exposed to that risk?* For example, the employer needs to answer this by referring to their risk assessment of the specific role and potential exposure to traumatic stress. If they had not previously done this, it will likely count against them. If they have not taken any action to reduce the risk or provide support to ameliorate the impact of a trauma, see the "Risk Assessments" section of this chapter.
- *If the employer knew of the risk, what, if any, measures (which were reasonably practicable to implement) could or ought to have been taken to reduce the risk of such injury to that employee?* For example, if the employer knew of the risk but their measures to reduce the risk of injury, such as trauma support, were either absent or not implemented as written in their policy, this would count against them. Employers may also need to refer to research carried out in the psychological and medical literature to demonstrate that their interventions were sound and based on best available evidence-based practice.
- *If the employer did not know of the risk, did that failure materially contribute to the development of that psychiatric harm?* For example, the employer would need to defend against the claim that not assessing the risk and not providing effective post-trauma support following the incident had led to harm to the employee.

Risk Assessments

Within the United Kingdom, the Health & Safety at Work Act (1974) specifies that an organization has a responsibility to proactively prevent psychological harm occurring to their employees through the work they carry out. The Management of Health and Safety at Work Regulations (1999) specifically requires organizations to carry out risk assessments to ensure psychological risks are minimized. Organizations would find it very difficult to avoid this duty especially where traumatic incidents frequently occur in the workplace during the course of an employee's work. However, organizations cannot provide totally risk-free environments, and therefore risk assessments should be carried out to cover:

- The activity or event that may give rise to harm.
- Who is at risk (including who might be at special risk).
- How likely it is that harm will occur.
- What are the realistically potential consequences?
- What are the procedures to control or reduce the risk?

Within an organization, the risk assessment would highlight the roles or locations where employees could be subject to a traumatic incident at work or where they work in a potentially dangerous or aggressive environment. In some instances the nature of the role means there is an increased risk of exposure to a traumatic incident in the course of work. Examples of roles specifically associated with risk of trauma include:

- Military personnel.
- Transportation workers (railways, maritime, aviation, and road).
- Emergency services (ambulance, fire, police, and forensic staff).

- Security service workers.
- Healthcare workers.
- Banking workers who work with customers plus retail employees.
- Offshore oil and gas industry employees.
- Employees working in heavy industry (e.g., chemicals).
- Nuclear workers.
- Postal workers.

However, it is important to recognize that in addition to these roles with higher risks, any worker could be exposed through experiencing physical attack, threat, armed raids, holdups, and/or verbal abuse, as well as even being terrorist victims or witnesses of terrorism.

Once a risk has been identified during the course of an individual's work, the organization needs to look at ways to materially reduce or control the risk (although this is not to say that the organization is specifically responsible for traumas related to acts of terrorism where this is unrelated to their work).

There are various levels in which risk management can be carried out (Rick *et al.*, 1998), which the following steps highlight:

Selection of staff: Are recruits able to cope with the type of work expected and are they aware of the risks in their role (self-selection here is important and "realistic job previews," where the potential recruits shadow a member of staff for a day or try it out themselves)? Do recruits have the necessary interpersonal skills to cope with situations that might lead to psychological trauma, especially in dealing with customers?

Adequate training of employees: Do employees need training to manage verbal or physical aggression? Are employees given "drills" on what to expect and do in the event of an incident?

Protective procedures: Are the supporting procedures to minimize the chances of an incident clear and regularly evaluated? Can the working environment provide a safe place for employees?

Support and post-trauma policies: Are there systems in place to support employees? Are these adequately promoted within the organization? Are these evaluated to ensure they remain "fit for purpose?" Given that incidents can be unexpected (e.g. 9/11 in New York or 7/7 in London), are the policies flexible enough to deal with this?

Managing Sickness Absence and Rehabilitation

Several existing frameworks provide a structure for developing the policies and practices that support effective sickness absence management. A recent example from NICE (2009; see "Resources" at the end of the chapter) is an evidence informed set of generic recommendations which can be adapted to the management of trauma or PTSD specifically and identifies the following key stages:

1. Initial enquiries from the employer.
2. Detailed assessment.
3. Coordinating and delivering interventions and services.

The NICE guidance also complements the recommendations in Dame Carol Black's (2008) review of the health of Britain's working age population, *Working for a Healthier Tomorrow*.

More detailed guidance for developing effective rehabilitation strategies for managing absence due to work-related stress (Thomson and Rick, 2008) are worth considering here. They detail the process of rehabilitation, which the following has been adapted from:

1. The employee who is on sick leave is contacted as early as possible by a line manager, HR, or OH.
2. Consent is obtained from the employee to enable a referral to be made for a health assessment as early as possible so that prognosis and treatment can be identified.
3. A rehabilitation plan is developed with the employee which typically includes a phased return to work.
4. Where necessary, the employee is provided with treatment which may include cognitive-behavioral therapy (CBT) or counseling if there is a common mild to moderate mental health condition.
5. A gradual return to work schedule is initiated and adjustments put in place as required.
6. Updated reports are provided by occupational health, with the employee's consent, regarding the success of the rehabilitation program with any advice given about further adjustments required within the workplace (our addition to the original Thomson & Rick model).
7. The employee's job needs to be adjusted to remove or reduce aspects of work which impedes recovery or which could lead to a relapse, and the employee's health and well-being as well as work output needs to be monitored during and after the return to work.

Thompson and Rick also provide a checklist for organizations to help them develop appropriate policies and practices (see "Resources" at the end of this chapter). The rehabilitation process and checklist apply equally to trauma-related absence.

Specific guidance on appropriate methods of assessment and therapeutic interventions are available from NICE guidance on common mental health disorders (NICE, 2011). With regard to PTSD, these guidelines specify:

- The need for referral to an appropriate practitioner for identification and assessment of any suspected problem.
- Referral for appropriate formal psychological intervention (i.e., trauma-focused CBT or EMDR).
- Psychosocial interventions based on informing employees about self-help and support groups and other local and national resources, and befriending or a rehabilitation program for people with longstanding moderate or severe disorders (NICE, 2011, p. 15).

Across all these frameworks, the pivotal role of the line manager becomes apparent such as in liaising with individual employees, referring to occupational health, developing a rehabilitation plan, and organizing flexible return to work and work adaptations. Line managers therefore have a central role to fulfill in enabling effective absence management and rehabilitation for trauma and PTSD. Recent work has recognized that to perform this role, line managers need specific and complex skills and this work has started to focus on the skill set, training, and support of line managers in rehabilitation. It is to this area that we turn next.

The Role of the Line Manager

Due to centralizing and setting up shared service centers, there is an increasing trend for processes and procedures related to the attendance, well-being, and conduct of employees to move from a specialist human resources (HR) department function to the line manager as the responsible agent (Larsen and Brewster, 2003). As we have seen in the "Managing Sickness Absence and Rehabilitation" section, the line manager has a pivotal role in the rehabilitation and return to work of an employee who is of sick. They are most likely to have frequent contact with the employee, have a good detailed understanding of the employee's work and the environment (including colleagues), and can provide ongoing feedback and monitoring of the employee's success (Pransky, Shaw, & McLellan, 2001). They need to be well acquainted with the policies which underpin the HR policies and procedures and should have access to advice from HR or OH, as needed, on specific cases.

In relation to traumatic stress the line manager is usually in contact with the absent employee following a traumatic event at work and during any consequent sickness absence. They would also be involved in supporting the employee back to work following a traumatic event and putting in place workplace adjustments such as gradually increasing their hours of work or providing mentoring support when going back to a place where the employee experienced a traumatic incident. The support and involvement of the employee's line manager are, therefore, crucial factors in effective rehabilitation following mental health issues, including where traumatic stress has occurred. In summary, research into the principles that line managers need to follow in the area of stress (Chartered Institute of Personnel and Development (CIPD), 2011; Dollard, Winefield, & Winefield, 1999; Donaldson-Feilder and Pryce, 2006; Pimental, 2002) equally applies to rehabilitation following a traumatic incident.

Policies and Practices

Trauma management policies

Written trauma policies need to consider not just the risks for those directly affected by the incident but also the many other people who can be indirectly impacted. Therefore, the policies need to take into account people who witnessed the incident, those first aiders where their input was not successful, close colleagues of those directly affected by the incident, families of those directly affected (especially where severe injury or death occurs), and managers who have a support function and may themselves be overwhelmed. Particular attention should be given to special circumstances such as hostage taking or where people felt particularly guilty about the incident or felt in some way responsible for what happened, since these experiences can create a bigger impact from a psychological point of view.

As mentioned in this chapter, the policy needs to respond to the risks identified in the organization and look for methods or interventions that reduce or control these risks. Having clear objectives stated at the beginning of the written policy is important so there is clarity about its purpose. Examples of such objectives are to:

- Prepare employees so that they know what to do if a traumatic incident occurs.
- Ensure support is provided to all those affected in a timely and effective manner.
- Highlight what line managers need to watch out for in their teams following a traumatic incident.
- Raise awareness of how all employees can take responsibility for themselves and colleagues through access to suitable post-trauma support.
- Facilitate an early return to work.
- Set in place ongoing review arrangements in order to test the effectiveness of what is done and make improvements where necessary.
- Educate managers and employees about the effects of trauma.
- Discharge the organization's legal duty of care to its employees.
- Take care of the organization's public relations following a work-related trauma and be seen as a caring employer.

A written trauma management policy should be stated in easy-to-read language and supplied to line managers so they understand their role in supporting each person impacted by the trauma. Ideally, this should be backed up by training which should include a section on attitudes to trauma and highlights the importance of good listening and communication skills so that the employee is encouraged to open up to the line manager about their difficulties.

Training of line managers

Training and awareness-raising sessions should be delivered to line managers which include the more practical elements of post-trauma support following the incident including checking on the well-being of employees, possibly accompanying employees to hospital, liaising with police or security services, assisting the employee with making statements, dealing with business aspects, and possibly in the short to medium term arranging cover to maintain operations.

Where a traumatic incident takes place, it will be helpful for a line manager to have a good understanding and awareness of how traumatic stress affects the individual, the symptoms of PTSD, and an understanding of what practical steps can be undertaken immediately following a traumatic incident at work including how to refer on for post-trauma support. They should be acquainted with the organization's policy in handling an incident at work and to ensure they can fulfill their role. Attitudes are also important since negative attitudes from line managers about trauma at work can result in further difficulties down the line both for the employer and for the employee.

Cunningham, James, and Dibben (2004) have highlighted that without training or awareness, line managers can become unsure about how they should communicate with employees, are unaware how they can put in place reasonable adjustments to support a phased return to work, and are unable to know how best to access support for their employee. For instance, a common question relates to the extent to which line managers should be directive about the benefits of an early return to work. They may feel reluctant to suggest to an employee, who perhaps is angry about how the organization did not protect him or her from a traumatic incident, that an early return to work would be recommended. The employee may also not wish to return and may try to avoid anything to do with work due to the "avoidance" symptom, described as one of the symptoms of someone with PTSD.

Training for line managers can respond to this issue by outlining how work generally has a beneficial psychological effect on individuals. Although the language would clearly need to be tailored by the line manager, they could talk about how work gives the employee a structure, a sense of belonging, a sense of status and meaning in society, as well as economic benefits. A return to work following trauma is even more important since it can help a psychological recovery in terms of "getting back to normal," seeing colleagues for social support, and moving away from a notion that "the world is a dangerous place." Clearly, this return needs to be handled carefully, and the employee may be nervous about his or her safety and question whether a repeat traumatic incident could occur. The line manager needs to be especially supportive here and to respond to the concerns that the employee has.

There also needs to be a recognition that some people will need a longer period off sick before being ready to return and that a small percentage might never be able to achieve this. Rehabilitation and returning to work are about doing things at the right time and involving the right people. Specialist services such as OH and counseling or trauma support can be brought in to provide advice to the line manager and have an important role in helping the employee to work through anxieties.

Management behaviors which support rehabilitation efforts

Research conducted for BOHRF (2006) and CIPD (2011) (see the resource section at the end of this chapter) has identified management behaviors of the line manager which support good rehabilitation practice and are helpful as they include positive as well as negative behavioral indicators. The following is a sample of these:

Whilst the employee is off, the manager:

- regularly communicates with the individual via telephone or email;
- regularly communicates work issues with the individual to keep him or her in the loop;
- focuses conversations more on the individual's well-being;
- is in touch with the individual's close colleagues with regard to their health;
- encourages work colleagues and other members of the organization to keep in touch with the individual;
- relays positive messages through family or friends;
- makes it clear that the individual should not rush back to work;

- makes it clear that the company will support the individual during his or her absence;
- reassures the individual that his or her job will be there for them upon returning to work;
- prevents the individual from pushing him or herself too much to return to work.

When the employee returns, the manager:

- gives the individual lighter duties or different jobs during his or her initial return to work;
- incorporates a phased return to work for the individual;
- remains objective when discussing return-to-work adaptations for the individual;
- explains the return-to-work process and procedures to the individual before he or she returns;
- explains any changes to the individual's role, responsibilities, and work practices;
- meets the individual on his or her first day back;
- makes the individual's first weeks back at work as low-stress as possible;
- is proactive in arranging regular meetings to discuss the individual's condition and the possible impact on his or her work;
- communicates openly;
- listens to the individual's concerns;
- understands that, despite looking fine, the individual may still be ill;
- appreciates the individual's wishes;
- has an open-door policy so the individual can always approach him or her with any concerns;
- adapts an approach that is more sensitive toward the individual;
- allows the individual to maintain a certain level of normality;
- is quick to respond to the individual via email or telephone when he or she has a concern;
- takes responsibility for the individual's rehabilitation;
- acknowledges the impact the individual's illness has on him or her;
- remains positive with the individual throughout his or her rehabilitation.

The manager avoids

- losing patience with the individual when things become difficult;
- displaying aggressive actions;
- questioning the individual's every move;
- going against the individual's requests for certain adjustments to be made to his or her work;
- making the individual feel like a nuisance for adding extra work to his or her schedule (list reproduced with kind permission from the British Occupational Health Research Foundation).

Line managers applying this framework need also to be realistic and balance it alongside the needs of the organization. For instance, where there is a clear lack of progress in the rehabilitation program, the line manager should obtain clear advice from occupational health and HR who will need to review whether the program is likely to be successful and if alternative options should be considered. Examples of these alternatives could be to revise the rehabilitation program, trial a new role, or, if all options have been attempted and failed, give consideration for retirement on health grounds or dismissal due to capability. Organizations that do not have ill health retirement options available may have insurance policies which pay for all or part of the employee's sick absence.

The CIPD framework also highlights that line managers need to understand their legal responsibilities and know how to follow the organization's policy and procedures; both of these points have been covered in this chapter.

Bringing It All Together: Rehabilitation Plans and Minimizing Trauma-related Sickness Absence

The costs of replacing employees who leave the organization, who go on long-term sickness absence, or who obtain ill-health retirement or are dismissed on grounds of capability are likely to be significant, and the objective of a comprehensive rehabilitation plan is to avoid such costs (Hogarth & Khan, 2004). When the employee returns to work, an employer who rushes the process or places the employee under too much pressure too quickly can create further sickness absence. The employer needs to be sensitive to the way the employee is welcomed back, especially if part of their role brings them back in contact with a location or place of work where the traumatic incident previously occurred or with coworkers who witnessed the incident. For instance, if the employee was verbally or physically attacked during the course of their work, it is important they are briefed on security aspects of their role so they know how best to respond if another traumatic incident occurs. This will also provide a sense of safety and "getting back to normal" which is important for the employee's psychological recovery. Where the employee has suffered repeated traumatic incidents at work resulting in sickness absence, specialist advice should be sought to comment on the appropriateness of the employee continuing in their role. For instance, a reasonable adjustment could be in providing a role where there is a lower risk of further exposure to traumatic stress or providing specific training in how to de-risk their current role for further psychological trauma through more effective security procedures or de-escalation techniques designed to minimize the severity of a traumatic incident. In addition, advice on what psychological treatment should be provided is another important aspect.

Therefore, advice should be sought on the appropriate timings of a return to work following trauma. This can be from the individual's doctor, although the organization may have the services of an occupational health advisor or workplace counselor who can perform this function and who should be more familiar with the specific working environment into which the rehabilitation is going to be attempted. The advice given should include a return to work date, adjustments that should be considered to support the return to work, along with any support that can be given by the line manager as well as from those involved in giving psychological talking therapies or practical assistance onsite. The advice should also set out how the employee should be monitored in terms of their behaviors and symptoms, and what the line manager needs to do if the employee does not meet the planned milestones as set out in the rehabilitation program. In this way, the return-to-work plan should not be seen as rigid. The timescales should be flexed according to what is experienced in practice with any revisions based on the principles contained in the advice from occupational health. In this way, the employee can be managed in a consensual way in light of the best medical evidence to meet both the employee's and employer's needs. To summarize, the employer's role – often delivered via the line manager – is to positively support a successful return to work, communicate with employees when they are absent, and, based on the advice of occupational health, implement a structured rehabilitation program (CIPD, 2004; Kendall, Linton, & Main, 1997).

The Conflicting Pressures on the Line Manager

As we have seen, the line manager is therefore vital in communicating how the organization will make reasonable adjustments to encourage the employee back to work and in following the occupational health advisor's practical and realistic steps for a gradual return to work.

However, the line manager will undoubtedly have other operational pressures on them and may not be very experienced or skilled in dealing with the rehabilitation of employees. This can result in the needs of the employee on a rehabilitation program moving down the line manager's priority list and important milestones being missed. The line manager may, therefore, enlist the support of a dedicated coordinator or "case manager." This can provide continuity of support as well as build up detailed knowledge of cases, end to end. The case manager could be an experienced operational manager in a large organization who "looks after" a number of cases in their jurisdiction, or it could be requested from the occupational health provider as a specialist service. The important point here is that if a case manager is appointed, they need to work closely with the local line manager so there is good ownership at this level.

For the more complex situations, case conferences can be used to bring together the various stakeholders so that all parts of the organization can discuss together the way forward. The line manager can outline what they have done to welcome the employee back and what adjustments outlined by the occupational health practitioner can be accommodated, the HR manager can advise on relevant HR absence management or disability policies, and the occupational health practitioner can clarify any of their advice which has previously been given. It can also be helpful to involve the employee and also an employee's supporter so that they feel involved in the process and can take ownership for their part in successfully returning to work.

Where the traumatic incident becomes the "last straw" for the employee, due to the final incident pushing them into a position of "not coping" due to the buildup of multiple traumas, or where the employee develops other conditions such as depression or anxiety, specialist advice from occupational health is important to accurately diagnose any mental health conditions. They may make a recommendation for psychological treatment within the workplace where the employer has made provisions for talking cures such as trauma-focused CBT. Alternatively, they may write to the employee's insurers or doctor to highlight the importance of putting in place appropriate treatment.

However, barriers can occur if the line manager has a negative attitude, such as believing that the traumatic incident was "not that much of a big deal" or that the employee was "acting up" as many others who experienced the same incident did not go off sick, as well as if the occupational health practitioner is risk averse:

> It is clear that at the start of an absence the majority of individuals express a desire to return to work. Some see themselves returning to work as part of the rehabilitation process from their illness. Others, sometimes reinforced by healthcare practitioners, develop the belief that they must be completely well before they can consider returning to work. The inevitable consequence becomes one of never being quite ready and thus they move into a state of perceived permanent ill health. (Wright, 2008)

Therefore, it is important for all within the organization to make efforts to make adjustments in the workplace to accommodate the employee, though these should be temporary with specific timescales for review. Examples would be where a postal worker is accompanied by a "buddy" for their first few days so that they regain their confidence in delivering letters and packages following a violent robbery on their walk, or a retail worker given paperwork tasks away from the shop front for the first week following an attempted robbery. Other examples would be temporarily providing the employee with extra support when they initially return to work along with a reduction in targets set. Reviews would measure progress against the goal of a full rehabilitation back to full duties, typically within 2–3 weeks, sometimes longer, depending on the nature of the initial traumatic incident.

Solution-Focused Interviewing

Part of the line manager's role is to focus on an early return to work and to champion the potential of rehabilitation since this turns the statement from "Work is making me ill" to the question "What do we need to change in order to help you return to work and perform successfully?"

Perhaps a useful way of seeing this switch is to consider the change from a problem-focused approach to a solution-focused way of thinking. For instance, de Shazer (1985) developed a solution-focused approach which was designed initially for talking therapy. This is a powerful way of operating and could be applied to the rehabilitation context; for instance between an employee off sick and their line manager or health professional in relation to a return to work. The following set of questions could be used to shift away from looking at problems and move to increase involvement by the employee jointly looking at solutions:

- What can we discuss in this meeting so that our conversation about returning to work will be useful? (Goal setting.)
- What will you be doing when the problem is solved? What will you notice? What would you be doing differently? (Future orientation.)
- What would be the smallest next step you can take to solve this problem? (Making big goals more workable.)

- How would others notice that you are making progress? What might your work colleagues or partner say you would be doing differently if things improved? (Expanding the possible solutions into the system.)
- What else can we do together to move this situation forward? (Eliciting cooperation from the employee.)
- Have you solved similar problems? If so, how did you solve them on that occasion? Who helped you and how did he or she help you? (Using exceptions and resources.)
- Are there moments in which the problem is less intense? What is different then? When is this happening? (Adding "nuances" instead of black–white opposition.)
- Has something changed since we last spoke about your views on returning to work? (Eliciting signs of spontaneous "pre-session" changes.)
- Now that you have achieved some progress, what is the next small step you could take? (Success building and breeding on success.)
- Imagine that you have returned to work; what will be different then? What will you do differently? What will your colleagues do differently? What will your line manager do differently? (Future orientation by visualizing a future in which the current problems are solved.)

The Challenge of Rehabilitation for Organizations

It has been outlined that the organization has a legal, business, and moral imperative to put in place reasonable adjustments to support the employee back to work, especially in cases of trauma at work. However, organizations sometimes struggle with this. Cunningham *et al.* (2004) found that where individuals were returning to work, relevant workplace adjustments were not always put in place or there were delays in making the changes and line managers did not adequately support the process. Di Guida (1995) found various reasons for this including a fear of change, potential cost increase, fear of employee re-injury, and lack of knowledge on how to implement a return-to-work plan effectively. This illustrates the journey that organizations need to make, and further research and cost–benefit analysis would be helpful to demonstrate how it makes good business sense to address these issues.

Conclusion and Summing Up

This chapter has been about the role of rehabilitation in supporting employees who have been affected by psychological trauma at work. The central role of the line manager is explored and the responsibilities of the organization highlighted, including the legal duty of care. In the United Kingdom, NICE (2005) has produced helpful guidance on the treatment of PTSD. It highlights how the first four weeks following the incident, "watchful waiting" should be put in place where the symptoms are mild, rather than prescribing psychoactive medication. After one month, treatment should be put in place that could include trauma-focused CBT or EMDR. Some people experience a delayed onset of the traumatic stress symptoms (estimated to be approximately 15% of cases).

Organizations that are developing a post-incident trauma policy or revising an existing policy need to take into account this guidance from NICE in both the four weeks following the trauma and the months following it. An organization needs to set out who is responsible for delivering the "watchful waiting" element as well as outline how treatment for traumatic stress can be effectively delivered. Being clear on how the line manager can refer for specialist support, which can include HR and OH services, is important. Being prepared for dealing with traumatic incidents is vital and the list of questions in the "Resources" section of this chapter can be used as prompts to check that an integrated policy is in place.

Resources

1. National Institute for Health and Clinical Excellence (NICE). (2009). Recommendations on managing long-term sickness absence and incapacity for work. Retrieved from http://www.nice.org.uk/PH19
2. *Questions that organizations can ask to help them build an integrated rehabilitation policy* (adapted from Thomson and Rick, 2008)

   Early contact by organization
   - What is "early" in terms of contacting the employee?
   - Who should make contact?
   - What guidance and support are available for the person making contact?
   - What happens if the employee is not willing to talk to the organization?

   Occupational health assessment
   - What is the organizational practice regarding referrals for a health assessment in cases of traumatic stress (immediately or if absence occurs following the trauma)?
   - How should line managers be advised on when to refer for a health assessment?
   - Who is best placed to conduct the health assessment?
   - Who is responsible for communicating the progress of the employee with their own doctor or occupational health?
   - Is the information available from the health assessment sufficiently detailed?
   - How will sensitive information about the employee be handled?

   Developing a rehabilitation plan
   - Who has overall responsibility for initiating rehabilitation plans?
   - Who has responsibility for discussing and agreeing the plan with the employee?
   - Aside from the line manager and employee, who else could be referred to?
   - What preparation can be done in advance (e.g., establishing what different return-to-work options are available)?
   - How are reviews managed, when, and by whom? Is a "case manager" required?

   Therapeutic interventions
   - Does the organization offer any form of psychological therapy?
   - When are these therapies used? For how long?
   - Who should be the provider? How can access to appropriate types of therapy be assured?

- How will the organization promote the long-term use of strategies learned through counseling?

Flexible return-to-work options

- How flexible are current return-to-work plans?
- How is the decision about initiating a return made?
- Who monitors the plan?
- When and how often is progress monitored?
- How are reviews and amendments decided?
- How does the organization react to plans that are not working?

Work adaptations

- How, and at what stage, are work adaptations considered for an employee?
- What range of work adaptations are available?

3. National Institute for Health and Clinical Excellence (NICE). (2011). NICE quick reference guide for: Common mental health disorders. Retrieved from http://www.nice.org.uk/guidance/CG123.
4. BOHRF. (2011). Manager support for return to work following long-term sickness absence – guidance. Retrieved from http://www.bohrf.org.uk/downloads/Managing_Rehabilitation-Guidance.pdf.

References

Black, C. (2008). *Working for a healthier tomorrow*. London: TSO.

Chartered Institute of Personnel and Development (CIPD). (2004). *Recovery, rehabilitation and retention: Maintaining a productive workforce*. London: Author.

Chartered Institute of Personnel and Development (CIPD). (2011). *Manager support for return to work following long-term sickness absence*. London: Author.

Cunningham, I. James, P., & Dibben, P. (2004). Bridging the gap between rhetoric and reality: line managers and the protection of job security for ill workers in the modern workplace. *British Journal of Management, 15*, 273–290.

de Shazer, S. (1985). *Keys to solution in brief therapy*. New York: Norton.

Di Guida, A. W. (1995). Negotiating a successful return to work program. *Journal of the American Association of Occupational Health Nurses, 43*, 101–106.

Dollard, M. F., Winefield, H. R., & Winefield, A. H. (1999). Predicting work stress compensation claims and return to work in welfare workers. *Journal of Occupational Health Psychology, 3*, 279–287.

Donaldson-Feilder, E. J., & Pryce, J. (2006, June 1). How can line managers help to minimise employees' workplace stress? *People Management*, 48.

Kendall, N. A. S., Linton, S. J., & Main, C. J. (1997). *Guide to assessing psychosocial yellow flags in acute low back pain: Risk factors for long term disability and work loss*. Wellington: Accident Rehabilitation and Compensation Insurance Corporation of New Zealand and National Health Committee.

Larsen, H., & Brewster, C. (2003). Line management responsibility for HRM: What is happening in Europe? *Employee Relations Journal, 25*, 228–244.

National Institute for Clinical Excellence (NICE). (2005). *Post traumatic stress disorder*. London: Author.

Pransky, G., Shaw, W., & McLellan, R. (2001). Employer attitudes, training, and return to work outcomes: a pilot study. *Assistive Technology, 13*, 131–138.

Rick, J., Perryman, S., Young, K., Guppy, A., & Hillage, J. (1998). *Workplace trauma and its management: Review of the literature.* Institute for Employment Studies Contract Research Report 170. Sudbury, UK: HSE Books.

Thomson, L., & Rick, J. (2008). An organisational approach to the rehabilitation of employees following stress related illness. In A. Kinder, R. Hughes, & C. Cooper (Eds.), *Employee well-being support: A workplace resource.* Chichester, UK: John Wiley & Sons, Ltd.

Waddell, G., & Burton, K. A. (2006). *Is work good for your health and well-being?* London: TSO.

Wright, D. (2008). Rehabilitation of mental health disabilities. In A. Kinder, R. Hughes, & C. L. Cooper (Eds.), *Employee well-being support: A workplace resource.* Chichester, UK: John Wiley & Sons, Ltd.

22

Healing the Traumatized Organization: An Exploration of Post-trauma Recovery and Growth in the Workplace Setting Using the Metaphor of the Nervous System as a Template to Highlight Collective Learning

Tony Buckley and Alison Dunn

Introduction

At the time of writing, it is over six years since the bombings took place on London's transport system on July 7, 2005. These events and their multilayered impact continue to have an intermittently high profile in the UK news with the inquest into the deaths of 52 people having recently concluded. Within Transport for London (TfL) itself, there has been much reflection, learning, planning, and development over these years, following the trauma that was experienced by the organization.

This chapter will apply a psychotherapeutic perspective to the organization's recovery process which will be illustrated through the use of case studies of some of the employees who played a key role during the events of 2005. There will be a specific focus on applying the understanding that neuroscience brings to individual trauma response and recovery in order to bring a new understanding to the organizational process.

Moving On: A Summary of Recovery and Development since 2005

Following the bombings, the transport system in London was recovered very quickly with all but the damaged sections of the lines operational on July 8. A number of

International Handbook of Workplace Trauma Support, First Edition. Edited by Rick Hughes, Andrew Kinder, and Cary L. Cooper.

exercises were undertaken both internally and externally in order to learn from the events and their aftermath and assist with ongoing planning for the future.

One of the developments that has taken place across TfL since 2005 is the establishment of an incident care team. This team consists of employees – volunteers who have been recruited, selected, and trained to provide immediate support to customers following an incident on the network. Their role is focused on providing immediate practical assistance to customers whose journey has been interrupted and may require help with their onward journey, with contacting friends and family or in larger incidents help with arranging accommodation, care for dependents, food, drink, clothes, and so on. Incident care teams were set up initially on airlines to provide support following air accidents (Coarsey, 2004), and their development was more recently extended to the railway network in the United Kingdom. TfL was in the process of setting up a team when the bombings took place in July 2005, and the very first volunteers worked at the family assistance center which was set up in London after the bombings to provide practical support to survivors, friends, and families.

The aftermath of the events also saw the setting up of interagency collaboration between the support teams working for the emergency services and TfL. A "Bronze" health group developed out of London's critical incident response arrangements which met regularly over a period of time to review the provision of support and share good practice. Relationships were developed during that time that underpinned the ongoing planning process. In the longer term this group became part of a trauma network set up as part of the British Association for Counselling and Psychotherapy's (BACP) workplace division, BACP Workplace. This network is now an active group meeting regularly in London sharing best practice in working with trauma and maintaining links and relationships between specialists involved in providing trauma counseling in the workplace.

TfL was also represented on the steering group that led the work to provide the NHS psychological support services. The work of this team in providing assessment and treatment to people affected by the bombings has been described by Brewin, Scragg, and Robertson (2008).

Within TfL's occupational health department, the counseling team has undertaken its own process of learning from the provision of the psychological support response to employees and managers across the London Underground (LU). In particular, the team's critical incident response plan was completely rewritten (Dunn, 2008). The plan is now reviewed and updated on a regular basis and everyone working in the counseling team takes part in a regular training to ensure that all are familiar with the processes outlined in the plan and their own role in any future response.

Since 2005, the importance of the psychological services within the organization provided by the in-house counseling team has been increasingly recognized. Our role is widely understood, and the part we play in responding to critical incidents highly valued by senior managers, local managers, and employees.

Since 2005, the role of volunteers belonging to our peer support program, the Trauma Support Group, has been expanded to include providing emotional first aid to colleagues following larger critical incidents as well as incidents affecting small numbers of employees – train suicide, for example. The training given to employees volunteering for the program has been expanded to reflect this new role. In any future incident, this will enable us to provide immediate onsite support to affected employees quickly and easily.

The in-house counseling team has continued to develop their understanding and skills at working with traumatized employees. Trauma continues to form a significant part of the regular workload for the team with employees experiencing trauma as a result of train suicide or incidences of workplace violence.

An additional service has also been introduced as part of the counseling service provision. A 24-hour telephone helpline is now available for employees, providing not only telephone counseling but also advice, information, and guidance on a whole range of issues. This service is provided in partnership with an employee assistance program (EAP) provider who will also be able to provide the team with immediate additional support should a further sizable incident take place on London's transport system.

The work to promote employee mental health has continued with the development and delivery of a mental health plan each year. The aims of the plan are to:

- Raise awareness of mental health issues and reduce stigma.
- Improve understanding and skill in responding to employees experiencing mental health issues.
- Minimize work-related stress as far as possible.

Data analysis and research underpin this work which is supported by a cross-functional steering group. Recent achievements include:

- Training for operational staff in building mental resilience.
- Production of a booklet and an e-learning package for employees on building mental resilience.
- Onsite support to teams experiencing significant stress issues.
- Delivery of an annual communications plan.
- Collaboration with colleagues in human resources on support through organizational change.
- Delivery of an annual mental health report at the director level.

Grieger, McCarroll, and Ursano (1996) conclude research by recommending a similar list of organizational interventions to prevent traumatic stress.

Training
Experience
Group or organizational leadership
Management of meaning
Management of exposure
Management of fatigue, sleep, and exhaustion
Buddy care
Natural social supports and caretakers
Education in disaster stress and strain
Education of health care providers
Screening (Grieger *et al.*, 1996, p. 449)

The Individual Recovery Process

The recovery process for employees followed a predictable path in that many of the affected employees either remained in work, or returned to work very quickly while being supported through their recovery process in different ways. A small number of employees experienced longer term and more disabling trauma reactions and were off sick for longer periods of time.

Research on the effects of the London bombings on TfL employees (Caddis, 2010) aimed to investigate the psychological effects on employees involved in order to mitigate such effects and help with planning interventions and resources should such an incident happen again. The research concluded that the incidence of PTSD among employees was within "normal" levels, that there was a short-term increase in sickness absence among employees for the first six months, and that the support provided by the in-house counseling team had a positive impact for employees.

> The effect of [counseling support] is seen in the return to baseline mean absence duration in the group attending [counseling], from an initial mean absence that was 3 times higher than those who did not attend counselling. A significant improvement was seen in the IES [Impact of Events Scale] before and after treatment. (Caddis, 2010, p. 90)

The Organizational Recovery Process

Organizations are living systems, which can be vulnerable – trauma can have a destabilizing impact (Bloom, 2011). Erikson describes the state of "collective trauma" with symptoms such as organizational dissociation, miscommunication, and helplessness (Erikson, cited in Bloom, 2011, p. 141). However, as Klein and Alexander (2011) state, the organization can also represent a source of help and healing – the challenge is for the organization to find ways to respond that can facilitate this healing (Tehrani, 2011).

Direct parallels of learning can be drawn between the individual's response to traumatic experience and the collective experience of our organization in terms of its response, recovery, and growth following traumatic events in a workplace context. In order to illustrate these parallels, research was conducted among employees whose perspective comes from their role in responding to the events and working toward organizational recovery. This included those involved with security, management of people, providing group or individual therapy, media, healthcare, and organizational leadership. Participants were given a short questionnaire and were invited to explore their responses in the aftermath of the events on July 7 commenting from both the individual and organizational perspectives. The questions focused around the following key therapeutic themes – common aspects of how individuals process trauma:

- behavioral responses;
- relationships;
- outlook and beliefs;
- learning and growth;

- resources and mental resilience;
- making meaning.

Key shared points of learning were identified from the responses. These were reviewed against the collective learning about trauma as portrayed in the therapeutic literature including the current neuroscience perspective. It is hoped that readers will be able to draw some useful conclusions about their own workplace environment in making sense of collective preparedness, response, and recovery from trauma in the context of an organizational setting.

An Organizational Nervous System Response

The nervous system has the capacity for both voluntary and involuntary responses to environmental stimuli and circumstances. Human beings have a remarkable survival response capacity which is mediated by the autonomic nervous system (ANS) when threat and danger are present: "It's a quick and dirty processing system" (Le Doux, 1998, p. 163). In ordinary life when there is no danger, the normal routine is for the brain to process all incoming sensory data via the "thalmo-cortical" system where we can take some adequate time to plan and act accordingly. When danger is suddenly present, however, "the direct thalmo-amygdala path is a shorter and thus a faster transmission route than the pathway from the thalamus through the cortex to the amydgala" (Le Doux, 1998, p. 164).

The amygdala reacts with split-second timing activating a cascade of neurochemicals and hormones to activate the sympathetic and parasympathetic nervous system which create adrenaline to mobilize emergency, survival-based, behavioral action systems including fight, flight, freeze, flop, or other highly adaptive responses. In the interest of expediency, this system completely overrides the ordinary daily functions of digestion, thinking, planning, and "social engagement system" (Porges 2011, p. 125). When the danger has truly passed, the individual can revert to ordinary functioning unless they get stuck in the traumatic response and may need help (via therapy or medical or other professional intervention) to recalibrate the nervous system.

Can organizations also be viewed as having two separate systems? On one hand is a response for ordinary, daily, business-as-usual (thalmo-cortex) functioning which is then, on the other hand, overridden in favor of a faster acting emergency system (thalmo-amygdala) activated to protect people and assets in extreme situations. People often can't think straight in dangerous situations. One reason for this is the cortical inhibition or suppression that takes place in favor of faster acting survival-based instinctual body responses. It is not possible to collect thoughts when the cortex is offline. As portrayed by the phrase "speechless with terror" it is very natural for people to be unable to think and speak properly in the midst of an emergency situation. Under extreme stress people can easily revert to habitual, procedurally learned responses such as withdrawal or aggression.

Do organizations have an effective autonomic response which will kick in at such times? This is likely achieved by advance planning and rehearsal before danger actually occurs (in peacetime) and developing slick systems as a result. This may include the development of emergency plans supported by training to test and implement systems, with repeated rehearsal to ensure that people are clear about their roles and fully enabled to carry them out. Such planning offers the assurance that even if people are frightened or in a state of anxiety about what is happening, they can still follow procedurally learned (auto-pilot) behaviors. A good example of this is how a train operator may calmly follow procedure to evacuate (detrain) the carriages in a situation such as power failure or derailment. Experience teaches us that the same is true for those staff whose role it is to provide psychological support services during and in the aftermath of trauma.

> Needing to have detailed plans and clear understanding from all about their roles should a similar event take place, that everyone has had training and is skilled and ready to respond.

The best-laid (and practiced) plans will provide some good security and confidence in organizational response to danger but will never completely cater for how people will instinctively and often heroically react in the heat of the moment.

> I am struck by some of the amazing stories of great courage, humanity, and personal resilience that emerged from the terror and horror of those events. I have a great belief in our ability to achieve great things at terrible times.

With regard to the events of July 7, 2005, it is well known that TfL front-line staff reacted spontaneously to rescue people, improvising emergency first aid with instinct and ingenuity, and disregarding their own safety and sometimes protocol to respond to the dramatic unfolding circumstances for which no training can fully prepare.

Le Doux (1998) describes the impulse to act "quick and dirty" (p. 163) as part of the survival response system, likening this to how you might react quickly to get out of the way when you think you see a snake in the corner of your eye, when later (cortical evaluation) you realize it was simply a curved stick. The faster acting system is a biological insurance policy in case of threat to life.

Conversely some members of the emergency services were inhibited ("immobilisation"; Ogden, Minton, & Pain, 2006, p. 104) at times from taking action due to protocol, having to wait for clearance from their chain of command. Where the natural response – urge to act – is inhibited, people may develop subsequent difficulties in coming to terms with the event(s) over time. Research shows that "helplessness" is a particular vulnerability factor in predicting subsequent difficulty such as PTSD. Herman (1992) reminds us of this process with regard to traumatized patients: "When neither resistance nor escape is possible, the human system of self-defense becomes overwhelmed and disorganised" (p. 34).

Pierre Janet (1925, cited in Ogden *et al.*, 2006) described the arrested impulses as truncated actions which can result in "everlasting recommencements" which will later need to be addressed in therapy.

The lesson here is about the value of advance planning, training, and rehearsal in a way that allows for and integrates a combination of a procedurally learned process which also allows for the unique "dirty work" which will inevitably be required on the day.

In thinking through the parallel between the nervous system and the organizational structure, it may be useful to identify which aspects of the individual's system may correlate to different parts of the organization. For example, within LU (London Underground) there is a department called the Network Operations Centre (NOC) which acts as the control and command center and communications hub during critical incidents. This office can be likened to the function of the amygdala – the brain's smoke alarm center which quickly mobilizes the autonomic response of the body by activating the release of chemical hormonal messengers from the pituitary gland (managers with capacity to decide on actions) to the adrenals (local managers) in readiness for fight or flight (actions of operational staff on the ground). The amygdala retains some projections to the thalamus (environmental input – external organizations and the media) and to the frontal lobes (new information or executive decisions from the head office), so actions and responses can in turn be modulated by the influence of this input as the situation changes. Consider which departments or aspects of your organization for which you can draw similar parallels to the communication of the nervous system.

One important question arising from such comparison is to ask, does this "organizational amygdala" have the capacity to calm down and return to homeostatic functioning when the crisis has passed, or does it remain overactive in a state of hyper- or hypoarousal? You may be able to think of examples whereby misconceptions and errors of judgment have resulted from a collective state of heightened autonomic arousal.

Institutionalized racism is perhaps a collective dysfunction mediated from a hyperaroused distrust of the other. Such shared negative perceptions can be the downside of empathy. "Groups of individuals often appear to catch the emotions of others, whether it be laughter in a movie theatre or hatred in a lynch mob." Hatfield, Cacioppo, and Rapson (1994, p. 204) suggest that certain careers or roles will show more or less susceptibility to "emotional contagion" (p. 179).

It may make sense for organizations to draw benefit from a "therapeutic" intervention to help minimize post-trauma dysfunction occurring. It is natural to initially have a fear-based reaction which presents as racial or cultural stereotyping. Intervention here might enable a phase of reflection and resensitization to diminish hyperarousal but also to help recalibrate the (organizations) autonomic nervous system in the post trauma climate. Within LU, therapeutic groups were organized for teams directly involved in the events of July 7. This setting enabled people to process events together which facilitated the understanding of different perspectives.

> The group as a whole has the capacity to bear and integrate traumatic experience that is greater than that of any individual member, and each member can draw upon the shared resources of the group to foster her own integration. (Herman, 1992, p. 126)

It is interesting to note the use of language surrounding trauma. Notice the shorthand which is employed which may serve to make the unthinkable more manageable. Huge impact events can be reduced down to simple words or numbers such as 7/7 or 9/11. In our context the experience of witnessing a train suicide is often called a "one under," which enables people to discuss it as part of everyday work conversations. Consider other shorthand phrases or euphemisms which may be used in various professions and organizations to enable emotional regulation.

> Creating and recalling a story requires the convergence of multisensory emotional, temporal, and memory capabilities that bridge all vectors of neural networks. (Cozolino, 2002, p. 113)

> As part of the ongoing regulation of affect, it is important for organizations to give some consideration to use of language and detail in relation to graphic trauma stories or images, whether these are limited or told in full, by whom, when, and why. Maintaining this dynamic tension between peoples' need to tell and the risk of retraumatization is how individuals and perhaps organizations are helped to remain within the "window of tolerance" (Ogden *et al.*, 2006, p. 71).

The Initial Response to Events

Having explored some parallels between the individual and organizational response to trauma, this chapter will now examine this theme further by comparing the initial response to events, the personal impact, relationships, and collaboration.

When shocking or traumatic events happen to us as groups or individuals, we are typically overwhelmed. Our priorities rapidly change. At such times, what people need most are other people. Note how the mobile phone networks were jammed with people trying to reach loved ones when the news about the bombings broke on July 7, 2005.

> My loved ones were contacted and contacted me.

People are "cast" into a new state of emotional arousal even when trying to take the news in.

> Surreal.

> Heightened confusion.

In the confusion there is likely to be a lot of fear, anxiety, panic, assumptions, misinformation, and yet an overwhelming need to find out more.

> At work, learning in bits and pieces about what had actually happened – having to plan very quickly while absorbing the enormous impact of what had taken place, and being aware of events (i.e., the rescue/recovery operation) still unfolding.

Notice in this comment the need to keep a sense of calm and logic for planning ongoing action against the potential for the looming emotional overwhelm. This is the point where we need the "mindful" aspect of brain active through the medial prefrontal cortex.

> Response flexibility is the capacity to pause before action. Such a process requires the assessment of ongoing stimuli the delay of reaction, selection from a variety of possible options, and the initiation of action. (Siegel. 2007, p. 142)

Just as an individual's body responds by diverting resources through shutting down cortical function and digestion in favor of increasing heart rate and respiration, similarly an organization has a diversion of resources in order to respond to the emergency and some normal "business-as-usual" functions may have to cease temporarily. This creates certain tensions where business continuity plans may need to run in parallel with emergency procedures. It is an important protective factor for some in traumatic situations to sense the capacity for (organizational) life to go on or/and to return to normal as quickly as possible.

When a national phone number is made available for public inquiries following an incident, calls come pouring in with anxious demands for advice and guidance. This is the time when those whose role is to provide support need absolute clarity with regard to expectations, roles, and actions. Therapists remit changes with the heightened awareness that along with employees, the managers and the company itself are also our clients.

> Communicating to contain anxiety as far as it was possible to do so.

Just listening to people and acknowledging what has happened for all are important. The strain of this can be considerable and may increase over time when the impact is also truly felt among the helpers.

> My experience of this was that the team was as rocked by the events as the organization itself.

The Personal Impact

Some stories can't be told. Note how many war veterans have marched proudly and silently year after year never speaking of their experiences. We must never underestimate the impact of trauma in any context.

> Immediately, my life changed completely – nothing seemed normal at home or at work. In common with other Londoners, travelling to work was scary for a while.

One member of staff expressed the difficulty of even beginning to fathom what was happening.

> The sheer horror of events could not be adequately relayed in print or celluloid.

The literature on trauma treatment shows that endless telling and retelling of such stories is inherently dysregulating, creating further re-occurrence of emotions, somatic symptoms, and intrusive imagery. Yet there is also a burning need to know more details of what happened (narrative integration) in order to try to make sense of the experience. It is important that organizations are aware of the psychoeducational facts and dynamics of these competing tensions which will either really help or seriously hinder recovery.

Overwhelming events are etched more deeply into memory which is commonly described by psychologists as flashbulb memory.

> When the amygdala lights up during threatening or aversive events the amount of activity predicts how well people later remember these experiences. (Schacter, 2001, p. 181)

The responses of the employees who took part in our research illustrate the impact for them personally and how events can fundamentally change people.

> I think it has also led me to stop believing that there is a god.

Others highlight the longer term (six years on) effects in terms of their own emotional and physical reactivity.

> The long-term impact as far as I am aware, is a stress response on hearing multiple sirens in central London, and a feeling of dread when I contemplate the possibility of being in a similar situation in the future.

For others, triggers include daily work environment of trains, stations, tunnels, platforms, or simply crowds moving up and down the escalators. In trauma processing, benign objects can become a conditioned stimulus. Memory researchers show this Pavlovian principle by pairing a "blue slide" with a "loud sound" stimulus. Participants show the same skin conductance reading when later seeing the blue slide without the sound (Schacter, 2001, p. 179).

Although compounded by the media hype surrounding these events we cannot discount how the anxiety can blend both people and objects together to also become a fear-conditioned stimulus.

> Whilst I do recall being wary of back-packs for a while after the bombings, or indeed back packs worn by people who "appeared Muslim."

Schacter (2001) describes such associations as "counterfactual thinking" (p. 166) and yet for some the experience of having worked in an environment where terrible things have happened may result in the opposite response of showing less fear for the future.

> I feel an increased awareness of terrorism but strangely I don't feel as threatened by potential bomb terrorism now as I did working in London during the height of the IRA conflict. I am not sure why it felt more threatening then.

Could this reaction in part come from the experience that the worst has already happened?

From all the responses we received, the biggest impact seems to be a new sense of profundity for people and the sanctity of life.

> Life is precious; it should not be taken or given away lightly.

Others showed some complex and contrary feelings:

> When events happen elsewhere, I feel an immediate sense of relief – thank goodness it isn't us – and then a sense of guilt at thinking that.

It is common for individuals to feel guilt (also counterfactual thinking) to varying degrees of intensity which is in part a retrospective way of regaining a feeling of agency or control perhaps in anticipation of future trauma.

> A loss of an internal locus of control is common for clients who have relied upon any of the immobilizing defences: freezing, submission, or submissive behaviours. (Ogden *et al.*, 2006, p. 105)

Communication

The responses demonstrated the value of communication between people as the key to all that happens in the aftermath of trauma. There are many interesting parallels that can be drawn when comparing individual and organizational coping strategies and recovery following a traumatic experience. It is clear that individual recovery is linked to how well grounded and secure we are in relationship to others before, during, and after the event. The connection between early-life security and individual resilience in coping with extreme experience is well documented in the attachment literature, but there are many other "resilience versus vulnerability" factors to consider (McFarlane & Yehuda, 1996, p. 155). What might the equivalent resilience and vulnerability factors be for an organization?

The same communication principle applies to group relationships on the level of a family, team, department, or whole organization (i.e., good relationships provide a sense of security that supports resilience). Cozolino (2006) points to neuronal firing and synaptic connectivity being remarkably similar from one individual brain, and likened it to the interbrain communication between networks of people:

> The brain is a living system. Neurons are, by their nature social; they shun isolation and depend on their neighbours for survival. (p. 39)

Intraphysic or intrapersonal refers to how the parts of the self communicate within one mind–brain system. This communication can be either free flowing dynamically, as with a person who is open with good integrative capacity, or can be restricted with compartmentalized disruption perhaps caused by avoidance, denial, repression, and dissociation.

Interpersonal refers to communication between people which can similarly be free or inhibited by the same facilitating or repressive forces. This can also be applied to communication within organizations such as the quality of relationships in meetings, between the various hierarchical layered structures, between departments, with unions, and of course between different organizations. Just as the individual brain has dynamic competing tension between excitatory and inhibitory neuronal activity, it is possible to view the communication systems of organizations as having similar function "dendritic branching" (Cozolino, 2006, p. 39), or pruning, to either increase or limit the spread of information. Much communication within and between organizations is facilitated by technology – for example, computer, radio systems, and telephone or video conferencing. On July 7, some critical problems were identified in radio signaling problems below ground (inhibitory) whereas other learning showed it was too much for some individuals to be exposed to events via TV screen monitors (excitatory). Although these are the tools of modern media, the quality of the communication resulting from their use also depends a great deal on the relationships of the users.

The regulation and flow of information both internal and external to the organization are keys to effective response to and recovery from trauma. As individuals we have systems for tuning in to our internal (bodily) and external environment via our senses. If we treat an organization as an organism with similar neural networks, our directors and company leaders who reflect and are equally in tune with both milieus would think and act mindfully in parallel with the integrative "resonance process" (Siegel, 2007, p. 42) circuits of the frontal medial cortex.

Collaboration

At times of extreme challenge we come to realize just how much we depend on one another. Much organizational resilience comes from available support and bonding through adversity.

We collaborated with other people in the organization and formed relationships that endured.

The events of July 7 and subsequent learning show a need to work closely and collaboratively with all emergency services and other key partners with a particular role to respond to trauma. As outlined in the introduction, workplace counselors in London working with trauma now have an effective collaborative network.

The sense that we were all in this together.

Of course, although these events took place on the TfL infrastructure, the sense of shared outrage crossed all boundaries in that such trauma is felt as an attack on the public, an attack on on Londoners, and indeed an assault on the national psyche with reverberations to all peaceable people and nations around the world. The outpouring of support from far and wide may significantly and positively impact our recovery and sense of pride in how we responded. One comment shows this sense experience as heightened.

Respect for our employees and how magnificently they responded.

The profile of the counseling department also came more sharply into view in that some members of senior management radically changed their perception (from seeing counselors as the "pink cardigan brigade") to deeper understanding of the real value of psychological support for employees.

Increased respect from senior people in the organization.

It is sadly often only in adversity that we stop to really appreciate what good people are capable of. The same may be said for organizations.

I think the organization has a gravitas and a respect that it didn't have previously.

This "gravitas" can be seen surrounding other traumatic events as highlighted in the media, typified by the heroic image of the New York fire-fighters after the 9/11 attacks.

Some companies and organizations may have a closed, noncollaborative communication style whereby they try to take care of all major issues internally, whereas others prefer to benefit from working collaboratively with external partners from similar or different industries. Our counseling and trauma team based within occupational health is collaborative by nature but we also have good links with external organizations for cross-referral and consultative support expertise. Collaboration with others outside the parameter of our own organization offers welcome perspective and insight at a time of great need. One provider who has a long relationship of offering us input on stress and trauma work was called upon at the moment of crisis.

I also think my "outsider" status helped me not feel too swamped by the impact and to remain detached enough.

In some ways for us this kind of relationship acted as an auxiliary cortex providing "interactive regulation" (Ogden *et al.*, 2006, p. 170) at a time of chaos and uncertainty. Following from the confusion and chaos, one operational manager reported that they just needed someone to sit them down and help them to think by asking:

Shall we just write a list of everybody who was due to be at work today?

In both of these examples, we see external support helping to activate the medial prefrontal cortex which supports mindfulness, a "top down processing" (Ogden *et al.*, 2006, p. 121) intervention to prevent "bottom up hijacking" (Goleman, 1995, p. 14) of the nervous system.

Recovery from Trauma

What would be a good indicator of trauma recovery for organizations? Would this be similar to that of an individual? Let us consider the typical list of reactions which are initially highly adaptive responses to a trauma but may later become the very debilitating symptoms blighting an individual's life and preventing him or her from moving on: hypervigilance, depression, problems with drugs and alcohol, emotional reactivity, irritability, mood swings, anxiety, sleep problems, and flashbacks. See McFarlane, Van Der Hart, and Van Der Kolk (1996, p. 421) for more extensive details of trauma symptoms.

Could aspects of the same list or any combination of these apply to what might in the longer term inhibit the organizations recovery and growth? In the aftermath of July 7 there has been some debate within our own organization and the media about how we can keep the traveling public safe in the future. The question has been asked as to whether there should be airport-type scanners at every entrance to the entire traveling network in the city of London. Media scare stories may heighten the fear about which type of people we should be most suspicious of traveling with. Can we trust backpackers or people of certain religious persuasion? Is this a temporary hypervigilance which should be transcended in the greater interest of recovery?

I was aware of exploring the anxieties of management who had reason to speak about cases that involved persons who appeared "similar to the bombers."

Conversely we have the perspective of a practitioner who feels empathy with individuals who are brought into this uninvited spotlight.

I journeyed with them far more intimately, as they explored their perceived difficulties. It was as if I were saying, "I see your humanity."

Summary: Learning from This Model of Understanding Organizational Trauma

Evolutionary biologists have shown there was a time in human history when we relied more on our subcortical (animal) brain to protect and defend against predatory groups. The subsequent development of the neocortex gives human beings discernment and the means of a much broader response repertoire including the ability to exercise control over emotional reactivity. This enables us to identify and localize threat without resorting to defensive judgments about (stereo)types of people or situations.

We maintain that therapeutic or psychological activity provided within the organization and aimed at supporting people before, during, or after a traumatic event is the equivalent to "Emotional intelligence" (Goleman, 1995, p. 163) necessary to act as the organization's "neurobiological regulator" (Schore, 2003, p. 256) which minimizes anxiety at all levels and optimizes a calm, balanced, and resilient approach to the challenges faced. For counselors, this work also comes in the form of psycho-educational advice to normalize, one-to-one counseling, stress-reduction programs, group work, and the wider contribution to the company's mental health strategy. The reader may have other examples of how emotional well-being is enhanced in their own organization.

Current research shows that the brain is continuously developing in response to environmental demands through "neural plasticity" (Cozolino, 2002, p. 296) including the voluntary and involuntary response systems. Organizations need to show the same change plasticity and integration of systems, procedures, and communication not only to recover from extreme experience but also to grow and flourish.

We suggest that the principles and learning from neuroscience can be a powerful lens through which to view organizational issues for people. Directors and company leaders can use strategies based on the "narrative integration" (Siegal, 2007, p. 309) function of the prefrontal lobes (executive brain) which have the capacity to moderate and calm the (survival mediated) emotional brain. This prevents a fear-based, emotional hijacking and perpetuation of traumatic symptoms being re-enacted in organizations.

> I'm very much aware of how a team can become divided and feelings of being "included" or "excluded" played out in interpersonal dynamics. The impact of such awareness upon me is that I have become more sensitized to the possibility of our team splitting.

Companies where people are exposed to traumatic experience need to invest in good psychological care for staff by preparing comprehensively for events and providing support during and after them. Good communication like synaptic circuits and systems of the brain should be "hierarchical, local, and divergent" (LeDoux, 2002, p. 50), and multidirectional in essence including laterally, horizontally and vertically, and internally and externally. Relationships should be understood as a key resilience factor on the same basis as empathic attachment "attunement" (Ogden *et al.*, 2006, p. 60) models enabling people to rise to the challenge when "bad stuff happens" and recover themselves well afterward. Further, it is important to trust that innate human responses and gestures are as valuable a contribution to trauma survival and recovery as are procedural systems.

As part of our role in providing psychological support to our organization, we in turn need to ensure we have good self-care, support, and supervision systems in parallel.

Our clinical supervisor paints a picture of the thousands of uniformed operational staff in TfL, whose archetypal role is to safely ferry us, on a daily basis, through the subterranean underworld of London via the tube network. They are our guides whose job is to deliver us safely to our destinations and do so with good cheer. The least we can give them in return is for us to also be a safe guide on the psychological journey through the subconscious in support of their recovery when bad things happen.

There is a very common theme to the question of what people have learned, and if there is any good coming from such horrific experiences, people seem drawn to powerfully reassert their bonds of that which unites rather than divides human beings.

> I hold that there remains at the "heart-of-humanity" a common unifying bond toward each other and existence.

Furthermore, some people described their outlook changing in their understanding and acceptance of "different" others.

> I am more tolerant of different cultures.

Employees show increased awareness of the dangers of harboring negatively biased perceptions of people.

> Clearer about unwillingness to make blanket judgments or assumptions about groups of people or faiths.

This has a particularly healing quality given that the attacks which happened on 7/7 that were driven by a fundamentalist religious ideology.

Furthermore, what began as a first "trauma reaction" as distrust to specific groups developed into an increased sense of wanting to know the "other" more.

> I journeyed with them far more intimately.

> It is as if I was saying I see your humanity.

In addition to this kind of learning is a sense of people at all levels of the organization really taking stock about priorities, a reassessment of what is really most important in life.

> The need to self-care and the importance of family and friends, a faith in humanity, are but a few of the things I have learnt.

And the spirit of togetherness come what may.

> Belief that humans can do things to help each other and show true spirit no matter what religion, creed, or culture.

These sentiments and changing views represent post-traumatic growth and an important meaning-making phase of traumatic recovery. This is sometimes described as a transpersonal element which is often linked to the meaning-making and integration phase of treatment.

> I believe that fundamentally people are good and that this goodness ultimately wins out. I believe that there is a meaning to everything and faith that what happens is meant to be – even if we can't make sense of it.

Nearly six years on, we can say that the learning for us has been vast and is ongoing. We are open to the recommendations of the coroner following the recent inquest which may represent an important stage in our recovery toward finding closure. There remains hope and optimism in the people affected and a looking forward to moving on.

In closing, a final piece of testimony seems to give us hope for the future:

> Belief that what is meant to be will be, and letting go of worrying about what is to come.

References

Bloom, S. (2011). Sanctuary: An operating system for living organisations. In N. Tehrani (Ed.), *Managing trauma in the workplace*. Hove, UK: Routledge.

Brewin, C. R., Scragg, P., & Robertson, M. (2008). Promoting mental health following the London bombings: A screen and treat approach. *Journal of Traumatic Stress*, *21*, 3–8.

Caddis, R. (2010). An investigation into the psychological impact and sickness absence on London Underground staff following the July 7 London bombings. Master's thesis, University of Manchester.

Coarsey, C. (2004). *Handbook for human services response*. Blairsville, GA: Higher Resources.

Cozolino, L. (2002). *The neuroscience of psychotherapy*. New York, NY: Norton.

Cozolino, L. (2006). *The neuroscience of human relationships*. New York, NY: Norton.

Dunn, A. (2008). Organisational responses to traumatic incidents. In A. Kinder, R. Hughes, & C. Cooper (Eds.), *Employee well-being support: A workplace resource*. Chichester: John Wiley & Sons, Ltd.

Goleman, D. (1995). *Emotional intelligence*. London: Bloomsbury.

Grieger, A., McCarroll, J. E., & Ursano, R. J. (1996). Prevention of post-traumatic stress consultation, training, and early treatment. In B. A. Van der Kolk, A. C. McFarlane, & L. Weisaeth (Eds.), *Traumatic stress: The effects of overwhelming experience on body, mind and society*. New York, NY: Guildford Press.

Hatfield, E., Cacioppo, J. T., & Rapson, R. L. (1994). *Emotional contagion*. Cambridge: Cambridge University Press.

Herman, J. L. (1992). *Trauma and recovery*. London: Basic Books.

Klein, S., & Alexander D., (2011). The impact of trauma within organisations. In N. Tehrani (Ed.), *Managing trauma in the workplace*. Hove, UK: Routledge.

LeDoux, J. (1998). *The emotional brain*. London: Phoenix.

LeDoux, J. (2002). *Synaptic self*. London: Penguin.

McFarlane, A. C., Van Der Hart, O., & Van Der Kolk, B. A. (1996). A general approach to treatment of postraumatic stress disorder. In B. A. Van der Kolk, A. C. McFarlane, & L. Weisaeth (Eds.), *Traumatic stress: The effects of overwhelming experience on body, mind and society*. New York, NY: Guilford Press.

McFarlane, A. C., & Yehuda, R. (1996). Resilience, vulnerability, and the course of posttraumatic reactions. In B. A. Van der Kolk, A. C. McFarlane, & L. Weisaeth (Eds.), *Traumatic stress: The effects of overwhelming experience on body, mind and society*. New York, NY: Guilford Press.

Ogden, P., Minton, K., & Pain, C. (2006). *Trauma and the body*. New York, NY: Norton.

Porges, S. W. (2011). *The Polyvagal Theory*. New York, NY: Norton.

Rothschild, B. (2000). *The body remembers*. New York, NY: Norton.

Schacter, D. (2001). *How the mind forgets and remembers*. London: Souvenir Press.

Schore, A. N. (2003). *Affect regulation*. New York, NY: Norton.

Tehrani, N. (2011). Building resilient organisations in a complex world. In N. Tehrani (Ed.), *Managing trauma in the workplace*. Hove, UK: Routledge.

23

The Management of Emotionally Disturbing Interventions in Fire and Rescue Services: Psychological Triage as a Framework for Acute Support

Erik L. J. L. De Soir

This chapter describes the impact(s) of emotionally disturbing and potentially traumatizing interventions and the way in which fire and rescue services personnel should be supported in the acute stage. Firstly, the controversy on the effectiveness of psychological debriefing will be revisited before shedding light on the way in which the detailed reconstruction of exceptional interventions, social sharing, open expression, and psycho-education may help fire and rescue personnel recover from emotional disturbing interventions. For some rescuers, early re-exposure to what they have been exposed to can be harmful. Therefore, this chapter will introduce the concept of psychological triage which may help peer support officers and mental health professionals manage their activities aiming at prevention, care, and after-care of the post-event sequelae. Using a psychosocial matrix, psychological triage will lead to a categorization in different groups based upon primary, secondary, or tertiary exposure. The management of emotionally disturbing events will be expressed in terms of primary, secondary, and tertiary prevention of post-event sequelae. Throughout this chapter, field experiences from the author's practice as a fire and rescue psychologist will be used to illustrate the experiences of personnel after exposure to shocking or disturbing interventions.

One of the secondary aims of this chapter is to introduce a new terminology with respect to stress and trauma, introducing the French concept of *effroi* (psychological terror) as a central feature of the potentially traumatic event.

In terms of trauma interventions, differences will be identified between sessions for psychological stabilization, emotional ventilation, and immediate recuperation. These sessions will be the stepping stones to further emotional and psychological uncoupling from an emotional disturbing intervention.

International Handbook of Workplace Trauma Support, First Edition. Edited by Rick Hughes, Andrew Kinder, and Cary L. Cooper.

Introduction

This chapter aims to generate some clarity surrounding the variety of effects of emotionally disturbing and potentially traumatic events. The traumatizing character of an emotionally disturbing event is always the result of a subjective interpretation of this event by the individual and not merely dependent on objective cues in the given event. In both the literature and the spoken language, there is too widespread a use of the term "trauma" these days; everything seems to become a trauma and the result is then that the involved victims develop a subsequent trauma after surviving one. As such, the causality between events and effects is very often unclear.

This conceptual lack of clarity also influences the practice of psychological support in fire and rescue personnel. The best illustration is the controversy on the effectiveness of psychological debriefing and whether or not it is effective. While techniques of psychological defusing and debriefing (Mitchell and Everly, 1993; Raphael, 1986) were originally developed to support professional (or professionally trained) caregivers – such as fire, rescue, police, or emergency services personnel – they have also been widely used (and researched upon) to support all kinds of victims of critical events. Although the definition of a critical incident is rather vague, the practice of psychological debriefing has rapidly grown in popularity. Researchers started to investigate whether or not single-session psychological debriefing also prevented post-traumatic stress disorder (PTSD) in primary victims. Both Critical Incident Stress Debriefing (CISD) – being an integral part of Critical Incident Stress Management (CISM) – and the latter concept of early intervention became container concepts of various kinds of interventions for various kinds of victims. Meanwhile, a whole disaster business (Deahl, 2000; Shepard, 2000) developed and professional caregivers or high-risk organizations (e.g., banks, petrochemical industry, rescue services, army, and police) were urged or legally forced to "do something" to support their personnel exposed to various kinds of emotionally disturbing and potentially traumatizing events. In De Clercq and Lebigot (2001) and De Soir and Vermeiren (2002), European trauma specialists offer an alternative view on stress theories and psychological trauma, introducing terminology other than the current concepts such as traumatic stress, acute stress disorder, and PTSD.

Psychological Debriefing: Positive or Negative Outcome?

In the literature (Rose & Tehrani, 2002), a number of methods of *psychological debriefing*, which generally correspond to a single-session and semistructured crisis intervention, applied shortly after a *traumatic event* and expected to prevent post-traumatic stress reactions (Bisson, McFarlane, and Rose, 2000), have been described. Dyregrov (1997) describes psychological debriefing as a structured group process where facts, thoughts, impressions, and reactions to a potentially stressful event – referred to as a *critical incident* – are explored and education on how to cope with reactions is provided. Originally, Mitchell (Mitchell & Everly, 1983) introduced his Critical Incident Stress Debriefing (CISD) as a group intervention for emergency personnel after exposure to secondary trauma (i.e., where emergency personnel witness or assist primary victims of a *traumatic event*). Most researchers ignored the fact that CISD is only part of an overall trauma support model commonly known as CISM and used a definition of debriefing to suit the study rather than the needs of involved fire and rescue personnel. Subsequently, the literature is inconsistent with the original objectives of CISM and uses various debriefing models that compete with the therapeutic interventions model or even *doing nothing*, described in the NICE Guidelines as *watchful waiting*. For example, Kenardy and Carr (1996) studied the effectiveness of psychological debriefing for helpers (emergency services personnel, counselors, and welfare officers) who responded to an earthquake but did not specify what type of debriefing was examined. The authors reported that there was no standardization of debriefing procedures, no knowledge of the extent to which the debriefing matched the CISD model, no way of determining whether or not any of the participants actually attended a debrief, and no assessment of whether the debriefing was appropriate to the level of stress and trauma experienced. Therefore, it may not be appropriate to use this evidence to suggest that psychological debriefing may be harmful (e.g., Kenardy, 2000; Raphael & Meldrum, 1995).

The debriefing literature has been criticized for a lack of randomized-controlled trials evaluating psychological debriefing, but these designs would involve (1) taking a group of people who have been exposed to a potentially traumatizing event, (2) randomly assigning them to a "debrief" group that attends a psychological debriefing session or to a "no-debrief" control group that does not attend a debriefing session, and then (3) contrasting the two groups on appropriate outcome measures. This process gives rise to a number of ethical issues. For instance, if researchers believe that psychological debriefing is beneficial, it is unethical to withhold the debriefing experience from the research participants in the control group. On the other hand, if researchers believe that psychological debriefing is harmful, then it is unethical to expose participants to the debriefing process.

Further, random allocation to debriefing or non-debriefing groups, with homogeneous teams of fire and rescue personnel who have been exposed to the same potentially traumatizing interventions and mandatory participation, would not be ethical either. Research participation should be voluntary and, as a consequence, the individuals who choose not to participate in the research may influence the results.

The lack of baseline data, clarity regarding response and dropout rates, and confounding factors should be taken into account when claiming that psychological debriefing might be potentially harmful. For example, despite using random assignment, individuals in the Bisson *et al.* (1997) study who were debriefed had, on average, more severe burns than the individuals who were not debriefed. This may have occurred because a number of people originally assigned to the "debrief" group left hospital before their debriefing was conducted (and were thus excluded from this study). Severity of burns might be expected to be a stronger predictor of trauma resulting from burns than a 45-minute discussion, in hospital, within days of hospital admission (Robinson, 2004). Similarly, in the Mayou *et al.* (2000) study, the victims of road traffic accidents who were debriefed had more severe injuries and spent more time in hospital, on average, than the accident victims who were not debriefed, also despite random assignment (Hobbs *et al.*, 1996).

The literature on the effectiveness of psychological debriefing shows a lack of conceptual clarity and consistency; it is unclear which models have been used, how the "debriefers" have been trained, and what the objectives of the "debrief" have been. It seems that any post-incident psychological intervention, often conducted in a single-session approach and for various kinds of trauma, has been called debriefing. The regretful conclusion is that these issues have often had serious implications on the implementation of support programs in police, fire, and rescue organizations.

The only conclusion that can seemingly be drawn from these studies is that it is not appropriate to offer one-off individual psychological debriefing sessions to victims of primary trauma while they are physically recovering from the trauma. It is not surprising that psychological debriefing – if it is still correct to call it that way – is not effective under these conditions. However, positive-outcome studies suffer from a number of limitations that need to be taken into account when evaluating scientific evidence in support for psychological debriefing: the effectiveness of psychological debriefing has to be proven in an ecological valid way, participants of psychological debriefing sessions are not assigned to these sessions in a random way, and other confounding factors (such as the way of being assigned to debriefing sessions, and the value of self-reports on the value of debriefing) exist.

It is clear that both negative and positive outcome studies on the effectiveness of psychological debriefing suffer from a number of limitations that need to be taken into account when evaluating their findings. Future research needs to be oriented toward a true understanding of the very specific world of fire, police, and rescue personnel.

Acute Reactions after Potentially Traumatic Events: Direct Victims and Significant Others

This section seeks to qualify an event to be *emotionally disturbing*, when this event is abrupt and shocking, and involves disturbing feelings of anxiety and/or depression, followed by guilt, shame, sadness, and/or rage. By its sudden impact, the event temporarily disrupts the emotional, physical, and/or cognitive equilibrium. Examples of these kinds of events include the painful or sudden death of a friend or a relative, witnessing severely injured or dead people, and other important losses. It is argued that these events are *shocking*, instead of being directly traumatizing, if they did not lead to a subjective and/or objective confrontation with death or if they did not involve a fight to survive during which the survivor(s) was (were) confronted with a state of psychological terror, frozen fright, and unspeakable experiences which are impossible to symbolize or verbalize, or in which there was a complete disruption between *signified* and *signifier*.

Traumatization can also emerge from identification with victims or on-scene contact with friends or relatives (or victims looking like friends or relatives) and especially children, always considered to be the ultimate victims. In other cases, such an event can also trigger earlier trauma and thus lead to post-traumatic sequelae, aggravating the already damaged psychological structure of the individual.

It is important to qualify an emotionally disturbing event as potentially traumatizing if this event also satisfies the following criteria: (1) the event is sudden, abrupt, and unexpected; (2) it involves feelings of extreme powerlessness, horror and/or terror, disruption, anguish, and/or shock; and (3) it implicates vehement emotions of anxiety and fear of death, due to (4) the subjective (feelings) or objective (real, direct) confrontation with death (i.e., the real or felt severe threat to one's physical and/or psychological integrity or the integrity of a significant other). What is considered to be central in this definition is the confrontation with death: the potentially traumatizing event confronts one with a world which is unknown, a world of cruelty and horror, the world of death in which certainties, norms, and values do not (seem to) exist anymore. The world of death is the world of the unspoken horror – *le néant* (the nothing), as French contemporary authors (Crocq, 2000; Lebigot, 2000) call it – in which everything becomes senseless, which is impossible to describe or to put into words, since humankind has no words or concepts to describe the real characteristics of death. This is the reality of survivors of terrible (industrial) accidents, wars, fires, explosions, earthquakes and floods, macro- or microsocial interpersonal terrorism, or severe threat. In this reality, overwhelming forces annihilate human values, norms, and/or life, with oppressive amounts of violence and power reducing the human being to "dust," leaving the survivors in a short but significant silence of emptiness, complete abandonment, and loneliness. This is typical for the immediate aftermath of trauma, in which victims *awake* again and try to get in contact again with *the spoken world of the living.*

In this description, the illusionary state of predictability and security, respect for the human being (or life) and its norms and values, and/or its basic assumptions and certainties about the world we are living in make place for a situation characterized by deep physical and/or psychological injury, irreversible damage, humiliation, and destruction beyond repair.

The overwhelming impact of this close encounter with death involves a typical situation of frozen fright and psychological terror which can be likened to the French concept of *effroi de la mort* as decribed by Lebigot (2000; De Clercq & Lebigot, 2001) and compared to the old Greek myth of Perseus by Crocq (2000, 2001).

Lebigot (2000) describes how a traumatizing event creates an embedment of oneself as dead into the psychic apparatus and how survivors lose their illusion of immortality. The traumatic moment is an exclusion moment too in which the language disappears, with an unspeakable moment of dereliction creating feelings of shame and abandonment.

In *Le Retour des Enfers et son Message* (*Coming Back from Hell and Its Message*), Crocq (2000) illustrates how hell is spontaneously evocated in the speech of the traumatized. Coming back from hell can perpetuate the remembrance of horror and misfortune.

Traumatizing events shake the very foundations of the human being. Beside feelings of extreme powerlessness and helplessness, and the overwhelming impression of deep penetration into one's own physical and psychological integrity, trauma survivors will have to cope with the potentially ego-destructive emotions of permanent uncertainty, (survivor) guilt, anxiety, shame, and loss of control. The more there has been severe physical injury, the longer the recovery and *working-through process* will last, and the more we can be pessimistic about the prognosis in the long term.

There is also the loss of *connectedness* with surrounding significant others and the life environment in general. As De Clercq and Lebigot (2001) state, trauma survivors have seen the reality of death (*le réel de la mort*) and lost the connection with the world of the living.

From here on, this article explores the various aspects of a model used to understand *the life threat and emotional-shock processing* in a chronological way. This interpretation – which finds its inspiration in the animal world (cf. the way animals act in a *predator–prey* context) – offers a simple parallel. It is important to carefully think about the different possibilities for immediate trauma support during these different stages of traumatization.

During the potentially traumatizing event (i.e., the *peri-traumatic stage*), direct victims act in a way which is very meaningful for their survival and comparable to what is found with animals when they are threatened by a predator, as expressed in the work of Nijenhuis (1999), who uses the animal trauma model to explain the successive trauma stages in trauma survivors (mainly in a context of sexual abuse). In most trauma accounts we can readily identify the next successive stages: (1) *immobility* and *total inhibition*: in nature, this kind of immobility often means "survival" and "escape from death," and *freezing* may happen in a state of *apprehension of danger and attempt to find the right or most adequate survival response*; (2) *flight*, if there is enough time and space for escape, otherwise numbness and freezing might return, or even the opposite reaction pattern, panic and senseless activation; (3) *fight*, for as long as the fight to survive has a sense and offers a chance to survive in the stage of the traumatization process; (4) *total submission* (i.e., the moment at which victims experience overwhelming power and violence of the predator, the perpetrator, technology, or simply nature; it seems as if they understand that fighting death has no more sense, and it is at that moment that *dissociative behavior* – alienation, depersonalization, anesthesia, analgesia, narrowing of attention, tunnel vision, out-of-body experiences, derealization, and so on – sets in, as if this would allow the victims to die without feeling pain or without even knowing consciously that they are soon to die); and, finally, the last stage in this traumatization sequence, if the danger or death threat disappears, is (5) *recovery, recuperation,* and *return of pain sensitivity*, partial consciousness of what happened, and widening of attention, that is, behaviors that are typical for a return to reality.

Van der Hart, Brown, and Graafland (2006) explain the core of psychological trauma as a failed integration of an event. For Crocq (2000) and Lebigot (2000), reality will never be the same again if one has seen "death" right into the eyes and has been confronted with the unknown, wordless, and unspeakable world of death. For a further analysis of this animal model of traumatization and an in-depth discussion of trauma and dissociation, and the disintegrating effects that trauma can have on the psyche and personality of victims, readers are directed to the theories of trauma-driven structural dissociation of the personality in Van der Hart *et al.* (1999), Nijenhuis and Van der Hart (1999), and Van der Hart, Nijenhuis, and Steele (2006).

After the return from death, survivors of wars, motor vehicle accidents, rape, assault, fires, hostage taking, and so on have to work through the fragmentary and wordless trauma sensations and will have to search for words in order to express trauma impact. For trauma survivors, the world will never appear to be the same again. They will have to go back into the *trauma labyrinth*, in search of a way to express what they lived through, in search of a story and a meaning which can reconnect them again to the world of the living – *the world of those who speak*, allowing them to reframe their world, reconstruct their basic assumptions and beliefs, and become one biopsychosocial whole again.

Different Stages in the Aftermath of Trauma: Acute Impact, Working-through, and Trauma Fixation

In the first stage – which is referred to as the *acute (or immediate) trauma stage* – in the immediate aftermath of trauma, immediately after living through the potentially traumatic impact, trauma survivors are confronted with a confusing mix of feelings of *disbelief*, *denial*, *relief*, and *despair*. These moments of disbelief and denial – during which survivors yearn for rest, recuperation, and safety – will be quickly disturbed and/or alternated by sudden and intrusive recollections and re-experiences of the traumatogenic event, during which the victim acts as if the event itself was recurring and the death threat has returned. The brain does not seem to make a difference between the original event and these intrusive recollections. Trauma survivors keep asking the same questions: *What happened? How did this happen? Who else is injured (or dead)? Why did this happen (to me, or to us)? Why now?* And *how will I (we) ever recover from this?* They are in a desperate need of information, still shaking from the event which just struck them, feeling the sequelae of the hyperarousal they needed in order to survive, still a bit disoriented and heavily affected by the close encounter with death. During this stage, trauma survivors have overwhelmingly material and practical needs. They keep asking themselves: *How will I eat? Where will I sleep? Who will pay for this? How can I tell to my relatives what just happened to me? How do I get home? What about my old sick mother, and how will she react? Will I ever find the energy and courage to go back to work after this?* And so on.

This initial stage will be followed by a *trauma working-through stage* – which is referred to as the *post-immediate* or *post-acute stage* – during which the trauma survivors will have to (1) accept what happened to them; (2) confront the negative emotions which are associated with these kind of events; (3) reach a daily life equilibrium again, or try to return to normal life activities; (4) work through their experiences; (5) search for a way to express and put into words their trauma experience; and (6) find a meaning and a story, in order to integrate what happened in their personal life story. Numerous models are offered in the current trauma literature, but most of them take more or less these different stages into account.

Most trauma survivors will have an urgent need to really understand what happened to them, and how it happened. This should be achieved through a detailed collective reconstruction of the event, taking all possible sources of information into account (television reports, newspaper articles, individual accounts and stories, etc.) and they will search for explication, understanding, compassion, recognition, and meaning. The longer they stay alone with these needs, the longer they will be haunted by vivid, intrusive, and/or weird re-experiences of the event, as if their minds look for understanding and closure of the event.

The intrusive recollections and re-experiences, while the survivors return to the hyperaroused states coupled with the repetitious reminders of the original event – and which are so typical for the *fight to survive* – alternated by moments (or periods) of denial and avoidance.

Finally, there is the third stage on the time axis of *trauma processing*, *assimilation*, and *accomodation* – or the *trauma fixation* or *chronification stage*; trauma survivors can get stuck in this stage, after several months during which they tried to cope with their experiences but reached a stage in which their initial fears and complaints worsened, becoming omnipresent and intense, and forcing them to invest nearly their complete quantum of daily energy to avoid the trauma-related symptoms or cope with the vivid, threatening re-experience attacks shutting down their ability to readapt to normal life again.

For one reason or another (e.g. previous trauma, concurrent life experiences, personality characteristics, and/or extremity of the event), the ***salutogenic*** (i.e., ***recovering***, ***health promoting***, and ***rehabilitating***) physical, emotional, and cognitive working-through of the trauma stops and urges for professional trauma care and therapy reduce.

Early trauma intervention and support may lessen the suffering for trauma survivors but will probably never prevent them from developing long-term sequelae or chronic PTSD. Once an impact has been traumatizing and there has been this overwhelming objective or subjective contact with death, the damage is done and nothing can revert this. As is described in the "Acute Reactions" section of this chapter, it is considered that, at least in some cases, there is a possibility for on-scene (peri-traumatic) primary prevention of post-traumatic sequelae, but these chances are rare and often unexploited.

Acute Reactions after Potentially Traumatic Events: Indirect Victims and Significant Others

An explorative field research into the experience of emotional distress by police, fire, and emergency medical services personnel brought an enormous amount of anecdotal data on how professional emergency responders manage stressful events in practice. These course-and-discussion sessions demonstrated in the first place that firefighters and crisis responders are ***doers rather than thinkers and talkers.***

During exercises with fire and emergency medical services personnel, it became increasingly clear that it is essential to know their world or, ideally, to be part of it to understand the way in which they react to traumatogenic events.

The firefighter or paramedic does not tolerate "busybodies" and does not want to feel a victim. He realizes that the borderline between success and failure, between saving and not being able to save, and therefore between being a hero or a "victim" is very thin indeed.

In the group, the following emotions usually surface; often overpowering impotence, a hated feeling of helplessness, a paralyzing grief about the human (and very recognizable) suffering of the victims, the intense guilt of not having been able to do more, and the anger generated by all this. This is not what they joined the fire brigade or the ambulance service for. The emotionally disturbing intervention may be considered as a difficult puzzle, from which the pieces have to be put back together again, to allow the involved rescuers to fully understand the context in which the intervention took place.

The acute psychological experience of a potentially traumatizing event is one of extreme powerlessness and loss of control. The victim loses his mouth as if his willpower were eliminated. At the same time this event causes a sudden and unexpected dislocation of the work and/or living conditions. Nothing will ever be the same again. There is always the threat of death or serious damage to psychological and physical integrity of the self or the other.

The caregiver can, in many instances, no longer maintain his image of the world. The basic assumptions and expectations about life are no longer valid. Everything, even in the practice of fire-fighting, becomes dishonest, unjust, unpredictable, and dangerous. There is danger behind every corner. Training no longer stands for controllability. Every intervention means "danger." Partners become afraid with every call up and so on.

During interventions, many emergency responders show a narrowing of attention, known in literature as the Easterbrook claim (Easterbrook, 1959). This narrowing of attention leads to a diminished capacity to take cues or information elements from the environment in which an event takes place (Bruner, Matter, & Papanek, 1955; Easterbrook, 1959; Eysenck, 1982; Mandler, 1975). It is therefore often very difficult for caregivers to come to a meaningful reconstruction of their intervention which resembles a giant puzzle from which they hold only a limited number of pieces. This makes it very difficult to come to a global image of the intervention. Yet it is indispensable to work through the event in a healthy way.

Caregivers often work on "automatic pilot." In this way most of the actions during the first instances of a traumatic intervention happen automatically, almost instinctively, and may seem unreal. Children are often "dolls" under such conditions, acquaintances "anonymous," injured or dead victims partly "dehumanized" through black humor to keep a psychological and emotional distance, and so on.

But the moment comes when the automatic pilot is promptly switched off. After the intervention, we know this phenomenon as the emotional *post-fact collapse*. During long interventions, one precise stimulus may surface to stop the automatic pilot, such as an image or impression the victim attaches to a relative, a teddy bear or child's doll, or other stimulus that pierces the hardness or armor of the caregiver. And from that moment onward, he starts to function mainly as a vulnerable individual. And he cannot keep this up for long. Once the intense experience is over and the danger averted, the caregiver in question gets an insight – albeit partly – into what has really happened and how it is acted. From that moment on, the *trauma video merry-go-round* begins. Because of the fragmented experience during the intervention, every caregiver starts to reconsider – read "ruminate on" – the events, wondering if he and his colleagues should or could not have done more. The more holes there are in the experience of the event, the longer the questioning process takes and the longer the mind ruminates on the experience.

The next part of this text sheds light on the psychosocial matrix as a framework for emotional triage of trauma victims. This emotional triage could be the starting point for the development of a series of support activities, from *psychological first aid* to *emotional and psychological uncoupling* (closely related to *psychological debriefing*) and *working through*, for the different victims' categories of a certain potentially traumatic event. It will become clear that the type of support has to vary as a function of the type of (potentially traumatic) impact or victims.

The Psychosocial Matrix for Crisis Psychological Support

The psychosocial matrix is a 3 × 3 matrix in which we find respectively in the rows and the columns: (1) the *primary*, *secondary*, and *tertiary victims*, belonging to one of these three categories depending on the type of potentially traumatizing impact suffered; and (2) the *primary*, *secondary*, and *tertiary prevention*, depending on when the trauma support takes place. The concrete realization of the complete framework for psychological (crisis) support consists, on the one hand, of a kind of psychological triage to sort out the different kinds of victims, dividing them in three different categories, and, on the other hand, the selection of the appropriate support technique, at the right moment and carried out by the right people, thus trying to realize an optimal fit between victims and the type of support they get. A third dimension might be added to the matrix, namely, the type of event: potentially traumatic (traumatogenic), potentially depressing (depressogenic), and potentially exhausting (exhaustogenic).

The *primary victims* in this model are the direct victims, those who have to be rescued and/or medically saved, who may have been directly confronted with the life-threatening potentially traumatic stimuli.

The *secondary victims* are the significant others, closely related to the primary victims or playing a significant role as bystanders, in the first rescue attempts (before the emergency services arrive), or providing the first assistance to the primary victims and their families. The *social tissue* of significant others – relatives, family members, friends, colleagues, and so on – creates a victim's dendrite of people who can be considered to be secondary victims. It would appear that, for each primary victim, there are approximately 10–15 secondary victims.

The *tertiary victims* are the professionally involved people, caregivers or law and order personnel – fire and rescue personnel, police, emergency medical services, and so on – who have been in direct contact with the primary and/or secondary victims.

With respect to the prevention, this trifurcated subdivision can be used, differentiating between primary, secondary, and tertiary prevention.

While, strictly speaking, *primary prevention* would include everything which is done to prevent the emotionally disturbing impact itself, we like to use a broader and perhaps less conventional definition of primary prevention, taking the whole series of activities of trauma education and preparation, training, and the creation of intervention models and structures, even considering the on-scene support along with the peri-traumatic first psychological support (cf. example of the tactics for victims aid by firefighters during the extrication and rescue of motor vehicle accident) to be a kind of primary trauma prevention.

If potentially traumatic or life-threatening events lead to hyperarousal states in which the victims have to fight for their lives and mobilize all possible animal-like survival mechanisms, sometimes going into dissociative behavior, these acute reactions may be predictive of later chronic trauma. Every action which can prevent these states of hyperarousal (i.e., every support lowering arousal in trauma victims, calming down, nurturing, etc.) and possibly avoid peri-traumatic dissociation, keeping the on-scene victims grounded, could be considered as primary prevention of long-term psychological trauma.

The immediate support, both on the scene of the accident or in temporary support centers on the field, carried out by the caregivers of fire and rescue or ambulance services, or even provided by volunteers from civil defense, Red Cross, or other services, is also considered to be primary prevention. *On-scene buddy aid* or *peer support* (the help for colleagues and from colleagues on the scene of the accident), then the *initial emotional and physical recuperative talk sessions* (sometimes described as *defusing*), are also considered to be *primary prevention of post-traumatic sequelae in tertiary victims.* These primary preventive support activities could be carried out by nonprofessional caregivers or peers.

Secondary prevention, in the post-immediate stage, essentially consists of (1) a quick and adequate detection of post-event psychological sequelae; and (2) a rapid and adequate intervention, carried out by the appropriate people, and all this at the right time. Secondary prevention targets the early detection of problematic responses or coping styles in victims, and a sufficient intervention tailored to the needs of the victims, in order to prevent these problems exacerbating and becoming chronic in the long term. It is considered that most *early intervention protocols* are a kind of secondary prevention (for tertiary victims).

These secondary preventive support activities could, in some cases, also be carried out by nonprofessional caregivers, as long as they work under permanent supervision of well-trained and professional mental health specialists.

Without going into too much detail, it would appear that the currently known models of *critical incident stress debriefing* or *psychological debriefing* have been designed as a secondary prevention for tertiary victims, *which should not be used to support or debrief primary or secondary victims.* The negative publicity surrounding these intervention techniques is not due to these protocols but to the incorrect use of support techniques with people who should not be re-exposed to their trauma again so quickly after the impact or after an insufficient physical, emotional, and psychological recovery period (as we will discuss in this chapter).

Tertiary prevention, finally, aims at the full professional curative trauma care, which can become necessary for the different categories of victims after several months during which these victims tried to cope with their experiences without any professional help. In this case, trauma victims can suffer from what is called PTSD in the DSM-IV (American Psychiatric Association, 1994).

Tertiary prevention could mean psychotherapeutical action from different perspectives, as there are (non-exhaustively) (1) (brief) cognitive-behavioral therapy, (2) psychoanalytically inspired trauma therapy, (3) (brief) eclectic therapy, (4) sensorimotor trauma therapy, (5) creative and/or arts therapy, (6) experiential (and/or existential) trauma therapy, (7) eye movement desensitization and reprocessing therapy, and (8) integrative trauma therapy.

In nearly all trauma models, the first stage of the therapy will aim to reduce and stabilize the current trauma symptoms and complaints, followed by a stage in which there will be a regurgitation of the trauma-related material, mostly using narrative exploration and cognitive reframing techniques, and, finally, in the last stage working toward integration of the (loss and) trauma in the personal life story of the survivor.

Primary Prevention of Psychological Trauma in Primary and Secondary Victims of Traumatogenic Events

As mentioned in this chapter, in most cases the acute (peri-traumatic) stage for the victims of a traumatogenic event could range from a few seconds to several hours. In many cases, victims will even need between 24 and 48 hours to "wake up" again from their *trauma trance* (*dissociative state*) or *tunnel* (cf. the forementioned dissociative responses and survival stages). When leaving these (sometimes functional) dissociative states, the primary victims start to slowly realize what they have experienced or how lucky they were to survive. They are still fearful that the threat will return and/or that in a repetitious way they will be aspirated back into their "traumatic tunnel" when the surrounding reality is still too cruel, extreme and/or overwhelming. The need to escape the reality remains. Primary trauma victims will only return back to reality very gradually and only when they perceive again a sense of safety, security, and stability in the surrounding environment.

For rescue workers and caregivers, it is very important to know how to guide and support the primary victims on their way back to reality, calm them down, help them ground themselves during and immediately after the rescue operations, and assist them in their first reorientation attempts. In particular, survivors who show dissociative responses need to be "grounded" in order to prevent them from staying overwhelmed by intrusive recollections of the event.

The first signs of post-impact recovery appear when victims start to search for information, about what happened, in the surrounding environment. This yearning for information in the immediate post-impact stage makes the primary victims very fragile and suggestible with respect to the first rumors about what happened.

The mental reconstruction of what really happened is very difficult for survivors since they all suffered more or less from a narrowing of their field of consciousness, focusing on peri-traumatic details which were relevant for their own survival or rescue. Lots of trauma-related, essentially preverbal sensations about speechless terror have been registered but need much more elaboration before they can be transformed into senseful traumatic memories. Therefore, psychological stabilization after traumatogenic events has to aim at physical recovery and cooling down but not at immediate verbal expression.

The on-scene support for primary, secondary, and tertiary victims could be executed along the same principles. The first psychological help in the peri-traumatic and immediate post-impact stage should aim to reduce the level of arousal and re-create basic security and safety around the traumatized victim. One could assume that the natural support a mother provides to a child in a state of anxiety, in trying to secure and to calm down, is a similar kind of support a traumatized victim needs.

Primary Prevention of Psychological Trauma in Tertiary Victims of Traumatogenic Events: First Assistance, Immediate Physical Recuperation, and Emotional Uncoupling

In what follows, the discussion of emotionally disturbing, shocking, or traumatizing interventions, in group and according to procedure, will be called psychological uncoupling (PU). It is, in fact, an individual or group oriented intervention – based on the commonly known psychological debriefing (PD) process – in which the most important elements of a past emotionally disturbing experience are treated shortly after the events. Lately psychological debriefing – mostly based on the elementary protocol of Critical Incident Stress Debriefing (Mitchell & Everly, 1983) – had been generally advised as the most appropriate stress-management technique for high-risk professions such as those providing aid in disasters, fire-fighters, military personnel, police personnel, and so on (Bergman & Queen, 1986; Dunning & Silva, 1980; Griffin, 1987; Mitchell, 1981; Raphael, 1986; Wagner, 1979; Yarmey & Jones, 1985). However, now there are a number of variants of the original Mitchell protocol of psychological debriefing widely used in psychological crisis intervention services. The problem is that in many cases outcome expectances of psychological debriefing have been too high and more recently specialists have questioned the effects of psychological debriefing (Van Emmerik *et al.*, 2002).

Firstly, the term "debriefing" can be misleading because many of its users do not even fully understand the meaning of it (can you debrief people who were not briefed in advance?) or believe that debriefings are simple because the term "debriefing" is very familiar to them from the point of view of "operational debriefing." Secondly, it is suggested that the outcome criterion – that is, the prevention of PTSD (in itself a debatable "condition") – may be the wrong one. The author's research (De Soir & Zech, 2011) indicated that participants from police services reported very positive effects after taking part in psychological uncoupling sessions (role clarification, reconstruction, enhanced understanding of the event, recognition, emotional support, etc.) but did not show less post-traumatic symptoms afterward.

In some cases, the testimonies of witnesses or direct victims can be essential in this reconstruction process. Very briefly, the various goals of PU would include *ventilating tensions and frustrations* (in many cases based upon the behavior of the press and "disaster tourists"); *normalization, comprehension,* and *legitimization of occurring reactions and feelings; creating a cognitive restructuration* (we hope to replace negative cognitions by positive ones in the course of the discussion), *creating a* (almost mythical) *bond among fellow caregivers*; and the *identification of those participants who may be supposed to run a high risk of problematic assimilation.*

Conclusions to the Implementation of Psychological Support for Fire and Rescue Services Personnel

From the perspective of the management of fire and rescue services, there are many reasons why it is important to provide quality support to rescuers who are constantly exposed to stressful and emotionally disturbing events. For example, providing adequate support may help to reduce absenteeism, lower the cost of compensation and litigation, and improve performance (Devilly & Cotton, 2003; Plant, 2000; Robinson, 2004). Accordingly, it is generally accepted that a healthy workforce is a more productive workforce (Devilly & Cotton, 2003).

In addition, most countries now have legal and moral responsibility to offer stress and trauma support to exposed personnel in risk organizations. Every organization has a legal duty of care to provide a safe working environment for employees, as dictated by occupational health and safety policy and legislation. This responsibility extends to protecting employees from possible psychological harm and suffering. However, the nature of fire and rescue interventions implies that it may be impossible for personnel to avoid exposure to stressful events.

These legal and moral obligations make it clear that it is unacceptable not to offer support after emotional disturbing and potentially traumatizing interventions.

Notes

1. The principle of triggering is one of the central problems in the working-through process of trauma victims. A psychological trauma is always characterized by a combination of several symptoms clusters, normally (1) the original potentially traumatizing event being a more or less direct contact with a life-threatening situation, (2) a cluster of symptoms in which the original event is re-experienced, (3) a cluster of symptoms in which the original event is denied or avoided, (4) a cluster of symptoms characterized by hyperarousal, and (5) a social dysfunctioning of the stricken individual. When trauma victims are confronted by various stimuli which make them remember or think about the original traumatizing event, these stimuli can *trigger* the same reactions (event dissociative responses) as the original event itself. The human brain does not seem to differentiate between the original event and the re-experienced events, leading to a potential neurobiological storm.
2. This anecdotal evidence was generated by visiting Belgium, Holland, France, Australia, New Zealand, Italy, Switzerland, Argentina, Mexico, and former Eastern European countries as a coordinator and trainer of fire-fighter and emergency medical stress teams. General exercises (a minimum of three hours) on the management of emotionally disturbing (potentially traumatizing) interventions in the fire-fighting and rescuing practice were held in more than two hundred fire brigades, ambulance services, and emergency medical departments. They consisted of three parts: an experience-oriented analysis of traumatic interventions, a practice-oriented discussion of real-life situations, and a theoretic (psycho-educative) placement of the mechanisms and phenomena under discussion.

References

American Psychiatric Association. (1994). *Diagnostic and statistical manual for mental disorders* (4th ed.). Washington DC: Author.

Bergman, L. H., & Queen, T. (1986). Critical incident stress. Part 1. *Fire Command*, 52–56.

Bisson, J. I., Jenkins, P. L., Alexander, J., *et al.* (1997). Randomised controlled trial of psychological debriefing for victims of acute burn trauma. *British Journal of Psychiatry*, *171*, 78–81.

Bisson, J. I., McFarlane, A. C., & Rose, S. (2000). Psychological debriefing. In E. B. Foa, T. M. Keane, & M. J. Friedman (Eds.). *Effective treatments for PTSD* (pp. 39–59). New York, NY: Guilford Press.

Bruner, J. S., Matter, J., & Papanek, M. L. (1955). Breadth of learning as a function of drive level and mechanization. *Psychological Review*, *62*, 1–10.

Crocq, L. (2000). Le retour des enfers et son message. *Revue Francophone du Stress et du Trauma*, *1*(1), 5–19.

Deahl, M. (2000). Psychological debriefing: Controversy and challenge. *Australian and New Zealand Journal of Psychiatry*, *34*, 929–939.

De Clercq, M., & Lebigot, F. (2001). *Les traumatismes psychiques.* Paris: Masson.

De Soir, E., & Vermeiren, E. (Eds.). (2002). *Les debriefings psychologiques en question.* Antwerp: Garant Uitgevers.

De Soir, E., & Zech, E. (2011). [Unpublished data].

Devilly, G. J., & Cotton, P. (2003). Psychological debriefing and the workplace: Defining a concept, controversies and guidelines for intervention. *Australian Psychologist, 38*(2), 144–150.

Dunning, C., & Silva, M. (1980) Disaster-induced trauma in rescue workers. *Victimology, 5*, 287–297.

Dyregrov, A. (1997). The process in psychological debriefings. *Journal of Traumatic Stress, 10*(4), 589–605.

Easterbrook, J. A. (1959). The effect of emotion on cue utilization and the organisation of behavior. *Psychological Review, 66*, 99–113.

Eysenck, M. W. (1982). *Attention and arousal: Cognition and performance.* Berlin: Springer-Verlag.

Griffin, C. A. (1987). Community disasters and posttraumatic stress disorder: a debriefing model for response. In T. Williams (Ed.), *Post-traumatic stress disorders: A handbook for clinicians* (pp. 293–298). Cincinnati, OH: American Disabled Veterans.

Hobbs, M., Mayou, R., Harrison, B., & Worlock, P. (1996). A randomized controlled trial of psychological debriefing for victims of road traffic accidents. *British Medical Journal, 313*, 1438–1439.

Kenardy, J. (2000). The current status of psychological debriefing: It may do more harm than good. *British Medical Journal, 321*, 1032–1033.

Kenardy, J., & Carr, V. (1996). Imbalance in the debriefing debate. *Bulletin of Australia Psychology* Society, *18*, 4–6.

Lebigot, F. (2000). La clinique de la névrose traumatique dans son rapport à l'événement. *Revue Francophone du Stress et du Trauma, 1*(1), 21–25.

Mandler, G. (1975). *Mind and emotion.* Huntington, NY: Krieger.

Mayou, R. A., Black, J., & Bryant, B. (2000). Unconsciousness, amnesia and psychiatric symptoms following road traffic accident injury. *British Journal of Psychiatry, 177*, 540–545.

Mitchell, J. T. (1981). *Emergency response to crisis: A crisis intervention guidebook of emergency service personnel.* Bowie, MD: RIBrady.

Mitchell, J. T., & Everly, G. E. (1993). *Critical Incident Stress Debriefing: An operations manual for the prevention of traumatic stress among emergency services and disaster workers.* Ellicott City, MD: Chevron.

Nijenhuis, E. R. S. (1999). *Somatoform dissociation: Phenomena, measurement, and theoretical issues.* Assen: Van Gorcum.

Nijenhuis, E. R. S., & Van der Hart, O. (1999). Forgetting and reexperiencing trauma: From anesthesia to pain. In J. Goodwin & R. Attias (Eds.), *Splintered reflections: Images of the body in trauma.* New York, NY: Basic Books.

Raphael, B. (1986). *When disaster strikes.* New York, NY: Basic Books.

Raphael, B., & Meldrum, L. (1995). Does debriefing after psychological trauma work? *British Medical Journal, 310*, 1479–1480.

Robinson, R. (2004). Counterbalancing misrepresentations of Critical Incident Stress Debriefing and Critical Incident Stress Management. *Australian Psychologist, 39*(1), 29–34.

Rose, S., & Tehrani, N. (2002). History, methods and development of psychological debriefing. In *Psychological debriefing: Profession Practice Board Working Party* (pp. 2–7). Leicester: British Psychological Society.

Shepard, B. (2000). *A war of nerves: soldiers and psychiatrists.* Jonathan Cape, London.

Van der Hart, O., Brown, P., & Graafland, M. (1999). Trauma-induced dissociative amnesia in World War I combat soldiers. *Australian and New Zealand Journal of Psychiatry, 33*, 37–46.

Van der Hart, O., Nijenhuis, E. R. S., & Steele, K. (2006). *The haunted self: Structural dissociation and the treatment of chronic traumatization*. New York: Norton.

Van Emmerik, A. A., Kamphuis, J. H., Hulsbosch, A. M., & Emmelkamp, P. M. (2002). Single session debriefing after psychological trauma: A meta-analysis. *Lancet*, *360*(9335), 766–771.

Wagner, M. (1979). Airline disaster: A stress debriefing program for police. *Police Stress*, *2*, 16–20.

Yarmey, M., & Jones, D. R. (1985). Secondary disaster victims: The emotional impact of recovering and identifying human remains. *American Journal of Psychiatry*, *142*, 303–307.

24

Working with Tsunami Survivors in South India: The Problem Lies in a Four-Letter Word

Sue Santi Ireson and Hash Patel

This chapter describes the work done in South India following the devastation of the 2004 tsunami. Most of the communities affected were small villages located on both the east and the west coasts of the southern tip of India. The villagers were predominately fishermen and those concerned with the fishing industry. The initial helping authorities consisted of those providing basic care; no agencies were providing psychological support.

The Indian Christian church was instrumental in recognizing the need for this and was beginning to organize localized support, but there were very few trained personnel who knew how to deal effectively with the post traumatic stress which was beginning to manifest itself in many of the survivors.

Our previous work had been developing counseling and psychotherapy courses at Bangalore University's Sampurna Montfort College. Montfort College provides education at the post-graduate and doctoral levels in the disciplines of psychology, counseling, and education. We felt privileged to be asked to assess and evaluate the need and feasibility of delivering specialist training in the treatment of post traumatic stress for the traumatized survivors of the tsunami and then to devise a suitable training course.

This chapter details the situation we found in the southern part of Tamil Nadu and the work we carried out. It describes the model we developed and ends with our recommendations for providing practical, material, and psychological support for people following catastrophic events (Figure 24.1).

Introduction

> Now death has shaken your faith, "*Why?*" "*Why must life be one of sorrow?*" "*Why?*" There are no pat answers. No one completely understands the mystery of death. Even if the questions were answered, would your pain be eased, your loneliness less terrible?

International Handbook of Workplace Trauma Support, First Edition. Edited by Rick Hughes, Andrew Kinder, and Cary L. Cooper.

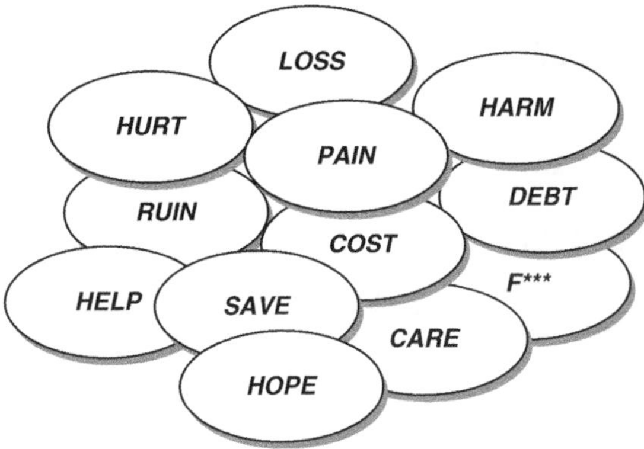

Figure 24.1 The mass of "four-letter words."

> "*Why*" may be more than a question. It may be an agonizing cry for a heart-breaking loss, an expression of distress, disappointment, bewilderment, alienation, and betrayal. There is no answer that bridges the chasm of irreparable separation. There is no satisfactory response for an unresolvable dilemma. Not all questions have complete answers. Unanswered "Why's" are part of life. The search may continue but the real question might be "*How [do I] pick up the pieces and go on living as meaningfully as possible?*"(Grollman, 2001)

The Impact of the 2004 Tsunami on the Southern Indian Coastline

In 2004, the Indian Ocean tsunami hit the southern coast of India, leaving in its wake more than 8000 people dead, 126 182 families homeless, and several thousand people injured. In total, the million people living in coastal fishing communities, such as in Tamil Nadu, were hardest hit (Shanthasheela, 2009).

Survivors experienced multiple intense stressors. Almost all had family or friends who had died. Those who survived had experienced the most intense fear as their lives and the lives of those they loved were threatened. In addition to this, many homes were destroyed, fishing boats and nets were lost, livestock was lost, money and personal papers were lost, and coconut plantations and gardens were lost or damaged by salination. The outcomes of the disaster are likely to be extremely adverse for a long time. As residents of Tamil Nadu struggled to understand what happened and to put their lives back together, helping agents were faced with multiple challenges as they attempted to provide much needed relief services (Asian Development Bank, 2006).

Poor-quality roads and a poorly developed basic infrastructure restricted access to these areas, causing an uneven distribution of relief supplies which resulted in resentment and frustration among survivors (Davidson, 2006; Sheth *et al.*, 2009).

One of the children in the Muttum village school drew for us a picture (Figure 24.2) of her traumatic experience of the tsunami wave overwhelming her village. Death resulting from sudden or traumatic accident or disaster raises an array of issues for the survivors – often somewhat different from those experienced after an expected or anticipated death. After a sudden loss, the grief response is often intensified since there is little or no opportunity to prepare for the loss, to say one's goodbyes, to finish

Figure 24.2 A drawing of the tsunami by Maya*, age 9.

*Name changed to protect anonymity.

unfinished business, or to prepare for the bereavement. Families and friends are suddenly forced to face the loss of loved ones instantaneously and without warning and often have to deal with multiple bereavements.

The tsunami shattered people's sense of order as they no longer could rely on the safe and predictable familiar world, but had to accept a new world full of chaos. Survivors of sudden loss will probably experience a greater sense of vulnerability and heightened anxiety and fear for themselves, their families, and their friends. It takes time for the family to reorganize. Families may be left feeling in a state of perpetual disarray with a lingering sense of unease and disorganization. Marital and other family relationships are very likely to become strained.

Additional problems arise if the grieving survivor was involved with the disaster or was physically injured. Memories of the disaster may dominate the person's mind. They may be overwhelmed with feelings of numbness, unreality, and fear. The bereaved person may suffer from "survivor guilt," wondering why they survived when others have died and believing that they could have or should have done more to prevent the tragedy.

Families may feel unable to fully grieve and reach closure in situations when there is no positive confirmation of the death, when the physical body has not been recovered, or when it cannot be positively identified. This can make it difficult to grasp the reality that the death has occurred as survivors continue to hope. Only when the reality is fully grasped can survivors move past the trauma to face the full realization and the pain of grief.

The search for meaning of the loss can challenge a survivor's religious and spiritual beliefs. Sudden losses in particular can precipitate an existential crisis as the survivor searches for meaning. They start questioning their internal belief system and values. Goals and plans which were important the week prior to the event suddenly may seem trivial in comparison. Survivors are forced to look at and re-evaluate their life priorities.

Traumatized survivors had no time to save loved ones or personal possessions. Tsunami survivors and refugees face specific and complex stressors and are at very high risk of developing clinical syndromes such as post-traumatic stress disorder (PTSD), major depression, substance abuse, and a range of social and emotional difficulties. Further experiences such as vehicle or workplace accidents, minor floods, and inter-ethnic violence might reawaken this trauma in later years (Tedeschi, 1999).

Our Brief

We were invited to provide some training in dealing with post-traumatic stress (PTS) by the Most Reverend Leon Tharmaraj, the then Bishop of Nagercoil, in the state of Tamil Nadu, and by Sampurna Montfort College in Bangalore (Bangalore University) where for some years we had been involved in their development of post-graduate counseling and psychotherapy programs. Our work was based in the southernmost tip of India in the Kanyakumari region.

Concerned Christian organizations in the affected areas had recognized the growing need for psychological help for the survivors of the Tsunami and contacted Sampurna Montfort College. They asked us if we could provide them with support, advice, and a training program. Although counseling psychology and psychotherapy are emerging professions in India there was a dearth of trainers to coordinate and underpin the work of counselors. The only accredited training organization in South India was Sampurna Montfort– established only in 1995 (Bangalore is some 750 km distance away from Kanyakumari). There was a great shortage of suitably qualified and experienced trainers who could work with the cross-disciplinary group, which is why we were asked to undertake this project. (As mentioned, we had been actively involved in developing counseling and psychotherapy courses at Montfort.)

Prior to developing the training, we visited the affected area. The region in southern Tamil Nadu in which we were going to work was predominantly Christian, but by some strange phenomenon, Hindu beaches nearby seemed to have escaped the tsunami leading to considerable confusion. Many people believed that it was as a result of their religious beliefs being faulty that they had been singled out. This meant that some individuals chose not to seek advice and guidance from their religious leaders, whom they no longer trusted.

We observed that practical helping agencies had provided basic shelter – food, water, and sanitation. Many people had to be housed in temporary huts built from corrugated iron, which were unbearably hot in the daytime. At that time there was no obvious plan for rebuilding the community, though people were very anxious as the rumor spread that the government would not allow them to rebuild near the sea. The fishermen found this particularly threatening and distressing as they believed they would be unable to predict the sea's moods in their time-honored way so they could know what to expect.

Few fishermen still had boats; the waves had smashed them. Farmers and smallholders found their land ruined by the salt water and their crops destroyed. In this normally hardworking and industrious community, the enforced idleness created further problems. Men in particular found it difficult to cope with their loss of role, so gambling and drinking increased significantly as displacement activities.

The survivors in the villages we visited were impoverished people with simple belief systems who were already quite disenfranchised. This disaster added to their emotional burden. As a general observation, the community was familiar with coping with very little in their lives, and we found many were able draw on their own inner strength using support from family and friends, agency workers, priests, and their religious and cultural practices and rituals (Figure 24.3).

A crisis training center had been set up by the Most Reverend Leon Tharmaraj in the Bishop's Palace at Nagercoil. One hundred and forty-seven Tamil-speaking professionals had been drawn from health care, social services, and religious institutions (along with eight people who had completed counselor training) from across South India. They were to be trained in active listening skills and in ways of identifying

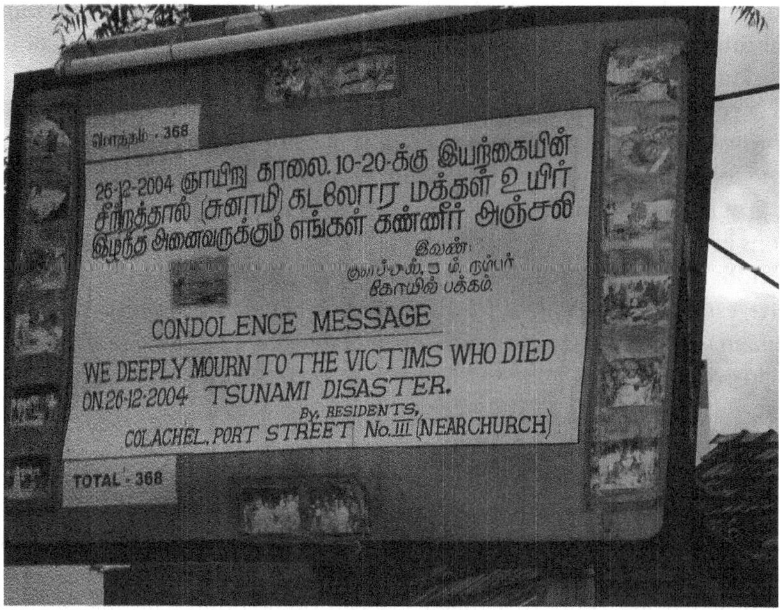

Figure 24.3 Memorial notice erected by survivors in Colachel village above one street's mass grave.

trauma symptoms, debriefing, and strategies for the management of PTS. The language barrier presented a major problem for effective training.

The Challenges Presented to Us

- *The concept of counseling in India.* Counseling has long been available in India but not in a form recognized in Western methodologies. Counseling in India tends to be advice giving, often from people who are seen as experienced and wise. We found no evidence of practical training in issues of loss and trauma having been incorporated into counseling and psychotherapy courses, so we had to devise a basic course, which could upskill both trained counselors and helpers from other professions.
- *Training methodologies.* From our previous experience of teaching at Montfort College we knew that teaching in India, even at the graduate and post-graduate levels, is very didactic; rote learning is the norm, and experiential exercises are seldom used. As Western trainers we value and use experiential trainee-centered approaches to consolidate learning. With this in mind, our training program was reorganized so we initially used a mostly "chalk and talk" approach. We then gradually moved the training program to a much more experiential, trainee-centered methodology.
- *Language problems.* Tamil and Malayalam are the local languages. English is becoming more used as a universal language in India, but the local people we were working with had maybe only a few words of English. This made it difficult to work in complex concepts without a translator. And the translator also had to understand the concepts himself.
- *Caste system.* We had an additional advantage in helping to open up the issues presented by working with different castes because Hash Patel (one of the authors of this chapter) is of Indian origin. Working across the caste system still presents problems, even today, and from a Western perspective it is difficult to imagine the depth to which this continues to affect even the most educated people. For example, we had noticed in some villages lower caste people being refused access by the fishermen to the water bowsers provided by Oxfam.
- *Lack of trained counselors.* As mentioned above, counseling psychology is a new profession in India, and there is still a shortage of trained counselors, supervisors, and trainers. The remoteness of the region meant that there was no network of counselors locally and they had to be drawn from a huge area of South India. This caused further language problems as some of the counselors did not speak Tamil or Malayalam, which put a lot of pressure on the three people who did speak these languages – and who had only the most basic of counseling training themselves.
- *Lack of theoretical understanding.* There was a lack of understanding by trainees of theoretical underpinnings of trauma work and somatized responses (panic attacks, flashbacks, etc.).
- *Lack of training resources.* We had access to only blackboard and chalk; there was a very erratic electricity supply that we could not rely on using computers, overhead projectors (OHPs), and so on.
- *Managing a very large training group.* The logistics of managing such a large group were difficult and challenging to us because we normally work with much smaller groups.

Figure 24.4 Brother Mathew SG, Hash Patel, Sue Santi Ireson, and Sister Joice SIC. Our team at one of the locations was accommodated in a family counseling center funded by Caritas (Switzerland) and the Diocese of Kottar.

Our Task

We had the task of upskilling the counselors and other helpers who had been hastily assembled through the network of the Catholic Church in South India. Their experience level of counseling was very varied with some who had not even been trained in the most basic of listening skills. Most had no experience of receiving supervision or therapy (Figure 24.4).

Working in the aftermath of a disaster requires helpers who are able to work comfortably in a wide variety of ways. They need to have a high level of interactive and social skills and be able to work with the widest range of needs: groups, families, children, marginalized groups, and so on. For some time after the crisis the helper's work may be more about making soup and listening sympathetically to people in fairly brief and informal encounters. It may also be about reiterating and clarifying information already given to the traumatized person. It is important to recognize that psychological support is optimally placed after setting up practical support such as shelter, food, water, and so on. The helper should know which helping agencies, nongovernment organizations (NGOs), and so on to contact for various different forms of help. It is confusing and may be distressing if the counselor seems not to know these important details.

Whilst of course it is essential to maintain the usual good boundaries and to offer the core counseling conditions of acceptance, congruence, and empathy, most encounters may not be the usual 50 minutes of face-to-face counseling in a quiet room. They may be much more fleeting and more public and may touch on things that may not normally be in a counseling session. The counselor can offer a great deal of comfort by just remembering people's names, and something of the issues they are facing. Simple questions about the person's welfare need not be looking for deep answers, but will quickly make a connection with the person, who may then wish to talk more.

For those who do want to talk to a counselor in more depth, a more private area needs to be available; however, counselors and other helpers should be available in the public areas at other times. This helps to demystify the helping process so that the survivors see the helpers as a friendly and helpful part of the community team.

The Training Program

We designed a six-day bespoke training program for this heterogeneous group of 147 people to respond to the urgent need for skilled workers in the field. Realistically we could only provide brief training in active listening skills alongside the teaching about the effects of trauma – psychologically, physiologically, and socially.

Using our trauma pathway model (Figure 24.5), we emphasized that the counselors must be able to work with anyone regardless of gender, religion, or caste. In India these three clearly define people's roles and status within communities.

Our training approach

Our training philosophy is based in a humanistic framework with the core conditions of unconditional positive regard, empathy, and congruence. We use an eclectic mix of techniques drawn from person-centered counseling, Gestalt therapy, art therapy, family therapy, transactional analysis (TA), and cognitive-behavioral therapy (CBT). We find a trainee-centered, experiential method of teaching and learning most effective. Our training program was designed to maximise learning since we were expected to deliver a useful framework for the trainees within a very short period of time. We intended that they would be able to offer "good enough" listening support to their clients and at the same time be able to normalize clients' experiences.

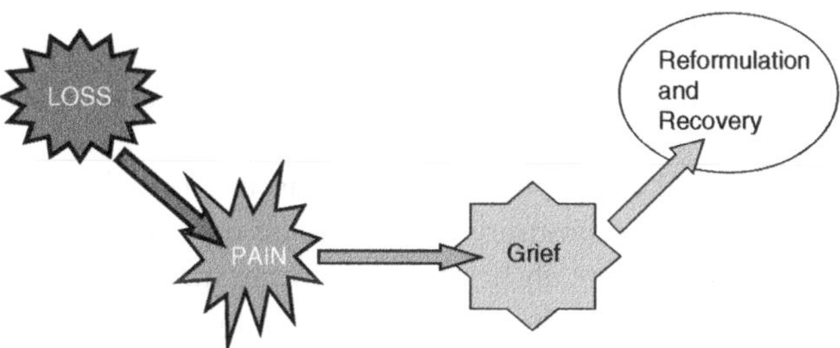

Figure 24.5 The trauma pathway model we devised and used for training.

We also intended that they would be able to support each other in an ongoing way as they continued to work in these often distressing circumstances.

Content of the training program

One of our main tasks in training was quickly to develop and ensure cohesion in the group, so that the group itself could provide support to its members. A secondary aim was to ensure that we left a group able to continue working successfully in this stressful environment after we departed.

We were presented with training this very large group in a very short space of time, so we decided to alternate between working with all of them and splitting them into two halves. Even so, the split groups of 70+ were still extremely large, so our strategy to contain this problem was to provide the following:

- Clear and well-managed boundaries.
- Highly organized schedule.
- Clearly focused tasks.
- Varied activities.
- Small- and large-group work.
- Demonstrations.
- Buddy system.

We had to get used to having an interpreter in the room, and also to the interpreter not always understanding what we meant, so sometimes we had to explain more to the interpreter before he could translate in a meaningful way. This sometimes meant that we felt we were focusing away from the group process and giving a rather stilted delivery. Fortunately, we had both experienced working with students whose spoken English was not always very good, or who found our English accents difficult to understand.

We started with group exercises to help consolidate the group as few of them knew each other and they were drawn from very different disciplines. We felt it was very important that they were able to work cooperatively and support each other. Our other requirement was to disseminate factual information about PTS.

We had written a leaflet entitled "Coping after the Tsunami" for the trainees. This was also translated into Tamil for them to distribute freely to their clients and other workers.

We had to regularly remind the trainees to be mindful that moving quickly to advice giving and problem solving for survivors could actually disempower the very people they were trying to help. We emphasized the points below and gave experiential exercises to help consolidate them:

- That the power of listening nonjudgmentally should never be underestimated as a healing tool.
- That counselors should avoid putting a Christian bias on what they said to their clients in this multifaith community (we had to be quite directive about this).
- That normalizing feelings is really helpful – reassuring survivors that the strange and upsetting feelings they experience after a disaster (as following other traumatic events) are normal – given what had happened.
- That counselors can help survivors to find effective ways of coping with their ongoing stress using resources within themselves and their family and social groups.

We endeavored to give as much practice in small groups as possible, so as to improve the counselors' listening skills in the context of the huge issues of loss in this post-traumatic period. We included lots of experiential exercises through role play to help to prepare the counselors for the immense emotional impact of the survivors' feelings of loss. We underpinned the work with loss theory developed by Elisabeth Kübler-Ross (1970) emphasizing that when someone is traumatized they suffer loss of their sense of safety, not to mention other losses that the tsunami survivors had experienced.

Our aim was not to try to provide a comprehensive counseling training program, but to put the emphasis on developing active listening skills to enable the helpers to be supportive and effective in debriefing the survivors of the tsunami. We emphasized the following:

- Our training was based on principles of crisis intervention.
- It may not solve all the problems presented during the brief time frame available.
- Sometimes it may be necessary to refer individuals for treatment after a debriefing.
- It is not therapy or a substitute for therapy (though it may be therapeutic).
- This kind of support can accelerate the rate of normal recovery in people who are having normal reactions to abnormal events.

It is important to recognize that there are a variety of ways in which people may express their distress following a traumatic event. Their reactions encompass the states illustrated in Table 24.1 or only some of these. There is no set pattern and no set

Table 24.1 The chart we developed to illustrate some of the physical, emotional, psychological, and behavioral symptoms possible following a traumatic experience

Physical	*Emotional*	*Psychological*	*Behavioural*
Nausea	Anxiety	Slowed thinking	Withdrawal
Upset stomach	Fear	Difficulty making decisions	Restlessness
Diarrhea	Grief	Difficulty in problem solving	Eating more
Chest pains	Emotional outbursts	Depression	Eating less
Change in speech	Feeling uncoordinated	Difficulty calculating	Increased alcohol use
Difficulty breathing	Feeling lost	Difficulty naming common objects	Addictive behaviors
Rapid heartbeat	Sadness	Loss of trust	Panic attacks
Muscle aches	Feeling abandoned	Confusion	Distressing dreams
Headaches	Reliving event	Flashbacks	Increased startle reflex
Chills	Feeling isolated		Poor attention span
Profuse sweating	Worry about others		Wanting to hide
Dizziness	Anger		Limiting contact with others
Sleep disturbances	Irritability		Disorientation, especially to time and place
Startled	Shocked		Displacement activities
More vulnerable to infection			Obsessive behaviors

time scale for these effects. Reactions to trauma have been identified and documented by many researchers such as Kübler-Ross (1970) and Murray Parkes (1996), who, whilst using different terms, are describing the same grieving process. It is very helpful for both survivors and workers to know that these responses, though distressing, are normal.

Sudden losses, like all losses, are likely to affect survivors in many different ways. One cannot compare loss. The greatest loss is the one that the grieving person is suffering. Each loss, whether sudden or not, creates its own unique issues. It is important to allow survivors to grieve in their own individual way.

Some identifiable symptoms of post-traumatic stress

Not everyone experiences the following symptoms. Some people may experience a few, and others may experience many. The severity of symptoms varies with individuals.

Alongside practicing the active listening skills, we taught students about the following key symptoms of post-traumatic stress:

- Shock
- Disbelief
- Regret
- Sadness
- Anger

Shock. People can "freeze" when they are shocked. They appear to have little emotional response, they may be unable to think coherently; they may find it difficult to retain information, and they may find it difficult to remember things they knew before or to do things they previously could do well. They may struggle to make sense of what is being said to them. The consequences of these normal effects of shock are that vital information may not get through to them to their detriment or to the detriment of others. People in shock may also need to be offered a high degree of immediate practical support as they may not be able to take care of either themselves or those dependent on them.

Disbelief. People may not believe that what has happened has actually taken place, and may continue to talk as if life has not changed. Trying to get them to "accept things" may be unhelpful to them, and indeed may further traumatize them. It is important to recognize that this is the person's way of coping with difficult information and they need to come to accept it in their own time.

For example, someone who has been traumatically widowed may talk as if their partner will be coming home quite soon. The listener should not challenge this in the early stages after the event (whilst at the same time not agreeing with it). On some level the person does know what has happened but needs to be given time and space to accept and acknowledge it. It is not unusual for someone to talk as if their loved one is not dead, but at the same time be listening to and contributing to funeral plans.

Regret. People may go back and back over the event in their minds almost as if by doing so they can come to a different outcome. They may say such things as "if only I had gone away that day," "if only I hadn't left my children out playing on their own," "if only I had noticed that something unusual was happening," and so on. Usually there is little difference that their actions would have made to the event, but they may need to be told that several times before they can accept it. Sometimes they have to accept that a relationship with someone has ended without the possibility of differences being sorted out.

Sadness. As realization of what has been lost grows, the person may experience great sadness. Helpers may have to cope with people who are in tears, which may feel difficult or embarrassing. They can empathize with the traumatized person and say things like "Of course you feel upset," "It's not surprising you feel sad after this terrible event," and so on. If counseling is available, this could be offered as the counselor may have more time to allow the person to explore how they are feeling than other helpers, who have other tasks to do.

Anger. People often feel that the most difficult emotion expressed following a trauma is anger. It is perfectly normal to feel angry when you have experienced a loss, whether it is bereavement, your way of life, your peace of mind, your sense of well-being, your identity, or even items precious to you. This is not always recognized and can be interpreted as meaning aggression or a threat to the person on the receiving end. Matters could then escalate to undesirable levels, with traumatized people left feeling misunderstood and mistreated.

The counselors might consider offering other helping strategies for management of people's emotions. Often anger is hardest to deal with, although it is a normal reaction to trauma, though of course it is not acceptable or helpful to allow uncontained anger or aggression toward helpers or others (Kinchin, 2005; Kübler-Ross, 1970).

Supporting the trainees

The group had bonded really well with very interesting cross-professional dissemination of information. For some, working with impoverished and marginalized groups was a new experience and we greatly appreciated the enrichment the experience brought by those who worked with these groups of people. Unfortunately, three members of the group were not able to complete their training, as they found the experience too stressful. However, all the others worked hard at understanding and practicing all the skills we had taught. By the end of our six weeks post-training, another eight people had felt unable to continue and had returned home. However, the rest of the group was in good spirits and doing very well.

To avoid burnout of our trainees, we spent some time focusing on the maintenance of good boundaries, asking them to use the group and a buddy system to look after their own emotional welfare. The buddy system involved pairing up with another trainee with the intention that they would support each other through the training and learning and afterward in the field. They were expected to exchange telephone numbers and to arrange to meet up on a regular basis.

Following the training course, the newly trained helpers were given placements throughout the region. We regularly traveled to all the centers where they were working, supervising and monitoring their work, and being available for support of some key workers within the village communities. The centers we worked in were all provided by Christian establishments where we were welcomed and given the warmest of hospitality.

Where possible we held support meetings for key members of the community; priests, teachers, community leaders, and so on to help them debrief – they were continuing to work, supporting and encouraging the community when they, too, had lost loved ones. In addition, we were sometimes able to act as a resource for NGOs and others working in the field, who were sometimes inexperienced and very distressed by what they saw.

Recommendations

Strategy

The support areas broke down into three distinct parts. Of course these overlap, but this kind of breakdown helped to focus both the NGOs and the survivors of the disaster. These areas were:

- Practical
- Financial
- Emotional

Practical matters

Practical help needs to be focused on people's basic needs (Figure 24.6, adapted from Maslow's hierarchy of needs; Maslow, 1968, 2011) for a safe place to be: many of those we encountered had lost everything and were in need of accommodation, food, and clothing.

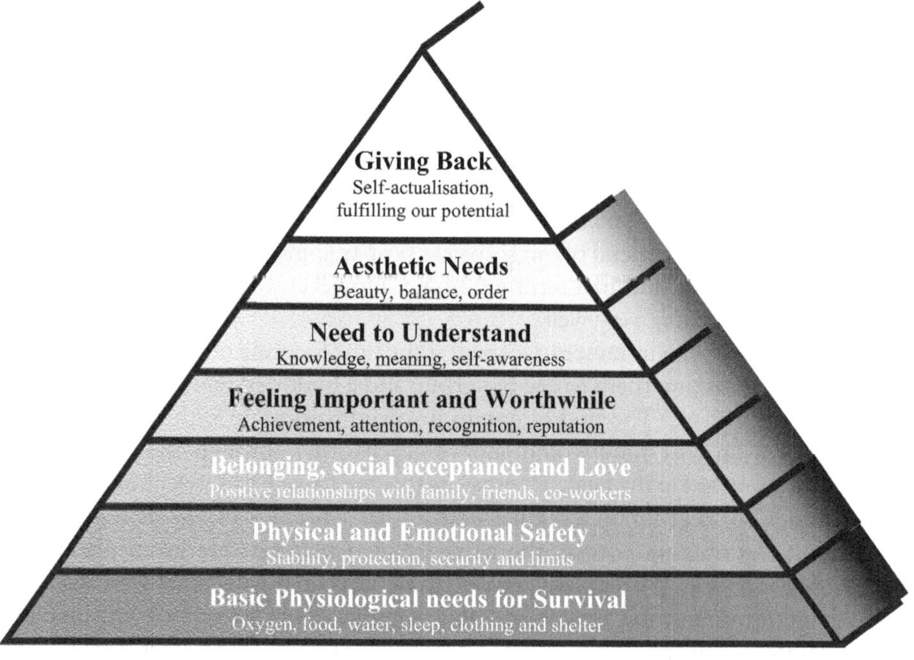

Figure 24.6 Maslow's hierarchy of needs.

Other basic essentials needed to be provided for, such as clean water, food, shelter, and so on. Oxfam and other international relief agencies were involved in this provision. People who have been traumatized may find it difficult to take in information so it must be explained clearly and simply what they will receive and where they can access it.

These essential requirements should have been mobilized very quickly, because many people had lost everything of material value and, most importantly, their loved ones. The confusion and slow response of national and regional government agencies resulted in nothing being available to these deeply shocked and traumatized people for quite some time. This greatly increased the emotional impact of the disaster, retraumatizing already traumatized people.

The quicker these essentials can be put in place, the better, so identifying contingency plans locally and nationally of who can mobilize these kinds of resources and authorizing them to do so are important from the onset of the disaster.

Financial matters

Both immediate and long-term financial support was necessary to those affected by the crisis: the purchase of tools and equipment in order to restart their means of livelihood needed to be available quickly in order to help recovery by allowing people to return to their familiar pattern of living. However, there was little or no consultation with the survivors which resulted in inappropriate and unfair distribution of precious resources. In one village five new fishing boats were made available by the government, but there were more than 50 fishing families whose boats were lost. This inevitably led to further stress, anger, frustration, and bitterness.

Emotional matters

Everyone affected by a trauma loses their sense that life is reasonably safe and predictable for a while (this may be momentary or for a long time). For those who are emotionally well supported with a good sense of their own worth in their world, the emotional effects of the trauma should be possible to deal with fairly quickly and they will probably already have good support networks that allow them to process their feelings and anxieties. For others, who have little personal support and/or whose lives have previously been affected by traumatic events, considerably more input may be required to support them and to help them recover from the disaster. Old traumas may be re-stimulated, which could make the person catastrophize and perhaps be unable to put this current event into context. They may find it difficult to trust those who are trying to help because previous experiences may have led them to believe that no one helps when trauma strikes (Paul, 2001).

In addition to this, people's regular routines and activities were thrown into total disarray, so here was not even a familiar structure for them to cling on to. We observed that in villages where children were returned quickly to a school routine they were adjusting more effectively.

Key helpers may need to be briefed about the effect of trauma on people. This can be done by meeting with them to brief them, followed up with simply and clearly written information which they can disseminate to other workers as appropriate. They need to be prepared for the range of people's emotional responses to trauma and crisis, ranging from anger to depression to helplessness. It should not be forgotten that the workers themselves may feel overwhelmed by the crisis and will need good support too.

It is important to keep things as normal as possible. Establishing routines and predictable tasks will help survivors to gain a practical perspective and to find some sense of psychological safety.

Immediate psychological support is optimally placed after setting up practical support such as shelter, food, water, and so on.

Trainers should be prepared with materials suitable to use in the most low-tech of conditions. We were easily able to access blackboards and chalk, but other resources were not really available.

For those who do want to talk to a counselor in more depth, a private area needs to be available; however, counselors and other helpers should be available in the public areas at other times. This helps to demystify the helping process so that the survivors see the helpers as a friendly and helpful part of the community team.

Conclusion

Creating order out of chaos

Catastrophes can bind people together to cooperate against any perceived threat. Response to a catastrophe can vary enormously from psychological freezing to high-level activity, and from withdrawing into long periods of sleep to severe insomnia.

In these situations, the trainer needs to be:

- Robust
- Resourceful
- Empathetic
- Adaptable
- Well informed
- Supported

Our recommendation is that trainers should work in pairs (or teams) in such highly demanding environments.

Everyone – survivors, counselors, and trainers – is connected through the common disaster experience. The psychological distance from the disaster is illustrated by Figure 24.7, where the most affected person is the survivor and then the counselor, who can be overwhelmed by the survivor's distress – something that peers and supervisors should be aware of and alert to. The trainer-supervisor is further removed, which helps to keep focus and objectivity.

There is a positive aspect to working in the aftermath of a disaster. The immediacy of the environment allows the counselor to enter a portion of the client's physical world, because they are sharing the environment with the client. Our experience of briefly visiting the disaster area before we formulated the program helped to focus us and enabled us to visualize the environment and the physical difficulties we would need to prepare our trainees to encounter.

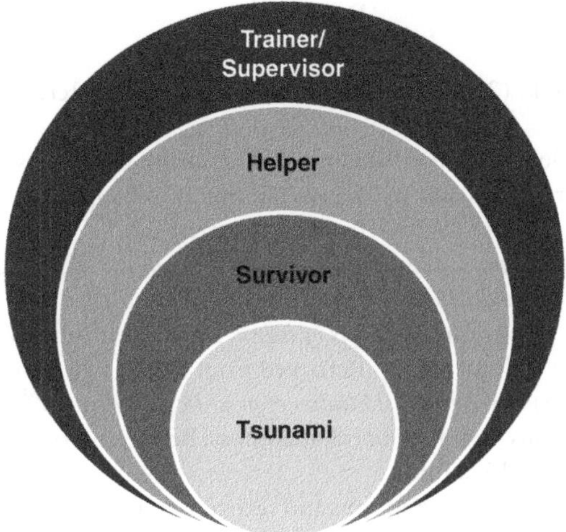

Figure 24.7 The ripple effect of the tsunami on survivors and helpers.

The negative aspect to working in this close environment can be an over-involvement and enmeshment with the client world. This factor needs to be guarded against and trainers should make sure that this is included in the training and ongoing supervision. Workers may find it difficult to deal with the privations of the most basic accommodation and rations.

The design of this kind of emergency training program needs to be:

- short;
- underpinned by basic theory;
- experientially based;
- varied activities;
- well supported.

It is important to incorporate a buddy system into this type of training program. This provides ongoing objectivity and support and helps ensure the counselor's well-being, which in turn provides a safer and more reliable resource for the survivors.

It should be assumed that there will be no resources available to the trainer in this sort of environment. Even when a disaster happens in better resourced areas, the trainer may not have access to what would normally be considered as essential for the delivery of their program – computers, OHPs, screens, photocopying, and so on.

Crisis work can be enriching and fulfilling for trainer and counselor alike since it meets an immediate need to respond to the plight of the survivors of the disaster.

None of our work in India is externally funded, nor is it affiliated to any particular organization. It has all been entirely self-financed and voluntary. Our choice to take our training programs to India has continued to enrich us in many ways. We are always really touched by the value placed on our work, the enthusiastic response of our trainees, and the warmth and hospitality we receive.

The task itself is the reward.

References

Asian Development, Bank. (2006). Tsunami: India – two years after. Retrieved from http://www.adb.org.

Davidson, J. R. T. (2006). After the tsunami: Mental health challenges to the community for today and tomorrow. *Journal of Clinical Psychiatry*, *67*(Suppl. 2), 3–8.

Grollman, R. A. (2001 , March). Why? *Journeys Newsletter*, 3.

Kinchin, D. (2005). *Post traumatic stress disorder: The invisible injury.* Didcot, UK: SuccessUnlimited.

Kübler-Ross, E. (1970). *On death and dying.* London: Tavistock.

Maslow, A. H. (1968). *Toward a psychology of being.* New York: John Wiley & Sons, Inc.

Maslow Project. (2011). [Home page]. Retrieved from http://www.maslowproject.com.

Murray Parkes, C. (1996) *Bereavement: Studies of grief in adult life* (3rd ed.). London: Pelican.

Paul, B. J. (2001). Reactions to sudden or traumatic loss. Retrieved from http://www.aarp.org/griefandloss/articles/15_a.html.

Shanthasheela, M. (2009). PTSD symptoms and recovery among different sectors of the people exposed to 2004 tsunami in Tamil Nadu, India. Retrieved from http://www.chikyu.ac.jp.

Sheth, A., Sanyal, S., Jaiswal, A., & Gandhi, P. (2006 , May). Effects of the December 2004 Indian Ocean tsunami on the Indian mainland. Retrieved from http://www.iitk.ac.in/nicee/RP/2006_Effect_EQSpectra.pdf.

Tedeschi, R. G. (1999). Violence transformed: Post-traumatic growth in survivors and their societies. *Aggression and Violent Behavior*, *4*(3), 319–341.

Further Reading

Doka, K. J. (1996). Sudden loss: The experiences of bereavement. In K. J. Doka (Ed.), *Living with grief after sudden loss: Suicide, homicide, accident, heart attack, stroke.* Washington, DC: Hospice Foundation of America.

Dyer, K. A. (2001 , October 7). 9-11: United in courage and grief. Why does my heart feel so bad? Retrieved from http://www.journeyofhearts.org/kirstimd/911_bad.htm.

Lendrum, S., and Syme, G. (1992). *The gift of tears.* London: Routledge.

Levine, P. A. (1997). *Waking the tiger: Healing trauma.* Berkeley, CA: North Atlantic.

National Association for Loss and Grief. (2011). Grief reactions associated with accidental or traumatic death. Retrieved from http://www.grieflink.asn.au/traumatic.html.

Rando, T. A. (1996). Complications in mourning traumatic death. In K. A. Doka (Ed.), *Living with grief after sudden loss: Suicide, homicide, accident, heart attack, stroke.* Washington, DC: Hospice Foundation of America.

Red Cross., (2001). When bad things happen. Retrieved from http://www.redcross.org/services/disaster/keepsafe/badthings.html.

Santi Ireson, S. (2001). *Working in general practice with somatisation of sexual abuse.* In K. Etherington (Ed.), *Counsellors in health settings.* London: Jessica Kingsley.

Shiraldi, G. R. (2009). *The post-traumatic stress disorder sourcebook.* Maidenhead, UK: McGraw-Hill.

Viorst, J. (1996). *Necessary losses.* London: Positive.

Williams, M. B., & Sommer, J.F. (Eds.). *Simple and complex stress disorder.* New York: Haworth Press.

25

Turning Training into Reality: Considerations When Training Teams for Deployment to Disasters

Gail Rowntree and Mark Akerlund

Overview and Introduction

Gail Rowntree and Mark Akerlund work as independent associates for Kenyon International Emergency Services (KIES) based in the United States and United Kingdom, respectively (see biographies at the front of this volume). Their roles are to support the families and team members who deploy to disasters worldwide by providing team welfare and support on site as well as completing debriefings and assessments for the team members when they return home.

Ensuring training and well-being of the teams are vital parts of their roles, and this chapter explores some of the core elements and specifics of training that need to be considered by organizations who may ask staff to support a disaster. The chapter will concentrate on the main areas for concern: what to train and who to train. Using two industry case studies the chapter offers firsthand learning to help organizations understand considerations for deploying emergency personnel. The chapter offers key learning points that have come from transferring theory into practice when supporting personnel who have deployed to a disaster and, finally, examines considerations for selection so organizations can plan an appropriate program that supports the deploying teams.

The Training Journey So Far

The academic and professional framework for training emergency personnel in relief and disaster operations has thus far been largely undertaken on an ad hoc basis. Until recently there were very few university-level courses that accommodated this growing industry and the need for formalized teaching worldwide. In fact prior to 1995 there were only two degree-level programs internationally. By 2005 there were over 40, which gives an indication of the increase in demand for these services and teaching. Most courses did not, however, follow an agreed curriculum (O'Connor, 2005). There continues to be a

International Handbook of Workplace Trauma Support, First Edition. Edited by Rick Hughes, Andrew Kinder, and Cary L. Cooper.

wide range of learning offered and there is some evidence to suggest that organizations and universities are attempting to develop a framework of best practice. Professional organizations, such as the Federal Emergency Management Agency (2011) in the United States, have recently created a web site specifically for emergency personnel and management, and the Emergency Planning Society (2011) in the United Kingdom is working toward chartered status. Both establishments are collaborating closely with educational institutions to ensure clarity and continuity in learning as well as offer a differentiator for emergency personnel and volunteers in recognition of their skills and experience in the field. This in turn promises to enhance the reputation of the industry.

To ensure the quality of the training programs and academic courses available, David Alexander (2003) rightly argues for an agreed set of measures to maintain quality. This has yet to be internationally agreed with most organizations keeping their training and learning in house. Currently opinion suggests that most organizational training pays little regard to academic theories or concepts, concentrating more on training needs analysis pertinent to the organization at the time.

Who Needs to Be Trained?

This question will of course vary from industry to industry, but essentially any personnel who will be required as part of their work duties or have volunteered to support an organization during a crisis situation will need some basic training in preparation. The crisis may take the form of operationally disruptive situations through to full mass disasters. For the benefit of this chapter the concentration will be on those personnel who may have to deal with mass disasters at the front line, that is, a humanitarian center, mortuary operations, or a crash site.

It is important to ensure that any personnel who may have to be exposed to front-line work have a full range of training to build a portfolio of skills and understand their own reactions to the situations as well as those working around them. Regular opportunities for updating these skills, whilst adding to them and sharing knowledge from past experiences and evaluation for improvements and development should form any part of a planned training regime.

Managers of emergency personnel need specific training around the challenges of delegation and decision making whilst on deployment. Often this group does not get the adequate training required to role model the behaviors that personnel rely on so greatly. Managers are required to make decisions whilst often lacking adequate information and in the spot light of the media and pressure from stakeholders. Understanding the impact of their behaviors and the support they can give their teams are key skills in which managers need to be trained. However, a lack of agreed best practice due to the various characteristics of each disaster makes this training unique to each organization.

What Needs to Be Trained?

There is little agreement in the disaster industry about what should be trained and as each disaster has its unique points it would be impossible to cover everything that is needed. However, there are some themes that run through each disaster that can be trained to allow for some of the more common experiences. The obvious areas to concentrate on for those personnel who will be front line are around the requirements of working within such a challenging environment. Personnel chosen to work in these areas are often qualified and usually experienced in their respective countries and associated institutions, but the conditions and equipment available can be different to what is expected and the processes and procedures may be confusing and complex due to international laws. Key areas for consideration when planning training for responding personnel are:

1. *How to work as a team on a crash site or disaster area*: Experiencing local weather challenges and the requirements for clearing a crash site can be taught in theory but having the real life experience is crucial. This can be achieved through mentoring or by observations in the field before allowing inexperienced personnel to undertake tasks. Use of video to show the processes and procedures can also be helpful during training.
2. *Processing human remains for identification*: Again, film and photography can be used for training, but observations by qualified personnel whilst working in a junior role will help with building confidence in both the processes used internationally, and also for working in challenging conditions when a full-scale mortuary may not be available.
3. *Processing personal effects – cleaning, processing, and identification and cataloguing*: Often this is where most personnel can start to gain experience in the field, whilst being taught by more experienced and skilled teachers. As most of this is conducted away from the crash site the conditions can also be better controlled, and training can be more structured.

   For those personnel who will be working within humanitarian centers, the essential elements for training should include:
4. *How to set up a humanitarian center (HAC)*: This can only really be trained in principle by using photos and mock-ups, deciding on what could be agreed best practice and cultural considerations at the time.
5. *Working with families of those persons directly affected (PDAs)*: These personnel are often experienced and skilled in this area, but the more administrative roles are a good starting point for new personnel who need to gain experience in the field but cannot yet work directly with families or survivors. A gradual introduction can be made over several deployments and measured accordingly by mentors and managers.
6. *Working with the authorities to support repatriation*: Specific skills such as languages and cultural knowledge may be required. Often these team members may not work directly with families at all but rather concentrate on the diplomatic and international bureaucracy that surrounds disasters. Of course, as they are based within the humanitarian center knowledge of how the center runs will be essential, but the role may not require further expansion of skills necessarily.
7. *Counseling of families of PDAs, injured PDAs, and deployed personnel*: Specific roles require qualified and skilled mental health or welfare professionals. These roles will be similar to the roles the personnel take in their everyday jobs and, although an

appraisal will be required, development is usually industry specific or kept within the confines of the specialism undertaken. However, it needs to be noted that as the disaster industry grows, these personnel can often be exposed to working on a crash site or with families and extra training and experience would be essential elements of the recruitment of these personnel.

Table 25.1 illustrates the three common areas of a disaster and the types of roles that many organizations employ within them. The table shows the prerequisite skills that deploying personnel should possess for the roles and finally the types of skills that can be trained in addition when planning ongoing training (or recruitment).

Hodgkinson and Stewart (1998) state that learning materials to give to students should include definitions or key jargon and a way of assessing knowledge and classifications of types of hazards to face. This can be aided by visual materials to

Table 25.1 Disaster types, mandatory training, and additional skills requirements

Area of deployment or role	*Requisite skill areas and qualifications pre-deployment*	*Specific deployment skills that may need additional training*
Mortuary operations	Funeral directing Embalming Repatriation Cosmetology Odontology Disaster victim identification (DVI) Data entry	Coping strategies Working in harsh environments Processes and procedures for international repatriation Working with agencies and stakeholders Technical equipment used for DVI Personal effects and technical equipment used for cleaning and restoring items
Humanitarian centers	Setting up and running humanitarian centers Languages and international experience Childcare qualifications Logistics and supply chain management experience, especially overseas Transport and accommodation Information technology (IT) systems Finance to control expenses and costs for the client and insurance companies	Specific checklists and procedures for setting up and closing down centers IT systems; how to configure equipment and ensure secure data Acquiring of equipment and knowledge of most common areas to concentrate on for purchasing or hiring Understanding of contracts and accounting Financial data required for insurance companies during a disaster
Welfare and team support	Qualified in assessment of clients Experienced in critical incident debriefing (CID) Qualified and recognized in area of expertise	Trauma psychology Post-deployment integration Trauma-specific psychiatric disorders Working in harsh environments

impart facts, data, and narratives from those who have direct experience for the field. They go further by suggesting group problem-solving activities, starting with simple problems and building to take on more complex ones, can be trained and, if possible, stakeholders and external agencies should be part of the learning process to aid relationship building. Working collaboratively and cooperatively should be natural to all interested parties, and by bringing together agencies and relevant organizations any possible conflict areas can be worked on in the classroom rather than on deployment (Davies & Wilkinson, 2004).

Using Scenarios as Exercises in Training for Front-line Personnel

Using scenarios in exercises is a popular and often useful way of replicating the situations and processes involved in disaster deployments. This can be especially helpful for those managers and emergency personnel who have little, or no, practical experience. Alexander (2003) suggests that the use of scenarios is essential for emergency personnel. These scenarios are also useful to develop the necessary skills required for working in the field, such as time management, cognitive mapping, mediation between local and virtual agencies and stakeholders, team management, and decision making whilst under pressure (Alexander, 2003).

A scenario for this chapter can be defined as a "Postdictive reconstruction of a past event or, more commonly, a hypothetical construction of future ones" (Forster, 1980, p. 147). This way the managers and emergency personnel can learn from an event by discussing the decisions made at the time and reviewing the time scales that these decisions were made alongside their impact or consequences. Scenarios can be used to enhance practical skills as well as knowledge and understanding of what is expected of personnel when dealing with stressful situations and understanding possible conditions or unique operational requirements.

Typical scenarios for exercises use a hypothetical (or past) event that requires the context of the situation, as well as the hazards, vulnerability, and risks. The time steps are frozen and the scenario suspends either at specific points during the course of the exercise or when interesting or particularly challenging points are reached. This allows for testing of assumptions and decisions made (Alexander, 2000).

Another element that should be considered when creating scenarios is to include a degree of media interaction, either by using real companies to take part or by using consultants' expert in this field. Hodginson and Stewart (1998) argue that understanding and working with the media can be a way to overcome simple miscommunications. Further they suggest that general public relations (PR) training and using real life media scenarios to discuss and review help the students to understand the media process during a disaster and support the development of those personnel whose role it is to interact with the media as spokesperson.

In the experience of the authors when training emergency personnel to prepare for deployment using specific drills and procedures in a similar way that airlines use them is particularly effective, to give an element of predictability. They should then be tested on a regular basis to allow for reviews. This allows trainers to test skills and team development, reaction times, decision making and communication skills and understand any gaps in training or selection that may need attention.

It is acknowledged by the authors that replicating the real emergencies can be challenging in a classroom environment; however, replicating the proximal pressure of decision making under certain conditions can help to test the teams' assumptions about a process or a reaction to a decision. In real deployments, managers have to make quick decisions in a pressured environment with long-lasting consequences and possibly without having the full information available. Scenarios give a taste of this and can test both practicality and personal characteristics.

Scenarios fill a gap between the training room and the practical experiences of the field. Students can apply theoretical knowledge into semipractical situations, and also test out different roles and responsibilities to change perspectives before the real event. An example is how Air Berlin uses scenarios to train their Gold Command team members.

Air Berlin's Experiences of Using Scenarios with A Command Team: Case Study

Air Berlin started using scenarios after they had a minor incident. Before then they had not used exercises or scenarios. Their use allowed the organization to bring in theory to practical exercises as well as using them to review processes and procedures from the real event. They found the combination very helpful. As Emergency Crisis Manager Kerstin Dettel explained in the following quote, it also highlighted the biggest gap for the Command Team. This is the highest level of decision making in a disaster and the team that usually sets up and runs the crisis management center, taking overall control of the operation and the organization during a disaster. The command team didn't realize the importance of training in the first place until it was included in the emergency procedures. When taking part in a scenario, the team members are required to read the manuals and to work through their own responsibilities. This may be the only time the team gets to concentrate on the requirements for the individual and the team.

> In the beginning the team was rather shy about taking part in scenarios and felt they were being asked to act, so they didn't generally like doing them, but it's amazing how quickly they forget they are playing a role and start to behave more naturally. This is when the learning really takes place. They stop playing the role and start being the role, which is great.

Dettel and the rest of the Board at Air Berlin are convinced that one of the main learning points is that the team is put under the same sort of pressure as they would be in a normal situation. This is true also for those in the team who have specific responsibilities.

> In the beginning of going through scenarios, quite a few felt self-conscious and not very confident in the roles. But as time goes by and the scenario plays out, they become more confident. In fact, when the center was activated you could hardly tell the difference between the exercises and the real thing. Scenarios actually helped us to question the processes and the procedures. When we had to open the center for a

real situation, it helped us all to see what was working and what we needed to do to make them better.

Dettel went on to explain how filming the scenarios gave everyone involved a way of seeing how their decisions and behaviors affected other stakeholders and agencies working with them. The power of filming them is that although initially it does put people under extra pressure when training, it allows them to see what they said and how they behaved, and after a while they forgot about the camera.

It was good to be able to show individuals what they had said (or had not said) and then the possible consequences of that in a real situation. It helps with communication and development and also acts as a mirror for certain behaviors that we may want to change. It's close to how they would react in real life. Also, having the media aspect live as part of the scenario added real value for the training.

Exercises using scenarios are a costly training program to use, and justifying them to board members can be challenging. Dettel overcame this by inviting board members to take part in the scenarios and actually make decisions about the direction of the exercise. It helped them to see the importance of training, and they saw firsthand the improvement and development of the Command Team.

Dettel went on to say:

I don't really have to justify cost too much as safety is so close to all of us. It is now quite normal to have these types of scenario-based exercises. The mind change happened over about six years. Having board members on the scenario exercises was a really positive statement. It backed up the corporate importance but also they saw how they were run.

Measuring the success of scenarios to see improvement is also a key point to help justify the cost of running them regularly. Air Berlin brings in observers who, with a specific checklist, focus on particularly points and behaviors throughout the exercise, but do not take an active role in them. They then see if certain indicators from both the processes and people skills have improved.

Finally, Dettel explained that the administration role is one of the key roles in the command center. Often not part of the decision making as they could be quite a junior person, they need to be experienced and able to work under pressure whilst not being intimidated by the senior team members.

We use people from operations control for this role. They have a good overview of the operation. They can also work under pressure. They get trained into the position. They don't have to make decisions, but they do have to be organized enough to offer information to the senior decision makers in a way that aids the process and communication.

The Challenge of Training Personnel for Deployment

Deploying as emergency personnel may not be the only role in an organization that volunteers have, or they have additional roles on top of their normal tasks, so the continuity of workload and pressure of balancing various roles can lead to personnel having less than the perfect amount of time to dedicate to training or updating their skills. Training and having staff out of the operation on a regular basis can be challenging and training for this area is often overlooked, or not given sufficient time or priority alongside other financial pressures that organizations face (Hosseini & Izadkhah, 2010). This can be overcome by frequent, short, sharp, and focused training sessions rather than irregular larger events that may disrupt an operation. A modular approach taking just one skill or process at a time in a bite-sized chunk may be sufficient for developing the basics for a team to deploy.

The cost of training needs to outweigh the time taken out of an operation; therefore, learning needs to be effective and personnel having the right program and trainers can maximize the learning for the participants (Wilson, 2000).

Training coping mechanisms can be a challenging element to any course as it often takes a prescriptive tone with areas deemed "positive" and other coping mechanisms seen as more "negative." The authors will not concentrate on this subject in detail in this chapter, but suffice to say that some sort of recognition of the different ways that deploying personnel cope with the experiences and tasks they face during a crisis situation should be included in any pre-deployment training. This should be offered by an experienced member of team welfare who has field experience and must include an appreciation of the demands of the jobs, an understanding of possible trauma, and personal coping strategies during the situation (Hodgkinson & Stewart, 1998).

Addressing possible maladaptive coping mechanisms, such as alcohol or drug taking, is essential for any training before personnel experience the pressures of the situation they may find themselves in. Some authors suggest that actually this group is well suited to employ personal coping strategies without too much intervention, but for new members of the team this may take a while to develop so training on what to expect and what they may experience in terms of feelings and reactions can still be helpful (Prati, Palestrini, & Pietrantoni, 2009).

Finally, having the right trainers with the right field experience as well as pedagogic and Andragogic qualifications can be a challenge as organizations may not have access to this cadre and organizations may find themselves training with well-meaning but ill-informed enthusiasts to fulfill the need to have regular training.

Specific Training for Managers Who Deploy

One of the most crucial areas for training is around how managers of teams need to behave and manage on deployments. These members of the disaster team are often looked to for immediate decisions and role modeling of coping strategies, whilst maintaining a supporting management role. It should be acknowledged at this point that one of the most difficult areas for a manager of a disaster team to learn and be comfortable with is being asked to make quick decisions with little information and delegate to people they don't know that well (Hosseini & Izadkhah, 2010). If team members are unwilling to leave their duties at the end of a shift (as is often the case due to a desire to be useful) and take a rest, it can be that this has been learned by the managers who have worked excessively long hours in a bid to be able to support their teams. The actions and behaviors of the managers during a crisis situation are more pronounced and important than during normal operations, and are core elements of any training given to managers (Hosseini & Izadkhah, 2010).

The understanding and knowledge of key policies and procedures required during a deployment are basic elements to start any training course for this group. This can help with giving managers a sense of control and familiarity in an environment when they actually have little control or familiarity with it. This will be of particular help to those managers who are experiencing a disaster situation for the first time. It allows for order and predictability of the operation during the early phases. Additionally, familiarity with disaster processes and procedures are an important area to understand as managers may have to manage teams they do not know well, or have not met before, and at the beginning phase of a disaster the control that a manager can bring can help a team to build relationships quickly. This can be more helpful in front-line operations (such as in a HAC) rather than a command and control approach (such as a crisis management center) (Suparamaniam & Dekker, 2003).

Deploying managers may benefit from how to set up crisis management centers and then run those centers by using scenarios or table-top exercises to learn the basic requirements. If they have specific roles in a disaster situation, then understanding the roles, the responsibilities, and how to work with other stakeholders is essential. Often, managers are given roles that sit outside their normal expertise and need to understand the boundaries of decisions making or the consequences of any decisions made in the bigger picture in terms of resources, costs, or possible conflict (Alexander, 2000).

Understanding and being able to cope with the stressors of a disaster is key if a manager is going to be able to support their teams who may be experiencing their own stress or trauma from the situation they find themselves in. Having a calm and self-aware manager is vital for an effective and efficient deployment team.

Box 1 Case Study of First-time Manager Deploying to Manage a Disaster Team

The name of the manager has been omitted to ensure confidentiality. The manager was deployed for an organization in a senior role in 2010

> Interestingly enough, a lot of the training I had has not been that structured in that I haven't really been on too many formal training courses. The only real training I had was my initial training when I first started. The rest I got from listening to other people on how to do things and what they experienced and what I might need to consider moving forward, so really I was dropped in at the deep end, which worked for me, but might not work for others.

The manager did understand the company policies and processes linked to managing a team in a HAC. His role meant he was required to become familiar with and set up a center, with support and mentoring, but all of this was theoretical and as is often the case the first time it was put into practice was on deployment and whilst managing a team.

The manager did feel ready to deploy because of a nursing background so in theory it was familiar territory. He stated his mind was in sharp focus and he felt comfortable and focused on the task in hand, so previous training and background helped:

> A lot of my concerns whilst I was there weren't necessarily about doing the right job or making the right decision, but I was concerned about perceptions and how other people thought I was doing. That added more stress. I had confidence to keep doing the right things, but I wasn't psychologically ready.

I had enough resources to help me do the job I needed to do. From a personal standpoint I really needed more experience, perhaps observing without being a manager or being a participant in a scenario exercise, because I could run the exercises but I had never been through one and I think I needed to know about how it felt. I hadn't linked everything together. The processes and roles were compartmentalized. Observing a deployment without having to be part of it would have been what I would have wanted. I know it's a luxury and I don't need it now but it would have really helped me get my head round what I needed to do.

Interestingly, having a mental health team member around helped as I was able to vent in private without being in front of the team, and (he) did tell me a few times I needed to take some rest as I was not making the best decisions or I was getting ratty. Sometimes I listened to him and sometimes I didn't, but having him there was a comfort.

Specific Skills Required by Trainers

As the area of disaster management is still in its relative infancy in terms of training and curriculum, it is essential that trainers both are suitably qualified to teach and also have field experience (Alexander, 2003).

It may be a requirement to have a cadre of trainers with a particular expertise and pedagogic or Andragogic experience to bring in a varied and flexible approach to any course, as one trainer is unlikely to be able to train a whole course purely from experience or expertise (Hosseini & Izadkhah, 2010)

Any program must, of course, be useful and relevant, but it may not always be necessary to have a course that automatically builds to a qualification. Training and the building of skills in the area may be more essential for particular participants such as emergency planners working for local councils who specialize in emergency services. To this end a modular-based program that personnel can then pick from to ensure their skills portfolio remains bespoke may be the best approach. In this way an organization can ensure all members of the team, regardless of experience or role, can get the individual training they need to be effective. This may also be a cost-effective way of having specific training without the cost of taking people out of an operation for training they may never need. As there appears to be a lack of academic agreement around specific managerial theories and concepts, it would be beneficial to concentrate on building practical experience and knowledge for those managers who expect to deploy.

Training to maintain and update skills and knowledge (such as processes and procedures, or health and safety) are areas that can be completed and evaluated using a blended approach through e-learning or using in-house experienced personnel, rather than external consultants. However, large-scale training using scenarios and exercises can be invaluable and for specific skills training then external consultants should be factored into any planning for budgets.

The Reality of Deployments: Understanding Team Welfare

The use of welfare teams is vital to the success of a deployment. Personnel face stressful circumstances and adjustment-related issues just prior to deployment, while on deployment, and after returning home from deployment. Careful consideration and attention should be given to all of these areas. It is critical that personnel have access to support throughout a deployment in both a formal and informal manner. This is also the one area of the subject that does follow predefined clinical models for both assessment of personnel and debriefing them, using the Critical Incident Debriefing (CID) model for trauma-based psychology (Jackson-Cherry & Erford, 2010).

Prior to a deployment, inexperienced personnel may face a great deal of uncertainty after being informed that they "might" be deployed. They then have to deal with the possibility that they will be leaving home and work for an extended period of time, and will have to transition all of their work, personal responsibilities, and so on, to others. Often emergency volunteer services such as working in a humanitarian center attract sensitive people who may need additional support when deploying for the first time as they be affected by the exposure to stressors (Cavanaugh *et al.*, 2008). This can be further complicated by time-zone changes when traveling to the site and the uncertainty of what they will be doing, what they might be exposed to, their housing conditions, and so on. Some personnel have been observed to arrive on a deployment with a "deer in the headlights" look, indicating that they are overwhelmed and will need time to adjust. This can be minimized if training includes helping them to understand that they are an important part of the process of disaster and that their efforts are essential (Miller & Garrett, 2009).

Perhaps the best method for utilizing clinical personnel on a deployment is through a "team welfare or well-being" approach. Terms such as "mental health," "psychotherapist," "counselor," and so on are quite common in some parts of the world, but may carry a negative stigma or connotation in other parts. For example, in North America it is very common to access mental health support services while it would be extremely uncommon in some Asian and/or Muslim cultures. Most other countries, religions, and cultures would fall somewhere between these points. Understanding and being comfortable with various terms are essential early knowledge for deploying personnel and their team welfare representatives.

Working in the field with deployed personnel provides several advantages for team welfare representatives. It provides an improved understanding of what the working conditions are like, what the personnel are being exposed to, and gives clinicians the ability to sense for themselves what others may be reporting on. Informal interactions with personnel from the various centers at the end of the day can be extremely beneficial in getting people to open up and talk about their experience in a casual environment along with other members from their team. Personnel seem to let their guard down more during nonworking functions and will often be more open about their feelings about what they have experienced. This also creates an opportunity to more closely observe persons in a social environment. It can be interesting to notice sociological trends about who dominates conversations, who is quiet, and so on. The team welfare representative should make an effort to ensure that all persons in a group are interacting and contributing to conversations. This can be particularly important during the first two or three days a person is on a deployment to make sure new persons arriving at the site are being introduced to others who they may not be working with and to reinforce a positive adjustment.

At the conclusion of a deployment, usually a day or two before returning home, the departing personnel will receive a debriefing for the purpose of helping them to process their experience, to enable them to provide feedback about the deployment, and to prepare them for the re-entry process and returning to their regular environment. This may include education about what to expect post-deployment and when to seek additional help if required (Norwood, Ursano, & Fullerton, 2000). The debriefing should concentrate on personal processing and helping persons to adjust back to their "regular" life. This can be a difficult transition and many people report being very surprised about how difficult the adjustment can be. The CID model is, therefore, adapted to not only review the mental health of the departing personnel but also gather vital feedback on the deployment to be incorporated into any post-deployment changes the organization may wish to make.

Some areas that CID concentrates on are how personnel become tightly bonded during a deployment and may experience a grief–loss reaction when leaving their colleagues with whom they have worked closely. Since their experience is so unique, others will not understand what they have been through. Personnel are therefore encouraged to remain in touch with one another through occasional phone calls or emails. It will help with the adjustment and to normalize the experience somewhat due to the added support. Personnel should be made aware that these relationships will largely fade over time.

Departing personnel should be educated about some of their potential reactions they may experience toward others after they return as well as how persons may respond to them. Some may idolize a team member or place them on a pedestal, calling them a hero or the like. Others may be completely avoidant because they don't want to get too close to what has happened or they feel emotional about what the individual has been doing. There can be a huge "disconnection."

Self-care will be very important during this stage, along with understanding that they should make sure that they are eating well, sleeping well, and not withdrawing and isolating themselves. Personnel can be encouraged to write about their feelings in a journal if necessary. They should be strongly encouraged to reconnect with current events, as many things will have happened in the world while they were away.

At the conclusion of the debriefing process, positive reinforcement should be given regarding the individual's contribution to the deployment. If there is a sense that any personnel may experience significant adjustment issues upon returning home, it may be appropriate to provide a follow-up call within the first two weeks. However, if continued difficulties are discovered at that time then a referral to a local clinician should be made for continued support.

Using Training for Selection of Personnel

Understanding how to recruit and select the most appropriate personnel for deployment is essential for effective and well-trained teams. Various psychometric tools can help and are considered an effective resource to help establish high performance from emergency teams (Subramaniam *et al.*, 2010), such as Belbin team profiling (Belbin, 2011) to understand how people might work together in a crisis team or the Myers–Briggs Inventory to determine specific characteristics sought for emergency personnel. It is important to review the organization's unique requirements when recruiting personnel to a disaster team. Of course, these tests alone will not give an organization a team that works to superhuman levels, but these tools can support the selection process.

Using scenarios to simulate specific situations can also be helpful for selection. It not only gives potential responding personnel a chance to experience (albeit in a remote way) the sights and sounds they may experience, but also allows recruiters to observe how the participants interact with each other and react to certain situations.

Some authors in this field suggest that most emergency personnel tend to be self-selecting due to their job choices and skills learned and, therefore, should not be compared to the "normal" population (Prati *et al.*, 2009), so recruitment processes should be mindful of assessing competencies such as coping with complex and unusual situations, assessment of core skills (especially for mortuary and humanitarian assistance teams), decision making, problem solving, as well as the supplementary use of personality or psychometric tests.

Conclusions

When considering the key points for training a disaster team, it can be seen from this chapter that a "one-size-fits-all" methodology may not be effective or cost efficient. Organizations should consider the requisite needs of their industry and the team skills that already exist.

Understanding what to train and who needs to be trained must be the core consideration of the decision makers before undertaking any kind of program. Ensuring that training standards are of the highest quality and that trainers have field experience or subject matter expertise must also be at the heart of supporting the needs of those personnel who may be asked to deploy on behalf of an organization. In addition the teams must have the framework in place to help them to transition back into everyday life without experiencing long-term effects.

The training provision within this industry is beginning to consider boundaries and frameworks to ensure higher standards and consistency, but as the authors have shown a full academic training program may not always be the most appropriate option for the needs of the organization. Until the academic world catches up with the mass disaster industry and creates tailored courses that meet the needs of potential students, a dynamic and flexible approach dependent on skills, experience, and requirements will always need to be the preferred option for training and preparation of deploying teams to disasters.

As this industry grows and the knowledge of what constitutes best practice expands, the need to understand what happens on a deployment and the most appropriate way to prepare the personnel and support them during the experience is paramount.

References

Alexander, D. (2000). Scenario methodology for teaching principles of emergency management. *Disaster Prevention and Management*, *9*(2), 89–97.

Alexander, D. (2003). Towards the development of standards in emergency management training and education. *Disaster Prevention and Management*, *12*(2), 113–123.

Belbin. (2011). BELBIN team roles. Retrieved from http://www.belbin.com.

Cavanaugh, J. C., Gelles, M. G., Reyes, G., Civiello, C. L., & Zahner, M. (2008). Effectively planning for and managing a major disaster. *The Psychologist-Manager Journal*, *11*, 221–239.

Davies, I., & Wilkinson, D. (2004). *Working for the training of trainer's courses to deliver crisis management courses.* Report, Mine Action and Disaster Management Centre. Cranfield, UK: Cranfield University.

Emergency Planning Society. (2011). [Home page]. Retrieved from http://www.the-eps.org.

Federal Emergency Management Agency. (2011). [Home page]. Retrieved from http://training.fema.gov.

Forster, H. D. (1980). *Disaster planning: The preservation of life and property.* New York, NY: Springer-Verlag.

Hodgkinson, P. E., & Stewart, M. (1998). *Coping with catastrophe: A handbook of post disaster psychological aftercare* (2nd ed.). New York, NY: Brunner-Routledge.

Hosseini, M., & Izadkhah, Y. O. (2010). Training emergency managers for earthquake response: Challenges and opportunities. *Disaster Prevention and Management*, *19*(2), 89–97.

Jackson-Cherry, L. R., & Erford, B. T. (2010). *Crisis intervention and prevention.* Upper Saddle River, NJ: Pearson.

Norwood, A. E., Ursano, R. J., & Fullerton, C. S. (2000). Disaster psychiatry: Principles and practice. *Psychiatric Quarterly*, *71*(3), 207–226.

O'Connor, M. J., Jr., (2005). From chaos to clarity: Educating emergency managers. Doctoral dissertation, University of Akron, OH.

Prati, G., Palestrini, L., & Pietrantoni, L. (2009). Coping strategies and professional quality of life among emergency workers. *The Australasian Journal of Disaster and Trauma Studies, 1*, 1–11.

Suparamaniam, N., & Dekker, S. (2003). Paradoxes of power: The separation of knowledge and authority in international disaster relief work. *Disaster Prevention and Management, 12*(4), 312–318.

Wilson, H. C. (2000). Emergency response preparedness: Small group training. Part 2 – training and learning styles. *Disaster Prevention and Management, 9*(2), 105–116.

Further Reading

Honey, P., & Mumford A. (1992). *The manual of learning styles* (3rd ed.). Maidenhead, UK: Peter Honey.

Kübler-Ross, E. (1969). *On death and dying*. New York, NY: Macmillan.

Miller, M. K., & Garrett, S. K. (2009). Improving disaster volunteer safety through data collection and skills matching. Paper presented at the 2009 Industrial Engineering Research Conference.

Ursano, R. J., McCaughey, B. G., & Fullerton, C. S. (2001). *Individual and community responses to trauma and disaster: The structure of human chaos*. Cambridge: Cambridge University Press.

26

Combating the Effects of Post-traumatic Stress and Other Trauma Associated with the Theatre of War

Walter Busuttil

Introduction

Psychiatric clinical services for service personnel are well developed in the British military. Peacetime military life and combat can cause a variety of mental health problems in some. The majority, however, cope well and never go on to develop mental health disorders.

Joining Up

Many are attracted to join the military out of a sense of adventure, for an outdoor life, for financial security, to learn a trade, and to gain meaningful qualifications. Many noncommissioned personnel are commonly recruited from lower socio-economic groups and from relatively deprived areas. The Royal Air Force and Royal Navy are more likely than the Army to recruit those who have better qualifications, and these individuals are more likely to be able to engage in trade-training opportunities that have more use in a later civilian life than those who join the Army. Many Army personnel are likely to be trained to be combat soldiers and to train in trades that are comparatively less likely to be useful in a civilian setting.

Some join up in order to escape poverty, poor housing, poor opportunities, and deprivation. Some will have been exposed to childhood adversity, childhood abuse, adoption, or poor role models, caregiving, and attachments. Some who join up find the adjustment too great to cope with, and a very small minority may develop brief psychiatric disturbances including time-limited stress-induced psychotic disorders which lead to an early exit from the military.

International Handbook of Workplace Trauma Support, First Edition. Edited by Rick Hughes, Andrew Kinder, and Cary L. Cooper.

Most who join the military family, however, adjust well and take the opportunity to reorder their lives and consequently make good attachments that remain intact even after leaving the military (Iversen *et al.*, 2005). However, for the minority pre-existing psychological attachment issues may be re-ignited when they leave the military (Busuttil, 2009; Iversen *et al.*, 2005; Novaco, Cook, & Sarason, 1983).

Significant numbers who join up come from military families or a military background. For these individuals, experience of civilian environments might be limited prior to joining up and even more limited on leaving the military. Many leave after having served for many years, making psychological adjustment to civilian life more difficult. The potential institutionalizing effects of the military environment are thus amplified.

While military life is for the vast majority extremely positive, it can also cause longer term problems relating to employability and work. This relates to issues concerning institutionalization, the development of mental illness through military service, and adjustment leaving the military environment with the re-ignition of attachment issues stemming from the primary (pre-military) families, or a combination of all of these factors.

Some who join the military even these days do not appreciate the implications of being exposed to a war theatre or even combat. They may not appreciate that military training can be a dangerous place and that exposure to occupational accidents can occur. Exposure to psychological trauma can in some lead to mental health difficulties.

The majority join up when they are young. For many this will be the first experience of leaving home and securing employment and work. The lower limiting age for the RAF, Army, and Royal Navy is 16 years (Defence Analytical Services and Advice (DASA), 2011). Joining at a young age is an important issue from a mental health point of view in that personality formation is still ongoing and is then shaped by military service and experiences.

The military environment is excellent at getting the best from people who might not initially be of the highest caliber. This is because military training and military service have been demonstrated to shift locus of control to internality as opposed to externality, and coping style to a problem-solving-focused coping style rather than an emotional-focused coping style. This means that those who join up and undergo military training are likely to become self-sufficient, resilient, and more able to counteract psychological threat (Mikulincer & Solomon, 1988).

The problem with this is that the very things that make the individual a better soldier will also make it more difficult for them to admit they have a problem and ask for help when there are problems. This is especially so in relation to mental health problems. Thus the qualities that make the individual a good soldier may also be the qualities that make the individual a bad patient.

In the British Armed Forces, there are approximately 180 000 service personnel with a turnover of some 25 000 leaving and being replaced every year.

Military Mental Health Care

Medical services offered by the military have two main roles. The first is to look after the individual's medical needs. The second is an occupational health role.

On employment, from the initial screening medical examination every service man and woman is awarded a coded medical employment standard (MES) which incorporates physical as well as mental health. This code maintains medical confidentiality and at the same time allows the executive to know whether or not the individual is fit to undertake a particular role. It also means that occupational health is monitored at all times. The MES ensures appropriate communication that each individual is fit to be employed in a particular role, including whether or not they are deployable to a war zone and engage in combat. Physical or mental ill health results in medical downgrading of the MES which might be temporary or permanent. In the latter case this may lead to loss of the chance of further promotion in the trade, or even medical discharge from the Armed Forces.

British Armed Forces military psychiatric service delivery is organized into 14 multidisciplinary Departments of Community Mental Health (DCMH) situated all over the United Kingdom, with similar DCMHs available overseas. Professionals include psychiatrists, mental health nurses, and therapists trained in cognitive-behavioral therapy (CBT) and eye movement desensitization and reprocessing (EMDR); psychologists; and social workers. Inpatient hospital services are contracted out to an NHS consortium. Military psychiatric hospital wards closed in the 1990s. In combat zones multidisciplinary field psychiatric teams comprising serving community psychiatric nurses primarily supported by visiting psychiatrists are deployed. In the military, access to high-standard culture-sensitive mental health care is rapid. Most of the clinical services are community based. Clinical audits indicate that approximately 5000 (or 4.5 referrals per 1000) (Rona *et al.*, 2007) of new mental health referrals present per year and that common clinical presentations include alcohol misuse, depression, anxiety, and adjustment disorders with low rates of post-traumatic stress disorder (PTSD) (4%) in the military population (McMannus, 2009).

Traumatic Exposure

The military can be a dangerous environment. Military training is made to be as realistic as possible with live weapon training being routine, and with exercises depicting realistic combat scenarios. Moreover the military operates a myriad of dangerous machinery. For these reasons, accidents can and do happen.

Exposure to military operations, including combat, peacekeeping, humanitarian missions, body and body part handling and recovery, emergency helicopter search and rescue, and other emergencies such as standing in for a civilian fire service on strike, can provoke traumatic stress disorders as well as other common mental health disorders, including depression and anxiety. Some allege that they have been exposed to bullying and harassment while serving in the military.

Military Operations

Operations can be directly threatening or indirectly threatening. Some operations involve intensive exposure to combat involving eyeball-to-eyeball contact, to the threat of insurgents who never take the appearance of an enemy force, or to the ongoing threat of improvised explosive devices (IEDs). Other operations may involve no direct threat but place the individual in traumatic situations. Examples include witnessing combat between factions, exposure to the aftermath of combat following ethnic cleansing, exposure to mass graves, civilian deaths including those of women and children, insurgency-led combat where insurgents masquerade as civilians, the use of the vulnerable including the mentally handicapped, and women and children as suicide bombers.

In most combat situations only a relatively small percentage of those involved are actually tasked in close-quarter fighting. Most are present in a supportive role although many of these are exposed to serious physical and psychological threats. Some perceive they are in danger from mortars or long-range missiles that never materialize. This waiting can lead to PTSD with more prominent hyperarousal symptoms and anxiety disorders. Close-quarter eyeball-to-eyeball fighting is more likely to lead to PTSD with prominent levels of re-experiencing symptoms and dissociation (Stouffer *et al.*, 1949).

Psychiatric Effects of Combat

Military psychiatrists have described war theatre mental breakdown as a combat stress reaction (CSR). A CSR incorporates features of acute stress disorder (ASD) as defined by the ICD-10 and DSM-IV; and is characterized by three phases of functional decompensation and symptom development (see Table 26.1). These comprise (Shalev, 1988a):

1. The *premonitory phase* commences before explicit exposure to psychological trauma occurs (pre-combat).
2. The *acute phase* is precipitated by exposure to a severe emotionally traumatic event (combat).
3. The *stabilization phase* develops over several days or weeks with an insidious onset. Often it is seen by the primary medical officer at the end of a military operation as the first manifestation of a CSR either in those who could handle the acute phase without medical help or in those whose CSR developed insidiously. Often this is the case in soldiers whose position of command has not "afforded" them the luxury of breaking down as long as active operations continued (Shalev, 1988a; Shalev & Munitz, 1993; Strange & Arthur, 1967). Another circumstance which predisposes to this type of reaction is the first contact with the family at home and especially the first leave after combat (Solomon, 1988). The condition should be distinguished from normal grief reactions which are also prevalent after combat. Good indicators of CSR are the persistence of intrusive memories, images, and nightmares (Bar-on, Noy, & Nardi, 1986; Feinstein, 1989; Shalev, 1988a).

The stabilization phase may lead to PTSD or spontaneous resolution. All clinical varieties of CSR, irrespective of their initial form, may result in PTSD. The clinical picture of acute CSR does not predict later PTSD (Shalev, 1988a, 1992; Solomon, Noy, & Bar-on, 1986a, 1986b), although the presence of CSR makes it more likely that

Table 26.1 Three phases of functional decompensation and symptom development (Shalev, 1988a)

Premonitory phase: Signs of CSR	*The acute phase: Psychiatric symptoms*	*Stabilization phase*
1. High arousal • Restricted field of interest • Inability to relax • Inability to shift attention • Inability to concentrate • Disrupted decision making 2. Emotional dysfunction including: • Irritability • Impulsive responses to stimuli • Uncontrolled emotional discharge • Diminished social interaction • Withdrawal and isolation • Loss of a sense of humor • Loss of affective adaptation to others • Sustained criticism and mistrust 3. Physiological manifestations of anxiety • Diarrhea, anxiety, and tremulousness • Weakness and cold sweating • Headaches and palpitations • Unexplained physical complaints serving as a pretext for any consultation	1. Cognitive impairment: • Dissociative states • Confusion and disorientation 2. Impaired stimulus response: • Hyper-reactiveness to stimuli • Inappropriate responses to minor events 3. Psychomotor symptoms: • Restlessness and agitation • Stupor and motor retardation 4. Affective symptoms: • Anxiety • Panic • Terror • Sadness • Guilt • Shame • Perplexity • Stupefaction • Shock 5. Conversion symptoms: • Paralysis • Blindness • Muteness	Symptom picture of this later reaction is midway between the acute phase of CSR and PTSD. 1. Affective symptoms • Depression • Guilt • Shame 2. Re-experiencing symptoms: • Intrusive thoughts • Vivid images of event(s) or scene(s) from the battle 3. Arousal • Sleep disturbance • Fatigue • Irritability

PTSD will develop in the short term (Mikulincer & Solomon, 1988) and remain present in the longer term (Solomon, 1989).

Research and CSR

Research studies on CSR have been conducted primarily by the Israeli military. Studies have shown that being wounded in action was not related to the degree of mental health improvement. A sense of isolation from the fighting unit was a risk factor in 40%

of subjects who developed CSR in one study, although it was concluded that although a relationship between unit cohesion and the incidence of CSR was established, causation was not proved. Exposure to a previous severe psychological trauma appeared to influence CSR symptoms, but the severity of these prior traumas was not related to the course of recovery. Pre-combat life stressors were unrelated to CSR intensity and were inversely related to improvement. The more intense the CSR, the slower the recovery and the smaller the percentage showing an improvement at follow-up. The intensity of the CSR in the acute phase was related to the recovery (Noy, 1991).

Personality attributes might be important. In the same study, 43% of CSR sufferers reported that they had never felt anger or fear prior to combat. These characteristics were interpreted as signifying the presence of "repressive personalities." When the study population was dichotomized into repressive personalities versus all the others, the repressive personality group showed less improvement, and this group was correlated with the presence and intensity of dominant symptoms of PTSD along with other comorbid symptoms. The implication was that comorbidity patterns can be determined by individuals' pre-morbid personality characteristics (Noy, 1991). This study suggested that two clusters of factors were relevant to the etiology and prognosis of the CSR. The first cluster related primarily to the etiology and included only situational factors such as battle stress and unit cohesion. The second cluster related mainly to the prognosis for recovery and included mainly personality variables, with personality factors probably determining whether or not recovery from CSR occurred as opposed to being included as part of the primary causal factors.

These conclusions have been substantiated by other studies. For example, in a controlled study aimed to predict cases of CSR on the basis of pre-war tests, it was found that psychologists could not predict who would become a CSR casualty; however, they were able to predict which CSR casualties would eventually go on to develop PTSD. These cases were characterized before the war by lower scores on cognitive organization: an external (as opposed to internal) locus of control and a high expectation of instrumental support (Barnet, Milgram, & Noy, 1987). Other studies add to the evidence that personality factors are operating in the recovery process of CSR. Emotion-focused coping has been shown as being more prevalent in soldiers who suffer from CSR and then go on to develop PTSD, as opposed to those who recover from CSR and do not develop PTSD (Mikulincer & Solomon, 1988; Solomon, Mikulincer, & Bebenishty, 1989). A poorer recovery from CSR has also been found to be associated with a history of exposure to life events immediately prior to exposure to the traumatic stressor (Solomon *et al.*, 1986a, 1986b).

CSR should be seen as a fluid state which can consolidate to more permanent conditions as well as spontaneous resolution. Indeed spontaneous resolution is much commoner than consolidation. CSR can resolve following early intervention comprising rest and basic cognitive therapy–based psychological interventions, or progress into long-term psychiatric illness including anxiety disorders, depression, and alcohol misuse as well as PTSD characterized by re-experiencing symptoms including nightmares, flashbacks, and intrusive memories; hyperarousal symptoms, including hypervigilance; physical and psychological symptoms of anxiety, including panic attacks and emotional numbing; and avoidance symptoms, including social withdrawal. As stated in Table 26.2, the main factor determining the development of CSR (as well as ASD and

Table 26.2 Risk factors for CSR (Modified from Noy, 1991)

Primary factors: Dose–response effect
1. Intensity of battle.
2. Unpredictability of stressor
3. Loss of social cohesiveness of combat unit

Secondary factors

A. Psychological deprivations
1. Passive role (air traffic and fighter controllers, cooks, and mechanics)
2. Lack of adequate (military) training for the actual role
3. Inability to sustain denial: over-exposure to casualties, atrocities, and death of friend or relative
4. Conflicts prior to combat, stress following separation from loved ones

B. Physical deprivations
1. Lack of sleep, food, and fluids
2. Physical exhaustion and illness
3. Poor weather conditions

C. Support deprivations
1. Leadership failure
2. Death or replacement of the leader
3. Isolation from basic unit
4. A new soldier in a unit
5. Lack of support from loved ones at home
6. Poor unit cohesion and esprit de corps

PTSD) is a dose–response effect: the more severe and prolonged exposure to psychologically traumatic stressors is, the more likely it is that mental breakdown and illness will develop (Shalev, 1988a).

Wounded in Action, Physical Injury, and Bravery: Personality and Mental Health Issues

Being wounded in action is seen as a badge of courage by many and by society in general. Nowadays, with forward-operating highly trained combat medics, rapid response medical teams include senior medical staff which are incorporated within helicopter crews as part of the evacuation procedure to forward-operating surgical teams; and combat zone hospitals are followed by aero-medical evacuation in well-equipped military aircraft back to sophisticated homeland military hospital facilities. Many soldiers who in the past would otherwise not have survived their severe wounds are surviving. Many who survive horrific injuries do so with long-term physical and functional impairments.

The literature relating to the long-term effects on mental health on those who have been wounded in action and who have survived sophisticated surgery close to the battlefield in the more recent wars and conflicts is lacking. Research from earlier wars demonstrates that being wounded can lead to long-term mental health disorders if recovery from physical wounds is limited and physical function is impaired by amputation, deformity, or chronic pain. In contrast, there is also evidence that being wounded in action may protect against the development of CSR and longer term combat-related psychiatric disorders. This almost certainly would be determined by long-term physical recovery and ability to function.

Uncontrolled studies of cohorts of World War Two CSR casualties reported high rates of over 90% who had "predisposing neurotic traits." Studies of groups of wounded soldiers also detected that 90% of them had so-called predisposing neurotic traits. Despite being subjected to the same stress of combat and the additional stress of being wounded, none of the wounded men had detectable CSR (Brill & Beebe, 1952; Noy, 1987, 1991). Controlled studies were first thought to prove differences in the personality traits of CSR casualties and comparable groups of wounded veterans (Solomon *et al.*, 1986a, 1986b). However, these findings could not be replicated in studies of much larger numbers of subjects, and it was thought that sampling errors in these first studies were the cause of the discrepancy (Noy, 1991).

Other controlled studies were conducted confirming that there were no significant differences in predisposing traits between CSR casualty groups and wounded groups who had not become CSR casualties (Gal, 1987; Noy, 1987).

Gal (1987) also showed that there was no detectable personality predisposition to bravery. Both Noy's and Gal's studies identified that stress and unit cohesion were the main variables which produced either valor or breakdown (and a high CSR rate) in combat. Thus, when a combat unit is under severe threat, cohesion and leadership foster bravery, while the collapse of the social network under these conditions increases the prevalence of CSR (Noy, 1991).

Mitigating the Effects of Operational Deployment and Combat

Preventing CSR and PTSD

The military has attempted to mitigate the development of CSR and longer term mental health disorders including PTSD by introducing measures employed pre-combat, intra-combat, and post-combat. Many of these measures are not new, and most have few clear preventative effects.

Pre-combat. Screening. One study demonstrated that eliminating recruits with a history of mental illness including ASD and PTSD at the outset reduced subsequent psychiatric ill health (Creamer *et al.*, 1993; McFarlane, 1989). In a different study, screening for mental health problems before deployment to a combat zone failed to mitigate subsequent morbidity with the intensity of combat exposure and increased group cohesion being more significant factors (Rona *et al.*, 2006, 2009).

Recruit training and locus of control. An external locus of control (as defined by Rotter's locus of control theory; Rotter, 1966) may hinder the recovery process from CSR and increase the risk progression to PTSD. A unique study which investigated the effects of recruit training in the US Marine Corps provides an insight into the intrinsic effects of initial military training on individuals' loci of control (Novaco *et al.*, 1983).

As a result of analysis of archive data which spanned one year it was found that approximately 49.3% of 205 recruits were discharged from basic recruit training within the first 17 days of joining after failing to adjust psychologically and behaviorally to the unfamiliar military system. Following careful observation, initial training procedures were identified as being directed at stripping the recruit of his individuality and ability to verbalize his emotions, at the risk of punishment. Virtually all psychological reward from the drill sergeant and other superiors was seen to involve negative reinforcement contingencies. This often led to acute stress reactions which resulted in referral for psychiatric screening in some cases. Analysis across different platoons (60–80 men make up a platoon) revealed that the main environmental factor that accounted for discharge was the drill sergeant. The higher attrition platoons did not produce higher performance soldiers (as judged on a variety of soldiering skills). Those training units with a high attrition rate during the training period continued to have a high attrition rate after graduation. Cognitive changes occurred as recruits progressed through training. With each achievement, confidence and an ability to take on new challenges increased. Those who were trained in low- or medium-attrition units became more internal in locus of control. This was especially true for recruits who began training categorized as externals and who had negative life experiences as indexed by the Life Experiences Survey (Sarson, Johnson, & Siegel, 1978) and by failure to complete high school. In contrast they found that training in high attrition-rate platoons resulted in shifts toward externality, and this was particularly true for those who began training as externals. It was postulated that these findings reflected the reinforcement contingencies in the training unit as engineered by the drill instructor team. This study demonstrates that it is possible to change an individual's locus of control depending upon the attitude of training instructors. While realistic training may be protective as far as the development of PTSD from CSR is concerned, it may also be detrimental in that changes in locus of control may be shifted externally hence increasing the risks of PTSD development. Unfortunately Novaco *et al.*'s study is descriptive and contains no statistical data. Undoubtedly this research needs to be replicated using methodologically sound techniques and follow-up should ideally extend longitudinally to the subjection of individuals to catastrophic stress situations, the development of CSR, and progression and nonprogression to PTSD.

The study also demonstrates the more general point that recruit training undertaken by any organization should ideally be monitored as far as the external or internal locus of control paradigm is concerned and techniques employed in training must be geared toward enhancing internalizer qualities and suppressing externalizer qualities. Training instructors, therefore, will require specific instruction themselves in order to implement techniques which increase the internalizer rates as opposed to the externalizer rate. Further research is urgently required in order to identify what training instructor qualities promote internal and external coping styles.

This study may also be seen as going some way to proving that training can shift one's locus of control and that, in military, if the shift to internal control can be achieved, then screening for and eliminating external controllers are less important or even unnecessary.

Realistic professional training. Realistic training allows enforcement of practical countermeasures to meet the demands of a threatening situation. Thus effective evasive action away from the threat or to attacking it directly by means of the weapons the soldier has been trained to use may be taken. For example, live fire exercises can help to reduce perceived psychological threat and the formation of supportive, cohesive groups (Busuttil, 1995; Novaco *et al.*, 1983); and being trained to dig a trench and know its optimal depth, protective cover, and camouflage will diminish the threat from shelling, making its effects more predictable and therefore less threatening or stressful (Stouffer *et al.*, 1949). Anecdotal reports of what soldiers say about their first encounter with the combat situation include comments such as "Nothing can prepare you for that." This is especially true if they have just seen one of their friends killed or wounded (Hendin & Pollinger Haas, 1984; Kormos, 1978; McManners, 1993). Being equipped with basic first aid knowledge of what to do if peers are wounded may help dispel some of their feelings of helplessness. Knowing that more expert help is available, that appropriate medically qualified back-up is close to the battlefield, and that speedy evacuation and specialist hospital intervention are available is very important as far as morale and motivation are concerned. In the Vietnam War, for example, the average transition time from being wounded on the battle field to being operated on in a forward hospital facility was three hours. It is postulated that the knowledge that this is possible helps soldiers to concentrate on the task in hand and to worry less about the fact that they might get wounded (Figley, 1978). This transit time nowadays is reduced. This knowledge is significant when one bears in mind that men who are wounded in combat usually die within the first two hours of being hit and that usually heavy bleeding is the main cause of these deaths. Immediate first aid, triage, and evacuation from the battle field by helicopter increase that chance of survival dramatically. This is the sort of "comforting" information that soldiers need to be told well before they are sent to a combat zone.

Stress inoculation training (SIT). SIT can increase resilience as demonstrated in nuclear biological chemical (NBC) warfare training exercises (Carter & Cammermeyer, 1989). It has also been found to mitigate the traumatic effects of body handling in police personnel (Miller, 1989).

Induction training. Induction training is used prior to exposure to a traumatic situation to reinforce SIT techniques. Briefings or lectures delivered by commanders or military mental health workers in collaboration with the chain of command are thought to enhance coping with expected adversity, enabling and facilitating open peer discussion (Srinivasan, 1993). However, pre-deployment briefings including mental health education adopted by the British military failed to reduce medium-term mental illness (Sharpley *et al.*, 2007).

Prophylactic medications. The use of potential prophylactic medications including SSRI antidepressants and beta blockers has been attempted, but no efficacious medication that reliably prevents CSR and PTSD has been identified (Foa *et al.*, 2009).

Intra-combat. Leadership, cohesion, and buddy–buddy care. The death of a respected leader and fragmentation of unit cohesion increases vulnerability of troops to develop CSR (Flannery, 1990). Good leadership includes example setting under high pressure, effective communication, and enforcing countermeasures to reduce deprivations such as lack of food and rest.

The buddy–buddy system has been in use in many armed forces across the world. This allows practical help including mutual equipment checks, drills and actions soldiers take routinely. This has been noted to enhance cohesion and mutual psychological support. Good buddy–buddy support has been shown to help those who develop CSR access treatment more quickly and enhance response to rehabilitation from CSR (Strange & Arthur, 1967).

Trauma Risk Management (TRiM). This is defined as a "proactive, post traumatic peer group delivered management strategy that aims to keep employees of hierarchical organizations functioning after traumatic events, to provide support and education to those who require it and to identify those with difficulties that require more specialist input" (Greenberg, Cawkill, & Sharpley, 2005, p. 2; see also Greenberg, Laughton, & Jones, 2008). Trauma Risk Management aims to equip nonmedical junior management personnel to manage the psychological aftermath of a traumatic incident, reduce stigma associated with mental health, and help early identification of those who decompensate psychologically following traumatic exposure. It can be seen as an extension of buddy–buddy support, and a peer support network within a hierarchical structure. It was initially implemented within the Royal Marines in the late 1990s. It is now being trailed by all three (British) military services. Evaluation of TRiM is awaited (Creamer *et al.*, 2009).

Proximity, immediacy, expectancy, and brevity (PIE B). These are essentially crisis intervention principles. They have been used since World War One in the treatment of CSR. The soldier who has decompensated will be treated close to the front line (proximity), immediately after he presents (immediacy), with the expectation he will return to the front line (expectancy), and within a brief period of time (brevity).

The aim is to replenish physical and psychological needs allowing sleep, intake of food and drink for a few days in relative safety. Minimal psychological intervention is attempted. Those who continue to be symptomatic are evacuated to a middle-zone facility. Failure to respond leads to evacuation out of the battle zone completely for longer term psychiatric intervention. It has been argued that this approach itself decreases the numbers afflicted by later PTSD. It has also been pointed out that no evaluation of those who are returned to the frontline has been carried out, including whether it is more or less likely for these to be wounded or killed in action (Solomon & Benbenishty, 1986).

Nowadays the principles of PIE B are blurred. Field mental health teams implement the principles, and these teams are usually deployed close to the battle front. The nature of modern conflicts means that fast-moving battles with little demarcation between front, middle, and rear battlefield zones are usually the norm. In addition, war theatre medical services include access to highly sophisticated surgery and intensive care, and rapid aero-medical evacuation procedures, meaning that the transition time from the front line to home psychiatric facilities can be very rapid indeed.

Psychological debriefing (PD) and critical incident stress management (CISM). During World War Two and the war in Korea, it was observed by General S. L. A. Marshall that soldiers who talked about their combat experiences shortly after returning from action functioned better as military units and psychologically than their counterparts who had not "debriefed" (Marshall, 1978; Shalev, 1988b). Psychological debriefing was based on the principle that it is helpful to talk in detail about facts, feelings, and sensations relating to traumatic experiences within a group format. This was said to help the individual to understand what had happened and to fill in "blank spots" from peers also involved in the traumatic event. Critical incident stress management incorporates the principles of PD and also includes peer support and psycho-education principles (Busuttil & Busuttil, 1997; Dyregov, 1989; Mitchell, 1983). There are controversies concerning one-off psychological debriefings especially with research studies supporting the notion that they might be harmful or at least ineffective. In 2000 within the UK military, the surgeon general banned the use of single-session PD. Following a Cochrane review (Rose, Bisson & Wessely, 2001) and the development of NICE guidelines in 2005, it was advised that single PD sessions should not be offered. The effects of phasic CISM remain unclear with many UK emergency services still using these techniques routinely with high client satisfaction and some research supporting positive benefits. Further evaluation is required (Hawker, Durkin, & Hawker, 2010; Tehrani, 2002).

Post-combat. Psychological decompression. Psychological decompression is a planned period of "time out" for those who have just left the combat zone. They are given time and space to "defuse" in a location well away from the combat zone, with rest and relaxation and psycho-educational briefings for a few days before returning home. No empirical evidence exists that this prevents subsequent mental health problems, although it is thought to enhance re-entry into the family (Hacker Hughes *et al.*, 2008). Support by the public for military personnel on return home, including overt support such as home welcoming parades, might reduce psychological reactions to combat (Figley, 1978).

Culture, alcohol, and PTSD. The military has a propensity to reflect a macho culture. It promotes the use of alcohol to socialize with a widespread "Work hard, play hard" attitude. There is a reliance on social activities revolving around the excessive use of alcohol even post deployment and combat (Hooper *et al.*, 2008; Iversen *et al.*, 2007).

Longer term problems associated with combat exposure include higher levels of alcohol misuse especially in the younger servicemen. In one British study, heavy drinkers (>30 units per week) were compared to light drinkers (<20 units per week). Heavy drinking was associated with current military service and with being unmarried, separated, or divorced. Heavy drinking was commoner in younger personnel who had been deployed to Bosnia. Those who drank heavily were more likely to smoke. Heavy drinking was associated with poorer subjective physical and mental health (Iversen *et al.*, 2007). Studies identify the range of comorbidity between PTSD and alcohol disorders to be 41–85%; this far exceeds the 19–29% prevalence of alcohol abuse or dependence in the general male population (Steindl *et al.*, 2003). Levels of alcohol misuse in British servicemen exposed to combat are at higher levels than comparative civilian populations, especially younger servicemen (Andrews *et al.*, 2009;Fear & Wesseley, 2009).

A recent study conducted on an initial sample of 25 000 personnel representing the 180 000 UK personnel who to date have deployed to Op Telec (Iraq) and Op Herrick

(Afghanistan) demonstrated that 19.7% suffered common mental disorders such as phobias and anxiety, with worrying rates of alcohol misuse as high as 13% (Fear *et al.* 2010).

This study demonstrated levels of PTSD in regular service personnel of 4% (for in-war-theatre personnel overall – this percentage was no greater than comparable nondeployed personnel), and increased levels of PTSD of 5% for Territorial Army (with compared nondeployed levels of PTSD at 1%) and 6.9% for front-line combat troops (Fear *et al.*, 2010).

Rates of PTSD in British personnel are low compared to US and Australian studies of military personnel deployed to Afghanistan and Iraq, with the latter having PTSD levels of up to 20% in combat veterans in some studies (Fear *et al.*, 2010). Longer term alcohol disorders were thought less likely in the US populations compared to British personnel. Reasons cited for the discrepancies in relation to alcohol and PTSD rates were thought to be due to cultural differences, with UK populations thought to be more likely to drink alcohol more heavily than US populations, and UK populations as being less likely to verbalize mental health problems compared to the US population. Other reasons were thought to be due to methodological differences of studies, as well as a lower dose–response effect following the implementation of "harmony guidelines," which in comparison limit the duration and frequency of deployments for the British soldier. It has been noted that rates of psychiatric problems increase if harmony guidelines are breached, with particular difficulties in those whose tours are unexpectedly extended (Rona *et al.*, 2007).

There is evidence however, from recent US studies that early prevalence rates of substance abuse disorders (SUDs) within 3–4 months of return from deployment are highly prevalent in US Iraq and Afghanistan military personnel with 20% of active duty personnel misusing substances, particularly alcohol and prescription drugs (Armed Forces Health Surveillance Center, 2009; Milliken, Auchterlonie, & Hoge, 2007). Using a two-item screen for alcohol misuse 3–4 months post deployment, researchers found that 27% of Army soldiers screened positive for alcohol misuse. Those who screened positive were more likely to have engaged in drinking and driving, riding with a driver who had been drinking, reporting late or missing work, using illicit drugs, being referred to alcohol treatment, and receiving a driving under the influence (DUI) offense (Santiago *et al.*, 2010). Similar findings were reported in a separate study with 25% screening positive for alcohol misuse 3–4 months post deployment (Wilk *et al.*, 2010).

A study conducted in UK veteran populations claiming war pensions showed that ex-servicemen were twice as likely to develop delayed-onset PTSD as civilians and that 36% of veterans who suffer from delayed onset PTSD developed it within the first year of leaving the services, suggesting that the loss of military support structures and adjustment to civilian life increase vulnerability. Many who developed delayed-onset PTSD reported major depressive disorder and alcohol abuse prior to PTSD onset with many reporting that they drank alcohol to excess in order to cope with mental health symptoms while still in the military (Andrews *et al.*, 2009).

Many ex-servicemen report that they were unable to present for help with mental health problems during their military service. Reasons given include fear of losing their career, a macho image, and stigma (Scheiner, 2008).

Suicide. Overall suicide rates in British servicemen and women are similar to civilian rates and are even lower in some age groups with one exception: suicide rates for ex-serviceman under 24 years are 2–3 times higher than their civilian counterparts. Reasons for this are unclear with causes suggested as including: pre-service vulnerability, trouble re-adjusting to civilian life, and exposure to more adverse experiences (Kapur *et al.*, 2009). These findings are not reflected in the US forces with much higher high rates of suicide in Vietnam War veterans reported (Kulka *et al.*, 1990) and higher rates in combat veterans from more recent wars including Iraq and Afghanistan (Kang & Bullman, 2008).

Military families. Marital relations and family life can be adversely affected by the cycle of peacetime detachments, leading to the intermittent husband syndrome (IHS). This syndrome was described in wives of oil rig workers, airline pilots' wives, and wives of military personnel (Busuttil & Busuttil, 2001). The IHS is characterized by symptoms of anxiety, depression, and sexual difficulties in the wives. It has been documented that unhappy wives are a common factor determining premature retirement in the husbands from the military (Iversen *et al.*, 2005).

Servicemen deployed to war zones may be exposed to threat of death. Studies have shown that a variety of psychiatric symptoms including anxiety, low mood, post-traumatic stress symptoms, and dysfunctional behaviors such as alcohol misuse and risk-taking behaviors can develop in spouses as a result such forms of separation. A process termed "anticipatory grief" is thought to take place. This is a process whereby a wife goes through all the stages of grief because she fears the loss of the life of her husband who has gone to war. If this grieving process completes, then on the husband's return, reunion and reintegration psychological mechanisms that reunite the couple and family might be compromised and this might lead to relationship breakdown. It is also noted that where there are children, the mother's psychiatric symptoms and function will be mirrored in any children she has (see Busuttil & Busuttil, 2001, for a review).

Historically British service families received primary care intervention through military general practices if they resided in family marital accommodation (married quarters). This used to be especially the case when posted overseas. As the draw-down of medical services has progressed and as more and more service families have opted to live within civilian communities rather than on military married quarters, primary care for service families has become increasingly rare. Civilian NHS Primary Care services need to become increasingly aware of these separation-related mental health issues.

It must be noted that rates of domestic violence are said to increase when servicemen return from combat operations. Evaluation of this issue has not been the subject of research, and this evaluation is sorely needed.

Leaving the Military

Transition and resettlement

Leaving the military is a time of vulnerability, and at this time psychological symptoms may increase due to the breaking of attachments with the "military family" and a need to engage with people in an alien civilian setting. Many for the first time since marriage find that they now spend much more time with their wives and children leading to spouses and children getting to know husband and father properly for the first time (Iversen *et al.*, 2005; van Staden *et al.*, 2007).

Some servicemen and women have problems when leaving the military, particularly when they are forced to leave against their wishes (e.g., if their contract has come to an end and despite wishing to remain are forced to leave). Others have disciplinary issues and may be ordered to leave. They thus leave under a label known as "services no longer required" (SNLR). Some may well be court-martialed and may have to spend time in a military prison such as the one based in Colchester, United Kingdom, before they then are discharged from the military when they have served their sentence.

Other military personnel apply for premature voluntary retirement (PvR). Sometimes this is because they cannot cope with military service due to undeclared mental health problems which they may have tried to mask (e.g., by self-medicating with alcohol). It should be noted that overt alcohol misuse may result in a discharge from the military on disciplinary grounds rather than for medical reasons.

Redundancy has become commoner within the military in recent years with some being made redundant and reacting psychologically to this as they do not wish to leave the military, others missing out on generous pay-outs, and yet others who are left in the military after the redundancy cycle has run its course, exposing them to increased work load and the potential to amplified occupational stress.

Recently much more attention is being given to the problems that arise following discharge, from the military, with the government through initiatives such as Fighting Fit (Murrison, 2010), and the Ministry of Defence (MOD) assisted by Third Sector Charities working to establish better discharge procedures, transition supports, and services to aid departure from the military; or retention where appropriate especially for those who have been severely wounded and who have physical disabilities.

At present the role of new facilities called Recovery Centres, funded in collaboration with the charity Help for Heroes, is currently being worked out. These centers, located across the country, will involve a new physical clinical rehabilitation service which will be a step-down from the military rehabilitation hospital based at Headley Court and other military rehabilitation services for those receiving rehabilitation due to severe physical injury. The psychological needs of those physically injured, including the effects of physical deformity and disability including psychiatric disorders and chronic pain, are catered for at Headley Court. It is hoped that the risk of this support and help being reduced on discharge from the military will be mitigated by the recovery centers which allow individuals to live there for a time period in order to finish their rehabilitation as well as provide a hub of interventions and welfare information via ex-service charities including mental health charities such as Combat Stress. The aim here is to allow a seamless transfer of care and support on leaving the military (see Chapter 29 for more information).

Conclusions

Military mental healthcare is highly developed in the British Armed Forces. It offers excellent standards of care that are theoretically easy to access.

Military life has the potential to precipitate stress disorders related to peacetime as well as to operations and combat exposure. Measures aimed at mitigating the impact of combat exposure are increasingly being put in place. PTSD is not the main clinical presentation. Research demonstrates that alcohol-related disorders as well as other nonpsychotic mental health disorders are commoner.

The further reduction of stigma and culture change evident within the British military will, it is hoped, encourage earlier clinical presentation allowing better clinical outcomes. However, many serving personnel may still not be accessing mental health care because of issues related to stigma, including a desire to retain a macho image and fears related to career loss. Some mental health problems become more evident once service personnel have left the military. The transitional period that service personnel have to go through on leaving the military has been examined and has become increasingly recognized as a period of vulnerability in the possible causation of mental health disorders. These issues have started to be addressed more effectively. Provision is also being made for those who have the potential to develop mental health disorders after suffering serious combat-related physical disability.

These issues will be explored further in Chapter 29, which deals with mental health issues in veterans.

References

Andrews, B., Brewin, C. R., Philpott, R., & Hejdenberg, J. (2009). Comparison of immediate-onset and delayed-onset posttraumatic stress disorder in military veterans. *Journal of Abnormal Psychology, 118*, 767–777.

Armed Forces Health Surveillance, Center., (2009). Alcohol-related medical encounters, active components, U.S. Armed Forces, January 2006–December 2008. *Medical Surveillance Monthly Report (MSMR), 16*(5), 6–9.

Barnet, I., Milgram, N., & Noy, S. (1987). Predisposition to CSR in pre-war projective tests. Master's thesis, Tel Aviv University. Cited in Noy, S. (1991). Combat stress reaction. In R. Gal & A. D. Mangelsdorff (Eds.), *Handbook of military psychology* (pp. 507–530). Chichester: John Wiley & Sons, Ltd.

Bar-on, R., Solomon, Z., Noy, S., & Nardi, C. (1986). The clinical picture of combat stress reactions in the 1982 war in Lebanon: Cross war comparisons. In N. A. Milgram (Ed.), *Stress and coping in time of war: Generalisations from the Israeli experience* (pp. 103–109). New York, NY: Brunner/Mazel.

Brill, N. Q., & Beebe, G. W. (1952). Some applications on a follow up study to standards for mobilisation. *American Journal of Psychiatry, 109*, 401–410.

Busuttil, W. (1995). *Interventions in posttraumatic stress syndromes: Implications for military and emergency service organisations.* Master's thesis, University of London.

Busuttil, W. (2009). *Psychometric data analyses and clinical audit data for combat stress 2005–2009* [Internal publication]. Leatherhead, UK: Combat Stress.

Busuttil, W., & Busuttil, A. M. C. (1997). Debriefing and crisis intervention. In D. Black, M. Newman, G. Mezey, & J. Harris Hendricks (Eds.), *Psychological trauma: A developmental approach* (pp. 238–249). Gaskill House, UK: Royal College of Psychiatrists.

Busuttil, W., & Busuttil, A. M. C. (2001). Psychological effects on families subjected to enforced and prolonged separations generated under life threatening situations. *Sexual and Relationship Therapy, 16*(3, Special Psychological Trauma Edition), 207–228.

Carter, B. J., & Cammermeyer, M. (1989). Human responses to simulated chemical warfare training in U. *S. Army reserve personnel. Military Medicine, 154,* 281–288.

Creamer, M., Burgess, P., Buckingham, W., & Pattison, P. (1993). Posttrauma reactions following a multiple shooting: A retrospective study and methodological inquiry. In J. P. Wilson & B. Raphael (Eds.), *International Handbook of Traumatic Stress Syndromes* (pp. 201–212). New York, NY: Plenum Press.

Creamer, M., Moreton, G., Lyng, H., Greenberg, N., & Richardson, D. (2009). Peer support and practical guidelines: Part 1. *In 11th European Conference on Traumatic Stress, European Society for Traumatic Stress, Oslo 15–18 June 2009* (Abstract 176). Oslo: European Society for Traumatic Stress.

Defence Analytical Services and Advice (DASA). (2011). [Home page]. Retrieved from http://www.dasa.mod.uk.

Dyregov, A. (1989). Caring for helpers in disaster situations: Psychological debriefing. *Disaster Management, 2,* 25–30.

Fear, N. T., Jones, M., Murphy, D., Hull, L., Iversen, A. C., Coker, B., *et al.* (2010). What are the consequences of deployment to Iraq and Afghanistan on the mental health of the UK armed forces? A cohort study. *Lancet, 375,* 1783–1797.

Fear, N., & Wesseley, S. (2009). Combat exposure increases risk of alcohol misuse in military personnel following deployment. *Evidence Based Mental Health, 12*(2), 60.

Feinstein, A. (1989). Posttraumatic stress disorder: A descriptive study supporting DSM-III-R criteria. *American Journal of Psychiatry, 146,* 665–666.

Figley, C. R. (Ed.). (1978). *Stress disorders among Vietnam veterans: Theory research and treatment.* New York, NY: Brunner/Mazel.

Flannery, R. B. (1990). Social support and psychological trauma: A methodological review. *Journal of Traumatic Stress, 3,* 593–611.

Foa, E. B., Keane, T. M., & Friedman, M. J. (Eds.). (2009). *Effective treatments for PTSD: Practice guidelines from the International Society for Traumatic Stress Studies* (2nd ed.). New York, NY: Guilford Press.

Gal, R. (1987). Combat stress as an opportunity: The case of heroism. In G. Belenky (Ed.), *Contemporary studies in combat psychiatry* (pp. 31–45). New York, NY: Greenwood Press.

Greenberg, N., Cawkill, P., & Sharpley, J. (2005). How to TRiM away at post traumatic stress reactions: traumatic risk management – now and in the future. *Journal of the Royal Naval Services, 91,* 26–31. Retrieved from http://ftp.rta.nato.int/public//PubFullText/RTO/MP/RTO-MP-HFM-134///MP-HFM-134-35.pdf.

Greenberg, N., Laughton, V., & Jones, M. (2008). Trauma Risk Management (TRiM) in the UK armed forces. *Journal of the Royal Army Medical Corps, 154*(2), 1213–1216.

Hacker Hughes, J. M., Earnshaw, N. M., Greenberg, N., Eldridge, R., Fear, N. T., French, C., *et al.* (2008). Use of psychological decompression in military operational environments. *Military Medicine, 173,* 534–538.

Hawker, D. M., Durkin, J., & Hawker, D. S. J. (2010, December). To debrief or not to debrief our heroes: That is the question. *Clinical Psychology and Psychotherapy, 18,* 453–463.

Hendin, H., & Pollinger Haas, A. (1984). *Wounds of war: Psychological aftermath of combat in Vietnam.* New York, NY: Basic Books.

Hooper, R., Rona, R. J., Jones, M., Fear, N. T., Hull, L., & Wesseley, S. (2008). Cigarette and alcohol use in the UK armed forces and their association with combat exposures: A prospective study. *Addictive Behaviours, 33,* 1067–1071.

Iversen, A., Dysen, C., Smith, N., Greenberg, N., Walwyn, R., Unwin, C., *et al.* (2005). "Goodbye and good luck": The mental health needs and treatment experiences of British ex-service personnel. *British Journal of Psychiatry, 186,* 480–486.

Iversen, A., Waterdrinker, A., Fear. N. T., Greenberg, N., Barker, C., Hotopf, M., *et al.* (2007). Factors associated with heavy alcohol consumption in the UK armed forces: Data from a health survey of Gulf, Bosnia and era veterans. *Military Medicine, 172*(9), 956–961.

Kang, H. K., & Bullman, T. A. (2008). Risk of suicide among US veterans after returning from the Iraq or Afghanistan war zones. *Journal of the American Medical Association (JAMA), 300*(6), 652–653.

Kapur, N., White, D., Blactchley, N., Bray, I., & Harrison, K. (2009). Sucicide after leaving the UK armed Forces a cohort study. *PLoS Medicine, 6*(3), 1–9.

Kormos, H. R. (1978). The nature of combat stress. In C. R. Figley (Ed.), *Stress disorders among Vietnam veterans: Theory research and treatment* (pp. 3–22) New York, NY: Brunner/Mazel.

Kulka, R. A., Schlenger, W. E., Fairbank, J. A., Hough, R. L., Jordan, B. K., & Marmar, C. R. (1990). *Trauma and the Vietnam War generation: Report of findings from the National Vietnam Veterans Readjustment Study.* New York: Brunner/Mazel.

Marshall, S. L. A. (1978). *Men against fire: The problem of battle command in future war.* Gloucester, UK: Peter Smith.

McFarlane, A. (1989). The aetiology of post traumatic morbidity: Predisposing, precipitating and perpetuating factors. *British Journal of Psychiatry, 154,* 221–228.

McManners, H. (1993). *The scars of war.* London: HarperCollins.

McMannus, F. B. (2009 , June 2–5). Accessing defence psychiatric services: Clinical pathways to the NHS and other agencies on leaving the military.Paper presented at Veterans Mental Health: Annual Meeting of the Royal College of Psychiatrists, BT Convention Centre. Liverpool, UK.

Mikulincer, M., & Solomon, Z. (1988). Attributional style and combat-related posttraumatic stress disorder. *Journal of Abnormal Psychology, 3*(97), 308–313.

Miller, I. (1989). Preparation for disaster: Trauma inoculation training for police disaster victim identification teams.Paper presented at the World Congress, World Federation for Mental Health, Auckland, NZ.

Milliken, C. S., Auchterlonie, J. L., & Hoge, C. W. (2007). Longitudinal assessment of mental health problems among active and reserve component soldiers returning from the Iraq war. *Journal of the American Medical Association (JAMA), 298*(18), 2141–2148.

Mitchell, J. T. (1983 , November). When disaster strikes: The Critical Incident Stress Debriefing process. *Journal of Emergency Medical Services,* 43–46.

Murrison, A. (2010). *Fighting fit: A mental health plan for servicemen and veterans.* London: Ministry of Defence.

National Institute for Health and Clinical Excellence (NICE). (2005). *Post-traumatic stress disorder (PTSD): The management of PTSD in adults and children in primary and secondary care.* NICE Guideline GC26. London: Author.

Novaco, R., Cook, T. M., & Sarason, I. G. (1983). Military recruit training: An arena for stress-coping skills. In D. Mietchenbaum & M. E. Jaremko (Eds.), *Stress reduction and prevention* (pp. 377–417). New York, NY: Plenum Press.

Noy, S. (1987). Stress and personality as factors in the causation and prognosis of combat reactions during the 1973 Arab–Israeli war. In G. L. Belenky (Ed.), *Contemporary studies in combat psychiatry.* (pp. 21–29). Westport CT: Greenwood Press.

Noy, S. (1991). Combat stress reaction. In R. Gal & A. D. Mangelsdorff (Eds.), *Handbook of military psychology* (pp. 507–530). Chichester: John Wiley & Sons, Ltd.

Rona, R. J., Fear, N. T., Hull, L., Greenberg, N., Earnshaw, H., Hotop, M., *et al.* (2007). Mental health consequences of overstretch in the UK armed forces: First phase of a cohort study. *British Medical Journal, 335*(7620), 571–572.

Rona, R. J., Hooper, R., Jones, M., Hull, L., Browne, T., Horn, O., *et al.* (2006). Mental health screening in the armed forces before the Iraq war and prevention of subsequent psychological morbidity: Follow up study. *British Medical Journal.* dol: 101136/ bmj.38985. 610949.55.

Rona, R. J., Hooper, R., Jones, M., Iversen, A., Hull, L., Murphy, D., *et al.* (2009). The contribution of prior psychological symptoms and combat exposure to post Iraq deployment mental health in the UK military. *Journal of Traumatic Stress, 22*(1), 11–19.

Rose, S., Bisson B., & Wessely, S. (2001). *A systematic review of brief psychological interventions ("debriefing") for the treatment of immediate trauma related symptoms and the prevention of post-traumatic stress disorder.* The Cochrane Collaboration (database online and CDROM) Oxford: Update Software. Updated issue 3, 2001.

Rotter, J. B. (1966). Generalized expectancies for internal versus external control of reinforcement. *Psychological Monographs: General and Applied, 80,* 609.

Santiago, P. N., Wilk, J. E., Milliken, C. S., Castro, C. A., Engel, C. C., & Hoge, C. W. (2010). Screening for alcohol misuse and alcohol-related behaviors among combat veterans. *Psychiatric Services, 61*(6), 575–581.

Sarson, I. G., Johnson, J. M., & Siegel, J. M. (1978). Assessing the impact of life changes: Development of the Life Experiences Survey. *Journal of Consulting and Clinical Psychology, 46,* 932–946.

Scheiner, N. S. (2008). Not "at ease": UK veterans' perceptions of the level of understanding of their psychological difficulties shown by the National Health Service. Doctoral thesis, City University London.

Shalev, A. (1988a). Psychiatric care of acute stress reactions to military threat. In R. J. Ursano, C. S. Fullerton, M. S. Dixon, & E. S. Doran (Eds.), *Groups and organisations in war, disasters and trauma* (pp. 107–119). Bethesda, MD: USUHS.

Shalev, A. (1988b). Event-oriented debriefing following traumatic exposure of groups: S. L. A. Marshall's approach. Presented to the annual meeting of the Society of Traumatic Stress Studies, Dallas, TX. In *Individual and community responses to trauma and disaster: The structure of human chaos.* London: Cambridge University Press.

Shalev, A. (1992). Posttraumatic stress disorder among injured survivors of a terrorist attack: Predictive value of early intrusion and avoidance symptoms. *Journal of Nervous and Mental Disorders, 180,* 505–509.

Shalev, A., & Munitz, H. (1988). Psychiatric care of acute stress reactions to military threat. In J. Ursano, C. S. Fullerton, M. S. Dixon,& E. S. Doran (Eds.), *Groups and organisations in war: Disasters and trauma* (pp. 107–119) Bethesda, MD: USUHS.

Sharpley, J. G., Fear, N. T., Greenberg, N., Jones, M., & Wesseley, S. (2007). Pre-deployment stress briefing: does it have an effect? *Occupational Medicine Advance Access.* doi: 101093/ occmed/kqm118.

Solomon, S. D., Noy, S., & Bar-on, R. (1986a). Risk factors in combat stress reactions: A study of Israeli soldiers in the 1982 Lebanon war. *Israel Journal of Psychiatry, 23,* 3–8.

Solomon, S. D., Noy, S., & Bar-on, R. (1986b). Who is at high risk for combat stress reaction syndrome? In N. A. Milgram (Ed.), *Stress and coping in time of war: Generalisations from the Israeli experience* (pp. 78–83) New York, NY: Brunner/Mazel.

Solomon, Z. (1988). The effect of combat-related post traumatic stress disorder on the family. *Psychiatry, 51,* 323–329.

Solomon, Z. (1989). Untreated combat-related PTSD – why some Israeli veterans do not seek help. *Israeli Journal of Psychiatry and Related Sciences, 26,* 111–123.

Solomon, Z., & Benbenishty, R. (1986). The role of proximity, immediacy and expectancy in frontline treatment of combat stress reaction among Israelis in the Lebanon war. *American Journal of Psychiatry, 143,* 613–617.

Solomon, Z., Mikulincer, M., & Bebenishty, R. (1989). Locus of control and combat-related post-traumatic stress disorder: The intervening role of battle intensity, threat appraisal and coping. *British Journal of Clinical Psychology, 28,* 131–144.

Srinivasan, M. (1993 , October 13). Mind at war. Army PTSD Video.Paper presented at the Defence Medical Services Military Psychiatry Conference, Royal Army Medical College, Millbank, London.

Steindl, S. R., Young, R. M., Creamer, M., & Crompton, D. (2003). Hazaderous alcohol use and treatment outcome in male combat veterans with posttraumatic stress disorder. *Journal of Traumatic Stress, 16,* 27–34.

Stouffer, S. A., Suchman, E. A., DeVinney, L. C., Star, S. A., & Williams, R. M. (1949). *The American soldier: Adjustment during army life* (Vol. 1). Princeton, NJ: Princeton University Press.

Strange, R. E., & Arthur, R. J. (1967). Hospital ship psychiatriy in a war zone. *American Journal of Psychiatry,* 281–286.

Tehrani, N. (Ed.). (2002). *Psychological debriefing.* Professional Practice Board Working Party. Leicester, UK: British Psychological Society.

van Staden, L., Fear, N. T., Iversen, A. French, C. E., Dandeker, C., & Wessely, S. (2007). Transition back into civilian life: A study of personnel leaving the U. K. *Armed Forces via military prison. Military Medicine, 172*(9), 925.

Wilk, J. E., Bliese, P. D., Kim, P. Y., Thomas, J. L., McGurk, D., & Hoge, C. W. (2010). Relationship of combat experiences to alcohol misuse among U.S. soldiers returning from the Iraq war. Drug and Alcohol Dependence, *108,* 115–121.

27

Trauma Counseling and Psychological Support in the People's Republic of China (PRC)

Xiaoping Zhu, Zhen Wang, and Tony Buon

Most literature and media coverage of trauma emanate from Western countries and from a Western perspective, while in China psychological trauma has not attracted much attention until recent years. The term "psychological trauma" appeared in local media reporting and in more widespread conversation in the wake of the Wenchuan earthquake in 2008.

Even in Chinese academic circles, the study of psychological trauma has not been given much attention until recently, and this has been a very gradual process. This chapter will briefly discuss the Chinese people's understanding of trauma from their traditional cultural point of view, methods of coping with trauma from a Chinese perspective, and the development and evolution of psychological and trauma counseling in China.

Chinese Cultural Background

Traditional Chinese culture is based on Confucianism, Buddhism, and Daoism which value the harmony and unity of human and nature; and, according to the traditional Chinese medicine which was developed out of this cultural background, diseases are deemed to stem from an imbalance of "yin" and "yang." Yin and yang comprise an ancient Chinese philosophy which is based on the idea that everything in the universe is formed and influenced by the combination of the yin and yang forces and the disorder of the "five elements": wood, fire, earth, metal, and water. These elements were used for describing interactions and relationships between phenomena.

This imbalance or disorder of yin and yang may result in the obstruction of the human body's energy for life, known as "qi" (also commonly spelled ch'i, chi, or ki). Qi is a fundamental concept of everyday Chinese culture, most often defined as "air" or "breath" and, by extension, the life force or spiritual energy that is part of everything that exists. This belief still deeply influences Chinese people's concept of disease and even their symptoms.

International Handbook of Workplace Trauma Support, First Edition. Edited by Rick Hughes, Andrew Kinder, and Cary L. Cooper.

According to traditional Chinese medicine, body and soul are linked together, so there is no obvious distinction between physical disease and mental illness. This concept holds that, in order to keep physical or mental health, people should avoid extreme expression of emotions and maintain the harmony of family and other social relationships. Traditional Chinese medicine regards the "visceral system" as the center where body and soul unite and internal organs as the manifestation of physical diseases. This may be one of the reasons why there are more somatic symptoms in Chinese patients involving depression (Kim, Li, & Kim, 1999).

Changing Coping Strategies and Social Development

The cultural background of Confucianism, Buddhism, and Daoism has a profound effect on the strategies that Chinese people adopt when they encounter disasters and trauma. When facing circumstances beyond their individual control, Chinese people of traditional views always hold the belief to "Do one's level best and leave the rest to God's will." Many events outside of human action are attributed to "God's will" or the "arrangement of fate." While this could be seen as a negative coping strategy, it effectively relieves feelings of shame, self-accusation, and helplessness in the face of trauma and, at least to some degree, reduces the negative effects arising from the trauma.

Traditional Chinese culture is based on agricultural culture, in which there is a phenomenon of "authority worship" and the strongly held belief that "Losers are always in the wrong." In such a belief system, people are not devoid of mercy and love, yet it's difficult for people who suffer trauma to fully express sorrowful emotion, because everyone else in the society believes that no matter how difficult the situation one faces, one should be strong enough to face it. To not do so would even see the person experiencing the disaster as being described as a coward.

In China there has traditionally been a strong social support system tied to the family or clan. In fact, in traditional Chinese people's view, people of the same surname belong to one clan, and these people share weal or woe (the population of the largest clan with the same family name in China is close to 100 million). In such a huge network based on the patriarchal surname, one's trauma is no longer regarded as personal business, but something that the whole clan will cope with together. In these circumstances it could be seen that the degree or impact of such trauma or disaster is diluted. This belief is evident in ancient Chinese stories recounting how immigrants experienced various misfortunes in other lands – because they were far away from their clans and lost the strong social support of the clan.

This situation had lasted for thousands of years until the founding of the People's Republic of China (PRC) in 1949. Along with the establishment of a new state structure this resulted in earth-shaking changes in the social structure. The sense of "clan" with feudal traditions was vigorously challenged, and new thoughts and concepts such as independence, equality, democracy, and freedom were gradually conveyed to the society by scholars and the government.

Yet such a transformation has not been always easy, especially during the first three decades after the founding of PRC. China suffered many natural disasters and social movements in which many people experienced severe psychological trauma. At that time, urban and rural residents were affiliated to different "units" and "production teams" for their different work places. As basic administrative organizations, these teams represented in a sense, government power, and when people experienced trauma, they often turned to the unit for help. Consequently, the role of clans in relieving trauma gradually declined. This transformed social support system could still cope with psychological trauma, but the units were less concerned with affection elements and more with rational elements. Therefore, during this period, people who suffered trauma became less inclined to express emotions, and many psychological pains were constrained and internalized. However, because of the authority of production teams (or other organizations which represent government) and people's confidence in these organizations, traumas which occurred in workplaces were generally dealt with through the organizations and, because of this, possibly in a positive way.

Nevertheless, the Great Proletarian Cultural Revolution had a severe psychological impact on many Chinese people. Many individuals and families experienced traumatic situations. Because of the sensitivity of this topic, many people who experienced psychological traumas during this period tried their best to avoid referring to it. This silence resulted in many individuals not dealing with the traumas for a long time. Today patients with psychological traumas arising from the Cultural Revolution can be seen in clinics of psychiatry and psychological counseling organizations in China.

Some scholars have put forward the theory of intergenerational transmission for the kind of psychological trauma attributed to the Cultural Revolution (Scharf, 2007), and a few Chinese scholars have tentatively made some investigations into this, but to date there is no published Chinese literature on this area. And while the intergenerational transmission of trauma is full of complexities, Kellermann (2001) reported that these children do show signs of stress and specific disturbances and demonstrate a higher vulnerability to post-traumatic stress disorder (PTSD).

Since 1979 China has undergone profound, even revolutionary change under the late Deng Xiaoping's reform program and "open door policy" (Liang, 1997). After China implemented the policy of reformation, the young modern society established in China changed greatly once again, and Western culture and ideology moved swiftly into China. In the China of this period, the importance of the individual and the attention paid to the individual by people gradually increased, and the population's mobility greatly increased. This saw the influence of units and production teams rapidly decreasing, and the psychological support and sense of security provided by them to people weakened synchronously. The influence and support system of clans, which had been weakened in the previous 30 years, continued to diminish.

China at this time did not have a social security system similar to that found in developed countries in the West, and this, together with the changes mentioned in this chapter, resulted in Chinese people finding themselves in a relative vacuum in relation to social support systems. People had to cope with more intense psychological stresses from life and work without sufficient psychological support. When people in this period experienced psychological traumas, they had to turn to other means for help because it was difficult for them to obtain enough resources from traditional psychological support systems. This resulted in the development of a Western-influenced psychological counseling services and Employee Assistance Programs (EAPs), particularly in major Chinese cities experiencing economic growth and an emerging middle class.

A Brief Introduction to the Development of Psychology in China

The development of indigenous psychology in China can be traced to the Pre-Qin Dynasty (221–206 BCE), and many psychological thoughts concerning body and mind relations, self-awareness, group mind, and individual psychology were clearly recorded in *The Book of Changes*, *The Book of Songs*, *The Classic of History*, and other early ancient books and records. From the Han Dynasty (206 BCE–220 CE) to Tang Dynasty (618–907), along with the combination of foreign Buddhism and the indigenous cultures, people attached a deeper level of attention to psychological problems. Buddhism emphasizes the inner world and has a strong connection with psychological phenomena, and it inspired many Chinese scholars to study psychological problems.

Achievements in science and technology during the Sung, Yuan, Ming, and Qing Dynasties (especially achievements in physiology in Chinese medicine) greatly influenced the development of psychological thought. However, though ancient scholars had explored human psychological activities, there was no real science of psychology in China at this time. The psychological thoughts were found in works of traditional philosophy, politics, history, ethics, education, medicine, military science, religion, and so on.

During the late period of Qing Dynasty (1644–1911), Western civilization was spreading out to the East and Western psychology gradually entered into China. The dissemination of modern Western psychology by Chinese scholars who went to Europe and America strengthened the combination of Chinese and foreign psychological thinking and accelerated the formation and development of a modern Chinese psychology. The provision of psychology courses in new-style schools and missionary schools also played a role in the early development of a Chinese modern psychology.

Around the May 4th Movement of 1919, with the catchphrase of "Science and Democracy," psychology aroused further attention in Chinese society and increasingly people went to Europe and the United States to learn psychology. After returning home, they were engaged in the teaching of psychology and began to introduce theories of various schools in Western psychology. A few psychologists also began to establish departments of psychology, psychology research institutions, and academic organizations related to psychology. There was also some limited academic exchange and the development of psychology journals and publication.

For example, at the beginning of the 1920s, colleges and universities in Beijing, Nanjing, and other cities began to offer psychology courses; Peking University founded the first psychology laboratory in China in 1917. After this, the development of Chinese psychology slowed down because of the unstable political situation in China and the influence of World War Two. After the establishment of PRC in 1949, Chinese psychology entered a period of renaissance. Following the development of economic construction and scientific undertakings, psychology research institutions were founded. The Chinese Psychological Society was re-established and psychologists from the whole country were gathered and organized to study the philosophy of dialectical materialism and the theories of Pavlov in particular. There was also a focus on Soviet psychology and attempts to transform Western psychology to suit Chinese thought. However, because of the influence of politics and society and other factors during the Cultural Revolution, psychology was eventually disparaged as idealism, and its development was greatly impacted upon and obstructed.

From the 1980s, psychology began to obtain a new life and this corresponded with the deepening of reform and opening up of the PRC. Western psychology began to be taken seriously again and with the economic development of the 1990s China saw professional institutes in various cities offering various kinds of psychology training courses through cooperation with overseas universities or psychology research institutes. An example of this was the Sino-German Senior Psychotherapist Continuous Training Program held by the German-Sino Academy for Psychotherapy jointly with Shanghai Mental Health Center, Hamburg University, Freud Research Institute of Frankfurt, and other organizations. This became the eminent training program in China at this time. It developed workshops in psychoanalysis, cognitive therapy, and family therapy, and many graduates of this program have become cornerstones in the Chinese psychotherapy field. Additionally, the Chinese-American Cognitive Behavior Therapy Training Program, the Chinese-Norwegian Psychoanalysis Training Program, the Chinese-American Trauma Therapy Training Program, and other training programs have been accepted gradually by the industry. Meanwhile, various kinds of training classes for the examination of Chinese psychological counselors sprang up like bamboo shoots in the 1990s; however, the quality of the training was mixed. In short, the current Chinese psychological counseling and therapy industry is going through a period of ongoing and rapid development.

On December 9 1997, *China Daily* reported on the expansion of psychological education. It stated that "more and more college students today have come to realise that healthy emotions and personalities are important for their lives and studies" (Cui, 1997, p. 2). It also reported that 70% of colleges and universities in Beijing and Shanghai provided psychological counseling services for students (Cui, 1997).

At present, Chinese psychology is taking its lead from the West, but people have begun to reflect on the cultural characteristics of Western psychology and attempted to see psychology from a cultural perspective. Research on the Sinicization of psychology has been given much more attention in recent years. Chinese psychologists emphasize the study of Chinese people's special psychology according to different characteristics of Chinese culture. Additionally, the participation of Chinese scholars in international academic exchanges on psychology has resulted in much development and advancement in the development of a Chinese psychological paradigm.

Trauma Counseling and the Development of Employee Assistance Programs in China

As mentioned above, after the implementation of reform and the general opening up of the PRC in the 1980s, significant and rapid changes took place in the Chinese social structure. Modern thought collided with traditional concepts, traditional values conflicted with external behaviors and an increasingly accelerated life and work rhythm, and the inevitable competition resulted in substantial pressures and stresses. Yet the simultaneous weakening of the traditional social support systems resulted in those with psychological problems not receiving support or assistance.

Together with the "one-child policy" implemented since the 1970s, a new generation of young people in China was growing up in an environment entirely different from that of their parents. Many of them were spoiled by their parents and received excessive attention from family members. This resulted in generational differences in relation to a number of issues between the young and the older generations, particularly regarding workplace conflict and workplace trauma. For example, in the face of a superior's unfair criticism, the employee from the older generation may often believe they should accept and endure such criticism, and they should be strong to face it even if feeling wronged. However, young workers would probably not accept this, and they would most likely point out the superior's mistake and safeguard their dignity. These young workers may even experience emotional disturbance in relation to this criticism.

For trauma in the workplace, the older generation does not have many options because they are used to "being strong to face them" and "believing that organizations will solve them adequately," while young workers are unlikely to accept the organizations' "adequate solution" when they are facing such traumas. Young Chinese people's mode of thinking is very different from their parents', but they have inherited the tradition of not being good at directly expressing their emotions. Therefore, young workers seem to develop greater psychological pressure and often appear lost in helplessness and despair when experiencing severe psychological traumas. They are aware that, psychologically, there is something wrong with them but they don't know how to resolve these psychological problems. The provision of trauma counseling to these young workers appears to be increasingly important.

During the 1980s and 1990s, the concept of psychological counseling was still new and strange to many people in China, while the actual demands of Chinese people for psychological counseling services were increasing. However, there is a significant shortage of specialist practitioners who are able to provide psychological counseling services in China. For example, we estimate that there are less than seven people providing psychological services persons for every million people presently in the PRC. In particular it is difficult for people who experience psychological trauma to obtain timely and effective psychological aid in the PRC. This contradiction between supply and demand encouraged, to some extent, the development of a psychological counseling and psychotherapy industry including EAPs during this period.

In 1997, Tony Buon and Xiaoping Zhu developed the first EAP in the PRC in Shanghai and began promoting the EAP concept throughout China (Buon, 1998). The Hong Kong Christian Service also provided EAPs in the then UK-administered Hong Kong from 1993 and reported "good usage" by Chinese people of their EAP (Fong & Lam, 1998). At the Asia-Pacific Conference of Employee Assistance Programs held in Hong Kong in March 1998, providers from many Asian countries also reported good usage of EAPs in their countries. Throughout the early part of the new century, many foreign EAP business began developing EAPs (to a limited degree) in China. In 2003, a not-for-profit organization, the Asia Pacific Employee Assistance Roundtable (APEAR), was formed to promote standards of practice and continuing development of EAPs in Asia. In 2005, a Chinese Government agency ran the First China International Forum on Employee Assistance Programs in Beijing. Today there are more than 20 providers of EAPs in China, both foreign and local, and these include organizations such as the China EAP Service Center, a local Chinese provider.

In fact, even before the development of these services in the 1980s and 1990s, Chinese enterprises were aware of paying close attention to staff's physical and psychological health. This was mainly reflected in the "ideological and political work" taken charge of by the Chinese Communist Party and government departments. There was also an emphasis on using methods of behavioral science in relation to staff management issues and the scientific ideological and political work after the implementation of reform and "opening up" during this time in Chinese history.

However, at the beginning of the twenty-first century many organizations began to pay attention to occupational psychological health and organizational development, and the concept of EAPs was introduced by some large-scale foreign-owned enterprises.

At first, some foreign-owned enterprises adopted internal EAP models or contracted with foreign EAP services to provide EAP services to their expatriate staff. But with an increase in the demand for EAP services, and the inclusion of Chinese staff in EAP contracts, a boom in EAP work began in earnest at the start of the twenty-first century. Services were generally provided to mainly foreign-owned enterprises that already had EAPs in their home countries. Some foreign companies who were the early adopters of the EAP model in China were Coca Cola, HP, Motorola, Cisco Systems, Alcatel, and Proctor & Gamble Co. However, at this time few Chinese-owned enterprises also developed EAPs.

Human biological makeup, the distribution of cognitive abilities, and the general features of the physical world are similar to all people. However people from different parts of the world have different sociocultural traditions and different experiences in their interactions with the world. These social variances may lead to different perspectives, from which they approach the same physical reality and how they conceptualize the world. These different sociocultural traditions and different experiences suggest that EAP counseling needs to be modified to suit the local market and be provided by psychologists who are part of that tradition and consequently share the same perspectives (Buon, 2000).

Over the past few years, there has been increasing belief in China that EAP services for national staff need to be provided by local (Chinese) psychological professionals. For this reason, a number of local EAP service agencies have emerged. This has also resulted in some Chinese enterprises starting to use EAPs or similar services. This includes organizations such as Lenovo Group, China Development Bank, China Mobile, and the Bank of China.

The Psychological Trauma Responses of Chinese People

Human beings have common psychological reactions in facing trauma, and Chinese people have no difference in this regard from people of other cultures. Chinese people experience normal post-traumatic psychological reactions. However, Chinese culture generally includes a strong volitional quality in facing traumas or difficulties and doesn't encourage the expression of emotions, especially for men. For example,

Chinese men will often believe that "A true man only weeps where he is touched to the marrow."

When hearing that someone has experienced a traumatic situation, it would be common for those around them to say "Pull yourself together." Therefore, in experiencing trauma, a person's fear or emotions are often not fully ventilated and a common coping strategy would involve suppression or isolation.

When a group experiences natural disasters or other traumatic events, this tends to result in different coping strategies or reactions. For instance, in facing disasters such as earthquake or flood, Chinese people tend to think "This is fate or God's will." They will tend to face such situations outside of their control with a mind-set of "Do one's level best and leave the rest to God's will." This reaction reduces guilt and makes it easier for people to accept the reality. Meanwhile, Chinese people hold a belief of "One who survives a great catastrophe is destined to good fortune for ever after," which can be a positive belief system in the treatment of trauma.

This may explain, at least in part, that after experiencing the Wenchuan earthquake on May 12, 2008 the actual incidence rate of post-traumatic stress disorder (PTSD) was reported as being 2.37% (Sun, Sun, & Li, 2010), far below the data reported after disasters in Western countries in spite of a huge number of disaster victims. According to a report on September 25, 2008, authorized by the Earthquake Relief Headquarters of the State Council of China, the death toll reached 69 227, with a further 374 643 injured and 17 923 reported missing (see also Sina, 2008).

Of course, during such natural disasters faced by groups, people will still experience negative emotions and individuals still tend to cope with these by means of suppression or isolation. Such coping strategies may be propitious to the mental adaptation of people involved in the short term, but later, psychological pain may reappear in other ways, including depression, anxiety, or even dissociative disorders. Therefore, it is essential to provide suitable trauma counseling to people who experienced trauma. Nevertheless, it is not easy to practice trauma counseling in China.

For most Chinese people, counseling of any type is a new concept, and with the disaster relief of traumatic accidents in the past, trauma counseling was not provided. This has resulted in many Chinese people rejecting assistance and psychological services. Additionally, because some professionals who provide trauma counseling haven't received sufficient training on trauma counseling, the risk of secondary trauma is high. This occurred during psychological intervention after the Wenchuan earthquake, which saw the avoidance of psychological intervention by many disaster victims.

However, in the 11.15 Large Fire which happened in Shanghai on November 15, 2010, the response was different. This fire resulted in 53 deaths, and psychological professionals involved in this crisis demonstrated a high level of professionalism. They established contact with victims' families and survivors of the fire, provided basic psychological support to survivors through non-intrusive approaches such as active listening and accompanying, and guided survivors to express their distress and fear with empathy.

It should be pointed out that there was a common phenomenon in trauma counseling after both the Wenchuan earthquake and the Shanghai fire disaster. For some people who demonstrated great fortitude or bravery in public or in group counseling, their sadness, fear, or other emotions would burst out suddenly once they came into an environment of individual counseling or had the opportunity to have one-to-one professional trauma counseling. This may be related to the subtle defense mechanism of repression and isolation in Chinese culture, and may also be the result of the fact that Chinese people do not want others to feel that they are weak or "not mentally strong," or that they don't want to "lose face." Nevertheless, a majority of people who experienced trauma are unwilling to receive psychological counseling; they will avoid the contact with psychological counselors because even simple contact means that they are not strong enough psychologically and it's difficult for them to "show their soft side" in public.

With the development of EAPs in China, some companies do utilize EAP providers for trauma counseling for traumatic events or work accidents. An example of this was the much publicized case involving the suicide of an employee in a manufacturing enterprise in China. A female employee of this company killed herself by jumping from the roof of a residence community. After learning the news, the Human Resources (HR) Department immediately informed the companies' EAP provider and this EAP provider established the trauma intervention program within the hour and assigned its staff to enter the workplace to carry out a field evaluation.

This employee, a recent graduate and recently engaged, was seen by other employees as an optimistic and outgoing young woman. Her suicide was a great shock to her colleagues and no one in her team could believe she had taken her own life. The first response of everyone was "It's impossible"; afterward, emotional reactions such as sadness, compunction, self-accusation, and bewilderment emerged among the team; some suffered sleep problems; and some team members felt demotivated and lethargic.

During the trauma intervention, the EAP professionals provided assistance with the depressive emotions of team members and enabled them to begin to accept their normal reactions to this incident through one-to-one sessions. This team had a good cohesion, and the team members appeared to trust and support each other. Following group counseling, they started discussions that demonstrated self-awareness and mutual understanding on the feelings of guilt and life values.

Since this traumatic incident did not take place in the company, it may have been expected that there would be less impact on other employees of the company. But in fact the event led to very significant reactions by other employees and management. Internal support systems developed and staff found they could support each other. Under a safe group-counseling environment, a wide range of employees received support and assistance.

Perspectives on the Development of Trauma Counseling in China

China is a country with a population of 1.3 billion, and it is experiencing economic development and social changes of an unprecedented scale. In China, various forms of trauma resulting from natural disasters or workplace accidents occur daily. While trauma counseling in China is at the initial stage of its development, we are convinced that trauma counseling will develop significantly over the next decade.

On one hand, following the continuous improvement of the economy, people will certainly demand more in relation to their psychological health along with their basic material demands; on the other hand, along with globalization and influence of Western culture, Chinese people's means of expressing emotions are gradually changing. Compared with their parents, young Chinese people can express their inner emotions more openly and appear to accept trauma counseling and other similar psychological interventions more easily. Government departments and an increasing number of private companies or organizations are paying greater attention to the psychological health of their staff and are much more willing to provide trauma counseling and other psychological services than they were even five years ago.

The development of trauma counseling in China must take into account the unique history and culture of the Chinese people and have what is often described as "Chinese characteristics." Among other things, the support of relevant Chinese government departments is particularly important. For example, the National Emergency Response Team for Public Health founded in China recently involved professionals in the fields of psychological crisis intervention and trauma counseling. We believe that this will play a demonstrative role for regional governments for trauma counseling to get attention, policy support, and even financial resources from the government. Furthermore, we believe this will allow trauma counseling services to extend from large-scale disasters and emergencies to the handling of general psychological traumatic incidents in workplace.

Nevertheless, the ongoing development of trauma counseling in China is stifled by a shortage of Chinese psychological professionals. It is presently a challenge to provide sufficient training to practitioners involved in trauma counseling so as to guarantee the quality and acceptability of trauma counseling. The development of trauma counseling in China needs to take into account the cultural applicability and the Sinicization of trauma-counseling technology. Perhaps Chinese psychological counseling experts will draw more influence from Eastern culture, rather than just accepting existing Western trauma-counseling methods. For example, Taoism advocates "Let it be," "Govern by doing nothing that goes against nature," and other thoughts that have possible implications for trauma counseling. Some western psychological counselors have even begun to look to the East for new therapeutic methods in their trauma counseling; this includes areas such as meditation technology, mindfulness-based cognitive therapy, and other methods with strong Eastern thought.

Psychological trauma is no longer a foreign concept in China, and people who experience trauma have gradually begun to seek professional psychological counseling and assistance, yet trauma counseling is still a new concept and the experiences of Western countries can affect this process of development. However, most importantly for the ongoing development of trauma counseling in China, we need to develop trauma counseling modes that both respect and integrate Chinese culture.

References

Buon, T. (1998). Sex, drink and drugs – can counselling win the battle? Managing your PRC workforce. *China Staff*, *4*(9), 15–19.

Buon, T. (2000). People's Republic of China (PRC). In D. Masi (Ed.), *International employee assistance anthology* (2nd ed., pp. 75–81). Washington, DC: Dallen.

Cui, N. (1997, December 8). Psychological education expands. *China Daily*, 2.

Fong, D., & Lam, K. (1998, 5–7 March). Workplace counselling: A study of its users and effectiveness. Paper presented at the Asia Pacific Conference of Employee Assistance Programme, Hong Kong.

Kellermann, N. P. F. (2001, Fall). Transmission of Holocaust trauma: An integrative view. *Psychiatry: Interpersonal and Biological Processes, 64*, 256–267.

Kim, K., Li, D., & Kim, D. (1999). Depressive symptoms in Koreans, Korean–Chinese and Chinese: a transcultural study. *Transcultural Psychiatry, 36*, Sep, 303–316.

Liang, C. K. (1997). [Personal communication]. Department of Psychology, Old Dominion University. Cited in Buon, T. (2000). People's Republic of China (PRC). In D. Masi (Ed.), *International employee assistance anthology* (2nd ed., pp. 75–81). Washington, DC: Dallen.

Scharf, M. (2007). Long-term effects of trauma: Psychosocial functioning of the second and third generation of Holocaust survivors. *Development and Psychopathology, 19*, 603–622.

Sina. (2008). Wenchuan Earthquake has already caused 69,196 fatalities and 18,379 missing (in Chinese). Retrieved from http://news.sina.com.cn/pc/2008-05-13/326/651.html.

Sun, Q., Sun, X., & Li, J. (2010). An epidemiologic investigation of anxiety disorder among Wenchuan earthquake victims. Paper presented at the 2nd National Anxiety Disorder Conference, Xian, China.

28

How Professionals can Help the Traumatized Organization

Pauline Rennie Peyton

Introduction

This chapter discusses the practical, hands-on ways in which traumatized organizations can receive help to help their staff. These include proactively creating a written policy in case a traumatic event occurs, calling in professionals after the event, and understanding the various ways in which these professionals can help not only those who have been traumatized but also human resources and management, who are often unsure about how to deal with unexpected and abnormal situations.

Pervasiveness of Trauma

Being asked to write this chapter has encouraged me to reflect upon my mental archive, and revisit the professional journey it catalogs, representing more than 25 years spent working with trauma in various roles: as a psychologist, psychotherapist, and counselor.

When casting the mind back over time, one almost by default returns to earliest memories, and so I recall my first experience. It was not a gentle entrance into the arena of trauma (not that it ever, by its nature, can be), but in retrospect I see that it was an invaluable one: I found myself looking into the eyes of train drivers into the paths of whose speeding trains suicidal people had thrown themselves in desperate efforts to end their lives. What I learned early on from these men is that while they all shared a similar experience, there were huge differences in the way they responded.

On one occasion I met with two drivers who, being together in the cab of a train, witnessed the same event. One of the men shrugged it off by saying, "That is life; these things happen. I'll be fine." I had no cause to doubt him, and believed that he was, indeed, "fine." He left the session after being told that he could return for treatment if he experienced any symptoms of trauma in the future. The other man, his colleague, was traumatized and showed evidence of PTSD. Ultimately, the first man did not need treatment and the second one benefited from it.

International Handbook of Workplace Trauma Support, First Edition. Edited by Rick Hughes, Andrew Kinder, and Cary L. Cooper.

These suicidal events are a lot more common than reports in the media might suggest. In another case, the train driver had coped well the first two times that this type of event had happened to him, but on the third occasion he was affected in a different way. This time he needed treatment, and kept asking, "Why me again?" Indeed, this was something that was difficult for him to grasp or for me to explain: why him again?

A tragic event of this sort traumatizes not only those who may have witnessed it – many others in fact suffer severe trauma, such as staff members who are called to the scene afterward to help get a train back into service. The idea of what had happened so traumatized one employee that he felt compelled to leave his job. He believed that he did not need treatment; all he wanted, he said, was to leave. I hoped he was right.

All an organization can do is make treatment available to their staff; they cannot force them to take it, and neither can they force employees to actually engage in the process even if they do attend. I had already spent a number of years working in the field of Critical Incident Stress Debriefing when I read about some researchers who were questioning its usefulness and even suggesting that it was counterproductive, especially where people were being forced to attend sessions. However, from both my own experience and that of many others, I knew that as long as people were agreeing to attend by choice, the work we were doing was not harmful to anyone and, moreover, was extremely useful to others.

One example sticks out as an excellent illustration. I was carrying out a critical incident debriefing in a country pub – a few days previously the chef had been attacked by a burglar armed with a machete. Before me was assembled a group consisting of all the people who had been working at the pub on the night in question – including the mother of one of the young waitresses – but excluding the chef who, according to the manager, had said he was "OK" despite having been stabbed and cut. It therefore did not surprise me that the chef was not present at this gathering.

The waitress's mother was saying that she couldn't stop thinking about how guilty she was feeling. She had been sitting in her car at the time of the incident and suddenly, on hearing a commotion, turned on her headlights to see what was happening. She saw a man carrying a machete and running off. Then she saw the injured chef. She was convinced that she had made the situation worse by "being nosey" and, had she not switched the headlights on, the thief would not have panicked and the chef would not have been hurt. She had only just finished speaking when the door opened and in walked a man with his arm in a sling and a slash wound, covered with a transparent dressing, down the side of his face. I welcomed him in, and he introduced himself as the chef and added quickly that he really wanted to speak. I let him go next, and the first thing he said was "I am so grateful to the person who turned their car headlights on; they probably saved my life. I know my injuries would have been far worse if it hadn't been for them." The waitress's mother may have never known this if they hadn't all been given the opportunity to debrief and talk about what happened. This experience demonstrated to me just how important such opportunities are – to all concerned.

Whether I have been debriefing after a bank robbery, an attempted raid on a building society, a convenience – store robbery, or many other events of this nature, what has always come out as significant to the people caught up in it is that their organization considers them important enough to pay someone like me to come in and offer them help, support, and treatment.

On the other hand, I have seen this effect completely nullified in cases where members of a lower level management team do not agree with the organizational policy of offering support – and make their opinion clear, whether consciously or not. One such example comes to mind. Several employees at a restaurant had been both physically hurt and traumatized by a violent robbery. During the debriefing, one of the line managers came in and said, "I am just grateful that no money was taken." This one comment totally obliterated any feelings of being taken care of that the staff had previously felt from the organization.

Timing as a Factor

As a practitioner rather than a researcher, I have come to realize that I concur with the view of Regel and Joseph (2010), who found that "early psychological contact is vital for the emotional and physical well-being of the patient in the early stages of recovery." Calling in a professional early after an event can enable the other members of staff to feel looked after. A colleague of mine describes this as "organizational nurturing."

Specific examples of successful early debriefing include one in which a seemingly fit and healthy man was working with his colleagues on a building site. Their day was a typical one, with everyone getting on with their jobs as usual, when suddenly, without any warning at all, one of the men collapsed from a massive heart attack and died. Within 72 hours the organization called me in, and I spoke to the managers and witnesses. Some staff said they were fine, while others were definitely shocked. One of the observations I made, and have seen subsequently as well, is that what affected these people may not necessarily have been the loss of the particular colleague but the manner in which they personalized the loss. For instance, the person who died may have reminded them of themselves or a family member. People say things like "He was younger than my dad," "He wasn't any older than me," "I have a son his age," and so on.

The managers in this case were particularly affected as they struggled to come to terms with the fact that everything had been so normal a few minutes before the man fell and no one had guessed that there was anything wrong with him.

Another example involved a well-loved and popular member of staff who came into the open-plan office where she worked, carrying her usual cup of coffee, smiling and greeting her coworkers. Within a minute she had collapsed on the floor, the victim of a stroke so sudden and severe that she was dead by the time her colleagues heard the crash and ran to her side. Her coworkers were devastated, and in the debriefing afterward they talked about wanting to hold their own memorial service to her, to organize a time when they could talk about her and remember her as they had known her and not as they'd last seen her.

Seeking "Closure"

Some colleagues want to achieve closure by attending a funeral or memorial service, whilst others want to remember the person as they knew them. This is not always possible owing to work commitments, religious reasons, or simply because sometimes the families of the deceased do not want colleagues to attend the funeral, especially if they believe that work-related stress had been a contributing factor to their loved one's death. In these cases, one-to-one grief counseling would help the person. Many grief counselors help people to create their own rituals to say goodbye to a colleague.

If possible, it is positive for colleagues to attend funerals; even if people don't socialize outside of work they still can, and often do, become close to each other. This is of course logical from a time perspective: sitting opposite someone for some 30–40 hours a week adds up to far longer than some people spend interacting with their partners, families, or friends. Some organizations have specific policies about whose funeral a member of staff may take time off to attend; in others, it is at the discretion of managers. In these situations, the human resources personnel often find it helpful to have the support of a professional whom they can just talk to about the options available and what the consequences in terms of staff morale might be.

Life events happen to us all, and we have no idea which will affect us or to what extent. However, what is clear is that they do not necessarily affect people at the time; this could happen later if the memory is triggered by something they read, or see in a film or on the news. Normalizing the fact that people can be impacted when these events happen – and that it is therefore normal to be feeling distressed – gives people permission to deal with feelings at the time and not block them out to be dealt with at some future date.

Bullying at Work

Another all-too-common event that traumatizes people in organizations is bullying at work. Unfortunately, it is not only those who *report* the behaviors to human resources, management, or their union representatives that constitute the true numbers of people who are bullied or harassed at work; for every incident reported, there are others that are not because while many people know they are feel bad, they believe there is something wrong with *them*, or they are simply too afraid to complain. Bullying behaviors can take myriad forms, from the one-off minor event to terminal, persistent, and cruel attacks. The impact of these behaviors can be traumatic not only to the individual on the receiving end but also to those witnessing them. When these behaviors are brought to the attention of human resources or senior management, it becomes another occasion when professional help is invaluable. The professional can offer an objective investigation into the issues and assist management by making recommendations about alternative ways of dealing with them. Bullying behaviors need to be stopped and not condoned (e.g., by doing nothing) within an organization; they otherwise have a habit of escalating rather than going away.

In cases of bullying and/or harassment in the workplace, a professional can guide human resource personnel through writing, updating, or implementing a policy. They can offer coaching or professional development skills to the person accused of the inappropriate behaviors, and they can counsel or coach the person who has been affected. In some severe cases, people traumatized at work need specialist trauma therapy.

Mediation

Another way in which professionals can help organizations is by carrying out mediation between members of a team who are in conflict. Mediation is the process of assisting people to find a common solution to a common problem. The mediator works as an objective facilitator whose aim is to enable people in conflict to reach their own solutions with a win–win outcome (Rennie Peyton, 2003). Bullying is always about the *impact* of behaviors on the victim and not the *intention* of the perpetrator; consequently mediation can sometimes give a traumatized individual the opportunity to explain just how they are impacted by the behavior of another or others. In addition, the presence of an objective, outside mediator can give an individual a sense of safety and provide an opportunity to be heard.

Team building is another way in which external professionals can help an organization in these cases – to help rebuild a fully functioning team that is able to move on from the events of the past. This can create an environment in which people can dump their grudges against each other and begin to work toward healthy, honest working relationships.

It is rare in an organization for a one-off event to be of sufficient magnitude to cause trauma; generally, a drip feed effect is more likely to have taken place whereby a person being bullied slowly and methodically over a period of time starts having disturbed sleep or even nightmares about the organization or the individual responsible. Consequently, they don't want to go into work, their health is affected, and they are no longer the same person in their home and social life. Frequently in these situations it is a family member, friend, or colleague who points out that the person has changed and that "something is wrong." The individual may have grown used to the gradual changes in themselves and feel so "down" about life that they cannot stop to figure out what the cause might be.

Extreme Trauma

So far the traumatic events I have discussed have been relatively (in the context of trauma) "everyday" occurrences such as robberies, bullying, injury, or even death at work. There are, of course, other more extreme events that can affect an organization, for instance kidnapping, torture, or other atrocities against a colleague who is working abroad. When these rare and deeply traumatizing events occur, a professional can help human resources or senior management make announcements to the staff. It shows greater consideration and caring for members of the organization that an internal announcement is made before people start gathering bits of information from gossip or the media. The kinds of comments I have heard include "No one bothers to tell us anything," and organizations clearly do not always understand the demoralizing effect of such inaction. Indeed, even if something is too sensitive to be disclosed in detail, a general announcement is still preferable to keeping people in the dark.

The Organizational Response

Calling in a professional when an event happens demonstrates to the members of the organization that their psychological health and well-being are important. It is the nurturing organization that increases its chances of fostering deeper loyalty and respect from employees.

Of course, organizations can get it wrong, too, such as when a genuine gesture of concern is not carried out with sufficient sensitivity. Organizations need to both understand and acknowledge that different members of staff will react differently to these events. Some people want to talk about them; others prefer to be left alone and talk to their families instead. I consider it important that the individual is asked what would be helpful to them, and told what help is available to them.

One organization I know has, according to people I have worked with there, "got it wrong" on many occasions. People who have been caught up in an unpleasant event are sent a bunch of flowers and are told that they are entitled to a counseling session paid for by the company, although this must take place after working hours or in their own time. Several people have told me how much they would have appreciated a phone call asking how they were, and what they needed, instead of – in the words of one of them – "a bunch of ##### flowers!" I appreciate that the organization concerned has a policy and wants to be seen as getting it right for their people, even if they were failing in some ways. On reading their policy I discovered that it did, as written, permit the sessions to take place during working hours. In this case I was able to help not only the victims by offering them counseling but also the organization itself in following its own procedures correctly.

Organizations sometimes fail to acknowledge that it is not just the incident itself but also knock-on effects that cause trauma. For example, a young woman covering a shift alone – only days after having been robbed at gun point – and having to lock up the store where this occurred, and then go home in the dark, suffered the knock-on effect. It is understandably terrifying for these young people who are working to pay for their education, to support children, or because they want to progress. I have heard secondhand stories of area managers who treat their staff as though there is something wrong with them for being afraid. This is where the area managers need training in dealing with people after a traumatic event, getting help to them, and enabling them to move on with their lives. But there is another dimension too: if people are not treated well after a traumatic incident, they are more likely to leave at the earliest opportunity, often with adverse economic consequences to the organization.

One of the difficulties that managers and human resources staff face is that they are expected somehow to be able to deal with all levels of human emotion and traumatic events. These key people need to be able to recognize not only when they are out of their depth and unable to cope, but also that these feelings are not necessarily inappropriate. They should be able to seek professional help.

Debriefing and "Pre-briefing"

In many organizations whose personnel travel on difficult assignments, such as to war-torn countries, it is part of the policy and procedure to debrief them on their return.

Like the train drivers referred to earlier, some of these people find it really beneficial while others think it a waste of time. Journalists and reporters covering major news items can witness, be involved in, or make films of traumatic events. For them it is often useful when professionals can carry out confidential debriefings, sometimes in groups or with individuals, which give them permission to express how they really felt without the fear of being seen as not part of a "tough team." It also gives the professional an opportunity to assess for PTSD and refer the individual for treatment as quickly as possible.

Another area of work that I have been involved in is preparing people to visit danger zones. Other professionals deal with all the techniques they can use to protect themselves physically and keep themselves safe. However, these people also need professional help to be prepared psychologically for what they are potentially going to face. I have found this to be particularly useful for those people who have already been, for instance, to the country involved and have particular images and disturbing memories. By dealing with these before they leave, a build-up of multiple traumas can be avoided.

Some organizations take the attitude that they should be prepared and trained in what to do if all or some of their staff is involved in a traumatic incident, and they consequently provide training programs in which human resources staff and managers work together to learn how to deal with a possible traumatic event. Professionals can also be invaluable to the organization when they are writing policies about what to do after a traumatic event. The input of an outside professional provides a perspective on matters that that may not have occurred to the management, such as the longer term psychological impact on staff members and the implications of that.

The External Specialist

Bringing a professional into the organization after a traumatic incident can serve as support for both the over-stretched human resources personnel and management. For example, after a serious train crash I was one of the people called in for people to talk to and off-load their emotions when the situation became too stressful. Another time, I was in the offices above a major London train station where staff were answering the constantly ringing telephones. One of the issues I dealt with was the staff's feeling hounded and bullied by the press. Just because people work in the rail industry does not mean they have a greater immunity to being traumatized by a rail accident that includes fatalities and serious injuries. This is something that the news-hungry press and public can forget whilst dealing with their own needs and emotions related to the tragedy. People were distressed at being asked for information that they did not have or could not give. In these situations, people anxious for information do not wait for a statement to be given from the press office; often they ring any number and make demands, sometimes loudly and aggressively. Bullying a junior member of staff to give information they do not know only makes an already tragic situation worse. This is where the professionals came in: we were able to take members of staff out of the highly charged situation, where the tension was exacerbated by television screens on which nonstop news bulletins were playing and replaying the event, and where the level of noise was almost unbearable, into a quiet room where they could talk to us.

In addition, we were there for management and human resources staff to talk to and download their intense feelings, thus enabling them to then go back and manage a difficult situation.

Organizational Preparedness

No matter how well trained the human resources people are in dealing with traumatic events that affect their staff – and how well written the policies – few organizations can be prepared on an emotional level for what happened on September 11, 2001. As a freelance organizational psychologist as well as a clinical practitioner, my first task was to go and "be with" a group of American executives who were sitting in a room in London, drinking coffee around a large conference table, and impatiently waiting for news. Many of their colleagues were at this stage "possibly dead"; these were the colleagues to whom many of them had been talking on the telephone when the planes struck. Many in the room were remembering how at that moment their lines went dead, terminating their calls. All that these men and women wanted was to go home and be with their families and, as one woman said, "wrap them in cotton wool." No one wanted to speak with me in private; they were happy to talk about their feelings and wanted to stay together as a group. One woman spoke about her feelings of guilt because, as a result of her being in London, her assistant in New York had taken her place at a conference being held in the World Trade Center. During this time I was aware that I couldn't make anything better and give them the good news that they craved, but I could be there to listen. I was also representing the organization's management team in its efforts to show they cared. I stayed with them until they were able to get clearance to fly out on their corporate jet and land in another part of the United States.

My next assignment at this time was to go into an organization that had lost 15 members of staff, and over 300 outside delegates, when the buildings fell. My being called into this organization was an example of the human resource director's ability to acknowledge that she was out of her depth and needed help to deal with this level of grief. The reception area was like a florist shop with flowers pouring in like rain from neighboring companies, competitors in the same field, clients, friends, and people offering support. This demonstrated to me that, culturally, if we don't know what to in a situation, we send flowers.

At first, people still had hope that there would be survivors when they were answering the telephone to the press, members of the staff or families of the delegates, old members of staff, and others. My role was varied; I worked with the chairman to write the announcement he was going to make to the staff informing them that no one had been rescued and that all of their colleagues on this assignment, and all of the other delegates, had perished.

I stayed in an office for days as people came in and spoke to me of their losses and fears. These were mostly young people who traveled and lived together all over the globe on company business – and their worlds were shattered. They had lost managers, friends, colleagues, lovers, and playmates – and so many of them all at once. They were also coming to terms with the fact that it could have been them who people were now grieving over. During my time with this organization I heard so much about the individuals who died, from so many people's perspectives, that I began to feel as though I had known them personally. This organization really understood how to take care of its staff as those who wanted to could come and see me at my consulting rooms for grief therapy at company expense.

A memorial service was organized for friends, family, and the staff to remember and say goodbye to those who had lost their lives. I was invited to be there, as people knew who I was and what my role had been, in case anyone wanted or needed my support. This I felt was beneficial not only to the staff; it was also good for me to have closure after a very difficult and draining piece of work.

Dealing from a professional perspective with the effects of an event in another country – whether a genocide in an African country, the tsunami in Asia, or terrorist attacks in the United States – does affect people differently than something on their own doorstep.

The terrorist attacks in London on July 7, 2005, had a wide-reaching and all-pervasive impact on so many people. People who live or commute to the city can identify with, and put themselves in the position of, those who were injured, traumatized, or even killed. Most people who live in, work in, or visit London at some time or another use the tube trains or busses. For some time afterward, I kept hearing people say, "I travel on that line" or "I use the busses all the time in London." What they are actually in shock about and really saying is that "It could have been me – me with those injuries, or me in the newspaper as one of the tragic victims of that event."

I was working very close to the stations where two of the bomb blasts occurred, and even closer to the square where the bus exploded, and this proximity taught me a lot about working with trauma. People were wandering around and not knowing which way to go; some were bloodied and disheveled, medical staff in scrubs ran about with equipment in shopping trolleys, and the emergency services were in evidence everywhere. People who the day before would have passed each other in the street now all had something in common. They were traumatized, and they wanted to contact their loved ones and make sure they were all right and then get home to them. People when traumatized want to return to "normality" as quickly as possible. What I heard frequently was "I don't know what to do," a sentiment I was able to identify with personally: wanting to help but not knowing how to made me feel redundant as it was clear that what people needed at that moment were medics and not psychologists.

The impact of this event was so strong that people were talking about it for months afterward, and I still see people stop to read the brass memorial plaque that marks the place where the bomb exploded with a look of horror and great sadness. Other people spoke of feeling that they had no right to be upset as they were not there. Yet, they were upset and of course they had every right to be.

As a professional I was finally able to feel useful when I set off on my first assignment, which was to debrief rescue workers in an office a mile away. The human resources department knew that they were out of their depth dealing with something this huge and called in professional help immediately. This had a sort of domino effect with everyone feeling that they had done what they could, including us professionals.

Conclusion

We have seen throughout this chapter the value of calling in professionals to do the jobs they are trained to do. The first step in this is acknowledging that help is needed. While some organizations have in place a professional support group or system of supervision for their personnel dealing with traumatic or potentially traumatic issues, sadly these are rare and it is up to the individual human resource worker or manager to seek help. Sometimes, just getting off site and visiting a professional for an hour can help them to put the situation and their reactions into perspective.

Every human being working in an organization has a history, and obviously each one is different. So even when a traumatic event takes place at work and affects the organization as a whole, individual's responses will be unique. In the course of their work they might experience or hear of situations that could trigger any of the following: feelings of upset; unpleasant memories from their past; disturbed sleep or disturbing dreams; difficulty in concentrating; feelings of being jumpy and easily startled; feelings of irritability, anger, tearfulness, and a heightened awareness of the potential dangers to themselves or others; and inability to enjoy their own life outside of work because of memories of their own or someone else's story. All of these are signs that they are not moving on from the situation and that they need to get professional help in letting it go. There is no stigma attached to being affected by a traumatic event; however, many people do not want to feel weak or that there is something "wrong" with them. The latter attitude is particularly prevalent in men who announce that they "should" be able to deal with these events. There is no logic to this as we are all human beings with human emotions and it is cultural beliefs and egos that get in the way of people seeking the help that they need.

"Caring for the carer" is an expression that many of us know and have our own interpretations of. However, how can those people working in the field of trauma, whether as a human resources manager dealing with staff issues – from someone who is distressed because the overtime they worked last week was not acknowledged in their pay packet, to a major disaster in which many members of their team have been critically injured or lost their lives – or otherwise, feel cared for? Many of these managers find themselves in these roles without training.

Those people who work in the emergency and armed services, and who most of their working lives are dealing with the mundane, are not exempt from being affected on occasions when something "gets under their skin," for instance if something happens that is too close to their own life events, such as if the injured child they're trying to rescue reminds them of themselves at that age or their own child.

Similarly, many people think that because professionals are trained in counseling, psychology, or psychotherapy, they are not prone to being affected by the traumas that they hear about. This is actually far from true, and it is up to professionals to be their own monitors. A trained professional can deal with many situations and just move on afterward; this can happen many times, and then suddenly one particular story or event can really affect them. Just as clients or patients respond uniquely based on their individual histories, so too is it with professionals. Which event or events will affect them in a traumatic way is not predictable. What is going on in their own lives currently, and the level of support they receive from others, can be factors.

For a professional, feeling that they are making a difference can have a huge influence on how they deal with their own feelings about a situation. Watching someone deal with, maybe grieve over, and then move on from a traumatic event might even have a healing effect on the professional.

Dealing with events as they happen in supervision prevents the individual being affected by multiple traumas. These do not have to be dramatic events; hearing about and working with a continuous stream of people's difficult and traumatic life events can affect the professional in ways that takes them by surprise. The value of having a group of colleagues who take the time to listen to and watch out for extreme stress signs in each other, whilst knowing they are receiving the same treatment, cannot be over-estimated. Traumatic events, even secondhand, need to be talked about and not anesthetized by the last glassful in a bottle of wine or stuffed down with a cream doughnut. As one professional, I can only say the following to others: talk about how it has affected you even if you think it is something that you have heard many times before. Then have your glass of wine or indulge your sweet tooth – but only because it is a treat and not a form of medication.

As professionals most of us have supervision which offers an invaluable space for us to talk about our work and the effect it has on us as human beings. We can have peer supervision, which is another opportunity – albeit with a slightly different focus – to talk about cases and possibly find different ways of working with a particular situation. Most of all, it facilitates the professional in getting help and support for themselves.

I advocate that those of us working in the fields of trauma at any level find ways to monitor our own psychological health. How often have you looked at the checklist you use to assess clients and patients and think about it in terms of yourself?

References

Regel, S., & Joseph, S. (2010). *Post-traumatic stress.* Oxford: Oxford University Press.

Rennie Peyton, P. (2003). *Dignity at work: Eliminate bullying and create a positive working environment.* New York: Brunner-Routledge.

29

Military Veterans' Mental Health: Long-term Post-trauma Support Needs

Walter Busuttil

Introduction

In Britain, a "military veteran" is defined as someone who has left the Armed Forces after having served for at least one day. When reviewing the literature it is important to make the distinction between this and a "combat veteran," who has been on military operations and has been exposed to combat situations.

In other countries such as the United States and Australia, the literature has focused on issues concerning combat veterans. This reflects the fact that national statutory clinical mental health services in these countries are geared toward treating those who have suffered mental injury during military combat or military operations, with others who have served in the military but who have not been involved in military operations and combat accessing less specialized care.

In the United Kingdom, since 1948, the National Health Service (NHS) has been responsible for the treatment of veterans' mental health. Until recently no specialist NHS provision has existed for veterans, and mainstream NHS clinical services have sought to treat any ex-service man or woman irrespective of whether or not they have been exposed to combat or military operations. Indeed the NHS has had no means of identifying those who have served in the military, and commonly this question of military service has never been raised or even given enough consideration in relation to the formulation of clinical mental health needs.

In the last three years, exciting new service developments have been initiated. These have been driven by the ex-service mental health Third Sector charity Combat Stress in conjunction with the Department of Health (DoH) and the Ministry of Defence (MOD).

In the British Armed Forces, there are approximately 180 000 service personnel at any one time, with some 25 000 leaving every year. There are some 5.5 million veterans and 7.5 million family dependents in Britain. While the vast majority of veterans adjust well to civilian life, some do not.

International Handbook of Workplace Trauma Support, First Edition. Edited by Rick Hughes, Andrew Kinder, and Cary L. Cooper.

Assessment of Need: Veteran Population Studies

Appropriate service planning for veterans demands that population studies assessing their mental health needs be conducted. The National Vietnam Readjustment Veterans Study (NVVRS) carried out in the United States in the late 1980s led to the setting up of comprehensive mental health services for veterans delivered by the Veterans Association (Kulka *et al.*, 1990). The NVVRS estimated that 50% of all who served in Vietnam (or 1.7 million) suffered partial or full PTSD at some time following military service and that 828 000 did so at the time of the study. Comorbidity rates were raised with 50% of PTSD sufferers also having at least one other diagnosable psychiatric disorder, with increased lifetime or point prevalence comorbidities for, most commonly, depression and alcohol and substance misuse. Divorce, unemployment, accidents including road traffic accident rates, and suicide rates were also increased (Kulka *et al.*, 1990).

In Australia, similar population studies were conducted, again in the Vietnam veteran population – around 60 000 Australians served with the Australian Defence forces in Vietnam. Once more clinical services were set up as a result of the findings, similar morbidities and comorbidities were found (see, e.g., O'Toole *et al.*, 1996).

It is estimated that there are 5.5 million veterans in the United Kingdom and 7.5 million first-degree dependents. Even small percentages of veterans with complicated mental health needs would amount to a difficult challenge to accommodate without adequate service planning and provision.

In Britain large population studies have not been conducted as yet, nor have they focused on combat veterans. In a small-scale study conducted in a matched population, national service noncombatant veteran and civilian population rates of mental illness were relatively the same (Woodhead *et al.*, 2011). Reliance on these findings can, however, be seen to be relatively misleading. A longitudinal study conducted in those deployed to recent and current war zones has been ongoing for the past few years. The main findings of this study, conducted by the Kings College Hospital Military Psychiatry research team on a population of military personnel who participated in the invasion and wars in Iraq (Operation TELIC) and Afghanistan (Operation Herrick), were presented in Chapter 26. It is pertinent to note at this point that in time it is hoped that this study will evolve into a longer term longitudinal study which includes veterans as the servicemen end their military careers. Approximately 25% of the original population are now veterans, having left the military (Fear *et al.*, 2010; Hotopf *et al.*, 2006; Iversen & Greenberg, 2009).

The Scottish government has recently funded a scoping study undertaken by the Robert Gordon University at Aberdeen in collaboration with Combat Stress, aimed at examining the feasibility of undertaking a veteran's population study. Another similar study is also currently ongoing in Cardiff assessing the needs of Welsh veterans.

Veterans in the Criminal Justice System

The picture here is again incomplete with mental health population studies urgently required. The National Probation Officers Association (NAPO) has calculated that there is a prevalence of 10% of veterans in the prison population (or 8500 individuals). In contrast the official government figures released by the Department of Analytical Services and Advice (DASA) cite a prevalence of 3% or 2500 individuals. The Veterans in Prison Association (VIPA; 2011) note that military veterans in prison represent the largest occupational group among offenders. NAPO cites that there are 12 500 further veterans on parole or subject to probation supervision – meaning that there are in excess of 20 000 veterans under correctional services control. NAPO states that common offenses are dominated by drug and alcohol misuse; there is a propensity to violence, particularly domestic violence; and many suffer from depression and PTSD. There is also said to be a higher rate of sex offending. This is not fully understood and requires further investigation. Small studies involving police desk surveys are now being conducted where people being booked for suspected offenses are for the first time being asked proactively whether they are currently serving in the military or whether they have served in the past.

The link between offending and mental health is not understood within the veteran population.

Treating Veterans' Mental Health Disorders in the United Kingdom

The Third Sector Charity Combat Stress originated in 1919 after the First World War mainly to provide respite care. In recent years the charity has been proactive in the media highlighting the plight of veterans suffering from mental health disorders. At the same time it has reorganized its clinical interventions, becoming the leading mental health provider for veterans' mental health and working in partnership with the DoH. In early 2010, Combat Stress signed a strategic partnership with the DoH and the MOD.

Combat Stress is the leading veterans' mental health charity. Some 100 000 veterans and their families have been helped over the years. Currently there are 4678 actively receiving mental health care in the community through outreach and welfare services, attending residential treatment services, or both, with some further 5887 "passive cases," those who either are in the process of accessing care or have already accessed care and for whom intervention is completed but discharge from care has not yet occurred. Referral rates have been increased by 72% over the past six years, with 1447 new inquiries for help received to the end of March 2011. Ten years ago, 300–400 new referrals were received each year, and the average age of the veteran was around 60, with a substantial number of World War Two veterans receiving care; now the average age is 42 years and falling (with a range of 20–93 years).

Around one half of all inquiries convert to clinical cases every year. There has been an increase in Iraq and Afghanistan veterans, currently presenting at a combined rate of approximately 160 new cases per year with just over 1000 unique individuals in all asking for help. Patients are offered a residential or, increasingly, a community outreach multidisciplinary team mental health assessment.

Approximately 50% are "self-referrals" with a further 10% from family and friends. It is usually the wife or "good woman" who instigates these referrals. Twelve percent come from other ex-service charities, and 5% from War Pensions Welfare Service; medical health and social services account for around 10%, with approximately only 3% of the total referrals coming directly from the NHS.

Ex-Army veterans account for 83%, with the remainder equally shared by ex–Royal Navy and ex-RAF veterans with a very small number of ex–Merchant Navy seamen. Female veterans account for around 3% of the total. Commonly, most will have served in at least one and usually multiple operational tours of duty with combat exposure being extremely common and with Northern Ireland operational service being the most prevalent.

The Combat Stress Clinical Population

Large new-patient clinical ($n = 604$) and psychometric audits ($n = 704$) (Busuttil, 2010) demonstrate that 92% of new clinical cases will have been exposed to multiple military-related psychological traumas, with 75% qualifying for a primary diagnosis of PTSD (representing 30% of all new inquiries for help that come to Combat Stress annually). Comorbid presentations, are high with 62% of these also presenting with depression and alcohol disorders including severe abuse and addiction. The clinical picture is further complicated as 52% of the veterans have other underlying issues including exposure to childhood trauma, neglect, and poor caregiving. Moreover, these veterans give histories of isolation, social exclusion, social withdrawal, unemployment including multiple episodes of unemployment (with 75% of new referrals being unemployed at the outset), multiple employment episodes and employers, inadequate housing, multiple house moves, periods of homelessness, episodes of being in prison or the criminal justice system, and multiple marriages and relationships. Many frequently present with behavioral disorders manifested by anger and outbursts.

Typically there is a significant delay post military discharge before a veteran presents to the charity. This time lapse has reduced from an average of 14 years to 13 years in 2010–2011. Many account for this delay by giving a history of being too macho or proud to declare their mental health symptoms while serving in the military or even after they have left the military. Some veterans give a history of having been fearful of declaring their mental health symptoms for fear of losing their job or reducing their prospects while in the military. Many therefore suffer in silence and find it difficult to acknowledge they even have problems. Hence may are "sent" to their general practitioner (GP) or to Combat Stress for treatment by their partners, some on an ultimatum and when their relationship might be at risk.

This delayed presentation means that chronic complex clinical presentations are the norm, with secondary functional deficits such as those mentioned. Some 80% of new clinical cases will have, however, tried to access help through the NHS, but for some reason this help has either been inadequate, or not been delivered if the individual has failed to engage.

Theoretical Complexities of Treating Veterans' Mental Health Disorders

Treating mental health problems in veterans is complicated because of pre-service difficulties including exposure to childhood trauma and poor family attachments. Military life itself can contribute through exposure to bullying, alcohol, psychological trauma, or family problems caused by cyclical deployment-related separations. Veterans are more likely to experience earlier onset of physical disorders as a result of occupational exposure including orthopedic problems with chronic pain and deafness (Busuttil, 2009a). Veterans suffering chronic PTSD have a higher than expected incidence of physical disorders including cardiac problems, cerebrovascular accident(s) (CVA[s]), and diabetes, with large US Vietnam combat veteran studies have demonstrated that such physical illnesses, and premature death, can develop 10 years earlier as compared to control groups of noncombat veterans (Boscarino, 1997, 2004, 2008).

Leaving the service and adjusting to civilian life can be a problem, as are help-seeking issues such as shame, stigma or guilt, and a macho image, thus preventing help-seeking behaviors and treatment engagement and compliance.

Many with comorbid alcohol disorders cannot be treated psychologically until undergoing detoxification. Seamless intervention between detoxification services and psychological trauma services is required for those who suffer comorbid PTSD and SUDs, in order to reduce chances of relapse back into substance misuse. The ideal is to follow detoxification with intensive psycho-education about PTSD and its interaction with alcohol and other substance misuse before further trauma-focused therapy is delivered (Busuttil, 2010; Hill & Busuttil, 2008).

It has been noted that many who have sought help from local services find that mental health professionals without experience of managing psychological trauma associated with the military service tend to focus on substance misuse, social problems, and anxiety and depression, while failing to tackle the root cause and seldom recognizing the problem as combat-related PTSD (Hill & Busuttil, 2008). Many have had problems with stigma and have been too ashamed to declare their military service, or they have lost faith in the NHS. Some have been told not to talk about their military service in a civilian group setting for fear that their traumatic experiences are too upsetting for other civilian patients accessing the same NHS services (Scheiner, 2008). Mainstream NHS services do not deliver bespoke veterans' services. Specialized psychological trauma services in the NHS are uncommon and National Institute for Health and Clinical Excellence (NICE) guidelines do not address the needs of chronic severe or chronic complex presentations of PTSD (NICE, 2005; United Kingdom Psychological Trauma Society, 2009). Clinical audits demonstrate that veterans accessing Combat Stress services are more likely to engage and remain in treatment, become plugged into the NHS locally to them, and remain engaged (Scheiner, 2008). Combat Stress clinical services do not aim to set themselves as an alternative to the NHS; rather, joint management between Combat Stress and the NHS is the norm, and Combat Stress aims to allow the veteran to regain confidence to re-engage with local NHS services if possible (Busuttil, 2009a).

Combat Stress Interventions

Many staff are ex-military; in particular, the regional welfare officers (RWOs) who are the first point of contact are all ex-military officers who share a common background, language, and culture with the veteran.

Combat Stress offers the option to self-refer, community mental health treatment and welfare, a national telephone help-line, access to telephone welfare advice, and clinical help for those already plugged into clinical services; as well as specialist evidence-based residential group and individual multidisciplinary treatments, carers' groups, and rehabilitation.

Treatment aims to plug the veteran back into his or her local NHS services. Joint working with other ex-service charities and work-retraining schemes is the norm.

Clinical Services

Combat Stress offers welfare support and diverse clinical interventions in the community as well as residential setting. It aims to work alongside NHS statutory services complementing statutory services, not replicating them, with the intention of plugging as many veterans back into mainstream NHS services as possible.

Combat Stress Treatment Models

Combat Stress operates three residential treatment centers located in Ayrshire, Shropshire, and Surrey, each with approximately 30 beds including some double rooms for carers to attend with the veteran. Each is run by a multidisciplinary team (MDT) comprising occupational therapists, psychotherapists, registered mental health nurses (psychiatrists, RMNs), nursing assistants, psychiatrists, and psychologists. The centers have a supportive therapeutic milieu which is sensitive to a military culture. Some but not all staff are ex-military. Peer support is encouraged, but this is directed by staff. Owing to the chronicity of many of the veterans' mental health illnesses, the main therapeutic aims are to improve function, maintain wellness, and treat psychiatric symptoms. A phasic rehabilitation model of care has been adopted in keeping with the recovery model (Foa *et al.*, 2009; Herman, 1992; NICE, 2005; UKTrauma Group, 2009).

Phase 1: Preparation. This includes outreach clinical and welfare assessment, multidisciplinary team referral meeting, information gathering from the GP and NHS team, and active preparation for initial assessment admission including NHS detoxification from alcohol and drugs.

Phase 2: Stabilization and safety. This includes a five-day MDT residential assessment, the prescription of appropriate medications, establishing trust and safety within a therapeutic milieu, and subsequent further admissions (current practice). It also includes care planning, liaison with GPs and other treating agencies, and the treatment of comorbid disorders such as chronic pain, depression, and alcohol dependence before trauma work is undertaken. Psycho-education groups are provided for PTSD, alcohol and illicit drugs, coping, anxiety, and anger management.

Phase 3: Disclosure and working through of the traumatic material utilizing individual trauma-focused therapies including trauma-focused cognitive-behavioral therapy (CBT) and eye movement desensitization and reprocessing (EMDR). Psychodynamic approaches are also used for attachment problems.

Phase 4: Rehabilitation and reintegration using occupational therapy (OT) interventions including normalizing activities of daily living and a healthy-living and OT-run exercise program, welfare support, and reintegration within society and work retraining.

Residential Treatment

Rolling treatment program

Historical funding streams through the war pensions agency, staffing levels, and configuration have influenced the ability to deliver treatment to so many patients. Until recently, Combat Stress essentially offered a one-size-fits-all clinical service which was incorporated within a mainly residential rolling treatment program.

This program comprised an initial RWO welfare community assessment, followed by a multidisciplinary clinical assessment within a residential center lasting for five days where a brief group program initiated the treatment process allowing basic psycho-education and coping skills to be taught; this was followed by up to three 10–15-day admissions over the following period of one year before care plans were reviewed. Veterans were able to access psycho-education, skills training, as well as trauma-focused therapies and rehabilitation. Many veterans received planned trauma-focused CBT delivered by Combat Stress staff while admitted as residential patients, or this was delivered within an outpatient setting or delivered through their local NHS as part of a coordinated joint management with Combat Stress clinicians. Joint working with NHS statutory services between residential admissions is the norm.

The rolling program has been upgraded over the past four years; however, the difficulty with it is that patients take time to be readmitted between residential periods and their access to trauma-focused therapies has been inconsistent and interrupted because of admission patterns. It is noteworthy that this admission pattern was initially based on funding streams of war pensions and not set up as part of a clinical intervention plan.

Rolling program: outcome audit and research evidence

Evidence through independent clinical audit studies demonstrated improved therapy and medication compliance, and symptom and functional improvements (Lee *et al.*, 2005).

A two-year follow-up clinical audit (n = 57) conducted in veterans suffering from severe complex clinical presentations (Busuttil, 2009b) demonstrated symptom reduction with 58% definitely better, 15.7% improved a little, and 1.8% discharged needing no further treatment; improvement in function (58% definitely better and 17.5%

improved a little), and improved employment rates from 26 to 47%; with reduced time in isolation, improved compliance with medication and uptake of psychotherapy, better engagement with NHS services, reduction and abstinence from alcohol and illicit drugs, and a better adjustment to civilian life overall, although this was not measured directly.

One study demonstrated deterioration in improvement between residential treatment admissions (Hart & Lyons, 2007). Thus the importance of maintenance of improvement in the community in between admissions was recognized. New clinical outreach services were set up for this and other reasons that will be discussed in this chapter.

Large exit satisfaction surveys ($n = 1681$) indicated very high levels of satisfaction with the residential services provided. An audit ($n = 49$) conducted using the Warwick Edinburgh Mental Well Being Scale (WEMWBS; Stewart-Brown & Janmohamed, 2008) measured over an average of 18 months at three treatment points demonstrated that high levels of mental well-being were maintained over time. Pre-treatment control group mean well-being scores were statistically significantly lower compared to post-treatment at all three follow-up points. Patients reported that they had easier engagement with Combat Stress and that treatment was conducted in a sympathetic and empathetic environment (Bellwood & Busuttil, 2009).

Recent Service Expansion

Combat Stress has recently been able to expand its clinical outreach services though charitable fundraising aimed at setting these services specifically. In addition, money awarded through National Specialist Commissioning (NSC) has been specifically aimed at funding an intensive residential program designed to treat those with the most complicated clinical presentations. This NSC funding has become available to veterans living in England. It is aimed at setting up a bespoke intensive residential treatment program devised on the same lines as the programs delivered by the Australian veterans' mental health services.

The residential rolling program is funded by the NHS in Scotland and by War Pensions Agency in other parts of Britain. Overall 60% of its overall funds are charitable.

Community Outreach Services have recently been upgraded from the RWO areas of operation and reorganized into 14 geographical regions across Britain and Ireland. Each region has a RWO, a community psychiatric nurse, and a generic mental health worker who is a psychotherapist with family therapy experience or a social worker. It is hoped that in the future there will also be access to local bespoke veterans' psychiatric and psychology sessional clinics operated by the NHS. This may be possible in the future through the setting up of Regional NHS and Armed Forces Forums. This is discussed further in this chapter.

Each outreach team is supported by a welfare support desk officer (WSDO) who coordinates the work of the team and who liaises and supports the veteran by telephone as well. The geography of the outreach clinical services is defined by the regions that RWOs have historically covered alone.

The RWO is a retired military officer, who, although not formally trained in mental health, has had in-house training and knowledge given to him about mental health problems. The RWO remains the main entry point into care for the veteran. This is because issues of stigma, shame, and other civilian barriers can be broken down by the RWO, who comes from the same culture and speaks the same language as the veteran.

Once referred, the veteran will receive an initial RWO home visit; the RWO also provides welfare advice, for example helping the veteran apply for benefits and access other ex-service charities as required. Currently if a mental health problem exists, then with the GP's written consent, veterans are offered a multidisciplinary residential five-day mental health assessment. This is conducted at one of three treatment centers. Some assessments can also be conducted on an outpatient and outreach basis.

Roll-out of the outreach services commenced 18 months ago. Early impressions are that those who are in most clinical need are more easily identified and more likely to engage in rehabilitation. This is especially so if they are affected by alcohol and drug misuse disorders and other comorbid presentations.

Early clinical audits show that the number of clinical patient community assessments has increased and has been performed with reduced waiting times. Many veterans had better access to evidence-based interventions within a community setting, using trauma-focused interventions delivered by Combat Stress staff, or veterans have been signposted to appropriate mainstream NHS services and have been maintained there. Others have been prompted and supported to remain engaged in treatment. Thus residential treatment has not been utilized unnecessarily. This is in keeping with research evidence from Australia, demonstrating that intensive treatments can be detrimental to those who do not actually need them, particularly if their psychiatric presentations are not severe. More family work and more carers' groups have been conducted, directed appropriately by a mental health professional.

Liaison between Combat Stress staff and NHS clinical services, including GPs, community mental health teams, psychology, and psychotherapy, has improved, allowing for better and more effective joint working. Education and awareness raising of NHS professionals in relation to veterans' mental health needs have also been enhanced.

It is thought that outreach services have helped to reduce stigmatization, and to allow veterans to ask for help more easily, especially if treatment is available locally. It is hoped that there will be a reduction in the time scale between service personnel leaving the military and asking for help with their mental health problems. Indeed, as mentioned, there has been a reduction from an average of 14 years to 13 years in 2010–2011, which coincided with the launch of six initial outreach services in 2009–2010. This reduction will, it is hoped, make it much easier to treat a far less complicated clinical presentation then otherwise would result.

It is hoped that in the future, outreach services will allow local veterans' group therapy to be conducted, including intensive-exposure group therapy programs such as those running in the United States for veterans who have treatment-resistant chronic PTSD (Ready *et al.*, 2008).

Support groups could also be increased in number. These have been running for some time in the community facilitated by a mental health professional. It should be noted that group meetings of veterans who are not directed by a mental health professional can in theory be detrimental to mental health in the long term; this research evidence was borne out after the setting up of so-called rap groups which evolved into unsupervised discussion groups in US Vietnam War veterans (Egendorf, 1975).

It is hoped that in the longer term, once bespoke psychiatric and psychological clinics are established, a range of stepped assessment, treatment, and management will be available to all veterans in an outreach setting as well as for those who require it in a residential setting. This would free residential treatment beds for those who need them most. It would also enable better working with residential treatment centers allowing joint residential, community outreach programs to be set up, following established practice from Australia and the United States. It is hoped that outreach services will augment and complement residential treatment programs, allowing their expansion, diversification, and development into a mix of residential and community-based treatments in keeping with best practice and the research evidence from Australia particularly. Residential programs that work in tandem with community outreach will decrease the need for residential treatment, allowing more intensive residential treatment to be delivered to those most in need.

One of the main aims of Combat Stress' current intervention is to help the veteran reaccess NHS mental health services wherever possible. In most cases joint clinical management between the NHS and Combat Stress is undertaken depending on the clinical presentation and the veteran's own wishes. Less commonly clinical management may be taken over by the NHS entirely. It is hoped that community outreach services will work with, liaise, champion, and educate the NHS local services. Outreach services still include welfare; this will ensure, for example, that help is delivered in accessing work retraining schemes.

Intensive Six-week Program

Content and its place within Combat Stress clinical services

The newly established six-week PTSD rehabilitation program which is to be funded by NSC fits into the overall recovery model adopted by Combat Stress. The program started to roll out at the end of 2011 and will last for six weeks. It will be run by two full-time-equivalent psychologists or CBT-trained nurse therapists; supported by consultant psychiatrists, RMNs, and other multidisciplinary treatment center staff; and appropriately supervised. Up to eight group members will participate. They will have access to all facilities in the treatment center including, for example, the activity center and gymnasium.

The work day will be from 0900 until 1700 hours during week days, with homework tasks and some therapy in the form of practical work being done at weekends. Two half days for family or carer involvement are also scheduled. The program will be manualized and audited for intratherapists' reliability. The program is based on the Australian veterans' rehabilitation programs (Creamer *et al.*, 1999) as well as other similar phasic PTSD rehabilitation programs run within the British military in the past (Busuttil, 1995) and for adult survivors of sexual abuse (Bloom, 1997; Busuttil, 2006; Cloitre *et al.*, 2002; Herman, 1992).

The main clinical components of such group programs comprise:

- Didactic and group psycho-education.
- Symptom management skills.
- Trauma-focused individual therapy.
- Group therapy, including trauma-focused work.
- Cognitive restructuring.
- Alcohol management.
- Problem solving.
- Family education and carer education linked to community career groups.

The program will be designed to encompass three blocks within each working day comprising:

1. Didactic psycho-education.
2. Individual trauma-focused CBT and psychiatric follow-up and ongoing assessment.
3. Skills training.

There is a large evidence base of over 4000 veterans suffering from chronic PTSD with comorbid presentations, including substance misuse and depression with social and relationship complications with significant comorbidity befitting from such programs. Patient selection is critical – they must be stable enough psychologically to be able to participate in the treatment without exiting therapy early.

The evidence is that outcome is best if there is:

- A mix of individual and group interventions.
- A mix of residential hospital, day center, and outreach.
- Inclusion of trauma-focused therapy and not just "rehabilitation" (Creamer *et al.*, 1999; Johnson, 1997; Johnson & Lublin, 1997).

Outcomes are generally as follows:

- A rule of thirds: one-third will do well, one-third get better, and one-third don't do so well – and need more input.
- Outcomes related to intensity of the program: this in turn is correlated with severity of the disorder; the higher the severity, the more intensive the therapy should be.
- If the patient has a mild disorder and an intensive program is delivered, then it is likely they may not improve and they might become worse.

It is anticipated that inclusion and exclusion criteria will include the following:

- Patients suffering from chronic moderate to severe PTSD with comorbid depression and/or other neurotic disorders such as anxiety, agoraphobia, and obsessive-compulsive disorder; alcohol and/or illicit substance misuse currently abstinent – stable.
- Exposure to multiple military traumas – two or more.

- Childhood attachment disorders or history of childhood abuse – maladaptive coping behaviors (but diagnoses of complex PTSD/borderline personality disorder, other personality disorder are exclusion criteria).
- Medication stable and complaint.
- Not currently psychotic.
- Dissociation under control – not dissociating.
- Not actively suicidal, homicidal, or violent; not sex offenders.
- Deliberate self-harm under control – stable enough to do therapy.
- Physical disorders stable and under control, including chronic pain.
- Absent brain damage or injury.
- History of social, family, and work dysfunction.
- Motivated: able to concentrate, do group work, and do cognitive therapy.
- Available for six weeks.
- Literate and intelligent – able to understand and take part in cognitive-behavioral therapy.

Joint Initiatives: National Health Service, Ministry of Defence, and Combat Stress Initiatives

The military covenant which the government subscribes to promises service personnel and the veteran that help will be given to those who have risked their lives in the service of freedom and the country. This has included priority treatment which in many occasions is not delivered by clinicians and services as clinical need usually takes preference over veterans' priority (Taylor, 2011).

Within the last 2–3 years, the MOD and NHS aided by Combat Stress set up six pilot sites across the country aimed at setting up diverse clinical services aimed at initiating treatment and at signposting veterans into mainstream mental health care. Evaluation of these services concluded that the inclusion of ex-military mental health workers and the inclusion of a mix of residential and outreach services produced the best outcomes (Dent Brown, 2010).

Other NHS initiatives include a directive to include veterans in the new Improving Access to Psychological Therapies (IAPT) services, where it is hoped this will allow better access to psychotherapies. The problem with this is that IAPT aims at treating noncomplex psychological presentations and, by and large, veterans present with complicated comorbid mental health disorders.

The MOD offers a mental assessment program (MAP) to veterans who have served on operations since 1982 (including the Falklands War) at St Thomas' Hospital in London.

For those reservists who have demobilized after deployment in an operational theatre since January 1, 2003, an assessment service is offered at Chilwell in Nottingham, United Kingdom.

The DoH funded up to seven Combat Stress badged nurse therapist–signposting positions within NHS Trusts in 2010. Up to three of these posts were appointed, but the situation was overtaken by the change in government in 2010 and the publication of the document *Fighting Fit*, authored by Dr Andrew Murrison MP. This important paper highlights a more cohesive approach to veterans' mental health nationally, including the setting up of a national helpline for veterans and their families which will be badged under the charity Combat Stress. Also included are better pre-release screening for mental health issues prior to exiting the military, follow-up health surveillance initiatives for those who have left the military, clinically supervised web chat rooms, and online treatments and support – currently involving a pilot study conducted in collaboration with the Tavistock and Portman NHS Foundation Trust under the Big White Wall support network which is being developed for service men and women, their families, and veterans and their families.

Also included in the document is the setting up of DoH-sponsored NHS/Armed Forces Regional Networks. Ten of these networks are currently in the process of being established. The aim of the networks is to ensure that veterans are at no disadvantage when dealing with the NHS. Each network will have a mental health subgroup, tasked with carrying out the recommendations of the *Fighting Fit* paper (Murrison, 2010). The model they are working toward differs between the regions, and standardization has not as yet occurred. The regions will follow PCT clusters once the strategic health authorities are dissolved. They will work in partnership with the MOD and Third Sector Charities including ex-service charities and in particular in relation to mental health with Combat Stress where the aim is to have seamless clinical community outreach services and residential specialized treatments made available nationally through Combat Stress and the NHS for veterans with mental health disorders.

The Future

In the future, it is essential that veterans' clinical services will be able to deliver diverse bespoke interventions comprising a mixture of residential and community-based services. Wherever possible, treatment should be offered locally to the veteran and residential interventions will be avoided. This philosophy is based on sound research findings from Australia that demonstrate that intensive interventions are required only for those patients who suffer from the most severe clinical presentations and that less intense interventions including community-based treatments are appropriate for those who are not severely unwell (Creamer *et al.*, 2002; Forbes *et al.*, 2008).

The research literature has demonstrated that most residential programs work best if they are part of a bespoke community outreach service (Creamer *et al.*, 2002; Forbes *et al.*, 2008). Combat Stress together with its NHS and DoH partners will aim to deliver programs comprising a mixture of residential and community-based treatment phases and will be targeted more specifically to individual needs. It is anticipated that these will include:

1. An intensive 5–6-week residential program for relatively recent-onset PTSD that is then augmented with community-based treatment interventions.
2. A PTSD substance misuse psycho-education program, which plugs into the program described in 1 and follows that pathway.
3. A PTSD treatment-resistant program which will commence in a residential setting but then will transfer to the community (Ready *et al.*, 2008).

4. An old-age program.
5. An upgraded rolling program (two-week admissions three times per year – with outreach maintenance).
6. A respite care program which will include summer camps which will allow the individual to have a holiday as well as offer respite to carers.
7. Carers' groups similar to those established in the Australian Veterans Mental Health Services are already offered in Combat Stress treatment centers as well as in the community. These will be expanded to be more widely run. The groups help, educate, and support the family members of the veteran accessing the care of Combat Stress.

Conclusions

Veteran's mental health needs are beginning to be better understood in the United Kingdom. Population studies are badly required but once conducted will help the ongoing establishment of appropriate clinical services. This will ensure better access into care, resulting in better rehabilitation of service men and veterans with mental health disorders and enabling better function, including increased employment rates.

References

Bellwood M., & Busuttil W. (2009). *Exit satisfaction surveys and Warwick Edinburgh Mental Well Being Scale (WEMWBS) outcome study.* [Internal publication]. Leatherhead, UK: Combat Stress.

Bloom S. (1997). *Creating sanctuary: Toward the evolution of sane societies.* London: Routledge.

Boscarino, J. A. (1997). Diseases among men 20 years after exposure to severe stress: Implications for clinical research and medical care. *Psychosomatic Medicine, 59*, 605–614.

Boscarino, J. A. (2004). Posttraumatic stress disorder and physical illness: Results from clinical and epidemiologic studies. *Annals of the New York Academy of Sciences, 1032*, 141–153.

Boscarino, J. A. (2008). A prospective study of PTSD and early-age heart disease mortality among Vietnam veterans: Implications for surveillance and prevention. *Psychosomatic Medicine, 70*, 668–676.

Busuttil W. (1995). Interventions in posttraumatic stress syndromes: Implications for military and emergency service organisations. Master's thesis, University of London.

Busuttil W. (2006). The development of a 90 day residential program for the treatment of complex post traumatic stress disorder, *Journal of Aggression, Maltreatment and Trauma 12*(1/2) 29–55.

Busuttil W. (2009). *Psychometric data analyses and clinical audit data for Combat Stress 2005–2009* [Internal publication]. Leatherhead, UK: Combat Stress.

Busuttil W. (2009). *Treatment outcome audit: Veteran cohort followed up between 2007–2009* [Internal publication]. Leatherhead, UK: Combat Stress.

Busuttil W. (2010, Spring). Veterans' mental health: The role of the third sector charity combat stress: Expanding community outreach services and bespoke residential treatment programmes. *BACP, (68)* 2–9.

Cloitre, M., Koenen, K. C., Cohen, L. R., & Han, H. (2002). Skills training in affective and interpersonal regulation followed by exposure: A phase-based treatment for PTSD related to childhood abuse. *Journal of Consulting and Clinical Psychology, 70*(5), 1067–1074.

Creamer, M., Forbes, D., Biddle, D., & Elliott, P. (2002). Inpatient versus day hospital treatment for chronic combat related posttraumatic stress disorder: A naturalistic comparison. *Journal of Nervous and Mental Disease*, *190*(3), 183–189.

Creamer, M., Morris, P., Biddle, D., & Elliott, P. (1999). Treatment outcome in Australian veterans with combat related posttraumatic stress disorder: A cause for cautious optimism? *Journal of Traumatic Stress*, *12*, 625–641.

Dent Brown, K. (2010). *An evaluation of six community mental health pilots for veterans of the Armed Forces: A case study series.* Sheffield, UK: University of Sheffield.

Egendorf, A. (1975). Vietnam veteran rap groups and themes of post war life. *Journal of Social Issues*, *31*, 111–124.

Fear, N. T., Jones, M., Murphy, D., Hull, L., Iversen, A. C., Coker, B., *et al.* (2010). What are the consequences of deployment to Iraq and Afghanistan on the mental health of the UK armed forces: A cohort study. *Lancet*, *375*, 1783–1797.

Foa, E. B., Keane, T. M., & Friedman, M. J. (Eds.). (2009). *Effective treatments for PTSD: Practice guidelines from the International Society for Traumatic Stress Studies* (2nd ed.) New York, NY: Guilford Press.

Forbes, D., Lewis, V., Parslow, R., Hawthorne, G., & Creamer, M. (2008). Naturalistic comparison of models of programmatic interventions for combat related post traumatic stress disorder. *Australian and New Zealand Journal of Psychiatry*, *42*, 1051–1059.

Hart, L. M., & Lyons, S. (2007). *Treatment effectiveness and client satisfaction in an inpatient programme for combat related chronic mental health disorders: A preliminary investigation* [Internal publication]. Leatherhead, UK: Combat Stress.

Herman, J. (1992). *Trauma and recovery: The aftermath of violence – from domestic abuse to political terror.* New York, NY: Basic Books.

Hill, D. M., & Busuttil, W. (2008). Dual diagnosis in service veterans with post traumatic stress disorder and co-existing substance abuse. *Advances in Dual Diagnosis*, *8*(1), 33–36.

Hotopf, M., Hull, L., Fear, N. T., Horn, O., Iversen, A., Jones, M., *et al.* (2006). The health of UK military personnel who deployed to the 2003 Iraq war: A cohort study. *Lancet*, *367*, 1731–1741.

Iversen, A., & Greenberg, N. (2009). Mental health of regular and reserve military veterans. *Advances in Psychiatric Treatment*, *15*, 100–106.

Johnson, D. R. (1997). Introduction: Inside the specialized inpatient PTSD units of the Department of Veterans Affairs. *Journal of Traumatic Stress*, *10*, 357–360.

Johnson, D. R., & Lubin, H. (1997). Treatment preferences of Vietnam veterans with post traumatic stress disorder. *Journal of Traumatic Stress*, *10*, 391–405.

Kulka, R., Schlenger, W. E., Fairbank, J. A., Hough, R. L., Jordan, B. K., Marmar, C. R. & Weiss, D. S. (1990). *Trauma and the Vietnam War generation.* New York, NY: Brunner Mazel.

Lee, H. A., Gabriel, R., & Bale, A. J. (2005), Clinical outcomes of Gulf Veterans' Medical Assessment Programme (GVMAP) referrals for Gulf Veterans with post traumatic stress disorder to specialised centres. *Military Medicine*, *170*(5), 400–406.

Murrison, A. (2010). *Fighting Fit: A mental health plan for servicemen and veterans.* London: Ministry of Defence.

National Institute for Health and Clinical Excellence (NICE). (2005). *Post-traumatic stress disorder (PTSD): The management of PTSD in adults and children in primary and secondary care.* NICE Guideline GC26. London: Author.

O'Toole, B. I. O., Marshall, R. P., Grayson, D. A., Schureck, R. J., Dobson, M., & French, M. (1996). The Australian Vietnam veterans health study: III. Psychological health of Australian Vietnam veterans and its relationship to combat. *International Journal of Epidemiology*, *25*, 331–339.

Ready, D. J., Thomas, K. R., Worley, V., Backscheider, A. G., Harvey, L. A. C., Baltzell, D., *et al.* (2008). A field test of group based exposure therapy with 102 veterans with war related posttraumatic stress disorder. *Journal of Traumatic Stress, 21*, 150–157.

Scheiner, N. S. (2008). Not "at ease": UK veterans' perceptions of the level of understanding of their psychological difficulties shown by the National Health Service. Doctoral thesis, City University London.

Stewart-Brown, S., & Janmohamed, K. (2008). *The Warwick Edinburgh Mental Well Being Scale (WEMWBS) user guide version 1* (Ed. J. Parkinson). Edinburgh: NHS Health Scotland.

Taylor, C. (2011). *The Armed Forces covenant.* Standard note SN/IA/5979. London: Government Publication.

United Kingdom Psychological Trauma, Society. (2009). [Home page]. Retrieved from http://www.ukpts.co.uk/site/trauma-services/.

Veterans in, Prison. (2011). [Home page]. Retrieved from http://www.veteransinprison.org.uk

Woodhead, C., Rona, R. J., Iversen, A., MacManus, D., Hotopf, M., Dean, K., *et al.* (2011). Mental health and health service use among post-national service veterans: Results from the 2007 Adult Psychiatric Morbidity Survey of England. *Psychological Medicine, 41*(2), 363–372.

30

Post-trauma Support: Learning from the Past to Help Shape a Better Future

Rick Hughes, Andrew Kinder, and Cary Cooper

Introduction

When this book was originally conceived, the aims were to provide a comprehensive account of traumatic stress within organizations, both on its impacts as well as on key techniques and approaches to support those affected. On reading the final manuscript, it is good to see the many varied approaches to the subject of trauma within organizations. There are many different manifestations of traumatic stress in the workplace, and there are many different solutions which have evolved to meet this need. It is also apparent that each organization needs to consider its unique situation and to design a system of support which is appropriate to its needs; a "one-size-fits-all" approach is unlikely to be effective, although there is considerable learning that can be made by looking at what has been trialed in different contexts.

The international perspective of this handbook is a vital ingredient since any organization exists within a social, political, and economic context which reflects the cultural perspective of that country. We can all clearly learn from each other, and the reader is encouraged to read each perspective provided with an open mind. The style of each chapter also reflects the particular background of the author which ranges from practitioner or researcher, on the one hand, to manager or trainer, on the other. Some chapters are more reflective in style as they pick out the themes and history of how traumatic stress manifests itself, whilst others are more case study oriented within the particular cultural perspective of an organization or geographical location. There are also chapters on concepts of trauma, academic and practitioner controversies, and new evolved models of practice. With this in mind, there is something for everyone interested in trauma within organizations.

The workplace is a "community" in its own right, bearing distinctive personalities and characteristics. Organizations may share a commonality of policies, procedures,

International Handbook of Workplace Trauma Support, First Edition. Edited by Rick Hughes, Andrew Kinder, and Cary L. Cooper.

and practices, but each organization will be a unique entity formed by the individuals working within.

The Workplace and Beyond

The familiar adage throughout this book is that an individual's trauma response is a normal reaction to abnormal events. Yet the communities in which we live serve as a wider backdrop posing their own collective responses and reactions. The workplace is a microcosm of the wider world, and few events that impact organizations remain there.

Within organizations, alongside trauma responses, there is an increasing interest to a linked subject that could be termed "well-being." This is generating greater interest in the public and private sectors (Kinder, Hughes, & Cooper, 2008) as well as at the government level, with the current UK prime minister advocating the "Big Society" as a means for collective responsibility, accountability, and wellness. The Foresight Mental Capital and Wellbeing Project (Dewe & Kompier, 2008) commissioned by the UK government sought out how to achieve the best possible mental development and mental well-being for the UK population in the future. Preparedness offers a means to plan for the future. But it is also from the past that the greatest lessons can be found.

However, lessons can only be learned when the truth can emerge, untainted by legal or commercial restraints. It can be advantageous for an organization to publicly announce details of their own investigations into traumatic events where there has been evidence of some culpability. Rather than hiding behind negligence and wrong-doing, a commitment to learn lessons can restore faith and trust amongst users and the public and, ultimately, shareholder confidence. BP's (2010) investigation into the 2010 *Deepwater Horizon* disaster has led to a 192-page report that represents a detailed overview into the event.

Public inquiries and other government-sanctioned or criminal investigations can also establish a cause and effect. The Bloody Sunday Inquiry, also known as the Saville Report, was established in 1998 by the then UK Prime Minister Tony Blair to establish the events surrounding the Sunday of January 30, 1972, where, in Derry, Northern Ireland, 13 people died and a similar number were injured at a civil rights march. Despite claims of a UK government "whitewash" of events and a delay of nearly 40 years into an understanding of what happened, the report concluded some UK military culpability (Saville & Toohey, 2010). Whilst the delay in exposing the truth has no doubt contributed to some distress for the family and friends of the deceased, a modicum of comfort will be derived from the revelation of the investigated facts.

Information by Media

The fast speed at which that modern media can now report on calamitous and catastrophic events gives the impression that traumatic events are on the increase. And the way in which the "drama" is conveyed in the media can whip up a frenzied response as was witnessed by the public outpouring of grief when Lady Diana Spencer (Princess Diana) died in 1997 – few knew her, yet a societal and shared empathy touched a nerve. Since then, 24-hour news channels seek to cater for what seems an insatiable appetite for information which, in itself, can contribute to a proxy traumatization for viewers, unable to do anything but get caught up in a collective involvement. Perhaps one of the biggest examples of this was the 9/11 disaster regarding the terrorist attack on the Twin Towers, a subject we shall return to later on in this chapter.

It could be argued that rather than seeking to convey an objective truth of an event, mass media manipulate users to drive ratings by sensationalizing, building an anxiety and fear amongst users that maintain a fixation to a given media outlet. In some cases it could be argued that proprietor, commercial, and political influence further filters a known reality which can exacerbate anxiety amongst those genuinely needing to understand and ascertain relevant, accurate, and appropriate information. Exceptions do emerge, and one example would be the film documentary by John Pilger (2010), *The War You Don't See*, where politicians, experts, and journalists admit to not having pushed sufficiently the Blair administration at the time to question the legality, legitimacy, and merits of taking military action in Iraq.

The Scale of Tragedy and Trauma

Sundnes and Birnbaum (2003) estimated that between 1951 and 2000, there had been 7312 major catastrophes, causing over 9 million fatalities and contributing to $961 895 000 worth of damage. This appears to exclude the vast number of smaller traumatic events that scar a more local workforce, community, or family.

The National Centre for Counterterrorism in their 2008 report approximated 11 800 terrorist attacks against noncombatants, resulting in 54 000 deaths, injuries, and kidnappings. Comparing to 2007, attacks had actually decreased by 2700 and deaths due to terrorism decreased by 6700. Southeast Asia, though, had the greatest number of casualties with 75 percent of the 235 high-casualty attacks (those that killed 10 or more people) (European Network of Victims of Terrorism (ENVT), 2009).

Northern Ireland

Returning to the theme of how organizations are influenced by their social context, we now take a reflective view on Northern Ireland and look at what we can learn from this setting before considering potential ways forward for trauma within organizations.

Northern Ireland has a history of civil conflict, known as the Troubles. It is recognized that this conflict not only involves large swathes of the Northern Irish and mainland UK communities but also represents a long-standing political and social struggle that goes back several centuries. Whilst this book seeks to establish current and "best" practice in effective post-trauma support, the consensus held by those involved in the support of current victims and mitigation of the long-term psychological ill health of people in Northern Ireland is to work toward the development of "better" practice, recognizing the developmental nature of improved service provision and delivery.

The Cost of the Troubles Study (Fay *et al.*, 1999) reported that from 1969 to December 3, 1997, 3585 people had been killed in Northern Ireland as a result of the civil conflict. If the United Kingdom as a whole, with its population of some 60 million, had experienced death pro rata as compared with the 1.8 million population in Northern Ireland, there would have been a total of 130 000 dead (Bloomfield, 1998). This pro rata comparison could be replicated at the locations of other large-scale disasters, including, but not limited to, the impact of 9/11 on New York in 2001, the thousands of small communities bearing the brunt of the December 2004 Asian tsunami, and the northeastern territories affected by the Japanese earthquake and tsunami in 2011.

Muldoon *et al.* (2005) claim that in Northern Ireland, during several decades of political violence, in addition to over 3500 deaths, there have been 16 000 charged with offenses, 34 000 shootings, and 14 000 bombings, suggesting that there will be few individuals, families, and communities who have not been directly or indirectly affected by the conflict.

The Voice of the Trauma Victim

Although models of trauma support are important to ensure they are effective, learning from "victims" of trauma helps to tell us about the effective interventions and how these can be further adapted and improved.

Taking the context of Northern Ireland, the community has endured some 40 years of civil conflict. Indeed, the extent to which this has affected the society is illustrated by Figure 30.1, which compares the lifetime prevalence of PTSD benchmarked against other countries (Ferry *et al.*, 2008).

Prevalence rates are well above the estimates from other comparable European countries (ESEMeD), including France, Spain, Belgium, the Netherlands, Italy, and

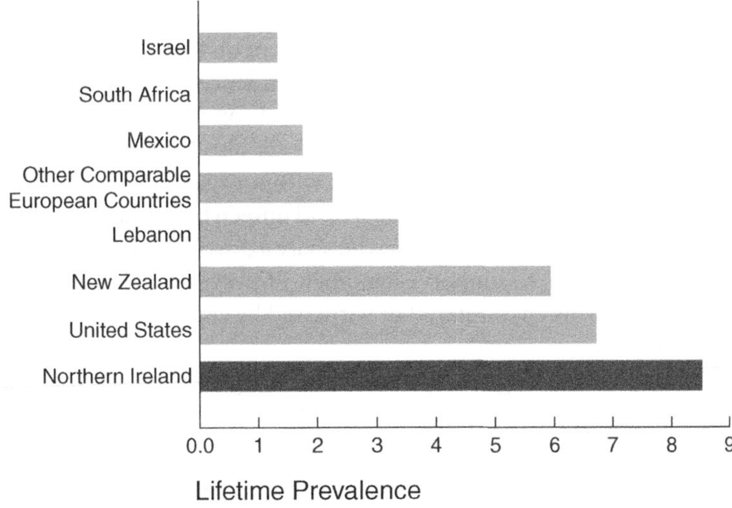

Figure 30.1 Comparison of lifetime prevalence of PTSD.

Germany; and other countries in which conflict has been a feature of their recent history (South Africa, Israel, and Lebanon). Indeed, the Commission for Victims and Survivors (CVS; 2010) suggests any comparison between Northern Ireland and South Africa, including the notion of working toward a South African–style Truth and Reconciliation Commission, is regarded as "uninformed and even naïve" (p. 38).

A victim-centered strategy is one that is likely to help aid recovery and to understand how best to provide the appropriate support interventions. One of the key elements to this is how the victims are able to be recognized and acknowledged.

"Recognition" occurs when someone feels that others have become able to see them as they see themselves, knowing them for the person they are or perhaps who they have come to be. Through recognition, victims feel they are properly known and understood by the rest of society. "Acknowledgment" occurs when the actions of others are informed by recognition of who the victim-survivor really is. Acknowledgment is the "working out" of recognition in the everyday lives of victims. A society that truly knows and understands the victims and survivors of its own conflict develops sufficient insight to orient services and make provision for its stricken citizens (CVS, 2010).

The ENVT, through the European Commission Directorate-General for Justice, Freedom and Security, published an influential chronicle of a recent symposium convened in its name. *The Voice of Victims in Europe: Testimonies* (ENVT, 2009) has given expression for people across Europe who share a common need to express themselves as victims of terrorism. The moving testimonies reflect a strong wish to share and to have a voice. And yet whilst these gatherings reflect a connection with terrorism, the stories could relate to almost any trauma in that there is loss, hurt, and pain.

Perceptions of Trauma

However, what is regarded as traumatic for one person may not be for another. This will be influenced by a range of variables including previous trauma, childhood development, and coping skills. Even amongst a population of traumatized individuals, some will cope better than others, and this is often dependent on the speed and availability of practical and psychological support. There is some evidence to suggest that large-scale traumatic events can foster a societal cohesiveness and community support which help some people to cope better (Gillespie *et al.*, 2002; see also Chapter 27, this volume). This appears to be consistent with experiences within organizations. An unexpected death of an employee, where this occurs off-site and unseen by colleagues, may generate a collective grieving that allows colleagues, who share a disconnection to the event but a connection to the deceased, to support each other. However, where death, injuries, or significant threat is actually encountered or witnessed by employees, a different impact might be created.

One's physical and psychological "distance" from or proximity to a traumatic event can also trigger a different set of experiences and subsequent support needs. Taylor and Frazer (1982) identified six different categories of "victims":

- Primary victims: those at the center of the trauma.
- Secondary victims: such as friends and family of the primary victim.

- Third-level victims: emergency, rescue, and first responders.
- Fourth-level victims: people within the community or workforce who offer to help.
- Fifth-level victims: those who may be indirectly involved.
- Sixth-level victims: people who could have been directly involved but were not.

The definition of a "victim" can present its own challenges. For instance, someone who is physically involved in a road traffic accident (RTA) may be less "traumatized" than another who witnesses the carnage. Similarly, the range of support services and compensation schemes open to "victims" might be based on a specified "label" that can exacerbate feelings of anger, resentment, and betrayal (CVS, 2010).

Post-trauma Support: The Debate

In the United Kingdom, the National Institute for Health and Clinical Excellence (NICE), a government body, is tasked with providing health and medical "advice" in order to set, maintain, and improve standards of practice. In 2005, they published a report on post-traumatic stress disorder (PTSD), *The Management of PTSD in Adults and Children in Primary and Secondary Care* (2005). This prescriptive document advocated four key pieces of "guidance."

For individuals who have experienced a traumatic event

1. Provision to that individual alone of brief, single-session interventions (often referred to as "debriefing") that focus on the traumatic incident should not be routine practice when delivering services.
2. Where symptoms are mild and have been present for less than four weeks after the trauma, watchful waiting, as a way of managing the difficulties presented by individual people with PTSD, should be considered. A follow-up contact should be arranged within one month. (NICE, 2005, p. 3)

Trauma-focused psychological treatment

1. Trauma-focused cognitive-behavioral therapy (CBT) should be offered to those with severe post-traumatic symptoms or with severe PTSD in the first month after the traumatic event. These treatments should normally be provided on an individual outpatient basis.
2. All people with PTSD should be offered a course of trauma-focused psychological treatment (trauma-focused CBT or eye movement desensitization and reprocessing (EMDR)). These treatments should normally be provided on an individual outpatient basis. (NICE, 2005, p. 3)

When this was published it could be seen as challenging the British Psychological Society's (2002) report on psychological debriefing and seemed to question much of what the "industry" had offered trauma "victims" in the past. The "watchful waiting" mandate both offered a dexterity and freedom to practitioners yet still called into question single-session interventions. Strict adherence to this line would suggest the end of mandatory psychological debriefing in the way it had been recognized and provided in some organizations in the past, yet there still appears to be support for its continuation, although definitions of what exactly was meant by "debriefing" has created some confusion.

This has, however, energized a further development of trauma support interventions, largely developed in, adapted for, and delivered at specific workplace environments. Some of the interventions are covered in greater detail elsewhere in this book and include Trauma Risk Management (TRiM) (Chapter 12), Support Post Trauma (SPoT), (Chapter 13), and Psychological First Aid (PFA) (Chapter 11).

Where these approaches or models have evolved, a body of evidence is required to prove their effectiveness but the very nature of achieving this can be challenging. De Soir (Chapter 23, this volume) argues that to establish a randomized control trial (RCT), one needs a control group. He suggests it may be unethical to withhold or delay treatment to a control group population who exhibit the signs of trauma akin to the trial group. However, Duffy *et al.* (2007) found, in their RCT of PTSD in the context of terrorism and other civil conflict in Northern Ireland, that in their 12-week study, the control or "wait" group, possessing a similar diagnosis for PTSD as the treatment group, did not show any signs of deterioration in their conditions and were offered treatment at the end of the trial period.

The NICE Guidelines (2005) advocate the use of trauma-based CBT and EMDR, though these should not be regarded as panaceas. It would seem that these options have found favor because of the number of RCTs that have been carried out which provide evidence of their effectiveness, yet it is argued that RCTs should not be the only way to merit the choice of intervention. Indeed. Andrews and Miller (Chapter 14, this volume) present a plausible argument against RCTs in favor of practice research networks (PRNs) where the weight of evidence-based practice informs what works and what does not.

Buckley and Dunn (Chapter 22, this volume) use the experience gleaned from Transport for London and the July 7, 2005, London bombings to explore trauma from the perspective of neuroscience. On a different level, Turnbull, Lwin, and McNab (Chapter 18, this volume) introduce the concept of mindfulness as further new conceptualization of how to treat trauma victims.

Other Manifestations of Trauma within Organizations

One of the authors was invited to provide defusing support to teams in an organization where a member of staff had committed suicide offsite. The most challenging aspect of this assignment was in relation to the high levels of emotion regarding how members of staff were able to (and, importantly, if they so wish), or can choose not to, find their own voice in reflecting the impact on them. A vacant room had been appointed as a quasi-shrine for the deceased. This allowed those who wanted a memorial site a place where they could "pay their respects" for the deceased. Other members of staff found this "memorial" intimidating, threatening, and inappropriate for the workplace. Their voice was important too.

Within this scenario, employees can focus their emotions on the organization, and sometimes these are projections where the organization can be blamed for the suicide due to work-related stress or bullying. Indeed, there are examples of organizations that have high rates of suicide where suicide notes are found linking their suicidal ideation to the workplace.

Suicide in the workplace can be traumatic in its own right and organizations need to think through how best to both prevent and deal with such occurrences (Kinder, 2006; Kinder & Cooper, 2009). Organizations would be advised to have clear risk assessment policies which include assessing the suicidal risk of an employee and the involvement of healthcare professionals. Joiner *et al.* (1999) identified a number of factors that point to an elevated risk of suicide:

- Past suicide attempts or behaviors.
- Current symptoms of suicide.
- Life stresses.
- Degree of psychological disturbance such as depression, anxiety, substance abuse, hopelessness, and helplessness.
- Degree of self-control in overcoming desire to take suicide actions.
- Lack of social support and alternative coping strategies.

The link between the reporting of suicide in the media and resulting spikes in suicide is worth noting here, and the World Health Organization (2000) has published guidance for the media to follow which highlights that detailed descriptions of the method used should be avoided, simplistic explanations of why a suicide occurs should not be given (i.e., it should not be blamed on work), and glorifications of those who have died by suicide should not be made. Describing the consequences of nonfatal attempts at suicide can act as deterrents (e.g., impact on friends and family, and brain damage). There is also reference to those people who are bereaved by suicide, and the guidance highlights the importance of offering support and psychological help to them in recognition that they will be impacted, again reinforcing the point that suicide impacts widely within the community.

This chapter has explored the impact of trauma within a society using Northern Ireland as a case in point. Of course, whilst reference is made to a society, the exact same parallel can be drawn with workplace communities. Spending half of our day in a workplace weds us to the cultures, values, norms, and systems of our organizations so we become embedded in an enforced new community where, for many, personal identity is heavily reflected by our work selves.

The manner in which workplaces cope with and orchestrate trauma support will have an enduring impact on all employees. Critical incident protocols should be in place to provide an operational strategy to deal with potentially traumatic events. By the very nature of them, they are unexpected and usually they cannot be anticipated time-wise. However detailed and meticulously prepared the protocols are, by the very nature of traumatic events, there will be things that do not work out the way they are planned.

It would appear to be prudent to have at the core of any critical incident management strategy an operational formula that allows a fair degree of flexibility and adaptability. Whilst incidents that may be more of the norm such as RTAs, onsite accidents, and suicides may fall into a more structured planning continuum, we also need to learn lessons through wide-scale indiscriminate actions of terrorism, such as 9/11 and, in 2011, the Norway shootings where over 75 people lost their lives.

Trauma in the Helping Professions

Those in the helping professions also need to take stock. Providing psychological support can be challenging in any situation, yet there are clear issues when delivering it in wars or in acts of terrorism where the people providing support may also be profoundly affected by the events that have unfolded. It is well documented that carers can experience compassion fatigue which reduces their own resources to cope and can result in a muddled response to the people they are aiming to support. It has also been observed that in various disasters, there has been a rush to "send in the counselors" which can add to the sense of confusion and bewilderment unless this response is carefully coordinated and the real practical as well as emotional needs of those affected are taken into account. Writers looking at the impact of trauma on the helper have highlighted some of these issues. These range from Tehrani's (2010) assessment of how in an organizational context the role of the human resources manager in dealing with emotion-laden material can affect their psychological well-being to Seeley's (2003) account of how the effects of 9/11 on mental health workers have had long-lasting effects.

These are worth looking at in more detail. For instance, Tehrani (2010) outlines the role of human resource professionals (HRPs) where "there is a cost to engaging with the stories and lives of distressed or traumatized people, which can result in the HRP experiencing symptoms of shock, disbelief and horror similar to those of the people they are trying to support" (p. 56). These negative experiences have been described as "compassion fatigue" (as described in this section) or "secondary trauma" (Figley, 1999) and are viewed as natural consequences of helping or wanting to help distressed people.

Tehrani then outlines that, given this emotion-laden role, HRPs look after themselves less well than other professional groups (e.g., when compared to counselors and occupational health advisors) and have less organizational support, such as supervision arrangements, in place (Tehrani, 2007). This could result in the significantly lower levels of positive attitudes and beliefs that were found with HRPs when compared to the other professions, "suggesting that HRPs were either less able to cope with their negative experiences or unable to translate their negative experiences into learning or feelings of competence" (Tehrani, 2010, p. 58)

Although it could be argued that HRPs are not trained to work therapeutically with distressing material and therefore these findings are not a surprise, Tehrani makes some important points about how organizations could benefit from standing back and checking that those providing support to others are not themselves harmed in doing this work. Other examples given in this book include the police, child protection workers, and air traffic controllers where traumatic material becomes part of an employee's experience as they carry out their responsibilities (Tehrani, 2010).

We have outlined how traumatic stress impacts those workers who are not specifically trained to act as psychological "experts," and we will now look at what happens to mental health workers when they are exposed to trauma and consider whether they are insulated from the effects of trauma by virtue of their training and experience.

At the time of writing, it is 10 years since 9/11, and repeats of the media footage and accounts of what happened on that fateful day serve as a reminder to us all of the enormous impact and shock that resulted not just to the community of Manhattan but also around the world. Seeley (2003) has written on the aftermath of the attack from the viewpoint of the mental health worker. She starts off by explaining how disasters cause extensive psychological harm even to those who were geographically distant from the trauma (Vlahov, 2002) and that the most pervading impact occurs where the trauma is caused by human acts (Norris, 2002). She highlights that the mental health workers who were called upon to support those affected were themselves needing to work through their own pain. Such workers operated "under great pressure, in scenes of unspeakable destruction, chaos, and horror," and they were faced with needing to deploy psychological help which went beyond the usual clinical boundaries. She makes the point that

> Transforming their usual professional models and roles to alleviate the suffering of others, their own suffering often went untreated, and grew worse after contacts with patients. Recognizing that those deeply hurt by the attack would not soon recover, and would need psychological treatment indefinitely, therapists resigned themselves to the fact that they would be reliving September 11 for years. (Seeley, 2003, p. 37)

She goes on to recount in graphic terms how mental health workers spoke "of being dazed, exhausted and numb; of spending days in their offices, where they would help patients voice their stories – where they would 'let her talk and listen to her, and listen to her, and listen to her' – and nights at home in tears" (p. 21). That emotional contagion was endemic and seemed to feed itself.

Reflecting on what she found, Seeley highlighted the need for "mental health preparedness" so that health services were organized and coordinated in advance rather than assembled following an incident amidst all the resulting chaos. She called for cultural sensitivity to be adopted so that the local support workers who are knowledgeable about the local population would be involved in this work and that training and research would be given in order to help the mental health worker understand what would help when (not if) the next disaster or terrorist attack occurs. Psychological protection for therapists involved in psychological support should not be overlooked lest the very people who are providing the help become at risk of harm themselves. She concludes her writing with a question to all those involved in providing psychological support by asking whether we are out of our depth in such a situation as 9/11 and stating that we should be honest and accept the limitations of our profession. Perhaps this highlights the need to bring perspectives from all different points of view and from different professions and organizations. Rudolph Giuliani, the New York mayor at the time of 9/11, called upon his fellow Americans within hours of the attack to "continue with your normal lives," to go to the cinema, to have a pizza, to walk in the park, and to respect your fellow neighbor, especially those who happen to be Muslim. Perhaps it is as simple as this: to carry on and not allow the terror to infect every part of our lives and nations.

Future Perspectives

We have explored in this last chapter the impact of trauma within our countries, our communities, and our organizations and perhaps we all have a unique response to this treatise as we explore the impact of trauma upon ourselves. The future of a more job-insecure workforce, the working of increasingly long hours (Burke & Cooper, 2008) as organizations try to do "more with less" resources, the excessive pressure to deliver during a recessionary period that could extend to a decade, the threat of terrorism in the workplace, and societal insecurities will mean that dealing with traumatic events is more likely rather than less in the future. As Franklin Delano Roosevelt wrote during the Great Depression, "True individual freedom cannot exist without economic security and independence ... People who are hungry and out of a job are the stuff of which dictatorships are made ... The hope of the Republic cannot forever tolerate either undeserved poverty or self-serving wealth" (quoted in Cooper & Wood, 2011).

We are experiencing these very events in the world today, which means a relatively unstable and potentially traumatic period, with its individual as well as organizational consequences, in the future. We hope that we have provided a platform in this book in pre-planning and developing appropriate critical incident management strategies and to do justice to the many varied and innovative ways in which traumatic stress has been tackled.

References

Bloomfield, K. (1998). *We will remember them: Report of the Northern Ireland victims commissioner to the secretary of state for Northern Ireland.* Retrieved from http://cain.ulst.ac.uk/issues/victims/docs/bloomfield98.pdf.

BP. (2010, September 8). *Deepwater Horizon: Accident investigation report.* Retrieved from http://www.bp.com/liveassets/bp_internet/globalbp/globalbp_uk_english/incident_response/STAGING/local_assets/downloads_pdfs/Deepwater_Horizon_Accident_Investigation_Report.pdf.

British Psychological Society. (2002, May). *Psychological debriefing: Professional Practice Board Working Party* (Ed. N. Tehrani). Leicester, UK: Author.

Burke, R., & Cooper, C. L. (2008). *The long work hours culture.* Yorkshire, UK: Emerald.

Commission for Victims and Survivors (CVS). (2010, June). *Dealing with the past: Advice to government.* Belfast: Author.

Cooper, C. L., & Wood, S. (2011, July 15). Happiness at work: Why it counts. *Guardian.* Retrieved from http://www.guardian.co.uk/money/2011/jul/15/happiness-work-why-counts.

Dewe, P., & Kompier, M. (2008). *Foresight mental capital and wellbeing report: Wellbeing and work: Future challenges.* London: Government Office for Science.

Duffy, M., Gillespie, K., & Clark, D. M. (2007). Post-traumatic stress disorder in the context of terrorism and other civil conflict in Northern Ireland: Randomised controlled trial. *British Medical Journal, 334,* 1147.

European Network of Victims of Terrorism (ENVT). (2009). *The voice of victims in Europe: Testimonies.* Retrieved from http://www.europeanvictims.net/images/blanca/ficheros/27_Report_11M_09.pdf.

Fay, M. T., Morrissey, M., Smith, M., & Wong, T. (1999). *The Cost of the Troubles Study: Report on the Northern Ireland Survey: The experience and impact of the Troubles.* Derry: INCORE.

Ferry, F., Bolton, D., Bunting, B., Devine, B., McCann, S., & Murphy, S. (2008). *Trauma, health and conflict in Northern Ireland: A study of the epidemiology of trauma related disorders and qualitative investigation of the impact of trauma and the individual.* Omagh: Northern Ireland Centre for Trauma and Transformation. Retrieved from http://www.nictt.org/attachments/article/22/NICTT%20THC%20in%20NI.pdf.

Figley, C. R. (1999). Compassion fatigue: towards a new understanding of the cost of caring. In B. Hudnall Stamm (Ed.), *Secondary traumatic stress: Self-care issues for clinicians, researchers and educators* (2nd ed.) Lutherville, MD: Sidran Press.

Gillespie, K., Duffy, M., Hackman, A., & Clark, D. (2002). Community based cognitive therapy in the treatment of post-traumatic stress disorder following the Omagh bomb. *Behaviour Research and Therapy, 40,* 345–357.

Joiner, T., Walker, R., Rudd, M., & Jobes, D. (1999). Scientising and routinising the assessment of suicidality in outpatient practice. *Professional Psychology: Research and Practice, 30,* 447–453.

Kinder, A. (2006, Winter). The last taboo: Suicide and sudden death. *Counselling at Work Journal.* Retrieved from http://www.bacpworkplace.org.uk/journal_pdf/acw_winter06_d.pdf.

Kinder, A., & Cooper, C. L. (2009). The costs of suicide and sudden death within an organization. *Journal of Death Studies, 33*(5), 411–419.

Kinder, A., Hughes, R., & Cooper, C. (2008). *Employee wellbeing support: A workplace resource.* Chichester: John Wiley & Sons, Ltd.

Muldoon, O., Schmid, K., Downes, C., Kremer, J., & Trew, K. (2005). *The legacy of the Troubles: Experience of the Troubles, mental health and social attitudes.* Belfast and Cork: Queens University Belfast and University College Cork.

National Institute for Health and Clinical Excellence, (NICE). (2005). *Post-traumatic stress disorder (PTSD): The management of PTSD in adults and children in primary and secondary care.* London: Author.

Norris, F. (2002). Disasters in urban context. *Journal of Urban Health, 79*(3), 308–314.

Pilger, J. (Dir.). (2010, December 14). *The war you don't see* [Television broadcast]. ITV1.

Saville, H. W., & Toohey, J. (2010, June 15). *Report of the Bloody Sunday Inquiry.* London: TSO.

Seeley, K. (2003). *The psychotherapy of trauma and the trauma of psychotherapy: Talking to therapists about 9–11.* New York: Center for Organizational Innovation. Retrieved from http://www.coi.columbia.edu/pdf/seeley_pot.pdf.

Sundnes, K. O., & Burnbaum, M. L. (2003). Health and disaster management: Guidelines for evaluation and research in the Utstein style. *Prehospital Disaster Medicine, 17* (Suppl. 3), 1–177.

Taylor, A. J. W., & Frazer, A. G. (1982). The stress of post-disaster body handling and victim identification work. *Journal of Human Stress, 8,* 4–12.

Tehrani, N. (2007). The cost of caring – the impact of secondary trauma on assumptions, values and beliefs. *Counselling Psychology Quarterly, 20*(4), 325–339.

Tehrani, N. (2010). *Managing trauma in the workplace: Supporting workers and organisations.* Hove, East Sussex, UK: Routledge.

Vlahov, D. (2002). Urban disaster: A population perspective. *Journal of Urban Health, 79*(3), 295.

World Health Organization. (2000). *Preventing suicide: A source for media professionals.* Geneva: Author.

Index

5HTT, *see* 5-hydroxytryptamine transporter
5- hydroxytryptamine transporter (5HTT) 282
9/11 attacks 71–3, 78–81, 306, 362, 476, 477, 482, 483
9/11 Commission 229

Abruzzo earthquake 99
absenteeism 75, 107, 146, 148, 199, 380
  impact of support on 209–10
  see also sickness absence
abuse
  and organizational adversity 162
  as predictor of PTSD 127
  child 99, 127, 251, 282, 416, 469
  domestic 140, 142
  emotional 244
  physical 244, 245, 247
  sexual 93, 213, 222, 244, 245, 246, 247, 373
  spousal 185
  verbal 87, 94, 106, 337
  see also alcohol abuse; drug abuse; substance abuse
abusive supervision 107
acceptance and commitment therapy (ACT) 288
accreditation
  of peer supporters 59, 60
  of private sector preparedness 230
ACT, *see* acceptance and commitment therapy
ACTH, *see* adrenocorticotrophic hormone
action research 78
active listening 388, 391, 393, 394
acute care 3, 9, 12, 14, 15
acute dependency model 6
acute impact 374–5
acute reactions 286, 371–7
acute stress disorder 6, 168, 174, 419
acute stress reactions (ASRs) 282–3, 424
acute trauma 6, 9, 167, 168, 242, 247, 263
adaptive cognitions 263
adaptive information processing (AIP) 259, 261–3, 269, 270
adaptive resolution 268
adjustment (psychological) 4, 14, 190, 462, 465
  difficulties 13, 14
  disorders 6, 124, 418
adjustments (workplace) 186, 335, 338, 339, 341–6
adrenal glands 282
adrenaline 354
adrenocorticotrophic hormone (ACTH) 282
adversarial growth 61
adversity 154–64, 315, 316, 362
affective regulation skills 280
affective states 315
affiliative system 285

International Handbook of Workplace Trauma Support, First Edition. Edited by Rick Hughes, Andrew Kinder, and Cary L. Cooper.

Afghan War 37, 45, 182, 428, 429, 459, 460
aggression
  and psychological debriefing 170
  nonphysical 105, 106, 117
  patient 91, 92
  physical 92, 200, 337
  tolerance of 106
  verbal 92, 200, 337
  workplace 105, 106
agoraphobia 124
Air Berlin 406–7
Air Florida crash 302
airline industry 80–1
Air Pacific Conference of Employee Assistance Programs 441
air traffic controllers 482
alcohol 35, 124, 142, 148, 248, 363, 388, 408, 430, 431, 459–65
  abuse 115, 298, 427, 428
  management 468
  misuse 418, 421, 427–30, 460
  see also drug abuse; substance abuse
alcoholism 112, 298
alienation 373
Allitt, Beverly 89
allostasis 133
American Association of Marriage and Family Therapists 215
American Civil War 8
amnesia 168, 243, 244, 250, 251
amygdala 98, 130, 354
analgesia 373
anesthesia 243, 250, 373
anger 77, 124, 146, 280, 375, 394, 395, 421, 456, 461
  management 464
animal model of traumatization 373
ANP, *see* apparently normal part of the personality
ANS, *see* autonomic nervous system
anthrax attacks 73, 76
anticipatory grief 429
anxiety
  and compassion fatigue 93
  and missile attacks 419
  and risk of suicide 481
  and stress inoculation 275
  and the media 476
  and the neurobiological regulator 364
  and trauma 127, 363, 443
  and traumatic injury 124
  diagnosis 344
  disorders 6, 39, 124, 125, 419, 421
  in British Armed Forces 418
  in days following accidents 174
  in spouses 429
  in veterans 462
anxiety management training 116
apartheid 140
APEAR, *see* Asia Pacific Employee Assistance Roundtable
aphonia 243
apparently normal part of the personality (ANP) 242, 244–8
appraisal 149, 276
armed raids 200, 209, 302, 337
arousal 77, 90, 129, 168, 259
Arthur Andersen 83
art therapy 391
Asia Pacific Employee Assistance Roundtable (APEAR) 441
ASRs, *see* acute stress reactions
assault 87, 88, 92, 94, 96, 106, 142, 168, 206, 222, 246, 261, 337
assault cycle 110
assertiveness 252
assessment
  clinical 464
  Combat Stress 463
  community 464
  for dissociation 253
  for EMDR 266, 267
  for resilience 89
  in CISM 53, 56, 58–9
  in E-EIM 175
  in psychological first aid 307
  in trauma support 147
  in TRiM 56
  of core skills 413
  of health 43
  of incidents 58–9
  of needs 459
  of risk 56, 59, 108, 110, 188, 336–7
ASSIST model 87, 95, 96–8
Atos Healthcare 95, 203, 204
atrocities 7, 13, 125, 422, 451
attachment
  and 5HTT genes 282
  and HPA axis responsiveness 282
  and resilience 360, 364
  family 462
  in army personnel 416, 417, 429
  need for 244

parental 244
problems 464
attention 260, 373, 376
Australian Centre for Posttraumatic Mental Health 59
Australian Defence Force 131
Australian Treatment Guidelines for PTSD 129
Australian Veterans Mental Health Services 471
authority figures 146
autonomic nervous system (ANS) 224, 278, 354
autonomic reactions 259
autonomic regulation 277
avoidance 77, 90, 123, 124, 146, 150, 171, 248, 259, 284, 285, 314, 341, 361, 421

BACP, *see* British Association for Counselling and Psychotherapy
BACP Workplace 351
Balkan Wars 12
bank robberies 302, 448
Barkham, Michael 215
barriers
to help seeking 184–7
to mental health care 39
Battlemind 56, 283
debriefing 56, 324
BattleSMART 319
Before, During and After (BDA) grid 56
Behar, Lior 309
behavioral disorders 461
behavioral health 39, 44–5
behavioral interventions 111, 117
behavioral problems 12
behavioral relaxation 43
behavioral repetition 40
behavioral skills 280, 323
behavioral systems 237
behavior-based training 323
behavior information 227–38
behavior modeling 323
Belbin team profiling 413
beliefs 62, 387
Ben Asdale trawler 193
benefit finding 61
bereavement 147, 386
BICEPS model of early intervention 11, 12
Big Society 475
Big White Wall 470
binge eating 288
biofeedback 278
bio-physiological effects of violence 91
Birnbaum, Elizabeth 77
Black, Dame Carol 334, 338
Blair, Tony 475
blame 92, 159, 190
Bloody Sunday Inquiry 475
Blum, H. Steven (Lt. Gen.) 228
body awareness therapies 223
body handling 418
body recovery 418
body scan 267, 268, 269, 287
BOHRF, *see* British Occupational Health Research Foundation
bonding 361
Book of Changes, The 439
Book of Songs, The 439
borderline personality disorder 245, 251
boredom 315
bouncing 111, 190
boundaries 395
Bradford City football ground disaster 49, 88, 303
braking techniques 278
bravado 187
bravery 422–3
breaking point 242, 243, 244, 246, 249, 251, 253
briefs 323
British Association for Counselling and Psychotherapy (BACP) 351
British Broadcasting Corporation 54, 187
British Foreign and Commonwealth Office 54
British Occupational Health Research Foundation (BOHRF) 204
Buddhism 284, 287, 436, 437, 439
buddy system 201, 300, 377, 395, 399, 426
Buerk, Michael 154
building security 79
bullying 74, 106, 299, 300, 418, 450–1, 453, 462, 481
burnout 92, 93, 236, 395
burn trauma 51
Bush, George W. 232
bystanders 72, 74, 377

calming 78, 147, 377
camaraderie 32, 139, 187
cancer 93, 287
Cantor Fitzgerald 71–2, 79
Cape Town 140

cardiac coherence 278
cardiac problems 462
cardiovascular disease 8, 131
Care First 307
care planning 463
Care Quality Commission 95
carer education 468
case monitoring 113
case studies 323
caste system 389
catharsis 170, 188
CBT, *see* cognitive behavioral therapy
Centre for Trauma, Resilience and Growth (CTRG) 58, 97
cerebrovascular accidents (CVAs) 462
CF, *see* compassion fatigue
chaos theory 155
chaplains 41, 42, 43
chat services 114
chemical plants 73
child abuse 99, 127, 251, 282, 416, 469
childhood trauma 125, 127, 461, 462
child neglect 87, 99, 244, 282, 461
child protection 99
  workers 482
China 436–45
Chinese-American Trauma Therapy Training Program 440
Chinese characteristics 445
Chinese medicine 437, 439
Chinese-Norwegian Psychoanalysis Training Program 440
Chinese Psychological Society 439
Chinese psychology 439–40
chronic fatigue syndrome 13
chronic pain 262, 287, 423, 430, 462, 463, 469
chronic stress 33, 278
chronic stressors 276, 314
CID, *see* Critical Incident Debriefing
CIPR, *see* critical incident processing
CISM, *see* Critical Incident Stress Management
Classic of History, The 439
classroom training 40
clinical assessment 464
clinical care 39
Clinical Outcomes in Routine Evaluation (CORE) 215, 218, 219
clinical outreach services 465
clinical risk management team 95
clinical trials 217
closure 267, 269, 449–50
Clothier Report 89
CMT, *see* compassionate mind training
coaching 450
cognitive appraisal 277
cognitive-based stress debriefing 323
cognitive behavioral therapy (CBT)
  and depression 116
  and guilt 149
  and PTSD 51, 116, 346, 479
  and rehabilitation 338
  and the rewind technique 221
  as tertiary prevention 378
  effectiveness 116–17, 167
  in British Armed Forces 418
  in Combat Stress treatment model 464
  in support for tsunami survivors 391
  in the workplace 116–17, 173
cognitive biases 83
cognitive change 288
cognitive construction 150
cognitive distortion 276
cognitive exercises 323
cognitive flexibility 43
cognitive interventions 324
cognitive networks 267
cognitive organization 421
cognitive reconstruction 150, 380
cognitive restructuring 116, 268, 468
cognitive therapy 149, 288, 421, 440, 469
  mindfulness-based 288, 445
cohesion 280, 326, 392, 421, 425, 444
  and coping 317
  and CSR 421, 423
  and morale 317
  following trauma 80
collaboration 361–3
collaborative training 324
collective efficacy 78, 147
collective trauma 353
Columbine High School shooting 302
combat
  and anxiety disorders 125
  and coronary heart disease 132
  and depression 125
  and domestic violence 429
  and motivation 315, 316
  and PTSD 13, 36–7, 125, 462
  and stress 6–13, 31–40, 44, 313
  and symptom management 187
  and trauma 31–40
  exposure to 13, 418

mitigating effects of 423–9
psychiatric effects 419–20
support during 34
use of psychotherapy during 34
violence of 36
witnessing 419
combat operational stress management program 44
combat readiness 317
combat stress 34, 36, 37, 40, 42, 44, 313
see also combat stress reaction
Combat Stress (charity) 430, 458–70
and the NHS 466
assessment 463
clinical population 461
clinical services 463
interventions 463
joint initiatives 469–70
recent service expansion 465–7
residential treatment 464–5
six-week program 467–70
treatment models 463–4
combat stress reaction (CSR) 6, 419–24
and cohesion 421
and depression 421
and early intervention 421
and pre-combat stressors 421
and PTSD 421, 424
and unit cohesion 423
etiology 421
personality factors 421
prognosis 421
recovery 421
research 420–2
risk factors 422
symptoms 123
combat stress teams 44
combat stress techniques 34
combat zone hospitals 422
command-and-control approach 409
command post blindness 228
Commission for Victims and Survivors (CVS) 478
communication 82, 235, 326, 341, 360–1, 364
and leadership 426
and the EBO 234
mass 155, 161
skills 316, 340, 405
strategies 80
with families 35
communications cycle 236
community assessment 464
community-based counseling centers 45
community health 40–5
community outreach services 465
comorbidity 125, 459, 461, 468
company-wide initiatives 114
compassionate autocracy 79
compassionate mind training (CMT) 285, 286
compassionate self-care 94
compassion fatigue (CF) 92, 144, 232, 236, 248, 482
compensation 121, 132, 380
competence 62, 92
Competency Checklist 58
complex disasters 227
complex dissociative disorders 253
Comprehensive Soldier Fitness (CSF) 41, 319
computer-based training 323
concentration 146, 259, 456
confidentiality 89, 114, 189
conflict resolution tactics 111
Confucianism 436, 437
connectedness 78, 147, 159, 372
consequence-focused training 110
conservation of resources (COR) theory 75
Consignia 201
constructionist self-development theory 93
constructive communication 252
consumer-related violence 91
continuity of care 44–5, 178
continuity plans 80
contract vulnerability 114
control 93, 150, 235, 267
perceived 36, 316
control groups 223
conversion syndromes 11
cooperative behaviors 75
coping 34, 61, 62, 75, 147, 276, 318, 344, 443
and adversity 161
and cohesion 317
and PTSD 319
assisted 234
emotion focused 421
in China 437
modeling 279
procedures 278
resources 78
role of EBC in fostering 236
skills 275, 280, 464, 478
training for 408

CORE, *see* Clinical Outcomes in Routine Evaluation
core skills 413
coronary arterial disease 132
coronary heart disease 132
corticotrophin releasing hormone (CRH) 282
cortisol 282
"cost of caring" 249
Cost of the Troubles Study 477
counseling
  after deaths at work 74
  and EAPs 113
  and PTSD 96
  and rehabilitation 338
  and residual stress 34
  community based centers 45
  for everyday stressors 41
  for factors complicating recovery 148
  for workplace violence 107
  in China 436–45
  in India 310, 388–99
  in Reuters America 81
  in RMG 207
  in the four-leg model 149
  in trauma support 147, 203, 351, 362, 364
  modes of delivery 114
  of deployed personnel 403–4
  of families of PDAs 403–4
  of PDAs 403–4
  profile of 362
  psychological debriefing as 52
  role of chaplains 42
  services 80
counseling centers 78
counterfactual thinking 360
countersnatches 206
court martials 430
CRH, *see* corticotrophin releasing hormone
crisis center 409
crisis intervention approaches 60, 115
crisis management 202–3, 207
crisis theory 14
crisis training center 388
Critical Incident Debriefing (CID) 411, 412
critical incident management 94, 295–311, 484
critical incident processing (CIPR) 55
critical incident protocols 481
Critical Incident Stress Debriefing (CISD)
  adjustments to 304
  and employee assistance programs 169, 303–4
  and psychological debriefing 48, 170
  and PTSD 78, 183
  and trauma management 17
  and workplace violence
  appropriate use of 378
  as group process 171
  difficulties 48, 50
  effectiveness 115, 182, 183
  follow-up in 53, 56
  for emergency personnel 369
  history of 168–9
  in mental health services 90
  use in EAPs 169
  see also psychological debriefing
Critical Incident Stress Management (CISM) 44, 48, 49, 283
  aims 52–3
  and education 53
  and military deployment 426–7
  and peer support 53
  and pre-crisis education 53
  and PTSD 52–3
  and TRiM 56
  as harmful 60
  assessment in 53, 56, 58–9
  commonalities in 55–6
  continued use of 63
  criticisms of 60
  follow-ups 53, 56
  origins 52
  planning in 56
  psychological debriefing in 53
  support in 56
  training in 57–60
  see also stress
cross-over violence 142
CSF, *see* Comprehensive Soldier Fitness
CSR, *see* combat stress reaction
Cultural Revolution 438, 440
culture
  and PTSD 427
  and resilience 158
  and responses to trauma 81
  as a resource 62
  as barrier to transfer of models 100
  Chinese 436–7, 442–5
  of emergency services 57
  organizational 83, 97, 318
  norms 108
  sensitivity to 235, 483

cumulative stress 33
cumulative traumatizing events 245, 246–7
CVAs, *see* cerebrovascular accidents
CVS, *see* Commission for Victims and Survivors
cyclical deployment 462
cynicism 315
Czech Republic 309

daily life activities 252
Dalai Lama 277
Daoism 436, 437
DASA, *see* Department of Analytical Services and Advice
day care centers 79
DBT, *see* dialectical behavior therapy
DCMH, *see* Departments of Community Mental Health
DDIS, *see* Dissociative Disorders Interview Schedule
DDNOS, *see* dissociative disorder not otherwise specified
death awareness 75
death reflection 75
deaths
  in the workplace 74–5
death threats 302
debriefing 280, 452–3, 479, 480
  group 303–6
  single session 189
  see also psychological debriefing
decision making 405, 406, 413
  value centered 43
Deepwater Horizon disaster 475
de-escalation techniques 111, 112, 343
deformity 430
defusing 53, 162, 202–3, 207, 369, 377
dehydroepsiandrosterone (DHEA) 282
delayed-onset PTSD 121–6, 130, 132, 133, 428
Delphi methodology 59
dendritic branching 361
Deng Xiaoping 438
denial of care needs model 5, 6
Department of Analytical Services and Advice (DASA) 460
Department of Homeland Security (DHS) 230
Department of Veterans Affairs Medical Centers 45
Departments of Community Mental Health (DCMH) 418
depersonalization 168, 244, 250, 251, 373
deployment
  and PTSD 38
  and team welfare 410–13
  mitigating effects of 423–9
  of peacekeepers 314
  training personnel for 408–10
  see also deployment stress
deployment-limiting medical conditions 43
deployment stress 34–5, 283
depression
  and CBT 116
  and combat 125
  and CSR 421
  and kindling 128
  and MBCT 288
  and post-traumatic stress disorder 461
  and PTSD 132, 459, 461, 468
  and risk of suicide 481
  and sensitization 128
  and the amygdala 130
  and trauma 121, 124, 127, 133, 363, 443
  diagnosis 344
  education about 43, 116
  etiology 121, 130, 133
  genetic predisposition to 128
  in British Armed Forces 418
  in Chinese medicine 437
  in tsunami survivors 387
  in veterans 459–62
  interpersonal psychotherapy for 173
  NICE guidelines 288
  support for sufferers 116
  triggers 131
  see also major depressive disorder
depressogenic events 377
derealization 168, 250, 251, 373
desensitization 267, 268, 269
detoxification 462, 463
Devon and Cornwall Police 195
DHEA, *see* dehydroepsiandrosterone
DHS, *see* Department of Homeland Security
diabetes 462
dialectical behavior therapy (DBT) 288
DID, *see* dissociative identity disorder
Disability Discrimination Act 1995 and 2010 186
disaster by design 231
disaster management 227
disaster mental health professionals 83
disaster preparation 78, 80
disaster-related behaviors 230–1

disaster response 44, 80
disaster teams 409–10
disaster types 404
disciplinary problems 12
discussion
  in psychological debriefing 58
  structured 60
discussion forums 114
disgruntled employees 72, 73
disruptive behavior disorders 127
dissociation
  and communication 361
  and EMDR 251
  and processing of emotional memories 124
  and SIT 276
  assessment for 253
  definition 240–1, 243
  double 214
  historical perspective 241–2
  importance 168
  in the workplace 240–54
  levels of 244–8
  maintenance of 248–9
  measurements 249–51
  organizational 353
  peri-traumatic 241, 249–50, 377
  predictive power 171
  role in acute trauma responses 167
  structural 241–53, 373
  symptoms 243, 250
  trauma-driven 373
  treatment 251–3
dissociative behavior 373, 377
dissociative disorder not otherwise specified (DDNOS) 244, 245
dissociative disorders 250, 253, 443
Dissociative Disorders Interview Schedule (DDIS) 251
Dissociative Experiences Scale (DES) 250
dissociative identity disorder (DID) 244, 245, 247
dissociative paralysis 250
dissociative responses 379
division of the personality 241
divorce 185, 246–7, 459
dizziness 242
domestic abuse 140, 142
  see also domestic violence; spousal abuse
domestic violence 429, 460
  and combat 429
  counselors 93
  see also domestic abuse
Dorset Fire Service 187
double agentry 93
double consciousness 241
double dissociation 214
doubling of the personality 241
downsizing 72
dreams 456
  see also nightmares
drills 40, 405
drive system 285
driving offenses 428
driving under the influence (DUI) 428
dropouts 217, 218
drug abuse 12, 142, 185, 466
  see also alcohol; drugs; substance abuse
drugs 12, 35, 141, 148, 248, 363, 387, 408, 460, 463–5
  see also drug abuse; substance abuse
DSM-IV 6
duality of persona 185
dual representation theory 149
DUI, *see* driving under the influence
duration of illness 129, 130, 131
duty of care 4, 96, 162, 302, 304, 380
dysphoria 174

EAPs, *see* employee assistance programs
early intervention
  and CSR 421
  and functional disorders 13
  and health promotion 63
  and prevention of disability 129
  and PTSD 13, 375
  and trauma prevention 31
  BICEPS model of 11, 12
  brevity of 11
  during World War Two 10–11
  dynamics 4
  for mercenaries 7
  immediacy of 11
  in recent wars 12–13
  in the military services 5–7
  in UN peacekeeping operations 12–13
  in violent societies 143–4
  models 3–15
  National Institute of Mental Health 173
  programs 42–3
  protocols 378
  psycho-educational model of 14
  role 167–79
early retirement 185

earthquakes 73, 99, 227
Easterbrook claim 376
EBCs, *see* emergency behavioral consultants
EBMS, *see* Emergency Behavior Management System
EBO, *see* emergency behavior officer
eclectic therapy 378
Edinburgh Early Intervention Model (E-EIM) 173–8
  as psychological treatment progam 178
  objectives 177
  routine use of 177
  stages 175–6
Edinburgh Psychological First Aid 173
education
  about depression 43, 116
  about self-injurious behavior 43
  about symptoms 267
  and CISM 53
  and emotional preparation 36
  and health assessments 43
  and mitigation of adverse effects 31
  and psychological debriefing 170
  and psychological first aid 306–8
  and resilience 306
  and stress 36
  cross-functional 43
  in cognitive behavioral therapy 116
  in employee assistance programs 308
  on normal responses to trauma 175
  pre-crisis 53
  pre-deployment programs 43
  psychosocial 266
  services 41
  web sites 14
E-EIM, *see* Edinburgh Early Intervention Model
effort syndrome 9
effroi 368, 372
ego strength 62
Eissler, K. R. 11
emergency action plan 79
emergency behavioral consultants (EBCs) 233, 236, 237
emergency behavioral management system (EBMS) 233, 237
emergency behavior officer (EBO) 231, 234–5
emergency communication systems 79
emergency escape procedures 79
emergency escape routes 79
emergency exercises 230
emergency management 234, 235
Emergency Operations Center (EOC) 81
Emergency Planning Society 402
emergency plans 79, 230, 354
emergency response structures 90
emergency response teams 72, 79
emotion regulation 252
emotion
  global sharing of 161
  sensitization to 155
  teleportation of 155
emotional abuse 244
emotional baggage 33–4
emotional contagion 356
emotional decompression 55
emotional engagement 315
emotional exhaustion 93
emotional intelligence 364
emotional neglect 244
emotional numbing 168, 421
emotional part of the personality (EP) 242, 243, 244, 245, 247, 248, 249
emotional preparation 36
emotional processing 34
emotional reactions 92
emotional reactivity 363
emotional regulation 98, 277
emotional shock processing 373
emotional support 94, 201, 202, 380, 397
emotional trauma 78
emotional triage 376
emotional uncoupling 379–80
emotional ventilation 368
emotion-focused coping 318, 421
emotions
  and coping 318
  and etiology of CSRs 8
  control of 318
  expression 442, 445
  importance in acute care models 9
  management of 395
  of disaster victims 397–8
empathic listening 111
employee assistance programs (EAPs) 295–311
  and alcoholism 112
  and CISD 169, 303–4
  and group debriefing 303–6
  and support 301–3
  and workplace violence 112–14
  compliance with clinical guidelines 296
  contracted 113, 114

employee assistance programs (*Continued*)
education in 308
history 298–301
in China 438, 440–2, 444
in different countries 309–11
in South Africa 74, 144
in the ASSIST model 95
in the Royal Mail Group 203
individual support mechanisms of 307
internal 113, 114
recent developments 301
resources 306
strategic role 308
training in 113
US Department of Defense 42
utility of 113
variability of delivery 114
employee-focused training 117
enforced passivity 280
enhanced understanding 380
Enron 83
environmental health hazards 31
environmental influences 11
environmental interventions 117
environmental mastery
environmental resources 324–7
environmental strengths 63
ENVT, *see* European Network of Victims of Terrorism
EOC, *see* Emergency Operations Center
EP, *see* emotional part of the personality
episodic memory 269
ethics 216, 219, 480
Ethiopia 154
ethnic cleansing 419
European Network of Victims of Terrorism (ENVT) 476, 478
evacuation
after wounding 425
and acute intervention models 10
and morale 12
from battle-line positions 10
of distressed personnel 187
of soldiers with self-control problems 13
of workplaces 79
orders 232
routes 80
to forward operating surgical teams 422
everlasting recommencements 355
evidence-based management 19–21
evidence-based practice 3, 7, 19–28
evidence-based risk factors 188
evidence-based support 199–212
evidence-based treatments 213, 217
evolutionary biology 285, 364
example setting 425
exercise 40, 150, 280
exhaustion 11, 93, 377
exhaustogenic events 377
expectations 108, 325
experimental manipulation 219
experts' opinions 59
exposure 288–9
direct 90
primary 368
secondary 368
tertiary 368
exposure-based therapies 275
external resources 315, 317–19, 327
external specialists 453–4
external stigma 185
extinction-based therapies 262
extreme trauma 451
eye movement desensitization and processing (EMDR) 251, 252, 259–70, 323, 378, 479, 480
and chronic pain 262
and memories 260
and NICE guidelines 201, 213
and PTSD 25, 261, 277, 346
and resilience 259, 263–4, 265
and SPoT 202
and workplace trauma 259–70
assessment for 266
client suitability 265, 267
description 260
eight-phase protocol 265–70
goals 262
in British Armed Forces 418
in Combat Stress treatment model 464
psychosocial education about 266
referral for 203
therapeutic effects 28, 260
three-pronged protocol 264–5

faith 41, 159
faith leaders 235
Falklands War 12
families
and resilience 319
and separation 43
and support 35, 41, 42, 43, 146–7, 318–19, 351, 401
and trauma management policies 339

as an external resource 317
as source of psychological first aid 308
cohesion 319
compensation in suicide cases 74
contact with 35, 235, 443
counseling after deaths at work 74
disorganization in 386
in China 437
problems in 462
relationships after disasters 386
family advocacy 41
family assistance centers 351
family education 468
family marital accommodation 429
family reunification centers 236
family therapy 391
famine 154
fatal road traffic collisions (FATACs) 191
fatigue 33, 34, 128
see also compassion fatigue
fear reduction models 6
Federal Emergency Management Agency 402
Federal Institute on Alcohol Abuse and Alcoholism 298
feedback
and psychological debriefing 58
from group debriefings 96–7
on levels of stress 236
on progress in treatment 217, 218–19
on therapeutic alliance 224
field mental health teams 426
field training 40
Fighting Fit 430, 469, 470
fight-or-flight response 98, 278, 282, 356, 373
financial matters 397
financial resources 81
financial support 397
firefighters 53, 72, 76, 240, 246, 314, 375, 380–1
first assistance 379–80
First China International Forum on Employee Assistance Programs 441
First Gulf War 13
first responders 232, 235, 240, 244
fitness 316, 317
"fit notes" 335
five elements 436
flashbacks 149, 243, 262, 363, 421
flashbulb memory 359
flexibility 41
flexible working 300
follow-up
in CISM 53, 56
in psychological debriefing 59
in TRiM 56, 190
follow-up services 113
Force Health Protection and Readiness program 31
Force Health Protection model 40
Foreign and Commonwealth Office 187
forensic nurses 90
Foresight Mental Capital and Wellbeing Project 475
forgetting 6
forward psychiatry 8, 10
four-leg model 148–51
Four Noble Truths 287
four-stage model of death awareness 75
Foxconn 73, 74, 80
fragmented information 83
France Telecom 73, 74, 80
fraud 162, 163
freezing 276, 373, 398
French Revolution 7
frontal lobe shutdown 278
frustration 380
fugue states 168
functional decompensation 419, 420
functional descriptive model 61
functional disorders, 13
functional incapacity 147
function checklist 98
functioning 77
funerals 449, 450
future orientation 345

GAF, *see* Global Assessment of Function
galanin 282
Galliano, Stephen 302–3
gambling 388
gangs 140
generalization effects 262, 269
generalized anxiety disorder 124
genes 282, 283
genetic strengths 62–3
Gestalt therapy 391
Gini coefficient 141, 151
Giuliani, Rudolph 79, 82, 483
Global Assessment of Function (GAF) 261
global risk profile 227
glucocorticoid resistance 127
GMP, *see* Greater Manchester Police

goal-directed behavior 173
goal setting 345
Goiânia 231
gossip 451
  see also rumors
Greater Manchester Police (GMP) 54
grief 147, 175, 176, 375, 385–6, 394, 412, 419
  anticipatory 429
grief counseling 450
grief therapy 455
grotesque images 279, 280
grounding 379
group cohesion 280
group debriefing 96, 170, 297, 303–6
group discussion 280, 287
group interventions 324
group living conditions 35
group processes 325
group risk assessment 188
group support networks 173
group therapy 468
group work 364
growth 60–3, 366
guard stations 79
guided imagery 267
guilt 91, 92, 149, 150, 194, 284, 286, 340, 360, 386
Gulf War 82
Gulf War syndrome 13
Guy, Keith 221

HAC, *see* humanitarian center
HAM-D, *see* Hamilton Depression score
Hamilton Depression score (HAM-D) 261
Han Dynasty 439
harassment 106, 418, 450
harmony guidelines 428
hatha yoga 287
hazardous chemical disasters 73
healing 78, 350–67
health and fitness programs 41
Health and Safety at Work Act 1974 336
Health and Safety Council 95
Health and Safety Executive (HSE) 157
health and safety teams 95
health professionals
  effects of psychological trauma on 87
  resistance to seeking help 99
health promotion 40–5, 61, 62, 63
health surveillance 31
health workers 55, 337
healthy behavior 147
heart-rate variability (HRV) 278
heavy industry 337
Help for Heroes 430
helplessness 150, 353, 355, 375, 481
help seeking
  and stigma 41, 184, 185
  barriers to 184–7
  for PTSD 186
  resistance among professionals 99
Herald of Free Enterprise ferry disaster 88, 167, 303
Herodotus 7
Hesnard, A. 9
HFA, *see* Hyogo Framework for Action
Hickley, Matthew 182
Hillsborough stadium disaster 88, 228, 302
hippocampus, sensitivity to stress 131
history taking 252, 253, 265, 267
Hofer, Johannes 7
holdups 337
Holmes, Geoff 307
Hong Kong Christian Service 441
hostage taking 200, 340
hostility 146
"hotspots" 284
hourglass model of psychotherapy research 223
house counseling team 353
housing 461
HPA, *see* hypothalamic pituitary adrenal axis
HRPs, *see* human resources professionals
HRV, *see* heart-rate variability
Human Givens Institute 215
humanitarian care 227
humanitarian center (HAC) 403, 404, 409, 410
humanitarian missions 418
human remains 403
human resources (HR) 95, 333, 339, 347
human resources professionals (HRPs) 482
human rights 108
human rights movement 12
humor 111, 376
Hurricane Ivan 232
Hurricane Katrina 228, 229, 232
Hyogo Framework for Action (HFA) 229
hyperarousal 377, 381, 419, 421
hyperlipidaemia 132
hypertension 131, 288
hypervigilance 129, 363, 421

hypothalamic pituitary adrenal (HPA) axis 125, 127, 282
hypothalamus 282
hysteria 9, 241–2

IAPT, *see* Improving Access to Psychological Therapies
ICAS 301
ICD-10 6
ICRC, *see* International Committee of the Red Cross
identification of bodies 173
identity alteration 251
identity confusion 251
IEDs, *see* improvised explosive devices
IES, *see* Impact of Event scores
IHS, *see* intermittent husband syndrome 429
imagery rehearsal 278
imaginable exposure therapy 116
imaginal exposure 51
imaginal reliving 51
immobility 373
immobilization 355, 360
immune activation 127
Impact of Event (IES) 261
  scale 209, 219
implanting 275
Improving Access to Psychological Therapies (IAPT) 469
improvised explosive devices (IEDs) 35, 318, 419
incident care team 251
incident response plan 351
incident site management 188
India 309, 310, 384–99
indirect victims 375–6
individual crisis intervention 48
individual debriefing 170
individual health 40–5
individual recovery 353
individual searches 79
individual therapy 288
induction processes 161
induction training 425
industrial accidents 227
influenza-like illness (ILI) 231–2
information 172, 201, 202, 235, 374, 397, 453
  about support 98
  and mitigation of adverse effects 31
  and recovery from trauma 361
  dysfunctionally stored 268, 269, 270
  for evidence-based practice 20–1
  fragmented 83
information networks 263
information-sharing models of early intervention 14
informed optimism 159, 163
inhibition 373
initial response 357–60
injuries 31, 231, 246, 372, 422–3
inoculation 274–90
insomnia 168, 398
installation 267, 269
Institute for Employment Studies 204
institutionalization 417
instrumental support 421
insubordination 10
insurgency 35, 419
integrated care model 211
integration 251, 252–3
integrative trauma therapy 378
intensive interventions 470
interactive regulation 363
interagency collaboration 351
intergenerational transmission 438
interhemispheric coherence 260
intermittent husband syndrome (IHS) 429
internal capacities 315–17, 323, 327
internal medicine 45
internal resources 62, 327
internal stigma 185, 186
International Center for Clinical Excellence (ICCE) 217, 224
International Committee of the Red Cross (ICRC) 53
International Employee Assistance Professionals Association (International EAPA) 298
international perspective 474
International Society for Traumatic Stress Studies (ISTSS) 170, 173
Internet 114, 235, 299
interpersonal conflict 316, 318
interpersonal networks 175–6
interpersonal psychotherapy (IPT) 173–7
interpersonal relationships 81
interpersonal skills 316, 324
interpreters 392
intervention
  behavioral 117
  cognitive 324
  distinguished from treatment 235
  environmental 117

intervention (*Continued*)
for large-scale trauma 30–45
group-based 148, 324
individual 148, 324
models of 144–51
post-event 112, 116, 183–4
post-trauma support 48–63
practice-based 324
primary 107, 110
secondary 107, 111
state-of-the-art 319, 323–7
tertiary 107
workplace violence strategies 107–112
see also early intervention
intimate partner violence 113
intrusive memories 177, 243, 244, 259, 347, 379, 419, 421
IPT, *see* interpersonal psychotherapy
IPT-Post Trauma (IPT-PT) 173–7
IPT-PT, *see* IPT-Post Trauma
IPT-Scotland 177
IPT-UK 177
Iraq War 37, 45, 427, 428, 429, 459, 460
irritability 77, 146, 363, 456
irritable heart 8
isolation 148, 420, 444, 461
ISTSS, *see* International Society for Traumatic Stress Studies
Italian Red Cross 87, 94, 99
Italian Society for Traumatic Stress Studies
Italy 99–100

job design 75
job dissatisfaction 74
job losses 72
job reassignments 72
job stress 74, 113
see also occupational stress
joining up 416–17
joint initiatives 469–70
Jones, Norman (Major) 181
journalists 182, 453
judgmental models 11
"juice" 35

Kanyakumari 387
Karolinska Institute 157
Kegworth plane crash 302
Kenyon International Emergency Services (KIES) 401
kernel statements 248
kindling 121, 126–9, 133
knowledge
and perception of control 36
as a resource 62
Korean War 12–13, 132, 426
Kubler-Ross, Elizabeth 393

labor–management conflicts 73
language
and trauma 356
inappropriate use of 160–1
problems 389
skills in 403
large-scale disasters 167
large-scale trauma 30–45
leadership 280, 318, 425
and healing 78
and meaning making 318
and recovery 82, 83
and resilience 317
and work motivation 78–9
attitudes of 186
development 324
responses to trauma 81
role of EBO in 236
training 11
learned resourcefulness 263
learning
organizational 97
transformational 159
learning-from-survivor model 7
legal requirements 335–6, 346
legal responsibilities 157, 380
legal services 41
legitimization of reactions 380
leisure time 35
liability 75–6, 131, 157
Life Experiences Survey 424
line managers, *see* managers
listening
active 388, 391, 393, 394
empathic 111
nonjudgmental 392
litigation 380
live fire exercises 418, 425
living conditions 318, 326
local responders 236
Lockerbie bombing 49, 167, 303
locus of control 360, 423–4
Loffler, Deborah 310
London Ambulance Service 182–3, 187
London Tube bombings 146, 148, 183, 250–2, 353, 455

London Underground 351, 356
long-range missiles 419
loss 385, 387, 393, 412
loss theory 393
Lutnick, Howard 71–2

Macy & Co. 298
major depressive disorder
  and childhood trauma 125
  and life events 126
  and PTSD 122
  and stress 125, 128
  duration of episodes 130
  etiology 125, 126, 128
  IPT-PT for 173–4
  neurobiology of 131
  recovery from 130
  tsunami survivors 387
  see also depression
malingerers 11
Management of Health and Safety at Work Regulations 1999 336
managers 333, 335, 338–41, 344–5;
  access to 114
  and deployment 408–10
  and psychological first aid 308
  and trauma management policies 339
  as staff-support champions 186
  behaviors that support rehabilitation 341–3
  of disaster teams 409–10
  role in incidents 207
  role in trauma support 82
  support for 162
  support from 211
  training 162, 408–10
  training of 145
man-made disasters 73, 83, 227
MAP, *see* mental assessment program
Marchioness pleasure boat disaster 49
Marine Corps 423
marital relationships 386
marital satisfaction 170
marriage enrichment programs 43
Marshall, S. L. A. 427
Maslow's hierarchy of needs 396
mass communication 155
mass disasters 402
mass emergencies (MEs) 227
mass graves 419
mass mobilization 7
mass trauma intervention 24
mastery 149, 150, 159, 267, 316
material resources 62
May 4th Movement 439
MBCT, *see* mindfulness-based cognitive therapy
McEntire, David 230
MDT, *see* multidisciplinary teams
meaning making 318, 366
media 172, 228, 235, 237, 405, 407, 451, 475–6, 481
  management 163
mediation 451
medical employment standard (MES) 418
medical model 8
medical training 80
medication 78, 148, 248, 425, 464
medics 240, 248, 422
meditation 277, 287, 290, 323, 445
meeting points 236
Memorial Hospital, New Orleans 232
memorial services 449
memorial sites 480
memories 128, 149, 170, 217, 252, 284, 419, 456
  and critical incidents 245, 247
  and EMDR 260
  engagement with 174
memory networks 259, 261, 262
Menezes, Jean Charles 155
mental assessment program (MAP) 469
mental disorders 182
mental health evaluation 78
mental health preparedness 482
mental health specialty care 45
mental skills 323
mercenaries 7
MEs, *see* mass emergencies
MES, *see* medical employment standard
Mexico 309, 310
military covenant 469
Military Mental Health Care 417–18
military operations other than war (MOOTW) 314, 317
military-political imperatives 7
military resilience 313–27
military stress 31–40
military training 417
Millennium Cohort Study 38
mindfulness 285–90, 480
  and individual therapy 288
  and the medial prefrontal cortex 363
  cultivation 287–90

mindfulness (*Continued*)
definition 288
practices 223
training 288, 289, 290
mindfulness-based cognitive therapy (MBCT) 288, 445
see also cognitive therapy
mindfulness-based fitness training (MMFT) 289
mindfulness-based stress reduction (MBSR) 287
Minerals and Management Services (MMS) 77
Ming Dynasty 439
mining disasters 72, 73, 76
miscommunication 236, 353
Mitchell, Jeffery 303
Mitchell model 58, 90
MMS, *see* Minerals and Management Services
mobile phone networks 357
model building 7–10
models
and diagnostic controversies 6
denial of care needs ("no-need" model) 5, 6
medical 8
moralist/judgmental 11
of acute care 8, 9, 14
of aftercare 8
of early intervention 3–15
personal conflict resolution 10
psychosocial 10
social network 10
mood swings 363
MOOTW, *see* military operations other than war
morale 7, 12, 80, 95, 170, 317, 326–7, 425
moralist models 11
mortality exposure 75
mortality triggers 75
mortars 419
mortuary employees 157
mortuary operations 404
motivation
following trauma 78–9, 83, 444
in combat situations 315, 316
in mental health services 90
in Vietnam War 12
of staff in challenging conditions 95
motivational resilience 233–4
motor control 250
motor incidents
Multidimensional Inventory of Dissociation (MID) 251
multidisciplinary teams (MDT) 463
multimedia 155
multiple traumas 457, 461
Munch, Edward 3
murder 72, 106, 140
Murrison, Andrew 469
Muttum village school 385
Myers–Briggs Inventory 413

Nagin, Ray 232
naïve optimism 159
name calling 106
NAPO, *see* National Probation Officers Association
narrative 61, 158, 159, 171, 364
National Centre for Counterterrorism 476
National Comorbidity Study 131
National Emergency Response Teams for Public Health 445
National Fire Protection Association (NFPA) 230
National Institute for Clinical Excellence (NICE)
and claims of clinical effectiveness 14
guidelines on depression 288
guidelines on EMDR 201, 213
guidelines on post-traumatic stress disorder 25, 95, 96, 170, 182, 187, 305, 337–8, 346, 462, 479
guidelines on psychological debriefing 24, 28, 48, 54, 57, 480
guidelines on TF-CBT 201, 213
guidelines on trauma management 24, 25, 201, 202, 203
guidelines on TRiM 187
guidelines on working conditions 182, 187
hierarchical model in guidelines 225
limitations of guidance 26
sources of evidence 24
National Institute of Mental Health 173
National Probation Officers Association (NAPO) 459
National Specialist Commissioning (NSC) 465
National Vietnam Readjustment Veterans Study (NVVRS) 459
Nationmaster 139
NATO 12
natural disasters 73, 83, 227, 437, 443

néant, le 372
needs analysis 111
negative affect 268
negative resilience 144
neglect
  child 87, 99, 244, 282, 461
  emotional 244
negligence 475
neocortex 364
Netherlands 325, 326
Network Operations Centre (NOC) 356
neural plasticity 364
neurobiological regulator 364
neurobiology
  and fitness 316–17
  of stress inoculation training 281–2
  of stress response 290
neuro-endocrine stress response 127
neuroimaging 262
neurolinguistic programming (NLP) 214
neuropeptide Y 282
neuroplasticity 278
neuroscience 223, 277, 480
New York Presbyterian Hospital 78
NFPA, *see* National Fire Protection Association
NHS/Armed Forces Regional Networks 470
nightmares 93, 149, 262, 419, 421
  see also dreams
NLP, *see* neurolinguistic programming
nonbureaucratic structure 81
nondirective therapy 26
no-need model, *see* denial of care needs model
nongovernmental organizations 390
non-hierarchical structure 81
nonjudgmental listening 392
nonphysical aggression 105, 106, 117
nonprofessional counselors 144
normalization 55, 96, 169, 175, 190, 276, 364, 380, 392
normalizing activities of daily living 464
Northern Ireland 12, 130, 215, 461, 476–7, 480, 481
Northern State Power Company 298
nostalgia 7
NSC, *see* National Specialist Commissioning
nuclear industry 337
nuclear power plants 73
nuclear submarines 73
numbing 77
nurturing 377
obesity 39
obsessive-compulsive disorder 128
occupational alcohol programs 298
occupational health (OH) 95, 333, 341, 347
  advisors 343
  physicians 200
occupational psychological health 442
occupationally related psychological disorders 30–1
occupational stress 32, 34–5, 37, 38, 74, 87, 113, 157, 158, 182
occupational therapy 10, 464
offshore oil/gas industry employees 337
OH, *see* occupational health
oil rig disasters 72, 73, 76, 77
Oklahoma City bombing 73, 79
old-age program 471
Omagh bombing 129
one-child policy 441
on-scene support 377, 379
on-site accidents 481
on-site training 78
open expression 368
open-door policy 342, 438
operational stress 34–5, 38
operational support 326
Op Herrick 427
Op Telec 427
optimism 41, 159, 163, 316
organizational adversity 162–3
organizational barriers 83
organizational causes of catastrophes 73
organizational challenges in traumatic events 78–9
organizational culture 83, 97, 318
organizational development (OD) 78, 160, 442
organizational diagnosis 80
organizational dissociation 353
organizational downsizing 72
organizational health 31, 40–5, 83
organizational learning 97
organizational liability 75–6
organizational nervous system response 354–7
organizational nurturing 449
organizational preparedness 454–5
organizational recovery 353–4
organizational resilience 33, 183, 187, 196, 361
organizational responsibility 75–6
organizational restructuring 72

organizational trauma 72–3, 76–8, 88–9
organizational violence 91
organization-specific guidance 26–7
orientation 235
orienting response 260
ORS, *see* Outcome Rating Scale
orthopedic problems 462
Outcome Rating Scale (ORS) 218, 219
out-of-body experience 168, 373
outreach 465, 466, 470
over-involvement 399
Oxfam 397

Paddington rail crash 167, 304
Page v. Smith 129
pain 430
  chronic 262, 287, 423, 430, 462, 463
  sensitivity 373
panic attacks 421
paralysis 243, 250
paramedics 240, 248
parasympathetic nervous system (PNS) 260, 278, 354
partial remission 130
participant selection 223
pathogenesis 63
patient aggression 91, 92
Patient Health Questionnaire (PHQ-9) 176
patriotism 79
Pavlov, Ivan 439
Pavlovian principle 359
PBC, *see* public behavior consultant
PDAs, *see* persons directly affected
peacekeeping 279–81, 313–14, 418
peer group champions 224
peer supervision 347
peer support 48–55, 58, 59, 146, 201, 319, 324, 351, 377
peri-traumatic dissociation 241, 249–50, 377
Peritraumatic Dissociative Experiences Questionnaire (PDEQ) 241
peri-traumatic prevention, *see* primary prevention
Perseus 372
personal adversity 160–1
personal attacks 159
personal conflict resolution model 10
personal growth 62
personal impact 358–60
personality 421, 422–3
  tests 413
personal protective equipment (PPE) 41
person-centered counseling 391
personification 253
personnel selection 413
persons directly affected (PDAs) 403
PFA, *see* psychological first aid
phase-oriented treatment 252
phasic rehabilitation 463–4, 467
phobias 213, 214, 248, 275
physical abuse 244, 245, 247
physical aggression 92, 200, 337
physical barriers 79
physical fitness 316, 317
physical recuperative talk sessions 377
physiological control strategies 36
PIE, *see* proximity, immediacy, and expectancy
Pilger, John 476
Piper Alpha oil rig disaster 88, 167, 303
PNS, *see* parasympathetic nervous system
Poland 309, 310
police 53, 55, 72, 129, 133, 185, 187, 195, 240, 246, 314, 380, 482
police desk surveys 460
Police Service of Northern Ireland (PSNI) 54
policies 339–43
  bullying 450
  for traumatic events 447
  in healthcare organizations 97
  managers' familiarity with 340
  on workplace violence 108
  post-trauma 337, 347, 339–40
political correctness 5
population expansion 229, 237
Portugal 309, 310
positive memory templates 260
positive psychology 41, 42, 60
positive reframing 323
positive reinforcement 413
post-deployment PTSD 34
post-fact collapse 376
post-trauma life stress 220
post-trauma policies 337, 347, 339–40
post-trauma recovery 350–67
post-trauma response 154–64
post-trauma support 48–63, 145, 458–71, 474–84
post-traumatic growth 60–3, 366
post-traumatic stress (PTS) 96, 154, 156
post-traumatic stress disorder (PTSD)
  acceptance as psychiatric disorder 302
  among veterans 38, 127, 459, 462
  and abuse 127

and alcohol 427
and allostatic load 133
and ASRs 282
and cardiovascular disease 131
and CBT 51, 116, 346, 479
and CISD 78, 183
and CISM 52–3
and combat 13, 36–7, 125, 462
and coronary arterial disease 132
and coronary heart disease 132
and counseling 96
and CSR 421, 424
and culture 427
and deployment 38
and depression 132, 459, 461, 468
and dissociation 245
and early intervention 13, 375
and E-EIM 174
and EMDR 25, 261, 277, 346
and family cohesion 319
and hypertension 131
and London bombings 353
and major depressive disorder 122
and missiles 419
and model building 6
and morale 317
and patient aggression 91
and peri-traumatic dissociation 249
and previous mental disorders 184
and primary structural dissociation 244
and psycho-education 464
and psychological debriefing 18, 22, 52–3, 60, 170, 178, 369, 380
and psychotherapy 78
and reactivity to stress 129
and residual stress 34
and self-critical thinking 286
and somatic symptoms 128, 277
and SPoT 27
and TF-CBT 25
and the ASSIST model 97
and the Millennium Cohort Study 38
and training 424
and triggers 128, 131
and TRiM 27, 188, 195–6
and unemployment 38
and workplace trauma 77, 335
as a dissociative disorder 253
as an adjustment disorder 6
as an anxiety disorder 6
as a treatable condition 38, 39
characteristics 77, 90, 259, 394–5
chronic presentations 462
comorbidity 459, 461
delayed onset 121–6, 130, 132, 133, 428
delay in receiving treatment 129–31
development of 259
diagnosis 126, 277
etiology 125, 130, 146
general guidance 25
health care utilization 124
help-seeking for 186
in emergency services 92
in firefighters 76
in peacekeeping troops 280
in spouses 429
intergeneration transmission 438
in train operators 75
in tsunami survivors 384, 387
in veterans 38, 121
legal obligations 133
line managers' understanding of 340
neurobiology of 131
NICE guidelines on 25, 95, 96, 170, 182, 187, 305, 337–8, 346, 462, 479
post-deployment 34
pre-deployment 38
predictors 220, 355
prevalence 39, 183, 240, 428, 443, 477–8
prevention 14, 123, 145, 313
recent onset 170
rehabilitation 467–9
remission 39, 168
role of traumatic events in 124
screening for 38
subsyndromal 123–4, 127
treatment 78, 129, 277, 284, 416–31
potentially traumatic events (PTEs) 182–3, 371–3
poverty 141, 142, 484
PPE, *see* personal protective equipment
practical support 201, 202, 351
practice-based evidence 221–3
practice-based interventions 324
practice-based knowledge 215
practice-based studies 223
practice research networks (PRNs) 213–24, 480
practitioner expertise 20, 22
Pragmatic Research Network 224
pre-briefing 452–3
pre-combat stressors 421
pre-crisis education 53

pre-deployment education and training programs 43
pre-deployment PTSD 38
premature voluntary retirement (PvR) 430
preparedness 454–5, 475
Pre-Qin Dynasty 439
presentification 253
pre-service difficulties 462
pre-trauma training 284
prevention
  and early intervention 31
  of disability 129
  of PTSD 14, 123, 145, 313
  of stress 40
  of suicide 42, 43
  of trauma 30–45, 83
  of workplace violence 108–10
  primary 39, 375, 377, 378–80
  secondary 42–3, 129, 377, 378
  strength based 42–3
  tertiary 377, 378
  training for 36, 110
preventive health orientation 62
preventive stress management 30
primary care 44–5
primary dissociation 251
primary exposure 368
primary interventions 107, 110, 114
primary prevention 39, 375, 377–80
primary structural dissociation 244, 253
primary traumatization 248
primary victims 376, 377, 378–9
Princess Diana 475
prioritization 83
problem solving 41, 280, 392, 413, 468
process debriefing 115
product champions 14
production setbacks 107
professional development skills 450
protection 31, 93–4, 147, 172, 175, 337
proximity, immediacy, and expectancy (PIE) 12
proximity, immediacy, expectancy and brevity (PIE-B) 426
pseudo-epileptic seizures 250
PSNI, *see* Police Service of Northern Ireland
PS-PREP, *see* Voluntary Private Sector Preparedness Accreditation and Certification Program law
psychodynamic approaches 464
psycho-education 14, 175, 188, 368, 463–4, 468
psychoform dissociation 250
psychological debriefing 97, 168–72
  aims 52–3
  and anxiety 27
  and CISD 48, 170
  and depression 27
  and feedback 58
  and military deployment 426–7
  and NICE guidelines 24, 28, 48, 54, 57, 480
  and psychological uncoupling 376
  and PTSD 18, 22, 27, 52–3, 60, 170, 178, 369, 380
  and resilience 323
  and retraumatization 48, 51–2
  and workplace violence 115
  appropriate use of 378
  as counseling 52
  as harmful 60, 170
  as therapy 52
  compulsory 53–4
  continued use of 63
  criticisms of 60
  definition 480
  difficult situations during 58
  discontinuation 54–5
  discussion in 58
  effectiveness 18, 21–3, 178, 182, 201, 204, 368, 369–71
  expectations of 380
  follow-up in 59
  for high-risk professions 379–80
  group 323
  in CISM 53
  origins 52
  role of EBC 237
  single-session 170
  stages of 169
  three-stage technique 56
  training in 56, 57–60
  value 167
  see also Critical Incident Stress Debriefing
psychological decompression 427
psychological disaggregation 241
psychological distance 398, 478
psychological first aid (PFA) 14, 44, 88, 96, 97, 172–3, 178, 283, 297
  and emotional triage 376
  for families of victims 308
  for managers 308
  in group settings 116
  in workplace environments 115–17, 480

nine-step model 172
recipients 308
TRiM as 187
value 167
psychological hardiness 323
psychological health services 31
psychological support 88, 172
psychological uncoupling (PU) 376, 379, 380
psychological well-being (PWB) 49, 61–3, 88
Psychological Well-being Post-traumatic Changes Questionnaire (PWB-PTCO) 62, 63
psychophysiological arousal 260
psychosocial education 266
psychosocial matrix 368, 376–8
psychosocial models 10, 11
psychosocial processes 9
psychosocial risks 162
psychosomatic disorders 11
psychosomatic reactions 8
psychotherapy
and PTSD 78
and residual stress 34
during combat 34
for factors complicating recovery 148
interpersonal 173, 174
in trauma support 147
narratives as key ingredient 171
psychotic disorders 416
PU, *see* psychological uncoupling
public behavior 231–2, 237
public behavior consultant (PBC) 234
public behavior evaluation (PBE) 235–7
PWB, *see* psychological well-being
PWB-PTCO, *see* Psychological Well-being Post-traumatic Changes Questionnaire

qi 436
Qing Dynasty 439
quality of life 186, 334

RA, *see* Reuters America
radiation 227, 231
rage factors 236
random control trials, *see* randomized control trials
randomized control trials (RCTs) 50, 216–20, 370, 480
rape 139, 168
Raphael, Beverly 172, 173
Raphael model 115
rapid eye movement sleep 260
rapid response medical teams 422
RCN, *see* Royal College of Nursing
RCTs, *see* randomized control trials
reaction times 405
realistic training 424–5
real-time responses 230
recent-onset PTSD 470
reconstruction 380
recovery 95, 148, 174, 363, 366, 373
and leadership 82, 83
and organizational culture 83
and resilience 314
and work 334
expectation of 187
factors leading to 81, 82
from adversity 316
from CSR 421
from major depression130
individual 353
organizational 353–4
post-trauma 82, 173, 174, 177, 350–67
recovery centers 430
recovery model 463
recruit training 320, 423–4
recuperation 368, 373, 379–80
see also recovery
Red Cross 87, 94, 99, 377
Italian 87, 94, 99
Reddy, Michael 301–2
redundancy 81, 299, 300, 430
re-experiencing 77, 90
referral 150, 172, 203, 413
reflection 75, 252
refugees 260
regional welfare officers 463–6
rehabilitation 31, 126, 333–48, 430
and CBT 338
and counseling 338
and management behaviors 341–3
and PTSD 467–9
effects of psychological debriefing on 23
in Combat Stress treatment model 464
in rolling treatment program 464
phasic programs 467
plans 338, 343–4, 339
rehearsal 80, 356
reintegration 31, 464
relapse 130

relationships
  after disasters 386
  and ongoing planning process 351
  and post-traumatic growth 61
  and resilience 361, 364
  breakdown of 146
  international 306
  interpersonal 81
  strengthening 237
  with employees 81
relaxation 9, 26, 43, 213, 277, 288, 289
religious faith 159
remission 39, 130, 168
Repatriation Medical Authority 125
repeated exposure 121–33
residential treatment 464–5, 468, 470, 463
residual stress 33–4
resilience 80–3, 158–9
  and adversity 161
  and attachment 360
  and collaboration 361
  and culture 158
  and early life security 360
  and education 306
  and EMDR 259, 263–4, 265
  and external resources 317
  and families 319
  and fitness 317
  and interpersonal conflicts 316
  and mastery 316
  and optimism 159
  and peacekeeping 313–14
  and post-traumatic growth 61
  and psychological debriefing 323
  and psychological first aid 306–8
  and recovery 82, 314
  and relationships 361, 364
  and stress 263
  and stress-modulating hormones 282
  and support 361
  and sustainability 314–15
  and the ASSIST model 98
  and TRiM 187, 196
  and vulnerability 360
  and well-being 315
  assessment for 89
  community 238
  creation of 161, 352
  definition 80, 263, 314–15
  factors associated with 158
  in mental health services 89
  increased by traumatic events 62
  knowledge of 234
  military 313–27
  motivational 233–4
  negative 144
  organizational 33, 183, 187, 196, 361
  programs 42–3, 235–5
  promotion of 49
  research 315
  role of EBC in fostering 236
  sources of 81
  stress-management approach 323
  training for 301, 313–27
Resilience XI Program 325–6
resistors 185, 186
resources
  and morale 326
  categories of 62–3
  coping 78
  environmental 324–7
  external 315, 317–19, 327
  financial 81
  for trainers 399
  internal 62, 327
  of EAPs 306
  redundant 80
respect 326
respite care 460, 471
restructuring 72, 74, 299
retirement 334, 343
Retour des Enfers et son Message, Le 372
retraumatization 48, 189, 190
return and reunion programs 43
return to work plan 344, 346
return to work schedule 338
reunions 173, 175
Reuters America (RA) 81
rewind technique
  and CBT 221
  description 213–14
  evaluation 215, 216
  evolution 214–15
"rewind-the-tape" analogy 190
ripple effects 234
risk assessment 56, 59, 108, 110, 188, 336–7
risk communication 236
risk-taking behavior 35, 429
RMG, *see* Royal Mail Group
road traffic accidents (RTAs) 221, 479, 481
robberies 107, 142, 345
Roberts, Peter (Capt.) 181
role ambiguity 93

role playing 58, 279, 323
role stressors 107
role tensions 90
role transition 176
rolling treatment program 464–5
Roosevelt, Franklin Delano 484
Royal Air Force 416, 417
Royal College of Nursing (RCN) 88
Royal Mail Group
  counseling offered by 207
  organizational changes 201
  support for work-related trauma 199–212
  use of SPoT in 27, 97, 202–3
Royal Marines
  compulsory psychological debriefing in 54
  TRiM in 56, 426
Royal Navy 26–7, 187, 202, 203, 416, 417
RTAs, *see* road traffic accidents
rules of engagement 13, 279, 315
rumination 61, 146, 376
rumors 172, 188, 236
  see also gossip

Salkovskis, Paul 223
Salmon, T. W. 9–10
Salmon principles 12
SAM, *see* situationally accessed memory149
Sampurna Montford College 384, 387
Sandman's formula 236
sati 287
Saville Report 475
SAV-T model 108, 109, 111, 112
scapegoating 159
Scream, The 3
screening
  for mental health problems 89
  for PTSD 38
  of employees with violent tendencies 108
  of soldiers 12
  of vulnerable people 183
  pre-combat 423
secondary dissociation 251
secondary exposure 368
secondary functional deficits 461
secondary interventions 111, 107
secondary prevention 42–3, 129, 377, 378
secondary structural dissociation 244–5
secondary traumatization 92, 232, 482
secondary victims 376, 377, 378–9
secure healthcare settings 87–100
security service workers 337
selection
  bias 216
  of staff 337
  of soldiers 12
self-acceptance 62
self-blame 91
self-care 252, 365, 413
self-compassion 283–6
  neural mechanism for 284
  training in 284, 290
Self-compassion Scale 284
self-control 13, 481
self-criticism 285, 286
self-destructive behavior 172
self-efficacy 41, 78, 147, 148, 236
self-esteem 159, 316, 334
self-harm 43, 249
self-help material 42, 202
self-management 288, 289
sense of coherence (SOC) 263
sensitization 121, 126–9, 133, 155
sensorimotor therapies 223
sensorimotor training 278
serial traumatization 264
"services no longer required" (SNLR) 430
sexual abuse 93, 213, 222, 244, 245, 246, 247, 373
sexual violence 140
shame 91, 190, 284, 286
Shanghai fire disaster 443
shelling 425
shell shock 8–9, 168
shooting incidents 72, 142, 482
sick leave 95, 148, 334
sickness absence 95, 334, 336, 337–9, 343–4
  see also absenteeism
sick role 148
SID, *see* sudden infant death
sight barriers 109
significant others 371–3, 377
"sin eaters" 93
single-session debriefing 170, 189, 201, 304
Sino-German Senior Psychotherapy Continuous Training Program 440
SIT, *see* stress inoculation training
situationally accessed memory (SAM) 149
skill-based interventions 323
skills
  acquisition 277
  affective regulation 280
  and mitigation of adverse effects 31

skills (*Continued*)
behavioral 280, 323
cognitive 280
communication 316, 340, 405
coping 280, 478
core 413
interpersonal 316, 324
mental 323
of trainers 410–11
problem solving 280
requirements 404
symptom management 468
skills training 252, 280, 464, 468
sleep problems 259, 363, 444, 456
SNS, *see* sympathetic nervous system 278
SOC, *see* sense of coherence
social care workers 55
social exclusion 461
socialization 155, 159
social media 42
social networks 10, 158, 423
social sharing 368
social support 8, 14, 25, 32, 34–6, 42, 49, 61, 78, 94, 146–50, 174, 183, 186–7, 201–2, 220, 248, 324, 437, 481
social support networks 14, 61
social withdrawal 421, 461
social workers 248
soldiers
and stress 313
dilemmas of 7
dissociation in 242
early intervention for 12
from ethnic minorities 12
high incidence of trauma in 240
internal capacities 315–17, 327
morale 7
nostalgia 7
psychological trauma 181
resilience 313–27
role of personality in dysfunction 10
screening of 12
secondary structural dissociation in 244
self-control problems 13
UN peacekeeping operations 279–81
vulnerability to CSRs 9
welfare of 8
see also military personnel
"soldier's heart" 8
Solomon Browne lifeboat 193
solution-focused interviews 345–6
somatic symptoms 437
and PTSD 128, 277
and stress 128
somatoform dissociation 250
Somatoform Dissociation Questionnaire – Peritraumatic (SDQ-P) 250
somatoform dissociative phenomena 250
South Africa 139–44
increased use of trauma management in 74
Soviet psychology 439
Spain 309–11
special operations 40
specialty clinics 39
spiritual counseling 34
spirituality 277
spontaneity, and recovery 81
SPoT, *see* Support Post Trauma
spousal abuse 185; *see also* domestic abuse
St. Andrew's Healthcare 88, 94–8, 100
stabilization 252, 267, 368, 419, 463
staff selection 337
staff support 89–90, 186
staff turnover 75, 95, 146
stages of trauma 374–5
stalking 107
startle response 129, 259, 456
status reports 232, 235
"staying with" 288
stigma
and clinical presentation 131, 431
and help seeking 41, 184, 185, 428, 462
and mental health care 39, 44–5, 222
and mental health terms 411
and outreach services 466
and the TRiM process 191
and traumatic events 456
external 185
internal 185, 186
reduction 352
strength-based intervention 319, 323
strength-based prevention programs 42–3
strengths
environmental 63
genetic 62–3
stress
acute 33
and adversity 161
and combat 6–13, 31–40, 44, 313, 421
and COR theory 75
and CSR 423
and deployment 34–5, 283
and education 36
and emotional engagement 315

and major depression 128
and mindfulness training 289
and organizational interventions 352
and peacekeeping 313
and psychopathology 132
and PTSD 129
and resilience 263
and resources 62–3
and somatic symptoms 128
and support 35, 36
and team cohesion 41
and the caring professions 87
and the work environment 143
background 247
buffers 36
chronic 33, 278
cumulative 33
definition 156
effects 33–40
feedback on levels of 236
from debriefing 178
group discussion of 280
in soldiers 313
life 174
management of 41, 236
manifestations in the workplace 474
military 31–40
monitoring 236
neurobiology of 281–2, 290
occupational 32, 34–5, 37, 38, 74, 87, 113, 157, 158, 182
post-traumatic 96, 154, 156
presentations about 287
prevention measures 40
protective factors 32, 40, 93–4
residual 33–4
sensitivity of hippocampus to 131
social 146
terminology associated with 31–2
traumatic 31–4-, 121–33, 177, 484
underlying mechanisms 156
see also Critical Incident Stress Management; post-traumatic stress disorder
stress inoculation training (SIT) 274–83, 290, 425
application and follow-through phase 278
as psychotherapy 275
conceptual education phase 276–7
evaluation 281
mindfulness as alternative 287
neurobiology 281–2
skills acquisition/consolidation 277–8
uses 279–81
stress management 157
see also Critical Incident Stress Management; preventive stress management
stress-modulating hormones 282
stressors
arising from roles 107
chronic 314
low level 33
operational 316
services to assist with 41
successful treatment with SIT 279
stress reactions 278
stress-reduction programs 364
stress-related behaviors 232
stress-related injuries 231
stress-related withdrawal factors 75
structural dissociation 241–53, 373
see also dissociation
structural violence 140
structured discussion 60
structured equation modeling 209
Subjective Units of Disturbance (SUD) scale 266, 268
subjective well-being (SWB) 62
substance abuse 93, 387, 428, 481
see also alcohol; drug abuse; substance misuse
substance misuse 459, 462, 468, 470
see also alcohol; drug abuse; substance abuse
substitute actions 248
subsyndromal PTSD 123–4, 127
SUD, *see* Subjective Units of Disturbance scale
sudden infant death (SID) 246
suicide bombers 419
suicides
among law enforcement personnel 185
among servicemen and women 429
compensations to families 74
data collection 43
of employees 73, 74, 80, 444, 480, 481–2
under trains 75, 302, 351, 352, 356
prevention 42, 43
prevalence 429, 459
Sung Dynasty 439
support
and depression 116
and EAPs 301–3
and fear-reduction models 6
and military deployment 35
and PTSD 375
and residual stress 34

support (*Continued*)
and resilience 361
and social network models of early intervention 10
and stress 35, 36
as a protective factor 32, 49
during combat 34
effectiveness of 210–11
emotional 94, 201, 202, 380, 397
evidence-based 199–212
family 35, 41, 42, 43, 146–7, 318–19, 351, 401
financial 397
for firefighters 380–1
for healthcare professionals 89
for managers 162
for professionals 87
for rescue-services personnel 380–1
for staff in secure mental health services 88–90
from house counseling team 353
from managers 59, 211
from peers 48–55, 58, 59, 146, 201, 319, 324, 351, 377
from social media 42
group-level 252
impact on absence 209–10
impact on symptoms 209–10
in CISM 56
in crisis situations 376–8
in the workplace 60
in TRiM 56
information about 98
instrumental 421
models 144–51
on-scene 377, 379
operational 326
organizational 170, 252
policies 337
post-trauma 48–63, 145, 458–71, 474–84
practical 201, 202, 351
psychological 88, 172
safety of 210–11
satisfaction with 208–9
social 8, 14, 25, 32, 34–6, 42, 49, 61, 78, 94, 146–50, 174, 183, 186–7, 201–2, 220, 248, 324, 437, 481
staff 89–90, 186
team 404
to organizations 365
trauma 146–51
types of 199
within Royal Mail Group 207
support groups 466–7
support networks 6
Support Post Trauma (SPoT)
and psychological first aid 14
and PTSD 27
in the Royal Mail Group 27, 97, 202–3
training for practitioners 27
and EMDR 202, 203
and TF-CBT 202, 203
in workplace environments 480
meetings 203, 207–8, 211
support programs 59
support teams 351
surveillance cameras 109
surveillance equipment 79
surveillance initiatives 470
surveillance of psychological disorders 31
survivor guilt 386
survivors
acute care of 14
and emotional trauma 78
effects of terrorist attacks on 72
needs of 14
of South India tsunami 384–99
practical support for 351
sustainability 314–15
SWB, *see* subjective well-being
swearing 111
sympathetic nervous system (SNS) 260, 278, 354
symptoms
after acute traumatic events 168
and incident type 209
and psychological debriefing 18
education about 267
impact of support on 209–10
management 14, 187, 252, 468
normalization of 150
of post-traumatic stress 77, 90, 259, 394–5
specific interventions for 25–6
system-wide interventions 78

table-top exercises 409
talking therapies 96
Tamil Nadu 385
Tang Dynasty 439
Taoism 445
tardiness 75, 107
target hardening 109, 110

taunts 281
Tavistock and Portman NHS Foundation Trust 470
team building 451
team cohesion 41
team development 405
team support 404
team welfare 410–13
tearfulness 456
technical causes of catastrophes 73
technical recovery 81
technological disaster 73, 227
technology 45
teleconferencing 81
telehealth 45
telephone counselors 93, 305
teleportation of emotion 155
terminology 31–2, 63, 411–12
terrorism 73, 231, 227, 229, 233, 476, 478, 480, 482, 484
  attacks 45, 76, 476
  in Northern Ireland 130
  research on effects 72
  threat of 484
tertiary dissociation 245
tertiary exposure 368
tertiary interventions 107
tertiary prevention 377, 378
tertiary victims 376, 377, 378, 379–80
testimonies 478
thalmo-amygdala path 354
thalmo-cortical system 354
Tharmaraj, Leon 387, 388
theft 200
theoretical understanding 389
therapeutic alliance 220, 224, 266
therapists 218
threat 111, 113, 185, 302, 337
Three Mile Island 80
three-stage technique 55, 56
Tiananmen Square massacre 302
tonic immobility 244, 276
Toronto 75
total inhibition 373
Toyota 83
training
  advance 356
  and emergency action plans 79, 354
  and mitigation of adverse effects 31
  and perception of control 36
  and prevention of trauma 36
  and PTSD 424
  and resilience 318
  and workplace violence 110
  anxiety management 116
  behavior-based 323
  by the CTRG 58
  classroom 40
  cognitive (knowledge-based) approach 319
  collaborative 324
  comprehensive programs 43
  computer-based 323
  consequence focused 110
  content 402–5, 414
  cost of 95
  design of program 399
  employee focused 117
  for coping 408
  for deployment to disasters 401–14
  for front-line personnel 405–6
  for managers 408–10
  for operational deployments 34
  for personnel selection 413
  for resilience 301, 313–27
  for self-compassion 284
  for SPoT practitioners 27
  for TRiM practitioners 26, 188
  in active listening 388
  in awareness and response 145
  in behavioral relaxation 43, 277
  in CISM 57–60
  in cognitive flexibility 43
  in defusing skills 162
  in EAPs 113
  in India 401–14
  in IPT-PT 177
  in leadership 11
  in mindfulness 288, 289, 290
  in psychological debriefing 56, 57–60
  in public relations 405
  in risk reduction 343
  in self-compassion 284, 290
  in strength-based skills 42–3
  in trauma support 145
  in US Marine Corps 423
  in value-centered decision making 43
  in violence risk 111
  induction 425
  mandatory 404
  medical 80
  methodologies 389
  military 40, 281, 416, 417
  of emergency personnel 401, 402

training (*Continued*)
of management 145, 162, 340–1
of peer supporters 59
of personnel for deployment 408
of recruits 320, 423–4
of therapists 288
of TRiM practitioners 188
of volunteers 144
on outcome measurement 224
on-site 78
positive effects on health 40
practice-based approach 319, 323
pre-deployment programs 43
pre-trauma 284
prevention focused 36, 110
qualities of 398
realistic 424–5
resources 389, 399
scenarios as exercises in 405–7
selection of trainees 402, 414
sensorimotor 278
skills required by 410–11
standards of 337, 414
trauma preparedness 144
see also skills training; stress inoculation training
train operators 75, 301, 354, 355, 447, 448, 453
transactional analysis 391
transformational learning 159
Transport for London (TfL) 350–1, 353, 355, 362, 365, 480
trauma advisors 94, 97
trauma awareness training program 100
trauma centers 87, 99
trauma fixation 374–5
trauma-focused cognitive behavioral therapy (TF-CBT) 25, 28, 344, 468, 479, 480
and SPoT 202
NICE guidelines on 201, 213
referral for 203
trauma impact 71–83
trauma inoculation 274–90
trauma labyrinth 373
trauma management
and critical incident stress debriefing 17
and psychological debriefing 17
for organizations 17–28
NICE guidelines on 24, 25
policies 339–40
trauma network 351
trauma pathway model 391
trauma preparedness training 144, 145
trauma processing 374
trauma responses 74–5, 475
of Chinese people 442–4
Trauma Risk Management (TRiM) 97, 181–96
and CISM 56
and organizational resilience 187
and potential psychological injuries 188
and psychological first aid 14
and PTSD 27, 188, 195–6
and resilience 187, 196
and stigma 191
as psychological first aid 187
assessment in 56
compulsory psychological debriefing in 54
definition 426
development 55–6
effectiveness 60, 195–6
follow-up 56, 190
implementation 188–91
in the Royal Navy 26–7, 187, 202, 203
in UK Army 319
in workplace environments 480
interviews 188, 190, 191
NICE guidelines on 187
origins 187
planning in 56
practitioner training 26, 188
real-life experiences 191–5
support in 56
training for practitioners 26, 27
use by BBC 187
use by Foreign and Commonwealth Office 187
use by London Ambulance Service 187
use by police 187, 195
use by Royal Marines 56, 187, 426
trauma support models 94–100, 144–51, 477
traumatic injury 124
traumatic stress 32, 121–33, 177, 484
traumatic tunnel 379
trauma time 243–4
traumatization 371, 373
traumatized organizations 80–2, 350–67, 447–57
trauma trance 379
TRC, *see* Truth and Reconciliation Commission
treatment

and large-scale trauma 30–45
delay 129–31
distinguished from intervention 235
plans 253
trench warfare 8
triage 307, 376, 425
triggers 75, 98, 122, 128, 131, 247, 259, 260, 262, 359, 381
TRiM, *see* Trauma Risk Management
triple comorbidity 125
troops, *see* soldiers
Truth and Reconciliation Commission (TRC) 140
tsunamis 227, 307, 477
survivors 384–99, 401–14
tunnel vision 373
Twin Towers attacks *see* 9/11 attacks
type I workplace violence 106–7, 109
type II workplace violence 107, 109
type III workplace violence 107

UK-EAPA, *see* UK- Employee Assistance Professionals Association
UK-Employee Assistance Professionals Association (UK-EAPA) 298
uncoupling 379–80
unemployment 141, 459, 461
among veterans 38
and PTSD 38
and quality of life 334
unfinished business 386
UNHCR, *see* United Nations High Commissioner for Refugees
Union Carbide, Bhopal 73, 76, 80, 83
United Kingdom Central Council (UKCC) 89
United Nations High Commissioner for Refugees (UNHCR) 53
United Nations peacekeeping troops 279–81
United Nations soldiers' stress syndrome 13
unsafe work behaviors 75
Utoeya shootings 63, 483

vagosympathetic regulation 277
Validity of Cognition (VOC) scale 266, 268
value-centered decision making 43
values 387
as a resource 62
VAM, *see* verbally accessed memory 149
ventilation 173, 380
verbal abuse 87, 94, 106, 337
verbal aggression 92, 200, 337
verbally accessed memory (VAM) 149
Vet Centers 45
veterans
definition 458
from the Gulf War 125
from the Korean War 132
from the Vietnam War 127, 131, 132
in the criminal justice system 459–60
mental health in 458–71
outreach programs 43
population studies 459
PTSD in 38, 121
research on return home 72
transition programs 43
unemployment among 38
use of rewind technique 215
Veterans Association 459
Veterans in Prison Association (VIPA) 460
veterans' priority 469
vicarious traumatisation 92, 93, 248
victims 371–3
categories 478–9
definition 479
indirect 375–6
primary 376–9
secondary 376–9
tertiary 376–80
Vietnam veterans 127, 131, 132
Vietnam War 12, 13, 127, 131, 132, 425, 429, 467
violence
and crime 141, 248
and previous military action 141
cognitive effects 91
cross-over 142
during armed raids 302
effects 91
in mental health services 89
in society 139–44
in South African townships 140
in the workplace 90, 105–17, 142–3, 200
indirect exposure 92–4
intimate partner 113
of combat 36
sexual 140
social dynamics of 139
social effects of 92
structural 140
threat of 35
types 91
violence risk training 111

VIPA, *see* Veterans in Prison Association
visceral system 437
visualization exercises 278
visual kinesthetic dissociation 214
VOC, *see* Validity of Cognition scale
Voice of Victims in Europe: Testimonies 478
voices 243
Voluntary Private Sector Preparedness Accreditation and Certification Program law (PS-PREP) 229–30

waiting list controls 216, 217, 220, 261
war neurosis 9
War on Terror 30, 40
War Pensions Agency 464, 465
Warwick Edinburgh Mental Well Being Scale (WEMBWS) 465
War You Don't See, The 476
watchful waiting 27, 145–6, 184, 201, 203, 346, 370, 479
web-based counseling 114
welfare support 464, 465
well-being 161, 170, 315, 326, 333–4, 411, 475
  and work 334
  of employees 83
  of staff in challenging conditions 95
  psychological 49, 61–3, 88
  subjective 62
wellness programs 113
WEMBWS, *see* Warwick Edinburgh Mental Well Being Scale
Wenchuan earthquake 436, 443
Western psychology 440
witnesses
  and emotional trauma 78
  and trauma management policies 339
witnessing combat 419
witnessing deaths 74–5, 185
witnessing traumatic images 36
women's health 45
work
  adaptation 186, 335, 338, 339, 341–6
  and health 334
  and recovery 334
  and self-esteem 334
  and well-being 334
  beneficial general effects 341
worker deaths 73, 74–5
worker injuries 73
working conditions 182
Working for a Healthier Tomorrow 338
working memory 289, 323, 376
working-through 372, 374–5
workplace
  accidents 74
  adjustments 186, 335, 338, 339, 341–6
  aggression 105, 106
  conflict 441
  resistance 146
  trauma 17–28, 163, 206, 441
  violence 90, 105–17, 142–3, 200
work-related injuries 121
worksite wellness programs 41–2
World Conference on Disaster Reduction 229
World Trade Center 71–3, 78, 81, 454
World War One (WWI) 8, 168, 242, 245
World War Two (WWII) 10–11, 181, 426, 439, 460
worried well 231
WSDO, *see* welfare support desk officer

xenophobia 141

yin/yang 436
Yuan Dynasty 439

Zeebrugge ferry disaster, *see Herald of Free Enterprise* disaster

Printed and bound by CPI Group (UK) Ltd, Croydon, CR0 4YY

07/07/2026

14916215-0001